Ion Specific Hofmeister Effects

Queen's College, Oxford, United Kingdom
3–5 September 2013

FARADAY DISCUSSIONS

Volume 160, 2013

RSC Publishing

The Faraday Division of the Royal Society of Chemistry, previously the Faraday Society, founded in 1903 to promote the study of sciences lying between Chemistry, Physics and Biology.

Faraday Discussions (Print ISSN 1359-6640, Electronic ISSN 1364-5498) is published 8 times a year by the Royal Society of Chemistry, Thomas Graham House, Science Park, Milton Road, Cambridge, UK CB4 0WF. Volume 160 ISBN-13: 978-1-84973-687-9

2013 annual subscription price: print+electronic £765, US $1428; electronic only £727, US $1356. Customers in Canada will be subject to a surcharge to cover GST. Customers in the EU subscribing to the electronic version only will be charged VAT. All orders, with cheques made payable to the Royal Society of Chemistry, should be sent to RSC Order Department, Royal Society of Chemistry, Thomas Graham House, Science Park, Milton Road, Cambridge, CB4 0WF, UK. Tel +44 (0) 1223 432398; E-mail orders@rsc.org

If you take an institutional subscription to any RSC journal you are entitled to free, site-wide web access to that journal. You can arrange access *via* Internet Protocol (IP) address at www.rsc.org/ip. Customers should make payments by cheque in sterling payable on a UK clearing bank or in US dollars payable on a US clearing bank.

PRINTED IN THE UK

MIX
Paper from responsible sources
FSC® C013604

Faraday Discussions documents a long-established series of *Faraday Discussion* meetings which provide a unique international forum for the exchange of views and newly acquired results in developing areas of physical chemistry, biophysical chemistry and chemical physics.

Co-sponsor

British Biophysical Society

Ion Specific Hofmeister Effects

Faraday Discussions
www.rsc.org/faraday_d

A General Discussion on Ion Specific Hofmeister Effects was held in Oxford, UK on 3rd, 4th and 5th of September 2012.

CONTENTS

ISSN 1359-6640; ISBN 978-1-84973-687-9

Cover
See Piotr Jurkiewicz, *Faraday Discuss.*, 2013, **160**, 341.
Clustering of cationic DOTAP lipids by Br^- anions. A single MD simulation snapshot with a view perpendicular to the membrane/water interface is presented, water molecules are not shown. Up to four cationic lipid headgroups (yellow) are bound by Br^- anions (green). Lateral inhomogeneities are induced and large hydrophobic patches, covered by hydrocarbon chains of lipids, are exposed (gray).

Image reproduced by permission of Dr Piotr Jurkiewicz from *Faraday Discuss.*, 2013, **160**, 341.

INTRODUCTORY LECTURE

PAPERS AND DISCUSSIONS

CONCLUDING REMARKS

ADDITIONAL INFORMATION

Introductory Lecture: Interpreting and predicting Hofmeister salt ion and solute effects on biopolymer and model processes using the solute partitioning model†

M. Thomas Record, Jr.‡[ab] Emily Guinn,‡[a] Laurel Pegram[a] and Michael Capp[b]

Received 14th November 2012, Accepted 16th November 2012
DOI: 10.1039/c2fd20128c

Understanding how Hofmeister salt ions and other solutes interact with proteins, nucleic acids, other biopolymers and water and thereby affect protein and nucleic acid processes as well as model processes (*e.g.* solubility of model compounds) in aqueous solution is a longstanding goal of biophysical research. Empirical Hofmeister salt and solute "m-values" (derivatives of the observed standard free energy change for a model or biopolymer process with respect to solute or salt concentration m_3) are equal to differences in chemical potential derivatives: m-value $= \Delta(d\mu_2/dm_3) = \Delta\mu_{23}$, which quantify the preferential interactions of the solute or salt with the surface of the biopolymer or model system (component 2) exposed or buried in the process. Using the solute partitioning model (SPM), we dissect μ_{23} values for interactions of a solute or Hofmeister salt with a set of model compounds displaying the key functional groups of biopolymers to obtain interaction potentials (called α-values) that quantify the interaction of the solute or salt per unit area of each functional group or type of surface. Interpreted using the SPM, these α-values provide quantitative information about both the hydration of functional groups and the competitive interaction of water and the solute or salt with functional groups. The analysis corroborates and quantifies previous proposals that the Hofmeister anion and cation series for biopolymer processes are determined by ion-specific, mostly unfavorable interactions with hydrocarbon surfaces; the balance between these unfavorable nonpolar interactions and often-favorable interactions of ions with polar functional groups determine the series null points. The placement of urea and glycine betaine (GB) at opposite ends of the corresponding series of nonelectrolytes results from the favorable interactions of urea, and unfavorable interactions of GB, with many (but not all) biopolymer functional groups. Interaction potentials and local-bulk partition coefficients quantifying the distribution of solutes (*e.g.* urea, glycine betaine) and Hofmeister salt ions in the vicinity of each functional group make good chemical sense when interpreted in terms of competitive noncovalent interactions. These interaction potentials allow solute and Hofmeister (noncoulombic) salt effects on protein and nucleic acid processes to be interpreted or predicted, and allow the use of solutes and salts as probes of

[a]*Department of Chemistry, University of Wisconsin, Madison WI 53706, USA*
[b]*Biochemistry, University of Wisconsin, Madison WI 53706, USA*

† Supported by NIH Grant GM47022 (MTR).
‡ Co-first authors.

interface formation and large-scale conformational changes in the steps of a biopolymer mechanism.

I. Introduction

A. Solute and salt ion series in aqueous chemistry and biochemistry

Solutes and salts are universally used *in vitro* to optimize or shift the thermodynamics and kinetics of biopolymer processes, including protein precipitation/purification, protein crystallization, protein folding, nucleic acid helix formation and RNA folding, and protein–protein, protein–ligand, and protein–nucleic acid interactions.[1–4] Dating back to Hofmeister's (1888) studies of the effectiveness of different salts as protein precipitants,[5] effects of high concentrations of salt anions, cations and nonelectrolyte solutes on many noncovalent protein (also nucleic acid, small molecule) processes in aqueous solution have been observed to fit the regular series listed in Fig. 1.[1,6,7] Solutes and salt ions at the right end of these series are the most stabilizing; that is, the most effective at driving protein folding, protein assembly and other noncovalent protein self-assembly and binding processes that remove large amounts of protein molecular surface from water. Solutes and salt ions at the left end of these series are most destabilizing, driving unfolding, disassembly, and the exposure of protein surface to the solution. In the assembly direction, all these processes replace protein–water and protein–solute interactions with protein–protein interactions. Generally, the protein surface buried or exposed (*i.e.* the change in water accessible surface area ΔASA) is primarily hydrocarbon. Approximately two-thirds of the surface buried in folding, and half of native protein surface, is aliphatic or aromatic hydrocarbon; typically about 90% of this hydrocarbon surface is aliphatic. Both folded and unfolded surfaces of proteins are about 15–20% amide oxygen and nitrogen.[8–10] Most of these amide groups are in the peptide backbone, with a O : N ASA ratio of ~2.4 : 1.

Addition of small nonelectrolytes such as glycine betaine (GB), proline, and trehalose and other sugars and larger polyols favors processes that remove protein surface from water.[11,12] Trimethylamine oxide (TMAO), a close chemical analog of GB, has

Solute and Hofmeister Salt Ion Series for Biopolymer Processes

	Solubilize, Unfold, Disassemble (Biopolymer ΔASA > 0)		Null Point (No effect)	Precipitate, Fold, Assemble (Biopolymer ΔASA < 0)
Proteins (Surface Composition 50-75% C, 50-30% O,N)				
Nonelectrolytes:	urea		ethylene glycol, glycerol	polyols, proline, glycine betaine
High [Salt] Hofmeister Series				
Anions:	SCN^-, I^-, ClO_4^-	NO_3^-, Br^-	Cl^-	Ac^-, Glu^-, F^-, SO_4^{2-}, HPO_4^{2-}
Cations:	GuH^+	NH_4^+	K^+, Na^+	NR_4^+
Nucleic Acid Duplex Formation (42%GC, Surface Composition ~30% C, ~70% O,N)				
Nonelectrolytes:	urea, glycerol, polyols, glycine betaine			
PEGs:	ethylene glycol (EG), diEG		PEG200-600	PEG>1000
High [Salt] Hofmeister Series				
Anions:	SCN^-, I^-, ClO_4^-, NO_3^-, Br^-, Cl^-		Ac^-, Glu^-, F^-, SO_4^{2-}, HPO_4^{2-}	
Cations:	GuH^+, NH_4^+		K^+, Na^+, NR_4^+	

Fig. 1 Solute and Hofmeister salt ion series for protein and nucleic acid processes. Solutes and salt ions are arranged from those that promote processes exposing biopolymer surface (blue) to those that promote burial of biopolymer surface (red) for effects on protein processes (top) and nucleic acid duplex formation (bottom). Approximate positions of the null (no effect) points in these series are indicated. Hofmeister salt ion series are those observed for folding or duplex formation at high salt concentration where Coulombic effects are minimized.

a similar effect. Though used as an osmolyte in some eukaryotic cells, TMAO is much less suitable than GB for *in vitro* biophysical and biochemical studies, because its aqueous solutions are quite alkaline and because it is chemically less stable than GB, decomposing into volatile trimethyl amine. Effects of solutes like glycerol and ethylene glycol on processes that bury or expose protein surface are typically small, neither favoring nor disfavoring them (hence the "null" (*i.e.* no large effect) designation for these solutes in Fig. 1). Urea and urea derivatives, on the other hand, favor processes that expose protein surface to water and hence destabilize protein assemblies and solubilize protein precipitates.[8,13]

Studies of effects of these small nonelectrolyte solutes on the stability of nucleic acid duplexes reveals a similar rank order[7,14–19] summarized in Fig. 1. However, all small nonelectrolytes examined to date destabilize nucleic acid duplexes, favoring melting and exposure of heterocyclic base surface to water. One of the least destabilizing is GB, which has no effect on the stability of AT base pairs, but destabilizes GC base pairs.[16] The surfaces buried in nucleic acid duplex formation and in protein folding or self-assembly processes differ greatly in composition: the nucleic acid base surface buried in helix formation is approximately two-thirds polar O and N and one-third nonpolar C, while that buried in protein folding is approximately two-thirds nonpolar C and one-third polar O and N.[7,9,20] Can these differences in surface composition explain the similar order but shifted null point of the series of nonelectrolyte solutes for protein and nucleic acid processes?

Another series of effects of uncharged solutes is the molecular weight (*i.e.* degree of polymerization) series from monomer to polymer, investigated most extensively for the series from ethylene glycol (EG) to PEG[21–25] (Fig. 1). For nucleic acid duplex formation, EG is destabilizing, but high molecular weight PEGs are stabilizing. Quantitative investigation of the series between EG and PEG provides an excellent method of separating the contributions of small solute effects (chemical preferential interactions) from those of large solute effects (formation of a flexible coil domain; excluded volume effects arising from the inability of the protein or nucleic acid being investigated to access the substantial volume of solution or interact with the chain segments in the interior of flexible coils). As discussed in more detail below, these physical effects arising from solute size are only observed for glycols larger than tri- or tetraEG. For protein folding, EG is a relatively nonperturbing solute, neither stabilizing nor destabilizing, while polymeric PEG can be stabilizing or destabilizing.[22,23,25] Pure excluded volume effects of a polymeric solute (which experimentally must be separated from the effect of segregating many segments of that polymeric solute in the interior of the flexible coil) are significantly larger for processes involving association of polymeric strands or protomers into multimers, aggregates or precipitates than for conformational changes without a change in strandedness.[21]

As in the case of large solutes, effects of salts also include both chemical and physical effects of interactions of salt ions with the biopolymers or model compound involved in the process of interest. These can include short-range chemical preferential interactions of the individual salt ions with the biopolymer surface buried or exposed in the process, chemical site binding of ions to initial and/or final conformations/states of the biopolymers, and long-range physical Coulombic interactions of these salt ions with the fixed charges on the initial and final conformations/states of the biopolymers or model compound. Coupling between these various modes of interaction is observed for polyions at low salt concentration, where Coulombically-driven accumulation of salt counterions can drive site binding or preferential interactions of the salt counterions with the polyion surface.[26,27] The term "Hofmeister effects" of salts usually refers to the weaker chemical (preferential interaction) effects of the individual salt ions manifested at relatively high salt concentration, as in Hofmeister's original studies of protein precipitation. For some polyion processes, "reverse" series and/or amplifications of Hofmeister effects of salt counterions are observed at low salt concentration where Coulombic effects of salts are significant, as discussed in subsequent sections.[26] The use of high salt concentrations and weakly charged (or uncharged) model compounds, biopolymers and surfaces minimizes Coulombic effects, allowing Hofmeister effects of salts to be most unambiguously observed.

Fig. 1 shows typical series of Hofmeister anions and univalent cations for protein self-assembly processes and for nucleic acid duplex formation. Anions favoring protein self-assembly include sulfate, phosphate, carboxylates including the physiological anion glutamate, and fluoride; cations favoring protein self-assembly include alkyl ammonium cations (NR_4^+), K^+ and Na^+. Most of these are strongly hydrated ions that prefer to interact with water, rather than with biopolymer surfaces. At the other end of these cation and anion series are ions that favor protein disassembly; these ions (*e.g.* thiocyanate, perchlorate and iodide anions and guanidinium cation) are less strongly hydrated and/or have more tendency to interact with biopolymer surfaces. Nucleic acid duplex formation exhibits the same rank order of Hofmeister anion and cation effects, but a very different null point that likely varies with base composition. No salt exerts a stabilizing Hofmeister effect on a 12 bp (7 AT, 5 GC) nucleic acid duplex, as shown in Fig. 5B below.[7]

B. Explanations of chemical effects of small solutes and salt ions

What is the molecular origin of these ubiquitous solute and salt ion series? Early interpretations of chemical effects of salts and solutes on aqueous processes were largely structural. Processes exhibiting these effects include the protein and nucleic acid processes discussed above, the model processes of dissolving organic compounds and disassembling micelles, as well as the process of transferring water from bulk water to the air–water interface. Salt and solute effects on these processes were attributed to their ability to affect the hydrogen-bonded structure of water. However, recent spectroscopic studies show that the effect of solutes and salt ions on water structure is confined to the first one or two layers of water of hydration of the solute or salt ion.[28]

One quantitative explanation of the thermodynamics of solute and Hofmeister salt effects invokes molecular-scale cavity formation.[1,29] The process of dissolving a model compound or a protein, or unfolding a protein, is divided into two steps: creating a cavity in water for the model compound or newly exposed protein surface, followed by making the interactions across that interface. To quantify the thermodynamics of cavity formation as a function of solute or salt concentration, surface tension increments are used; these quantify the effect of solute or salt concentration on the free energy of transfer of water from bulk to surface, per unit area of surface created. At a semi-quantitative level, surface tension increments of Hofmeister salts are consistent with effects of Hofmeister salts on processes that expose or bury primarily hydrocarbon surface.[1] Most fundamentally, this means that Hofmeister salt effects on surface tension (*i.e.* salt effects on the transfer of water from bulk to the nonpolar interface with air) and on hydrocarbon solubility (*i.e.* salt effects on the process of exposing hydrocarbon surface to water and hydrating this nonpolar molecular surface) have similar thermodynamic origins. However, this correlation does not necessarily mean that cavity formation and molecular-scale surface tension effects are the fundamental physical phenomena responsible for effects of Hofmeister salts on processes exposing hydrocarbon surface to water.

The solute partitioning model (SPM) is an alternative model for the interpretation of chemical (preferential interaction) effects of nonelectrolyte solutes and Hofmeister salts on all aqueous processes.[8,9,30–34] Recent molecular thermodynamic analysis of both surface tension and hydrocarbon solubility data using the SPM, detailed below, interprets solute and salt effects on surface tension and on solubility of uncharged model compounds quantitatively in terms of a surface hydration b_1 ($H_2O/\mathring{A}^2$) and a solute partition coefficient K_p. K_p, the ratio of local to bulk solute concentration, quantifies the competition between the solute or salt ions and water for a molecular or macroscopic surface, and the extent to which the solute or salt ion accumulates at the surface, replacing local water, or is excluded from that surface, leaving it hydrated. For most Hofmeister salt ions, K_p values quantifying accumulation ($K_p > 1$) or exclusion ($K_p < 1$) from the air–water interface and molecular hydrocarbon surface in water are similar, thereby providing a quantitative explanation of the parallels between salt effects on surface tension and hydrocarbon solubility in terms of ion partitioning between local and bulk water, not cavity formation.[33,34]

Timasheff recognized that the interactions of Hofmeister salts with proteins, like those of nonelectrolyte solutes, were preferential (chemical) interactions between the salt (or solute) component (3) and the protein component (2).[3] His data are presented as preferential interaction coefficients Γ_{32} and chemical potential derivatives $\mu_{23} = d\mu_2/dm_3$ (where $\Gamma_{32} = -\mu_{23}/\mu_{33}$) and discussed qualitatively using both the cavity model and the local–bulk model (the precursor of the SPM). Tanford[35] and Schellman[36,37] developed weak binding models based on exchange between the solute/salt ion and water to discuss preferential interactions of urea and other solutes and salts with proteins and interpret μ_{23} values. Tanford[38] and more recently Bolen and collaborators[39,40] analyzed solubility increments quantifying the effects of urea and other nonelectrolytes on the solubility of amino acids and dipeptides to obtain side chain and backbone transfer free energies, which Bolen and collaborators have used in conjunction with a surface area analysis to predict effects of these solutes on the standard free energy change for protein unfolding and other protein processes.[39,40]

The SPM-based analysis reviewed here quantifies the competition between a solute or salt ion and water for any molecular or macroscopic surface, using structural input (water-accessible surface area (ASA) and composition). This analysis yields two levels of useful information. Interaction potentials, obtained from osmometric and solubility data for model compounds and from surface tension increments for the air–water interface, quantify the interaction of a solute with a unit area of a particular functional group, atom, or surface. These interaction potentials are very useful for interpreting or predicting interactions of solutes or salts with biopolymer or other surfaces, as well as interpreting or predicting *m*-values or changes in surface area (ΔASA) in protein or nucleic acid processes.[7,8,12] From these interaction potentials, local–bulk partition coefficients quantifying the accumulation or exclusion of the solute from the vicinity of this group or surface are obtained. These partition coefficients for solutes like urea, GB, and salt ions and functional groups like amide and anionic oxygens, amide and cationic nitrogens, and aliphatic and aromatic carbons, are consistent with expectations based on competitive hydrogen bonding capabilities of the solute and water to interact with polar groups, and also provide evidence for the significance of cation-π and π–π interactions between solute and aromatic groups on model compounds.[8,12]

The SPM-based analysis not only predicts and interprets the interactions of solutes and salt ions with the functional groups of biopolymers, but also is applicable to predict and interpret effects of small solutes and salts on the surface tension of water and on all biopolymer and model compound processes in water.[7,8,12,33,41] The fundamental premise of this analysis, in common with the backbone and side chain transfer free energy analysis,[38–40] is that these contributions to solutes and salt effects are independent and additive. For interactions of two larger solutes, each with multiple functional groups, especially when presented on constrained surfaces, deviations from additivity are more likely. Deviations would indicate that favorable (complementary) or unfavorable nearest neighbor effects are significant, as in forming arrays of noncovalent interations in a ligand–biopolymer or biopolymer–biopolymer interface. Interactions of larger or more highly charged solutes with biopolymers may include significant contributions of physical (*e.g.* excluded volume, Coulombic) effects as well as chemical (preferential interaction) effects.[7,21]

C. *m*-Values: experimental thermodynamic inputs to characterize solute and Hofmeister salt effects on biopolymer and model processes

Solute and Hofmeister salt effects on aqueous processes are characterized by *m*-values, which are derivatives of the standard free energy change $\Delta G^\circ_{obs} = -RT\ln K_{obs}$ with respect to solute or salt concentration:

$$m\text{-value} = \frac{d\Delta G^\circ_{obs}}{dm_3} = -RT\frac{d\ln K_{obs}}{dm_3} \tag{1}$$

In eqn (1), K_{obs} is the equilibrium constant for the process expressed in terms of equilibrium concentrations of products and reactants, and ΔG°_{obs} is the corresponding standard free energy change. Interactions of solutes or salt ions (component 3) with biopolymers or model compounds (component 2) in aqueous solution affect the activity coefficient of component 2.

m-Values quantify solute and Hofmeister salt effects on standard free energy changes, not only for biopolymer self-assembly or disassembly processes, but also for model processes that change the exposure of nonpolar or polar surface to water, including dissolving a model compound in water (where K_{obs} is the solubility and the *m*-value is proportional to the Setchenow coefficient[1,42]), micelle formation (where K_{obs} is the reciprocal of the critical monomer concentration (CMC)), and transferring water from bulk to the air–water interface, where the surface tension is the free energy change per unit area and the surface tension increment (STI) is the *m*-value.

II. Examples of *m*-value plots illustrating effects of solutes and Hofmeister salt ions on aqueous biopolymer and model processes

A. Effects of small nonelectrolyte solutes on protein unfolding and protein stability

Fig. 2 plots the logarithm of the equilibrium concentration quotient (observed equilibrium constant) K_{obs} for unfolding the DNA binding domain of lac repressor, a small (51 folded residues) single domain globular protein, as a function of urea and GB concentration; $\ln K_{obs}$ varies linearly with solute concentration over a wide range. For this very small protein, addition of 2.1 M GB reduces K_{obs} for unfolding by tenfold; addition of 3.0 M urea increases K_{obs} by tenfold. Slopes of these plots, when multiplied by $-RT$, yield urea and GB unfolding *m*-values. Solute *m*-values are generally independent of solute concentration over a wide range; typically molar scale plots are more linear than molal scale plots, and both molar and molal scale plots have the same initial slope.[9,13] For small nonelectrolyte solutes like urea and GB, these *m*-values are entirely determined by chemical (preferential) interactions.[3,8,12] Urea favors protein unfolding because it is locally accumulated in the vicinity of the protein surface exposed in unfolding (ΔASA); the composition of this surface is approximately 65–70% hydrocarbon (C) and 15–20% amide (O, N). GB favors folding because it is locally excluded from these surfaces, which prefer to remain hydrated. Urea unfolding *m*-values are proportional to ΔASA of unfolding (Fig. 3), a key finding of Myers *et al.*,[13] which is predicted and interpreted using the SPM.[9,20] From SPM-based analysis of *m*-values one also obtains partition coefficients K_p which relate the local concentration of the solute (here urea, GB) near the surface of interest to the bulk solute concentration. For the surface exposed in unfolding, the composite K_p value for urea is greater than unity (accumulation) while that for GB is less than unity (exclusion).[8,12]

B. Effects of varying the degree of polymerization of the solute (ethylene glycol to polyethylene glycol) on nucleic acid helix melting

Fig. 4A summarizes the effects of the solute series from EG to PEG on K_{obs} for melting of a 12 bp DNA duplex. All PEG concentrations are expressed on a

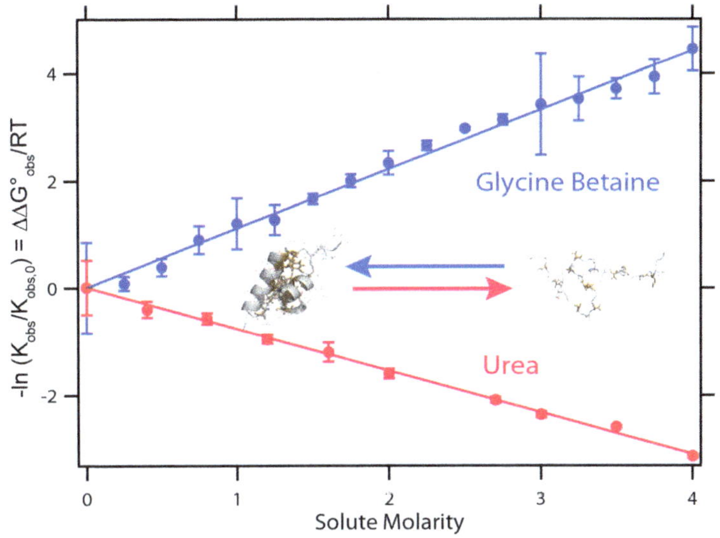

Fig. 2 Effects of urea and glycine betaine (GB) on protein unfolding. Plots of $-\ln(K_{obs}/K_{obs,0}) = \Delta\Delta G^{\circ}_{obs}/RT$ *vs.* solute molarity for effects of urea (25 °C) and GB (59 °C) on unfolding of the lac repressor DNA binding domain (lacDBD).[75] K_{obs} is the unfolding equilibrium constant in the presence of solute and $K_{obs,0}$ is the unfolding equilibrium constant in the absence of solute; slopes are *m*-values/*RT*. Inset Figure shows the folded structure (one monomer of dimer in PDB 1OSL) and one unfolded structure (determined by ProtSA[76]) of lacDBD with the surface exposed on unfolding highlighted in yellow.

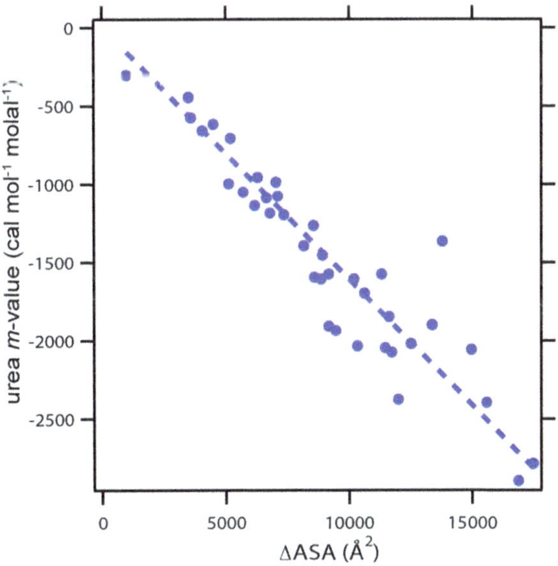

Fig. 3 Myers, Pace, and Scholtz[13] plot of urea *m*-values for protein unfolding as a function of ΔASA of unfolding. Protein data set and ΔASA determinations (for a model of the urea-denatured state where the protein backbone is in an extended β (all trans) conformation and side chains are rotated to avoid steric clash) from Guinn *et al.*[8] Unfolding *m*-values determined at temperatures ranging from 0 to 40 °C with most at 20–25 °C (for conditions see Hong *et al.*[61] and references therein).

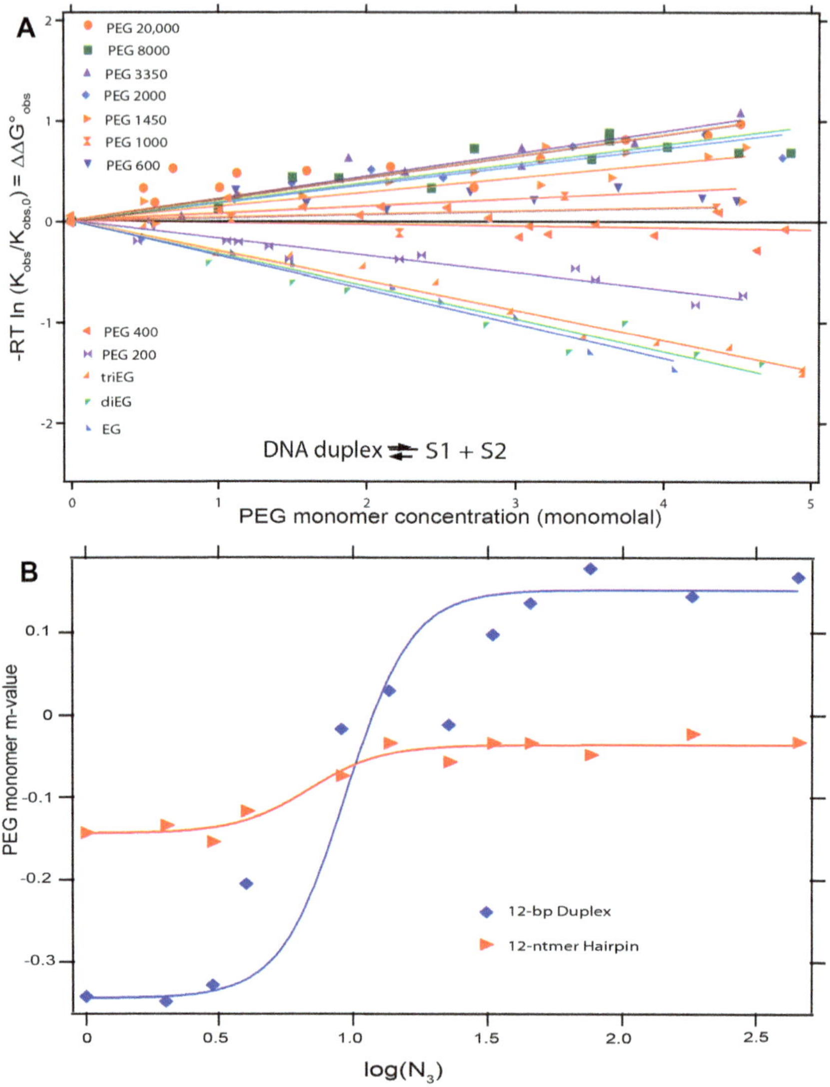

Fig. 4 Effects of polyethylene glycol (PEG) degree of polymerization (N_3) on DNA duplex and hairpin melting. (A) Plots of $-RT\ln K_{obs}/K_{obs,0} = \Delta\Delta G^{\circ}{}_{obs}$ *vs.* PEG monomer concentration (monomolal) for the series from ethylene glycol to PEG 20 000 on DNA duplex melting at 40 °C, where K_{obs} is the melting equilibrium constant in the presence of PEG and $K_{obs,0}$ is the melting equilibrium constant in the absence of PEG. (B) PEG monomer *m*-values (slopes from A) for melting a 12 bp DNA duplex and 4 bp hairpin plotted as a function of log(N_3) where N_3 is the degree of polymerization of PEG. Figure adapted from Knowles *et al.*[21]

monomolal scale; use of this PEG monomer concentration scale not only allows data for the entire series to be represented clearly on one plot but also illustrates the very different effects of low, moderate and high molecular weight PEG on the duplex melting process. Low molecular weight PEGs favor melting; for EG, diEG and triEG, monomer *m*-values are the same, indicating that the favorable chemical (preferential) interaction of the DNA surfaces exposed in melting with the two –CH_2OH end groups of diEG and triEG (and with the two halves of EG) is the same as with the –CH_2OCH_2– interior groups. As the PEG molecular weight is increased from

PEG 200 (tetraethylene glycol) to PEG 1450, the favorable effect on melting per PEG monomer, which is the same for EG, di- and tri-EG, is first reduced and then replaced by a strong unfavorable effect on melting (*i. e.* duplex stabilization) of higher molecular weight PEG. Above PEG 1450, this stabilization becomes independent of PEG molecular weight. A summary plot of the variation in PEG monomer *m*-value with the number of residues N_3 in the PEG molecule (log scale) is shown in Fig. 4B.

For a 4 bp hairpin helix, where melting is not accompanied by strand separation, qualitatively similar but smaller effects of increasing PEG molecular weight on the PEG monomer *m*-value are observed (see Fig. 4B). Monomer *m*-values of EG, diEG and triEG for hairpin melting are all the same, but are somewhat less than half as large in magnitude as for the 12 bp duplex, consistent with an interaction with the ASA exposed in melting, which is less than half as large for the 4 bp hairpin as for the 12 bp duplex. For the 4 bp hairpin, the high PEG molecular weight plateau occurs at a *m*-value near zero; the favorable chemical interaction of functional groups of low molecular weight glycols in this series (EG, di- and triEG) with the nucleic acid base surface exposed in melting is largely eliminated by a physical effect of high molecular weight PEG, but not replaced by a helix-stabilizing effect as in the case of the 12 bp duplex.

Two large-solute effects become significant for PEG oligomers larger than tri- or tetraEG. One is excluded volume; the flexible PEG polymer excludes the DNA strands from the volume of solution occupied by the chain molecule. This effect is much more important for 12 bp duplex melting, with strand separation, than for 4 bp hairpin melting, which only involves a change in shape and not a change in strandedness. The other is sequestration; the large fraction of PEG monomers which transiently are located in the "interior" of the chain molecule cannot interact chemically with the DNA surface exposed in melting, so the chemical effect of EG and PEG is gradually eliminated with increasing PEG molecular weight. Knowles *et al.*[21] interpret these behaviors quantitatively. Fig. 4B demonstrates that for these systems only chemical (preferential interaction) effects of the solute are observed up to a solute molecular weight of approximately 200 (tetraEG).

C. Hofmeister effects of salts on model processes that expose an uncharged, nonpolar surface to water

Fig. 5A and B provide classic examples of Hofmeister effects, where Coulombic effects are largely absent and the surfaces involved are entirely nonpolar. Surface tension differences for sodium and guanidinium salts, plotted *vs.* salt concentration in Fig. 5A, are free energy differences per unit of surface area for the transfer of water from bulk to the nonpolar air–water surface in a salt solution as compared to pure water. These free energy differences increase linearly with salt concentration and yield salt-specific *m*-values for this transfer process. Logarithms of solubility ratios for dissolving benzene in water and in salt solutions are plotted *vs.* salt concentration in Fig. 5B. These plots are also linear with slopes that are interpreted as salt *m*-values/RT for this benzene transfer process. Most (but not all) salts increase surface tension and reduce benzene solubility. The cation and anion series deduced from these surface tension and hydrocarbon solubility data are the same as the Hofmeister series of cations and anions determined for protein solubility and other protein processes (Fig. 1), though the null points of many of these series differ as discussed below. Analysis of these and other data for these model transfer processes demonstrates that the cation and anion of the salt make additive, independent contributions to the *m*-value, and provide values for the amount of local water at the air–water and molecular hydrocarbon surface and for the partition coefficients of individual salt ions between bulk water and this local water.[33,34]

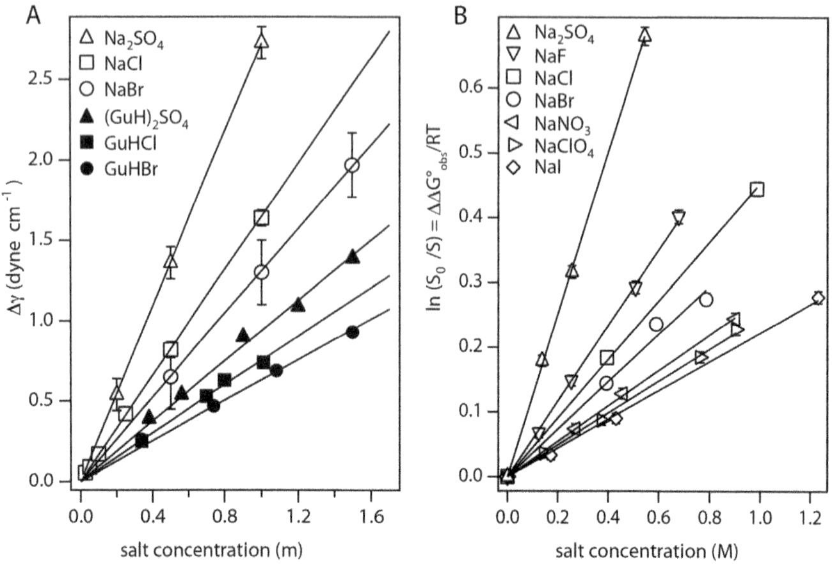

Fig. 5 Hofmeister salt effects on model processes involving nonpolar surfaces. Panel A: Increase in surface tension of water ($\Delta\gamma$) as a function of the molality of Na^+ and GuH^+ salts of SO_4^{2-}, Cl^- and Br^- (see Pegram *et al.*[34] for literature references; constant temperatures for each series in range 15–30 °C). Slopes are surface tension increments (STI) (eqn (13)) characterizing effects of these salts on the free energy of transfer of water from bulk to surface. Panel B) Effects of Na^+ salts on the logarithm of the relative solubility of benzene ($\ln(S_0/S) = \Delta\Delta G^{\circ}_{obs}/RT$, where S is the solubility in the presence of salt and S_0 is the solubility in the absence of salt) at 25 °C as a function of salt molarity for the anion series SO_4^{2-}, F^-, Cl^-, Br^-, NO_3^-, ClO_4^-, and I^-. Slopes are m-values/RT for the effects of these salts on transfer of benzene from the pure liquid to water. Figure reused with permission from Pegram and Record.[34]

D. Salt effects (Coulombic, Hofmeister) on protein and nucleic acid melting and protein aggregation

1) Melting. Fig. 6 summarizes effects of a wide range of Hofmeister salts on the logarithm of the equilibrium constant K_{obs} for unfolding the lac repressor DNA binding domain (Fig. 6A; *cf.* Fig. 2) and for melting a 12 bp DNA duplex (Fig. 6B). At low salt concentration (<0.5 molal), Coulombic effects are large and cause the stability ($-RT\ln K_{obs}$) of the folded and helical forms to increase with increasing salt concentration. Plots of $\ln K_{obs}$ *vs.* ln[salt] are linear in this low salt range (data of ref. 7; not shown), with slopes which are similar for all salts of the same valence, as expected for a Coulombic effect.[43] This Coulombic effect is particularly pronounced for the 12 bp DNA oligoanion; transition from the duplex to two separate strands reduces the number of negatively charged phosphates from 22 (duplex) to 11 for each melted strand and also reduces the axial charge density from 2 phosphate charges to less than 1 per 3.4 Å. The 51-residue DBD has a net charge of +2, with 7 positive and 5 negative charges which presumably become more separated in the unfolded form and cause its stability to increase more modestly with salt concentration in this range.

Above 0.1 M salt (lacDBD) or 0.5 M salt (12-mer DNA) where Coulombic effects are expected to be small, large salt-specific differences in K_{obs} and in its salt derivative (*i.e.* m-value/RT) are observed. Salt series for both processes follow the rank order of the Hofmeister series, with very different null points. The stability of lacDBD increases with increasing salt concentration for all salts investigated except GuHCl, but the stability of the DNA duplex is not increased by any salt in this range. These very different null points of the Hofmeister salt series for lacDBD

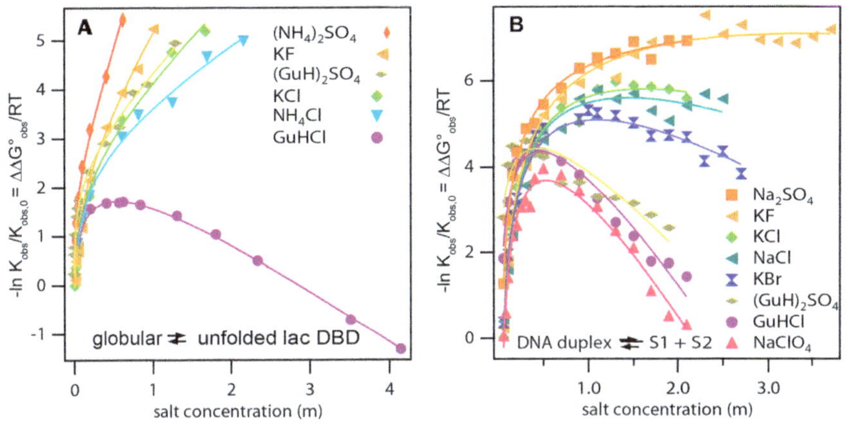

Fig. 6 Effects of Hofmeister salts on protein unfolding and DNA melting. Panels show the effects of low and high concentrations of salts spanning the Hofmeister series on $-\ln(K_{obs}/K_{obs,0}) = \Delta\Delta G^{\circ}_{obs}/RT$ for (A) lacDBD unfolding at 37 °C and (B) DNA helix melting at 40 °C. K_{obs} is the unfolding or melting equilibrium constant in the presence of salt and $K_{obs,0}$ is the reference equilibrium constant in low-salt buffer. Fitted curves allow a separation of Hofmeister and Coulombic effects of these salts at high salt concentration, as described in Pegram *et al.*[7] Figure adapted with permission from Pegram *et al.*

and the 12 bp DNA duplex must be a consequence of the very different compositions of the surfaces exposed in melting (ΔASA). The lacDBD surface exposed in unfolding is primarily (~70%) nonpolar C and only (~30%) polar N, O, while the DNA surface exposed in melting is primarily (~70%) polar N and O, and only ~30% nonpolar C.

2) Protein aggregation and phase separation. Fig. 7 (Zhang & Cremer[26]) shows the dramatic and complicated effects of different sodium salts with anions from the middle to the solubilizing end of the Hofmeister series on the cloud point transition temperature T_{cp} for liquid–liquid phase separation and aggregation of lysozyme upon cooling at pH 9.4 where it is positively charged. Derivatives $dT_{cp}/d[salt]$ measure the ratio of the salt m-value to the enthalpy change for aggregation. At high [salt], where Coulombic salt effects are minimized and Hofmeister effects

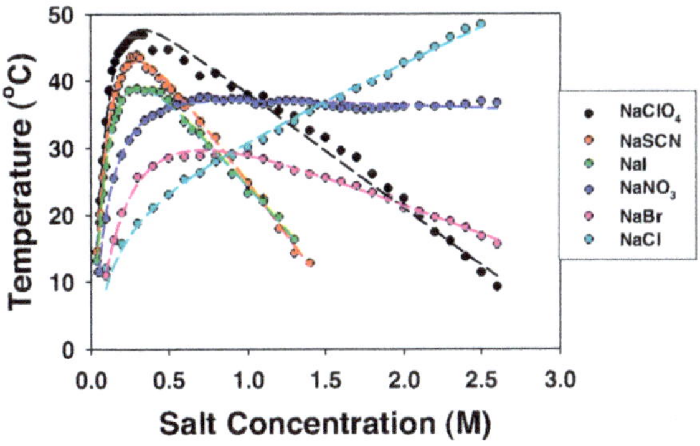

Fig. 7 Reverse and normal Hofmeister salt effects. The cloud-point temperature of the cationic protein lysozyme (pH 9.4) as a function of anion type and concentration (Zhang and Cremer[26]), showing high salt (normal) and low salt (magnified, reversed) Hofmeister salt effects on liquid–liquid phase separation and aggregation. Figure reprinted with permission.

of these sodium salts are dominant, T_{cp} varies linearly with salt concentration. These high-[salt] slopes follow the normal protein Hofmeister anion series, (positive for Cl^-, null point near NO_3^-, and increasingly negative for Br^-, ClO_4^-, and SCN^-). At lower salt concentration (below 0.4 M), a reverse order of highly magnified effects of all these sodium salts is observed, where SCN^-, I^-, and ClO_4^- favor aggregation much more than Cl^-.These magnified, reverse-order Hofmeister effects of salt counterions (here anions) at low [salt] are the result of Coulombically driven anion binding (most significant for SCN^-, least for Cl^-) to positively charged lysozyme at low salt concentration, reducing its charge and favoring aggregation and phase separation.[26]

III. Analysis of solute and Hofmeister salt effects

A. Interpretation of *m*-values in terms of chemical potential derivatives μ_{23}

1) *m*-Values for biopolymer processes: relationship to $\Delta\mu_{23}$. Interactions of solutes or salt ions (component 3) with biopolymers (component 2) in aqueous solution affect the activity coefficient (f_2) of the biopolymer, giving rise to significant derivatives $d\ln f_2/dm_3$. These activity coefficient derivatives are to an excellent approximation the same as reduced chemical potential derivatives $(1/RT)d\mu_2/dm_3 = \mu_{23}/RT$. Because the observed equilibrium constant K_{obs} for the biopolymer process is defined in terms of product and reactant concentrations and not activities, K_{obs} depends on m_3 and this dependence (i.e the solute *m*-value) is determined by the dependence of f_2 on m_3, or equivalently by μ_{23}:

$$m\text{-value} = -RT\frac{d\ln K_{obs}}{dm_3} = RT\Delta\frac{d\ln f_2}{dm_3} = \Delta\frac{d\mu_2}{dm_3} = \Delta\mu_{23} \quad (2)$$

Chemical potential derivatives μ_{23} are closely related to preferential interaction coefficients (dialysis or Donnan coefficients) $\Gamma_{32} = -\mu_{23}/\mu_{33}$ where $\mu_{33} = d\mu_3/dm_3$. For an uncharged solute $d\mu_3/dm_3 = (RT/m_3)(1 + \varepsilon_3)$ where $\varepsilon_3 = d\ln f_3/d\ln m_3$ represents an often-small correction for solute–solute nonideality (*i.e.* ε_3 is small in magnitude in comparison to unity) that can be approximated by its value in the absence of the biopolymer or model compound species 2. Experimentally determined values of Γ_{32} and μ_{23} may be compared with theoretical (*e.g.* Poisson–Boltzmann) or computational (*e.g.* Monte Carlo, molecular dynamics) predictions of the radial distribution of solute 3 in the vicinity of solute 2 using Kirkwood–Buff theory and integrations over the local excess or deficit in the radial distribution of solute 3.[44,45]

For biopolymer processes where the solute effect is characterized by a *m*-value = $\Delta\mu_{23}$, we assume additivity of free energy contributions from these short-range interactions so that the *m*-value can be interpreted as the value of μ_{23} for interaction of solute (3) with the biopolymer surface area exposed (or buried) in the process (*i.e.* the ΔASA).[9,20] The extensive data base of urea *m*-values for protein unfolding reveals that these *m*-values are proportional to ΔASA for this homologous series of surfaces (varying in total area but with very similar compositions (65–70% hydrocarbon (C), 15–20% amide (O,N), the remainder other polar (O, N). This proportionality of urea *m*-values to ΔASA, first established by Myers *et al.*[13] and interpreted using the SPM by Courtenay *et al.*,[9] is shown for an updated set of small, single domain globular proteins lacking disulfides in Fig. 2.

2) *m*-Values for model processes; relationship to μ_{23}. *m*-Values also quantify solute and Hofmeister salt effects on standard free energy changes of model processes that expose a molecular or macroscopic surface to water. These processes, their solute or salt *m*-value, and the relation to μ_{23}, include:

1) Transfer of model compounds from the pure liquid or solid phase to water, quantified by the solubility m_2^{ss}:

$$\text{solubility } m\text{-value} = -RT\frac{\text{dln}m_2^{ss}}{\text{d}m_3} \approx \mu_{23} \tag{4}$$

In eqn (4), the solubility m-value, determined at constant a_2, is equal to μ_{23} (at constant m_3) only for sparingly soluble model compounds for which the activity coefficient γ_2 is not a function of m_2 even at the solubility limit. Osmometry is a good alternative to quantify μ_{23} for interactions of solutes with highly soluble model compounds, as reviewed below.

2) Transfer of amphiphiles from a micelle to water, quantified by the critical monomer concentration CMC:

$$\text{micelle dissociation } m\text{-value} = -RT\frac{\text{dlnCMC}}{\text{d}m_3} = \Delta\mu_{23} \tag{5}$$

where $\Delta\mu_{23}$ is interpreted as μ_{23} for the interaction of solute (3) with that portion of the surface of the amphiphile (2) that is buried in the micelle and exposed to water in micelle dissociation.

3) Transfer of water from bulk to the air–water surface, quantified by the surface tension γ:

$$\text{surface } m\text{-value}/A = \frac{\text{d}\gamma}{\text{d}m_2} = \text{STI} = \alpha_2 \tag{6}$$

where α_2 is a solute–surface interaction potential, quantifying the direction and strength of the solute–surface interaction per unit area of surface (see below). STI m-values, while for a two component system, contain the same quantitative information about the interaction of the solute with the air–water surface (relative to interactions with water) as do values of μ_{23} for a molecular surface in water.

3) Determination of μ_{23} for solute-model compound interactions by osomometry. Values of μ_{23} quantifying preferential interactions of solutes and Hofmeister salts with soluble, non-volatile model compounds are obtained directly from the difference ΔOsm[46,47] between the three component osmolality (Osm(m_2, m_3)) and the two component osmolalities (Osm(m_2), Osm(m_3)), determined by vapor pressure osmometry[8,11,12,15,48] or isopiestic distillation:[49]

$$\Delta\text{Osm} = \text{Osm}(m_2, m_3) - \text{Osm}(m_2) - \text{Osm}(m_3) = m_2 m_3 \frac{\mu_{23}}{RT} \tag{7}$$

The slope of a plot of ΔOsm *vs.* the concentration product m_2m_3 is therefore equivalent to a free energy derivative m-value/RT quantifying the effect of one solute on the chemical potential of the other. As an example, Fig. 8A shows osmolality measurements on a constant-molality K_2Oxalate solution as a function of the molal concentration of GB added. The black curve is the osmolality Osm(m_2,m_3) of the three-component GB–K_2Oxalate aqueous solution. The red curve is the sum of the osmolalities of the individual two-component solutions (variable GB Osm(m_3) (blue curve); constant K_2Oxalate Osm(m_2) (green line)). The difference between the black and red curves is ΔOsm (eqn (7)). Fig. 8B shows the plot of ΔOsm *vs.* the product m_2m_3 for all concentrations of GB and K_2Oxalate investigated; the plot is linear and yields (from the slope) a constant value for the chemical potential derivative $\mu_{23}/RT = 0.55\ \text{m}^{-1}$ for the interaction of GB and K_2Oxalate ($\mu_{23} = \mu_{32}$).

Vapor pressure osmometry (VPO) and solubility measurements are complementary methods for quantifying interactions of a solute or Hofmeister salt with model compounds or folded (native) biopolymers. VPO is most useful for relatively soluble, nonvolatile compounds while solubility is most useful for sparingly soluble compounds. Eqn (7) for analysis of VPO data is not applicable to situations where both the solute and the model compound or biopolymer are salts.

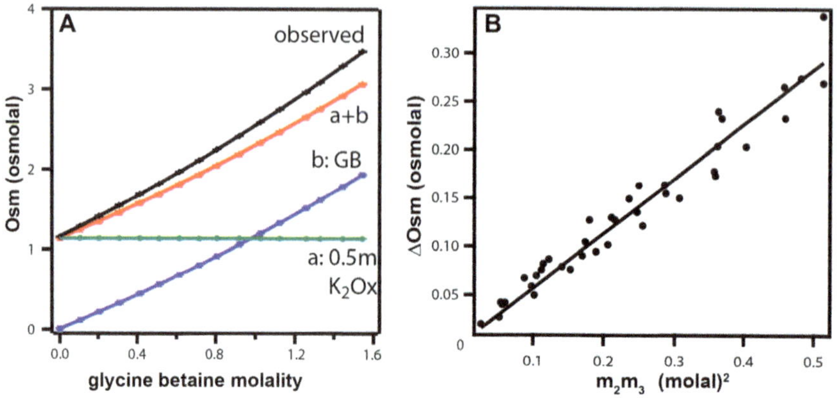

Fig. 8 VPO determination of ΔOsm and μ_{23} for the interaction of glycine betaine (GB) with K_2Oxalate. (A) Determination of ΔOsm for a series of experiments at 25 °C in which the molality of K_2Oxalate is fixed at 0.5 M and the molality of GB is increased. Green line (a) is the osmolality of 0.5 M K_2Oxalate in the absence of GB. Blue curve (b) is the osmolality of GB solutions as a function of its molal concentration. Red curve (a + b) is the sum of the osmolalities of these two component solutions: 0.5 M K_2Oxalate (a) and variable GB molality (b), plotted against GB molality. Top black curve is the observed osmolality of the three component solutions (0.5 M K_2Oxalate, variable GB) plotted *vs.* GB concentration. The difference between the red and black curves is the excess osmolality ΔOsm; this is positive and increases with increasing GB concentration. (B) A plot of ΔOsm against m_2m_3, the product of molal concentrations of K_2Oxalate and GB, using K_2Oxalate data from Capp *et al.*;[12] slope is μ_{23}/RT (eqn (7)).

B. Analysis of chemical potential derivatives (μ_{23}) and *m*-values ($\Delta\mu_{23}$) using surface areas and the solute partitioning model (SPM)

Guggenheim proposed a two-state (local, bulk) model of water near the air–water interface[31,41] and others proposed an analogous (local, bulk) model of water in the vicinity of a molecular surface.[30,50] From these precedents, we developed the solute partitioning model (SPM) to describe the competitive short range interactions of solutes or salt ions with water for an interface or surface. In the SPM, the number of water molecules per Å^2 of interface or molecular surface is designated b_1, and a microscopic partition coefficient K_p quantifies the local accumulation or exclusion of the solute from the local water at the interface or surface.

1) Analysis of μ_{23} for interaction of a solute or Hofmeister salt with a surface. From a molecular thermodynamic analysis using the SPM, the chemical potential derivative μ_{23} quantifying the interaction of a nonelectrolyte solute with a homogeneous molecular surface (*e.g.* aromatic or aliphatic hydrocarbon) is predicted to be proportional to ASA, with a proportionality constant α which is determined by b_1 and K_p. If the molecular surface is chemically heterogeneous, then b_1 and K_p are the average hydration and solute partition coefficient of the different functional groups that make up the surface.

$$\frac{\mu_{23}}{RT} = \alpha_{3,2}\text{ASA} = \frac{-(K_p - 1)b_1(1 + \varepsilon_3)\text{ASA}}{55.5} \tag{8}$$

where the microscopic partition coefficient $K_p = m_3^{\text{loc}}/m_3^{\text{bulk}}$ and the solute(3)–solute(3) self-nonideality correction term is $\varepsilon_3 = \text{dln}f_3/\text{dln}m_3$. The proportionality constant $\alpha_{3,2}$ defined in eqn (8) is called an interaction potential because it quantifies the strength and direction of the preferential interaction of the solute with the surface, is independent of ASA, and to a first approximation is independent of solute concentration. At solute concentrations above 1 M, some dependence of the quantities K_p, b_1, and ε_3 on m_3 is expected.[9,32] If the solute is completely excluded from the surface, so that $K_{p,3} = 0$, then measurement of $\alpha_{3,2}$ provides a determination of the hydration of the surface b_1.

Where the solute is a salt with ν ions per formula unit (ν_+ cations and ν_- anions; $\nu = \nu_+ + \nu_-$), SPM analysis yields the following expressions to interpret the chemical potential derivative μ_{23} for the Hofmeister (noncoulombic) interaction of the salt component and the individual salt ions with the surface:[31,33]

$$\frac{\mu_{23}}{RT} = \alpha_{3,2}\text{ASA} = \frac{-(\nu_+K_{p,+} + \nu_-K_{p,-} - \nu)b_1(1+\varepsilon_\pm)\ \text{ASA}}{55.5}$$
$$= (\nu_+\alpha_{+,2} + \nu_-\alpha_{-,2})\text{ASA} \qquad (9)$$

In eqn (9), the individual ion interaction potentials are $\alpha_{+,2} = -(K_{p,+} - 1)b_1(1+\varepsilon_\pm)/55.5$ and $\alpha_{-,2} = -(K_{p,-} - 1)b_1(1+\varepsilon_\pm)/55.5$ and the individual ion partition coefficients are $K_{p,+} = m_+^{\text{loc}}/m_+^{\text{bulk}}$, $K_{p,-} = m_-^{\text{loc}}/m_-^{\text{bulk}}$. The term $\varepsilon_\pm = \text{dln}f_\pm/\text{dln}m_3$ corrects for self-nonideality of the salt component (3). If both cation and anion of a salt are completely excluded from the surface, so that $K_{p,+} = K_{p,-} = 0$, then measurement of $\alpha_{3,2}$ for that salt provides a determination of the hydration of the surface b_1.

2) Analysis of *m*-values ($\Delta\mu_{23}$) for the effect of a solute or Hofmeister salt on a process with a change in surface area ΔASA

From the SPM-based eqn (8), the *m*-value quantifying the effect of a nonelectrolyte solute on a process is interpreted a

$$\frac{m\text{-value}}{RT} = -\frac{\text{dln}K_{obs}}{\text{d}m_3} = \Delta\frac{\mu_{23}}{RT} = \alpha_{3,2}\Delta\text{ASA} = \frac{-(K_p - 1)b_1(1+\varepsilon_3)\Delta\text{ASA}}{55.5} \qquad (10)$$

where $\alpha_{3,2}$ is the interaction potential for the surface specified by the ΔASA.

From eqn (9), the *m*-value quantifying the noncoulombic (Hofmeister) effect of a salt on a process is interpreted as

$$\frac{m\text{-value}}{RT} = \Delta\frac{\mu_{23}}{RT} = \frac{-(\nu_+K_{p,+} + \nu_-K_{p,-} - \nu)b_1(1+\varepsilon_\pm)\Delta\text{ASA}}{55.5}$$
$$= (\nu_+\alpha_{+,2} + \nu_-\alpha_{-,2})\Delta\text{ASA} \qquad (11)$$

The SPM-based analysis therefore predicts that solute and Hofmeister salt *m*-values are (a) independent of solute concentration, in agreement with experiment (see for example Fig. 2), and (b) proportional to ΔASA for homologous series of biopolymer surfaces with same composition (same K_p and b_1), like those exposed in unfolding globular proteins (Fig. 3).

3) SPM-based interpretation of surface tension increments

Completely analogous expressions to eqn (8) and 9 are obtained to interpret surface tension increments STI of nonelectrolytes[51] and salts.[31,33,41] For nonelectrolyte solutes,

$$\text{STI} = \alpha_2 = \frac{-(K_p - 1)b_1(1+\varepsilon_3)}{55.5} \qquad (12)$$

and, for salts,

$$\text{STI} = \alpha_2 = \frac{-(\nu_+K_{p,+} + \nu\ _-K_{p,-} - \nu)b_1(1+\varepsilon_\pm)}{55.5} = -(\nu_+\alpha_+ + \nu_-\alpha_-) \qquad (13)$$

Measurements of STI for solutes and salts which are completely excluded from the air–water surface allow determination of b_1, the amount of local water per unit area of that surface.

C. Dissection of solute–surface interaction potentials α for heterogeneous molecular surfaces

For interactions of a solute or salt (component 3) with the chemically heterogeneous surface of a biopolymer or a small model compound, different functional groups are expected to interact differently with the solute and exhibit different values of K_p (and possibly b_1). Given the short-range character of these interactions, we proposed that the observed μ_{23}/RT is the sum of contributions $\alpha_i\mathrm{ASA}_i$ from interactions of the solute with the different types of biopolymer surface (functional groups):[8,12,21]

$$\frac{\mu_{23}}{RT} = \sum \alpha_i \mathrm{ASA}_i \tag{14}$$

where each α_i for the interaction of solute 3 with a functional group on solute 2 is related to a local partition coefficient for partitioning of solute 3 between bulk water and the local water of hydration of that functional group:

$$\alpha_i = \frac{-(K_{p,i} - 1)b_1(1 + \varepsilon_3)}{55.5} \tag{15}$$

as in eqn (8). Values of μ_{23}/RT for interactions of a solute with inorganic salts and electrolytes as model compounds include the contribution of the inorganic salt ions:

$$\frac{\mu_{23}}{RT} = \sum_i \alpha_i(\mathrm{ASA})_i + (\nu_{\mathrm{ion}})\beta_{\mathrm{ion}} \tag{16}$$

where β_{ion} is the contribution to μ_{23}/RT from the solute-ion interaction, and ν_{ion} is the stoichiometric number of that ion per formula unit of the salt.

A "two way" breakdown of μ_{23} values into interactions of individual functional groups on solute 3 with functional groups on solute 2 is currently being investigated (Knowles *et al.*, in preparation):

$$\frac{\mu_{23}}{RT} = \sum\sum \alpha_{ij}\mathrm{ASA}_i\mathrm{ASA}_j \tag{17}$$

Data is being obtained to test the dissection of μ_{23} values for model compounds, and the buildup of μ_{23} values or *m*-values for biopolymer processes from the interactions of the solute with the individual functional groups on the biopolymer surface involved in the process. We are systematically determining α_i values for the interaction of urea, GB, Hofmeister salts and other small solutes of biochemical significance (proline, glycols, polyols, sugars, TFE, *etc.*) with different biopolymer functional groups from model compound data. From the resulting database of α_i values, it is straightforward to predict or interpret values of μ_{23}/RT for the interactions of these solutes with native biopolymers, and to predict or interpret *m*-values for biopolymer processes. Hence solutes serve as very useful probes of interface formation and large-scale coupled conformational changes in the steps of biopolymer processes, including open complex formation and transcription initiation by RNA polymerase at promoter DNA.

IV. Case studies: quantifying the interactions of GB, urea, Hofmeister salts and salt ions with functional groups of proteins and nucleic acids and the hydration of these groups; applications to protein and nucleic acid processes

A) Introduction

Here we summarize the determination of solute–surface interaction potentials (α-values) and partition coefficients K_p for the interactions of GB, urea and Hofmeister salts with the functional groups of proteins, and the hydration (b_1) of some of these groups, from the database of model compound interaction data (μ_{23} values).

Quantitative thermodynamic information about these very important noncovalent interactions of biochemistry and aqueous chemistry was not previously available. Interaction potentials for urea and GB with polar and charged groups make good chemical sense in terms of competitive hydrogen bonding. With these α-values (or, equivalently, K_p values) and structural information about the composition and amount of protein surface exposed or buried in a protein process, solute or Hofmeister salt m-values for the process can be predicted. Alternatively, from experimental measurements of solute and/or Hofmeister salt m-values for the steps of a protein process or mechanism, information about the amount and composition of the protein surface exposed or buried in interfaces and coupled conformational changes in those steps can be obtained. In this way solutes and Hofmeister salts can serve as very useful probes of steps involving interface formation or large scale conformational changes in the mechanism of action of protein molecular machines.

B) Determining the hydration b_1 of different groups and surfaces

To use solutes as probes or to predict or interpret solute or Hofmeister salt m-values, only α-values (and ASA information for a prediction) are required. The SPM-based analysis provides justification for the use of ASA information to interpret m-values or values of μ_{23} and obtain α-values, but the SPM is not otherwise needed for this more empirical analysis. To interpret α-values and obtain solute partition coefficients (K_p) using the SPM-based analysis, it is necessary to determine the hydration b_1 of each functional group and type of surface. Lower bound values of the hydration b_1 of some functional groups and surfaces can be determined by assuming complete exclusion ($K_p = 0$) of the solute or salt exhibiting the largest positive α-value (*cf.* eqn (8), (9), (12), (13)). Na_2SO_4 and sucrose have the most positive surface tension increments and surface α-values, indicating that Na^+, SO_4^{2-} and sucrose are the most highly excluded cation, anion and nonelectrolyte from the air–water interface. Na_2SO_4 has the largest unfavorable effect of any salt on solubility of benzene and toluene and the largest aromatic hydrocarbon α-value,[33] indicating that Na^+ and SO_4^{2-} are the most highly excluded cation and anion from aromatic hydrocarbon surface. α-Values for interactions of the excluded solute glycine betaine with anionic and amide O surface are more positive than for interactions of GB with any other functional group.[12]

From Na_2SO_4–aromatic hydrocarbon solubility m-values and Na_2SO_4 and sucrose STI values, analyzed using eqn (9), (12), and (13), lower bound values of b_1 of 0.18 $H_2O/Å^2$ to 0.20 $H_2O/Å^2$ are obtained for the amount of local water at the molecular surface of aromatic hydrocarbons in water and at the air–water interface. This experimentally-derived lower bound on the amount of local (hydration or surface) water corresponds to two layers of water at bulk density, an eminently reasonable result. MD simulations predict complete exclusion of Na^+ and SO_4^{2-} ions of Na_2SO_4 from approximately this amount of water at the air–water interface (see Fig. 12 below).[52,53] Recent spectroscopic studies show that the effect of solutes and salt ions on water structure is confined to the first one or two layers of water of hydration of the solute or salt ion.[28] From analysis of the GB data, similar lower bounds on b_1 are obtained for anionic carboxylate and amide O groups, though anionic phosphate O may be more hydrated. We therefore use a hydration $b_1 = 0.18$ $H_2O/Å^2$ to calculate solute partition coefficients K_p for all types of protein surface. Recent MD simulations of water at the surface of the villin headpiece subdomain (HP-36) conclude that the thickness of "biological water" is uniform for different segments of the protein (~5 Å).[54]

C) Why glycine betaine (GB) is a strong osmoprotectant and protein stabilizer while urea is a protein denaturant: opposite preferential interactions with key protein functional groups

Glycine betaine and urea have opposite effects on protein self-assembly processes (*cf.* Fig. 1). Addition of GB favors the assembly direction (folding (*e.g.* Fig. 3), precipitation),[55] which removes protein surface from water, while urea favors disassembly (unfolding (*e.g.* Fig. 3), dissolving),[2] which exposes protein surface to water. But GB does not drive nucleic acid helix formation, having little or no effect on the stability of AT-rich DNA and destabilizing GC-rich DNA,[14–16,18] while urea is destabilizing for all base compositions.[17,56,57] GB is an exceptionally good solute to reduce water activity and boost osmolality *in vivo*,[58] in concentrated protein or nucleic acid solutions,[14,15,59] and more generally *in vitro*,[12] while urea is not.[15,20] What are the thermodynamic and molecular origins of these very different interactions of GB and urea with biopolymers, and very different effects of GB and urea on protein and nucleic acid processes? What information is needed to predict or interpret GB and urea *m*-values (*i.e.* differences in μ_{23}; eqn (2)) for protein processes?

1. Obtaining μ_{23} values for interactions of urea and GB with functional groups of proteins by osmometry and solubility assays. To determine the interactions of a solute with the functional groups and types of surface of proteins and nucleic acids using the analysis of section III above requires measurement of the preferential interactions of these solutes with model compounds displaying various combinations of these functional groups and surfaces. For proteins, these types of surface include: nonpolar (aliphatic, aromatic) hydrocarbon (C), polar (amide, hydroxyl) and anionic (carboxylate) oxygen (O), and polar (amide) and cationic (ammonium, guanidinium) nitrogen (N).

Preferential interactions of small solutes like GB and urea with these functional groups of proteins are short-range (noncoulombic) and relatively weak, because they are competitive interactions, in which the solute and the functional group can interact with water or with each other. A reasonable starting point, motivated by the data, is to assume that the free energy consequences of these preferential interactions are additive, and that the preferential interaction of GB, urea or any other small solute or salt ion with a protein surface can be built up as the sum of its interactions with individual functional groups on that surface, quantified using model compound data (*cf.* eqn (16)). Large datasets (μ_{23} values) quantifying the interaction of urea and GB with model compounds displaying one or more protein functional group have been obtained by VPO, solubility, and phase partitioning studies, and deduced (for hydrocarbon surfaces) from micelle formation assays.[8,12]

Fig. 9 shows VPO data for the interaction of urea and GB with sufficiently soluble model compounds plotted as ΔOsm *vs.* m_2m_3 as in Fig. 8B. (From eqn (7), the slopes of these plots are μ_{23}/RT.) Addition of urea lowers the chemical potential of all nonelectrolyte model compounds studied (*i.e.* $\mu_{23} < 0$; Fig. 9A panels a and b), indicating that urea interacts favorably with most protein functional groups. Interactions of urea with Na^+ and K^+ salts of organic ions (Fig. 9A panels c and d) are less favorable, indicating most simply that preferential interactions of urea with these cations are unfavorable. On the other hand GB raises the chemical potential of most model compounds studied (both nonelectrolytes and electrolytes; Fig. 9B), indicating that preferential interactions of GB with most protein surface types are unfavorable. GB exhibits favorable preferential interactions only with Na-Benzoate, NaCl, and ArgHCl, indicating most simply that at least its interactions with aromatic hydrocarbon surface and Cl^- are favorable. The VPO data sets for interactions of urea and GB with nonelectrolyte solutes in Fig. 9 clearly demonstrate the same trends as those observed for the effects of these solutes on biopolymer processes (see Fig. 2). For both urea and GB, VPO data sets were augmented by solubility determinations of μ_{23} for sparingly soluble model compounds.[8,12]

2. Analysis of urea- and GB- model compound μ_{23} values into interactions with individual functional groups of proteins. Values of μ_{23} for interactions of urea and GB with model compounds were dissected, using eqn (16), into individual contributions from interactions of these solutes with each model compound functional group and inorganic ion. The resulting α_i and β_{ion} values are shown in Table 1. Since the

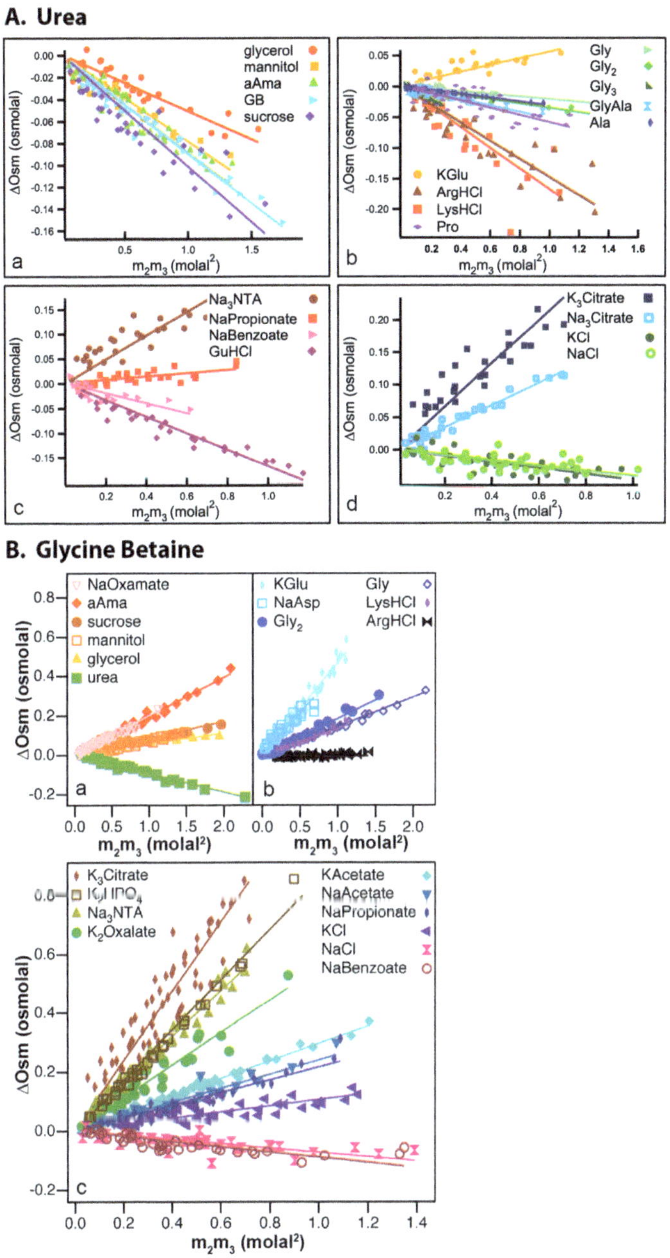

Fig. 9 Interactions of urea (panel A) and GB (panel B) with model compounds. Excess osmolalities ΔOsm (eqn (7)) determined by VPO at 25 °C are plotted *vs.* m_2m_3, the product of molal concentrations of the model compound and urea or GB. Slopes are μ_{23}/RT. VPO data for nonelectrolytes are grouped in panels (a) and (b) for urea and panel (a) for GB; the other panels show VPO data for salts. Adapted with permission from Guinn *et al.*,[8] Capp *et al.*[12]

model compounds studied differ greatly in surface composition, and the number analyzed (45 for urea, 27 for GB) greatly exceeds the number of α_i and β_{ion} values fitted for, these values are well-determined. When these α_i and β_{ion} values are used to predict μ_{23}/RT values for interactions of urea and GB with all the model compounds in the training set, very good agreement with the experimental values is obtained.[8,12]

Table 1 Interactions of urea and glycine betaine with functional groups of proteins (α_i and K_p values)[a] and inorganic ions (β_{ion} values)[b]

	$10^3\alpha_i/(m^{-1}\ \text{Å}^{-2})$		SPM K_p	
Surface type (*i*)	Urea	GB	Urea	GB
Aromatic C	−0.89 ± 0.05	−2.3 ± 0.4	1.28 ± 0.02	1.62 ± 0.11
Amide O	−0.87 ± 0.18	2.8 ± 1.0	1.28 ± 0.06	0.24 ± 0.27
Carboxylate O	−0.40 ± 0.15	2.9 ± 0.2	1.13 ± 0.05	0.22 ± 0.06
Amide N	−0.32 ± 0.23	−2.0 ± 0.7	1.10 ± 0.07	1.54 ± 0.19
Hydroxyl O	−0.25 ± 0.06	0.1 ± 0.2	1.08 ± 0.02	0.97 ± 0.06
Aliphatic C	−0.11 ± 0.05	0.3 ± 0.3	1.03 ± 0.02	0.92 ± 0.08
Cationic N	0.18 ± 0.16	−1.2 ± 0.4	0.94 ± 0.05	1.32 ± 0.11
Inorganic Ion	$\beta_{ion}/(m^{-1})$ [c]			
Na^+	0.104 ± 0.013	0		
K^+	0.146 ± 0.014	0.05 ± 0.02		
Cl^-	−0.170 ± 0.021	−0.04 ± 0.04		

[a] α_i defined by eqn (14) and K_p defined by eqn (15), all values determined at 25 °C. [b] Guinn *et al.*[8] [c] β_{ion} defined by eqn (16).

Using the SPM-based molecular thermodynamic interpretation of α values (eqn (15)), and $b_1 = 0.18\ H_2O/\text{Å}^2$ (as discussed above), local–bulk partition coefficients K_p are obtained for the distribution of urea and GB between each functional group or surface type and bulk water. Comparison of these partition coefficients (K_p) for urea and GB (Table 1) provides molecular insight into the competitive interactions involving water, solute and model compound functional groups that determine these K_p and α values. For polar functional groups, hydrogen bonding interactions appear most important; urea and GB differ greatly in their interactions with these O and N moieties because urea has multiple hydrogen bond donors and acceptors while GB has only hydrogen bond acceptors.[8,12]

Table 1 quantifies the extent to which the functional groups of proteins prefer to interact with urea and GB, as compared to interactions with water. The strongest preferential interactions of urea are favorable interactions with amide O and aromatic C; interactions of urea with anionic and hydroxyl O, amide N and aliphatic C are also favorable but significantly weaker, while the urea-cationic N interaction is unfavorable. Urea accumulates to a moderate extent in the vicinity of both amide O and aromatic C (local concentration ~28% greater than bulk) and to a lesser extent at anionic O (~13% above bulk), amide N (~10% greater than bulk) and hydroxyl O (~8% greater than bulk). Very modest accumulation of urea is observed at aliphatic C (only ~3% greater than bulk). Cationic N surface (and Na^+ and K^+ ions) prefer to interact with water than with urea; as a result the local concentration of urea in the vicinity of cationic N is 6% less than its bulk concentration. Cl^-, on the other hand, exhibits a favorable preferential interaction with urea. The favorable or unfavorable character of the urea–ion interaction is consistent with expectations based on the surface composition of urea (~2/3 partially-positive amide N, ~1/3 partially-negative amide O).

By far the strongest preferential interactions of GB are its highly unfavorable interactions with anionic and amide O. As discussed above, GB is highly excluded from two layers of water at the surface of these oxygens (local concentration less than 25% of the bulk GB concentration). GB is weakly excluded from aliphatic C (local concentration ~8% less than bulk) and hydroxyl O (local concentration ~3% less than bulk). For urea, all four of these interactions are favorable. GB accumulates to a very significant extent in the vicinity of aromatic C and amide and cationic N (local concentrations 62%, 54% and 32% greater than bulk). Comparison of β_{ion} values shows that net interactions of GB with inorganic ions (Na^+, K^+, Cl^-) are relatively weak by comparison with urea–ion interactions.

3. Molecular interpretation of K_p values for urea and GB accumulation or exclusion in the vicinity of functional groups of proteins. A "two-way" dissection (see eqn (17)) of μ_{23} values for interactions of urea and GB with model compounds into contributions from interactions of individual functional groups on both solutes has not yet been accomplished. Nevertheless a qualitative interpretation of K_p values for interactions of urea and GB with the polar and charged functional groups of proteins is possible, leading to very reasonable deductions about the competitive hydrogen bonding capabilities of these solutes and water. For example, the favorable interactions of urea with amide O and anionic O indicate that the hydrogen bonds formed with amide-NH_2 groups of urea as the donor and these oxygens as acceptor are more favorable (in part because there are more ways of forming them) than the corresponding hydrogen bonds with water, resulting in accumulation of urea at these groups. GB, on the other hand, lacks a hydrogen bond donor and so has no way to compete with water to interact with these amide and anionic oxygens, which therefore prefer to remain hydrated, largely excluding GB.

The observed accumulation of GB in the vicinity of amide and cationic N indicates that the anionic carboxyl oxygens of GB are a better acceptor of hydrogen bonds from these nitrogens than is the polar oxygen of water. Urea is much less accumulated at amide N than at amide O and is somewhat excluded from cationic N. These results indicate that while urea–NH···O=C-amide hydrogen bonds are quite favorable, relative to hydrogen bonds to water, urea–NH···N-amide hydrogen bonds are unfavorable relative to interactions with water. Amide hydrogen bonds in α-helices and β-sheets are of course –NH···O=C–. The shift from a modest favorable interaction of urea with amide N to a slightly unfavorable interaction of urea with cationic N indicates that protonating the nitrogen disfavors urea–NH···$^+$NH (cationic N) hydrogen bonds (and eliminates the possibility of hydrogen bonds where urea N is the donor) by more than it favors (cationic N)$^+$NH···O=C(urea) hydrogen bonds.

Both urea and GB are accumulated at aromatic C; for both solutes this interaction is the strongest favorable preferential interaction. For GB, this quite favorable interaction is almost certainly a cation–π interaction of the trimethyl ammonium group of GB with the aromatic ring.[12] The more modest accumulation of urea at the aromatic surface presumably arises from hydrogen bonding or partial-cation π between amide N groups on urea and the aromatic π-system, and/or π–π stacking interactions between the π-system or urea and the ring.[60] Urea and GB exhibit weak favorable and unfavorable interactions with aliphatic C. The interaction of water with both aromatic and aliphatic C surface is unfavorable, relative to water–water interactions (the hydrophobic effect), and the interactions of urea and GB with these hydrocarbon surfaces need to be interpreted relative to that interaction.

4. Application of interaction potentials (α values) or K_p values to interpret and predict effects of urea and GB on protein processes (folding, DNA binding). Interaction potentials (α values) or K_p values in Table 1 allow one to interpret and predict urea and GB *m*-values for steps of protein processes in terms of the amount and composition of the surface (ΔASA) involved in the process. Values of μ_{23} for interactions of urea with folded and unfolded protein surfaces can also be predicted from ASA information. Because urea has been widely used to determine protein stability by extrapolation to zero urea concentration, a large collection of urea *m*-values is available for analysis.[13,61]

a) Protein folding. Using the α values in Table 1 and ΔASA values calculated using an extended chain model, we predicted *m*-values for unfolding a set of well-characterized globular proteins and for dissociation of lac repressor from lac operator DNA and found good agreement between predicted and experimental values (Fig. 10). Since these unfolding processes form a homologous series with similar compositions of the protein surface exposed in unfolding and differing primarily in the total ASA exposed, a systematic offset of predicted from experimental *m*-values would indicate a systematic error in the ASA analysis or an error in one or more of the α-values, but no such systematic deviation is observed, and the scatter of the data is about the line is similar to that observed in Fig. 3. Only one isothermal GB *m*-value has been determined; for unfolding of the lac DNA binding domain (Fig. 2) the experimental *m*-value is 790 cal mol^{-1} molal^{-1} at 59 °C, similar to the predicted *m*-value of 690 cal mol^{-1} molal^{-1} at 25 °C. While this 12% difference between predicted and experimental *m*-values is not outside of the uncertainty in both values, it is interesting to observe that predicted and observed values of the quantity *m*-value/*RT* differ by only 3%. If the stabilizing effect of GB were entirely entropic, as would be predicted to be the case for a completely excluded solute, then *m*-value/*RT* would be independent of temperature while the *m*-value itself would be more temperature dependent.

b) Protein interface formation and coupled conformational changes. The α-values in Table 1 are very useful to predict and interpret effects of interface formation and coupled conformational changes in protein processes. Effects of urea and GB on binding of lac repressor tetramer to 40 bp lac operator DNA are shown in Fig. 11. Logarithms of normalized binding constants ($\ln(K_{obs}/K_{obs,0})$), determined from urea and GB titrations of poised lac repressor–lac operator mixtures, are plotted as functions of urea (A) and GB (B) concentration at 25 °C. In this binding interaction, not only is an extensive interface formed between two DBD of lac repressor and 20 bp of operator DNA, but also two protein interfaces form as the two flexible hinge regions that tether the DBD to core repressor fold and the DBD-operator assembly then interacts with the core repressor.[61–63] Formation of these three interfaces buries 6900 Å^2 of protein and DNA surface.[12] From structural data on the amount and composition of the protein and DNA surfaces buried, we

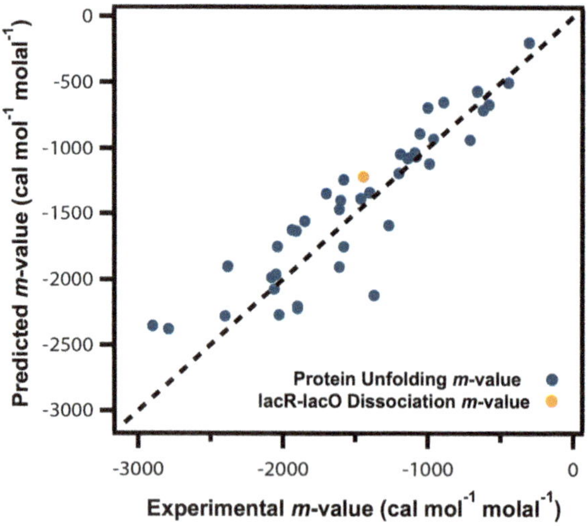

Fig. 10 Predicting urea *m*-values. Comparison of predicted and experimental *m*-values for effects of urea on protein unfolding and dissociation of lac repressor from operator DNA. Predicted values calculated from eqn (14) using urea α-values (Table 1) and published amounts and composition of the ΔASA.[8,12] Dashed line represents equality of predicted and experimental values. Experimental unfolding *m*-values are from Fig. 7 and lacR–lacO dissociation *m*-value is from Fig. 11. Predicted protein unfolding *m*-values are from Guinn *et al.*[8] and predicted lacR–lacO *m*-value is from Fig. 11.[12,61]

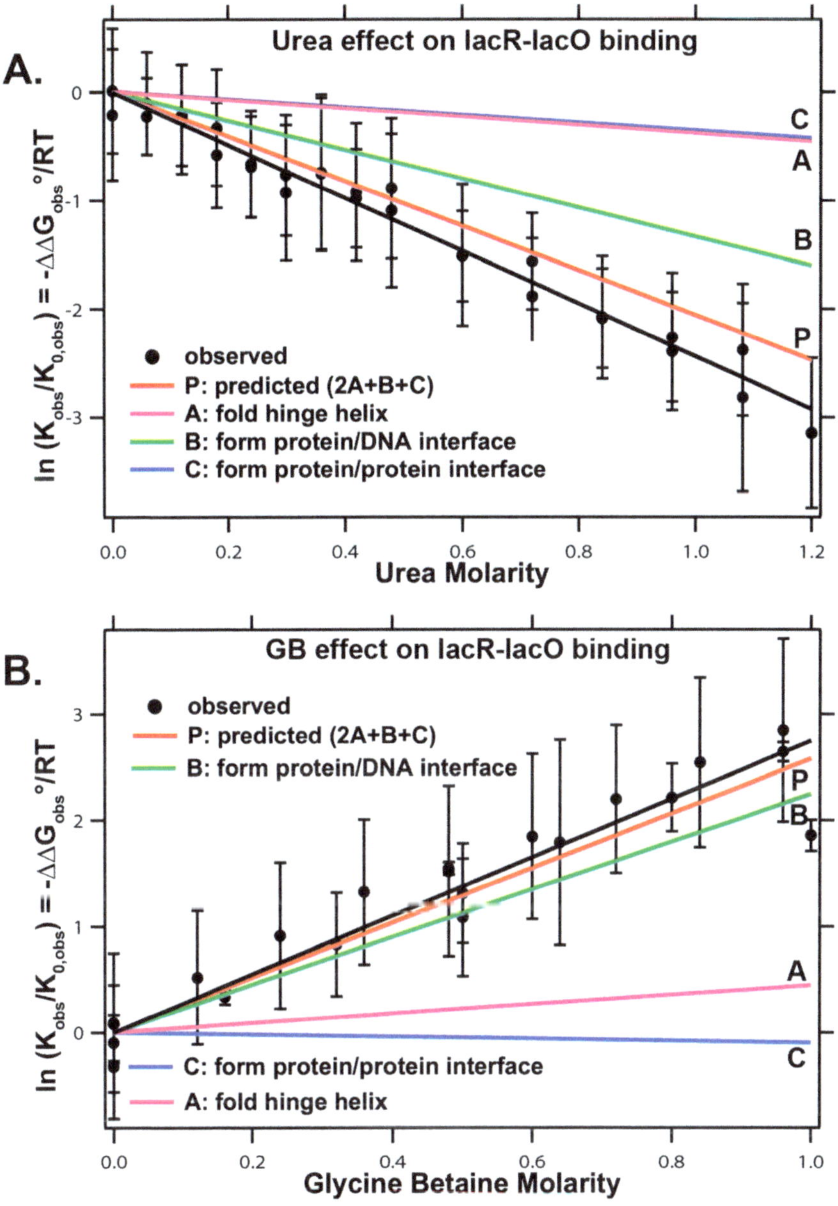

Fig. 11 Effect of urea and GB on binding of lac repressor protein to lac operator DNA. Comparison of observed and predicted effects of urea (panel A) and GB (panel B) on the lac repressor–lac operator binding constant K_{obs} relative to its value K_0 in the absence of solute. (Experiments performed at constant salt molality at 25 °C.) Predicted contributions to the dependence of $\ln(K_{obs}/K_0)$ on solute concentration from the three interfaces formed during this assembly process are also shown. Experimental lacR–lacO binding data in Panel A from unpublished data (M. W. Capp); binding data for panel B and ASA calculations are from Capp *et al.*[12]

predict an overall urea m-value = 1.24 ± 0.19 kcal mol^{-1} m^{-1} and GB m-value = 1.54 ± 0.84 kcal mol^{-1} m^{-1}, in agreement within the experimental uncertainty with experimental values of 1.42 ± 0.21 kcal mol^{-1} m^{-1} and 1.66 ± 0.24 kcal mol^{-1} m^{-1} respectively (Fig. 11). (The large uncertainty in the GB prediction reflects the uncertainties in the GB α values for aliphatic C and amide O (Table 1). Investigation of interactions of GB with additional model compounds will reduce these uncertainties.)

Formation of the repressor–operator interface buries 630 Å² of anionic DNA phosphate O surface, at which urea accumulates strongly and from which GB is strongly excluded.[12,56] Formation of this interface is predicted to contribute −1.3 m^{-1} (62%) to the urea *m*-value and 2.2 m^{-1} (85%) to the GB *m*-value. Folding the hinge helices, which buries 520 Å² of amide surface, is predicted to contribute −0.4 m^{-1} (19%) to the urea *m*-value and 0.4 m^{-1} (15%) to the GB *m*-value. Formation of the core repressor–DBD interface is predicted to contribute −0.4 m^{-1} (19%) to the urea *m*-value and to make no significant contribution to the GB *m*-value as a result of relatively small and compensating effects (Fig. 11). Therefore, while urea and GB are predicted to detect primarily lacR–lacO interface formation, formation of the two protein–protein lacR interfaces also contributes significantly (38%) to the urea *m*-value.

c) Predicting the most important interactions responsible for urea and GB effects. Dissection by functional group of the effects of urea and GB on different protein processes shows quantitatively why urea is a denaturant and GB is a stabilizer. Table 2 shows predicted contributions of different surface types to urea and GB *m*-values for unfolding a globular protein, melting an α-helix (which exposes a higher proportion of amide ASA) and for the interaction with native protein surface. All values in the table are for 1000 Å² of ASA.

For protein unfolding, interactions of urea and GB with amide O, aliphatic C and aromatic C surface exposed in unfolding are predicted to make the largest contributions to the *m*-value (Table 2). Interactions of urea with amide O and aromatic C are similarly favorable and moderately strong: because the ΔASA of amide O significantly exceeds that of aromatic C in unfolding, amide O is predicted to contribute

Table 2 Predicted contributions to an *m*-value or μ_{23} at 25 °C (cal mol⁻¹ molal⁻¹) for exposure of 1000 Å² of protein surface[a]

	Process								
	Globular protein unfolding			α-Helix melting			Surface of native protein[b]		
Surface type (*i*)	ΔASA	Urea	GB	ΔASA	Urea	GB	ASA	Urea	GB
Amide O	120	−62	199	480	−247	796	100	−52	166
Carboxylate O	40	−9	69	100	−24	172	160	−38	275
Hydroxyl O	30	−4	2	0	0	0	20	−3	1
Amide N	50	−9	−59	90	−17	−107	40	−8	−47
Cationic N	40	4	−28	0	0	0	130	14	−92
Aliphatic C	650	−42	115	330	−21	59	530	−35	94
Aromatic C	70	−37	−95	0	0	0	20	−11	−27
Predicted *m*-value or μ_{23} per 1000 Å² of surface:	−159	203		−309	920		−133	370	
Experimentally derived value:	−184[c]	224[d]		−598	ND		−124	776	

[a] Guinn *et al.*[8] [b] From osmometric data for interaction of solutes with native BSA.[14,20] BSA is anionic with a net charge of approximately −17 at the pH of these experiments,[74] so the BSA component includes at least the corresponding number of Na^+ ions. Interactions of solutes with these cations are not included in the predictions. Inclusion of these interactions would not significantly affect the predicted GB μ_{23} value, but would make the urea μ_{23} value less favorable (Table 1). [c] Urea globular protein *m*-value from average of Fig. 7 dataset for urea scaled to 1000 Å². [d] GB globular protein *m*-value from lacDBD data in Fig. 2 scaled to 1000 Å². Data taken at 59 °C but prediction determined from 25 °C α_i values.

~40% and aromatic C ~20% of the unfolding *m*-value. While the α-value for urea and aliphatic C is only marginally different from zero, its contribution to the ΔASA is largest and this interaction is predicted to contribute ~25% of the *m*-value. GB has both favorable and unfavorable interactions with protein functional groups; the net interaction with both folded and unfolded protein surface is predicted to be unfavorable, resulting in exclusion of GB. For unfolding, the unfavorable interaction of GB with amide O is predicted to be approximately as large as the overall GB *m*-value. The next two largest contributions are predicted to be from interactions of GB with aliphatic C and aromatic C surface; these are about equal in magnitude and opposite in sign.

For melting of a series of α-helices, urea *m*-values appear to be proportional to ΔASA, but with a very different proportionality constant than that observed for unfolding of globular proteins; the urea *m*-value/ΔASA ratio for α-helix melting[20,64] is more than three times larger than for unfolding of globular proteins. Using the α-values of Table 1, the *m*-value/ΔASA ratio for α-helix melting is predicted to be twice as large as for unfolding of globular proteins (Table 2), in semi-quantitative agreement with experiment. The predicted GB *m*-value for α-helix melting is also significantly larger (~4.4×) than the predicted globular protein unfolding *m*-value; data is not yet available to test this prediction. Table 2 shows that these differences arise primarily from the much larger proportion of amide ASA exposed in α-helix melting, and the strong favorable (urea) and unfavorable (GB) interactions with amide O surface, as previously proposed.[20]

The native protein surface displays a smaller proportion of hydrocarbon surface (55% of total) than is exposed in protein unfolding (72%) and so displays a larger proportion of polar and charged surface. GB interacts unfavorably with native protein surface due to its significant unfavorable interaction with polar and anionic oxygen surface, making GB a good osmoprotectant because its chemical potential is raised in the presence of protein surface increasing its contribution to the solution's osmolality. Urea on the other hand interacts favorably with native protein surface due to its favorable interaction with most surface types and so will be a poor osmoprotectant because its osmolality will be lowered in the presence of protein surface. However, we predict urea to be an effective protein solubilizer because its favorable interactions with native protein surface will increase protein solubility in water (GB's unfavorable interactions will lower protein solubility).

5. Comparison with other approaches.

a) Transfer free energy analysis. In the transfer free energy analysis of Auton and Bolen,[39,65] preferential interactions of urea and other solutes with amino acids and small peptides were determined from solubility measurements and dissected to quantify transfer free energies of the 20 amino acid side chains and the peptide backbone unit from water to one molar solute, closely related to μ_{23} values. From these data they concluded that urea denatures proteins primarily because of its favorable interaction with the peptide backbone. On the other hand, from the SPM-based α-value analysis described above, Guinn *et al.*[8] find that favorable interactions of urea with amide O (mostly backbone), aliphatic C (mostly side chain) and aromatic C (all side chain) make the most significant contributions. Predictions of the transfer free energy analysis agree well with *m*-values for urea and other solutes for protein unfolding and other protein processes.[40] In the transfer free energy analysis of protein unfolding, the average accessibility of backbone and individual side chains for two models (extended, more compact) of the unfolded state is used to calculate the ΔASA; Guinn *et al.* used an extended chain model of the unfolded state. Left to be resolved are some significant differences between values of μ_{23} for interactions of urea with the subset of five amino acids determined by osmometry and those calculated from the solubility data of ref. 39. In addition, values of μ_{23} predicted for interactions of urea from the best fit set of model compound α-values disagree significantly with those calculated from the solubility data of ref. 39 for

10 of the 18 amino acids and the cyclic dipeptide investigated.[8] By contrast predicted values of μ_{23} for amino acids and the cyclic dipeptide agree in 9 of 11 cases with experimental values of μ_{23} from osmometry and solubility in the data set of ref. 8.

b) Molecular dynamics simulations. While preferential interactions represent the strength of solute interactions with biopolymer functional groups relative to water they do not give the molecular origin of these interactions. Insight into the noncovalent interactions between solutes and biopolymers is provided by a statistical thermodynamic analysis (most simply the SPM, more generally Kirkwood–Buff (KB) integrals). A Kirkwood–Buff integral relates the radial distribution function of species 2 around species 3 to the chemical potential derivative μ_{23}.[45] Hence radial distributions of one solute in the vicinity of another predicted by simulation can be compared with values of μ_{23} from thermodynamic experiments by Kirkwood–Buff integrals.

Because solute effects are relatively subtle, MD simulation parameters must be carefully chosen to properly describe them. In fact, the standard Charmm force field actually predicts accumulation of GB at triglycine surface where it should be excluded.[66] In recent studies, Smith *et al.* have successfully applied atomistic simulations to obtain KB integrals for a number of binary and ternary solution systems; they have also used the KB integrals to improve atomistic force fields for small molecules such as urea.[44]

Using the KB force field, simulations of urea with the trp cage miniprotein[67] and polyglycine[68] indicate that urea is significantly hydrogen bonded to the protein backbone and that the number of hydrogen bonds to the backbone increases with urea concentration. Horinek *et al.* computed that in 60% of hydrogen bonds with polyglycine urea is acting as a donor, presumably to amide O, consistent with the relatively strong favorable preferential interaction of urea with amide O we observe. Of the computed 40% of hydrogen bonds where urea acts as an acceptor, urea O is the only acceptor seen supporting our interpretation that amide NH–O hydrogen bonds are favorable while NH–N bonds are not.[8,68] Urea interactions with hydrocarbon are also found to be significant in simulations since interactions with trp cage side chains, which contain significant aromatic and aliphatic hydrocarbon surface, contribute 60% to the urea *m*-value[67] (compared to 55% predicted from our preferential interaction data). Simulations of interactions of urea with a RNA hairpin show that urea destabilizes through hydrogen bonds to the bases as well as π–π stacking interactions with the heterocyclic nucleic acid ring, indicating that urea could be stacking on protein aromatic rings as well.[60]

c) Spectroscopic studies of urea–amide interactions. Evidence for hydrogen bonds of urea with amide groups has been obtained by FTIR and NMR spectroscopy. Sagle *et al.*[69] used FTIR to study urea's hydrogen bonding to the amide O of poly(*N*-isopropylacrylamide) (PNIPAM). When urea is added, the amide I absorption band shifts to a frequency indicative of CO–HN hydrogen bonding and collapsed state of the polymer is destabilized, presumably because urea hydrogen bonds to PNIPAM better than water does.[69] These spectroscopic measurements were used to calculate the fraction θ of PNIPAM amide O hydrogen bonded to urea as a function of urea concentration. This fraction θ can be predicted from the SPM K_p value:[8]

$$\theta = \frac{n_3^{\mathrm{loc}}}{n_3^{\mathrm{loc}} + n_1^{\mathrm{loc}}} = \frac{K_p m_3}{(m_1 + K_p m_3)}$$

The fraction of PNIPAM amide O that are hydrogen bonded to urea agrees with the SPM prediction within error below 4 molar urea.

Lim *et al.* studied the effect of urea, GuHCl and polyols on acid and base catalyzed amide hydrogen exchange (HX) using 1D NMR spectroscopy. They observed that urea slows the rate of amide HX of the diamide aAma, presumably by hydrogen bonding with amide N and so inhibiting HX. However, GuHCl had very little effect on the HX rate. Analysis of model compound solubility data (Fig. 14) shows that GuH^+ accumulates at amide surface. From the discussion above, this interaction appears to involve hydrogen bonding of GuH^+ to amide O and not to amide N. The absence of an effect of GuH^+ on exchange of the amide N is consistent with this interpretation. The polyols sorbitol and glycerol also do not affect the HX rate, indicating that they are excluded from the amide surface.[70]

D) Molecular interactions responsible for the Hofmeister anion and cation series and their different null points for different aqueous processes

What is the molecular basis of the Hofmeister series of cations and anions observed for such diverse processes in aqueous solution as unfolding proteins, dissolving hydrocarbons or proteins, and transferring water from bulk to the air–water interface? Do the Hofmeister ion series of Fig. 1 primarily reflect the very different tendencies of different ions to interact with a macroscopic or molecular nonpolar surface in water[1,2,71] or are other types of surface or combinations of surfaces (like the methylene and amide groups of the peptide backbone[7,33,72,73]) involved? Why are the null points of these series different for model processes like increasing the amount of air–water surface or dissolving hydrocarbons than for protein unfolding and other protein processes, and different yet again for nucleic acid melting? For protein, nucleic acid and other polyampholyte or polyelectrolyte processes, when and to what extent can Coulombic and Hofmeister effects of salts be separated? For processes involving polyions, how can amplified or reverse Hofmeister effects of counterions at low salt be explained?

1. Analysis of μ_{23} values and surface tension increments for interactions of Hofmeister salts with molecular and macroscopic surfaces to obtain single ion K_p values. Analysis of the extensive literature data for surface tension increments (STI; *i.e.* air–water surface free energy *m*-values per unit area) and hydrocarbon solubility free energy increments (*i.e.* solubility *m*-values for hydrocarbons), like those displayed in Fig. 5A and 5B, using an SPM-based analysis (eqn (10) and 11)[31,33,41,51] provides answers to the above questions, and provides a quantitative database to predict and interpret Hofmeister salt effects. For these homogenous nonpolar surfaces, interaction potentials (α-values) quantifying the preferential interaction of the salt component (the composite effect of the salt cation and anion) per unit area of the surface are obtained from (STI)/k_BT for air–water surface, where k_B is

Table 3 Hofmeister salts: Comparison of interaction potentials (α_i) for molecular hydrocarbon surfaces with reduced surface tension increments (STI/kT)

		$\alpha_i \times 10^3/m^{-1}Å^{-1}$ [a]		
Salt	STI/$kT \times 10^3/(m^{-1}Å^{-1})$	Aliphatic C	Aromatic C	Amide (O,N)
Na_2SO_4	6.7 ± 0.1	6.0 ± 0.1	5.9 ± 0.2	−6.8 ± 0.5
$(NH_4)_2SO_4$	5.6 ± 0.3	—	3.8 ± 0.1	−3.3 ± 0.3
KF	4.4 ± 0.3	3.4 ± 0.1	2.4 ± 0.1	−4.6 ± 0.3
GuH_2SO_4	2.3 ± 0.5	2.2 ± 0.3	0.3 ± 0.5	−5.8 ± 1.0
NaCl	4.2 ± 0.4	2.2 ± 0.1	2.0 ± 0.1	−4.2 ± 0.2
KCl	3.9 ± 0.3	2.0 ± 0.1	1.7 ± 0.1	−4.2 ± 0.2
KBr	3.3 ± 0.1	1.7 ± 0.1	1.2 ± 0.2	−4.1 ± 0.4
$NaClO_4$	0.5 ± 0.1	0.9 ± 0.1	1.1 ± 0.1	−7.0 ± 0.3
NH_4Cl	3.4 ± 0.4	—	0.8 ± 0.3	−1.2 ± 0.8
GuHCl	1.8 ± 0.6	0.3 ± 0.1	−0.8 ± 0.2	−3.7 ± 0.4

[a] α_i values determined at 25–30 °C.

Boltzmann's constant, and from μ_{23}/ASA for molecular hydrocarbon surface. These are listed in Table 3 for the complete set of Na^+ salts where both types of data (STI and solubility increments) are available, and for a few other cases of interest. Table 3 shows that the rank orders of salt interaction potentials for these two nonpolar surfaces are quite similar, but that interaction potentials for the air–water surface are systematically larger (less favorable preferential interaction of the salt) for air–water than for the aromatic hydrocarbon–water surface.

These trends of surface interaction potentials are clarified when interpreted to obtain microscopic local–bulk partition coefficients (K_p values) obtained from a SPM-based analysis. To determine K_p values requires a determination of the amount of local water b_1 per unit surface area obtained from the STI or solubility increment for a completely excluded salt. MD simulations indicate that Na_2SO_4 is completely excluded from the air–water interface, unlike other salts and acids.[52,53] Fig. 12 shows distributions of Na^+ and SO_4^{2-} near the air–water interface predicted from molecular dynamics simulations.[52] Both Na^+ and SO_4^{2-} are completely excluded from a surface region of comparable thickness (5–10 Å) to that deduced from the SPM-based analysis.[34,41] This supports the assumption that Na_2SO_4, the salt with the largest positive STI and largest effect on hydrocarbon solubility, is completely excluded ($K_p = 0$) from the local water at both these surfaces. As discussed earlier, this assumption yields $b_1 = 0.18 \pm 0.01$ H_2O/A^2, corresponding to approximately

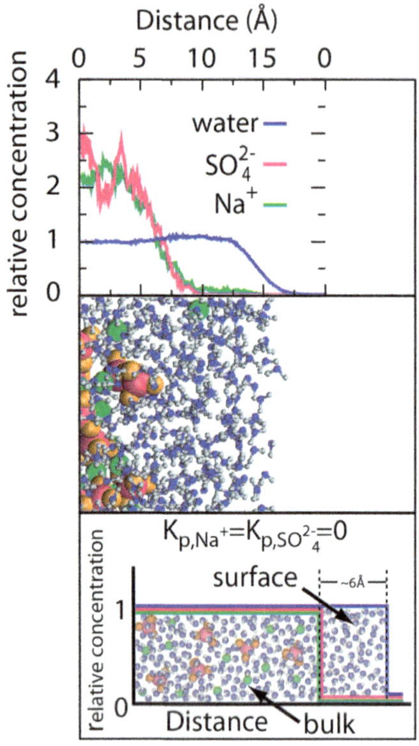

Fig. 12 Ion distributions (panel A) and molecular snapshot (panel B) of the air–water interface from molecular dynamics (MD) simulations of Gopalakrishnan *et al.*[52] Relative concentration means the local concentration at each point of water and ions relative to the average concentration for the entire simulation cell. The complete exclusion of Na^+ and SO_2^{4-} from the surface water in panels A and B agrees well with the treatment of Na_2SO_4 in the SPM as a completely excluded reference salt (panel C). Adapted with permission from Pegram and Record.[34]

two layers of water at bulk density. If Na_2SO_4 were not completely excluded ($K_p > 0$), then b_1 for these surfaces would of necessity be larger than 0.18 $H_2O/Å^2$. We assume that $b_1 = 0.18$ $H_2O/Å^2$ for other molecular surfaces; at present anionic phosphate oxygen appears to be the only surface where more than two layers of local water are present.[12]

Table 4 and Fig. 13 summarize partition coefficients K_p for individual salt ions at the nonpolar air–water surface and at the molecular surface of aromatic and aliphatic hydrocarbons in water. To within the experimental uncertainty, these single-ion K_p values for both cations and anions are independent of salt concentration up to approximately 1 molal or higher. Values of K_p for cation and anion are additive, as predicted by eqn (9), and are independent of the salt investigated. For both the air–water nonpolar surface and the molecular nonpolar surface of aromatic hydrocarbons (and aliphatic hydrocarbons, where data is available), these K_p values define cation and anion series which closely follow the cation and anion Hofmeister series for protein processes listed in Fig. 1. However, the null points ($K_p = 1$) for both cations and anions differ for the different surfaces.

This dissection of individual Hofmeister ion K_p values indicates that all Hofmeister cations are excluded from the air–water interface. (The proton is accumulated but is not part of the Hofmeister cation series for protein processes.) Cations like GuH^+ and NH_4^+, which are at the accumulated, destabilizing end of the protein Hofmeister series, are moderately excluded from the air–water interface. Concentrations of GuH^+ and NH_4^+ in the two layers of water at the air–water surface are predicted to be approximately 67% and 25% as large as their bulk concentrations, respectively; no Hofmeister cation yet investigated is more accumulated at the air–water interface than GuH^+. Anions from the favorably-interacting (accumulated) end of the protein Hofmeister series (SCN^-, I^-, ClO_4^-) also accumulate at the air water interface. Local concentrations of SCN^- and ClO_4^- are predicted to exceed their bulk concentrations by 1.6 and 1.8 fold, respectively, while I^- is predicted to be less dramatically accumulated (1.2 fold greater surface concentration compared to bulk). Acetate, which typically ranks with other carboxylates and with F^- near the excluded end of the Hofmeister series for protein processes, accumulates at the air–water surface. We hypothesized that accumulation of acetate allows its methyl end to be at the surface while the carboxylate end remains in water.[41]

Table 4 SPM partition coefficients ($K_p = m_3^{loc}/m_3^{bulk}$)[a] of individual Hofmeister cations and anions[b]

	K_p[c]				K_p[c]		
Cation	Air/Water	Aromatic	Aliphatic	Anion	Air/Water	Aromatic	Aliphatic
GuH^+	0.67	0.99	0.70	SCN^-	1.64	1.80	—
NH_4^+	0.25	0.49	0.28	I^-	1.18	1.62	1.54
Li^+	0.08	0.28	0.23	ClO_4^-	1.77	1.54	—
K^+	0.12	0.11	0.08	NO_3^-	0.98	1.47	—
Na^+	0.00	0.03	—	Br^-	0.86	1.40	—
				Cl^-	0.69	1.23	1.23
				H^-	0.53	0.92	—
				SO_4^-	0.00	0.00	—
H^+	1.50	0.61	0.59	OH^-	0.58	1.03	—
				Acetate-	1.30	—	—

[a] K_p defined in eqn (9). [b] Pegram and Record.[33,41] [c] Uncertainties in these K_p values typically ±10–20% for $K_p > 0.3$ and >20% for $K_p < 0.3$.

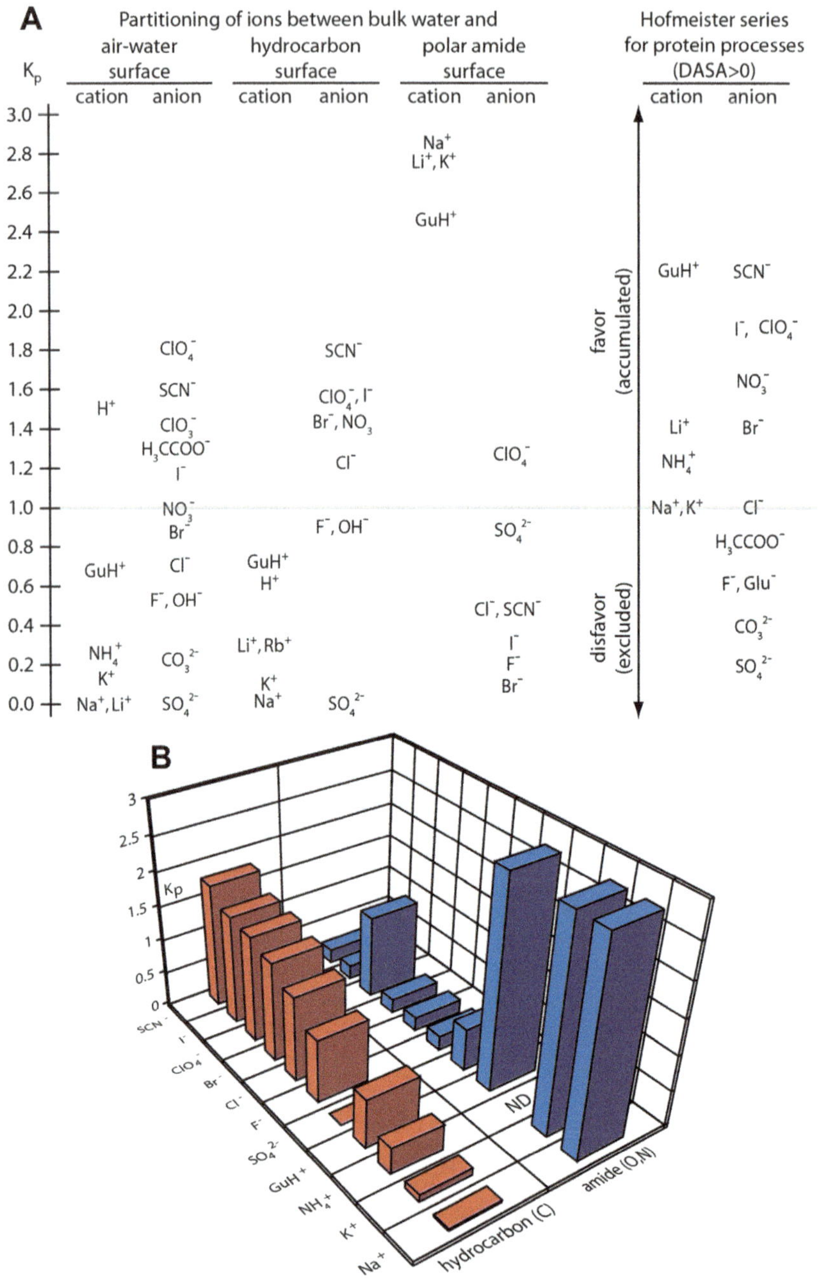

Fig. 13 Single ion partition coefficients for Hofmeister cations and anions at nonpolar and polar surfaces. (A) Rank orders of cation and anion partition coefficients $K_p = m_{ion}^{local}/m_{ion}^{bulk}$ quantifying local accumulation ($K_p > 1$) or exclusion ($K_p < 1$) of these ions in the vicinity of the macroscopic air–water surface and the molecular surface of hydrocarbons and amide groups in water. The null point of each series is at $K_p = 1$. On the right is an approximate alignment of the ion series for protein processes (Fig. 1) with these K_p series. (B) Projection bar graph displaying K_p values for interactions of key Hofmeister cations and anions with molecular hydrocarbon (red) and amide (blue) surfaces (from panel A left), arranged according to the Hofmeister series order for effects of these ions on protein processes (Fig. 1). Both panels adapted with permission from Pegram and Record.[33,34]

Hofmeister anions accumulate to a greater extent at the surface of aromatic hydrocarbons (and also aliphatic hydrocarbons, where data is available) in water than at the air–water surface. For interactions with aromatic hydrocarbon surface, the null point of the Hofmeister anion series is F^-. This is a very different null point than those observed for air–water (NO_3^-) and protein (Cl^-) surfaces. F^- is highly excluded from the air–water surface ($K_p = 0.53$) and is net excluded from protein surface. For Hofmeister cations, the differences are smaller and may not be as significant as for anions. For interactions with aromatic hydrocarbon surface, GuH^+ is at the null point ($K_p = 1$) of the cation series. Both GuH^+ and NH_4^+ interact more favorably with aromatic than with aliphatic surface. Extents of exclusion of GuH^+ and NH_4^+ from aliphatic and from air–water surface are similar. Since GuH^+ and NH_4^+ are the cationic functional groups of arginine and lysine side chains of proteins, their interactions with aromatic and aliphatic hydrocarbon surfaces are relevant in many biological contexts. The finding that GuH^+ is much less strongly excluded from hydrocarbon surfaces than are alkali metal cations is the principal explanation for why GuHCl is a protein unfolding agent and NaCl or KCl are not, as discussed below.

2. Interactions of Hofmeister salts with amide-containing model compounds: deducing interactions of cations and anions with amide groups. To interpret or predict Hofmeister salt effects on the thermodynamics of protein processes, individual thermodynamic contributions of interactions of anions and cations with amide and other types of polar biopolymer molecular surface as a function of salt concentration are needed. Literature data exists to accomplish part of this goal; more experiments are needed to complete this analysis. Salt effects on solubility of oligopeptides and amide model compounds follow the classical Hofmeister series, but the null point is shifted from that observed for molecular hydrocarbon or air–water surfaces. Fig. 14 shows literature data for a series of end-capped oligoglycine peptides

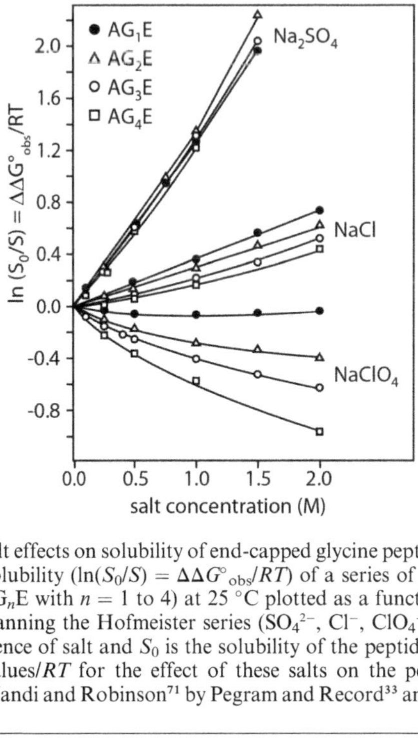

Fig. 14 Hofmeister salt effects on solubility of end-capped glycine peptides. Values of the logarithm of the relative solubility ($\ln(S_0/S) = \Delta\Delta G^\circ_{obs}/RT$) of a series of glycine peptides (acetyl glycine ethyl esters; AG_nE with $n = 1$ to 4) at 25 °C plotted as a function of concentration of Na^+ salts of anions spanning the Hofmeister series (SO_4^{2-}, Cl^-, ClO_4^-). S is the solubility of the peptide in the presence of salt and S_0 is the solubility of the peptide in the absence of salt. Initial slopes are m-values/RT for the effect of these salts on the peptide transfer process. Figure redrawn from Nandi and Robinson[71] by Pegram and Record[33] and used with permission.

(N-acetyl(glycine)$_n$ethyl esters, or AG$_n$E, with surface compositions ranging from 13% to 28% amide O,N, 75% to 66% aliphatic C, and 12% to 7% ester O surface as n increases from 1 to 4). While all Na salts reduce the solubility of benzene (Fig. 5), salts at the favorably interacting end of the Hofmeister series (*e.g.* $NaClO_4$, NaSCN) increase the solubility of some of these peptides. In addition, for all salts, the solubility m-value, or μ_{23}, becomes less positive (or more negative) as the amount of amide surface increases (Fig. 14). Robinson *et al.* concluded that the amide is "salted in" while hydrocarbon groups are "salted out", and that these groups contributed additively to the solubility m-value.[71]

Pegram and Record dissected values of μ_{23} obtained for each salt and peptide assuming additive contributions from the end groups, methylene carbons and amide groups.[33] End group contributions could only be assessed semiquantitatively from solubility data on the highly soluble model compound ethyl acetate; errors introduced from use of these data do not appear large, and the contribution of the end groups of course is constant as n is varied in a series of peptides. Partition coefficients K_p quantifying the distribution of the salt component in the vicinity of the amide group were obtained by application of the SPM-based analysis to these data; these are tabulated for the Na-based series of Hofmeister salts and the AG$_n$E series of peptides in Supplemental of ref. 33.

Separation of salt-interior amide K_p values into single-ion K_p values for the cation and anion of the salt is more ambiguous than for salt-hydrocarbon K_p values, because no salt component or salt ion is completely excluded from the amide group the way that Na_2SO_4 is completely excluded from aromatic hydrocarbon or air–water surface. To obtain single ion K_p values for amide interactions, use was made of the classical observations that NaCl and KCl typically exhibit little if any Hofmeister (*i.e.* non-Coulombic) effect on unfolding transitions of globular proteins and that Cl^-, Na^+, and K^+ ions are thus inferred to be centrally positioned in individual anion and cation series derived from unfolding data.[7,33] The surface composition of the AG$_4$E peptide is similar to that of the surface exposed in unfolding a globular protein (approximately 2/3 nonpolar C and 1/3 polar N,O). Indeed, for this peptide, overall partition coefficients K_p for NaCl and KCl, calculated using the same hydration $b_1 = 0.18$ H_2O/A^2 as determined for hydrocarbon and amide O data, are equal and close to unity ($K_{p,NaCl} = K_{p,KCl} = 0.97$).

For this AG$_4$E peptide, overall single ion partition coefficients of 0.97 were assigned to Na^+, K^+ and Cl^-. From this assignment and single ion partition coeffcients for hydrocarbon surface (Table 4), estimates of single ion partition coefficients for amide (O,N) surface are obtained. For both cations and anions, these predicted single ion K_p values make good chemical sense, as discussed below. Fig. 13 compares single ion partition coefficients for air–water, aromatic hydrocarbon, and amide surfaces; Fig. 13 is a summary bar graph representation which places single ion partition coefficients for hydrocarbon and amide surfaces in the order predicted by the protein Hofmeister series. Both Figures show clearly that the rank order of the Hofmeister anion and cation series for protein processes result from the rank order of preferential interactions of these ions with hydrocarbon, not amide surfaces.

The chart of K_p values in Fig. 13 emphasizes that single ion partition coefficients for amide surface are remarkably different for cations and anions, and partition coefficients for all cations and some anions are remarkably different for amide and hydrocarbon surfaces. The SPM-based analysis was required to determine these striking single ion behaviors, which had not been previously appreciated. In particular, all cations investigated (including Na^+, K^+, GuH^+) are highly accumulated at the amide surface. Presumably part of the explanation for this is that $\sim$¾ of the surface of an interior amide group (*e.g.* amides of the peptide backbone) is partially-negative O. Na^+ and K^+ are highly excluded from the hydrocarbon (and air–water) surface, while GuH^+ is randomly distributed between the aromatic hydrocarbon surface and bulk water and slightly excluded from the aliphatic hydrocarbon surface. Indeed, as discussed below, the difference in denaturing abilities of these chloride salts results from the highly unfavorable (protein stabilizing) interaction of Na^+ and K^+, but not GuH^+, with hydrocarbon surface. If the assumption used to generate single ion K_p values for the AG_4E peptide were incorrect, this would shift the cation and anion K_p values but would not affect the relative rankings nor the conclusion that cations are highly accumulated and that most anions are excluded from interior amide groups of the peptide backbone of AG_nE peptides and proteins.

3. Prediction of Hofmeister salt effects on biopolymer processes. From these ion partition coefficients for nonpolar and polar amide surfaces, one can now interpret or predict, for the first time, most of the non-Coulombic effect of Hofmeister salts on amide or peptide solubility, protein unfolding, and other protein processes in terms of structural information (the amount and composition of the ΔASA in the process).[7,33] For example, protein unfolding exposes surface (ΔASA) which is ~1/3 polar (divided equally between amide groups and other polar or charged surface) and 2/3 nonpolar; therefore, quantitative analysis of the thermodynamic consequences of accumulation (or exclusion) of salt cations and anions near both nonpolar and polar surface is of key importance for the interpretation of Hofmeister salt effects on protein unfolding. Because most anions are excluded to a similar extent from the amide surface (Fig. 13), the Hofmeister anion series for unfolding is determined by whether the anion is accumulated (like iodide, perchlorate) or excluded (like fluoride, sulfate) near a nonpolar surface. Most cations are excluded from nonpolar surfaces but strongly accumulated at polar amide surfaces and as a result are nearly neutral in driving unfolding. Guanidinium cation is much less excluded from a nonpolar surface than other cations, but equally accumulated at an amide surface; this behavior provides a novel quantitative explanation of its effect as a protein denaturant. The Hofmeister order of anions and cations arises from the extent of accumulation or exclusion of these ions near a nonpolar surface. Although interactions of cations and anions with polar amide surface do not exhibit the Hofmeister order, favorable cation–amide interactions are crucial for determining whether a given Hofmeister salt will favor or disfavor a biopolymer or model process in which both nonpolar and amide surfaces are exposed. For example, DNA duplex melting exposes a surface which is only ~35% hydrocarbon with the rest being primarily polar N and O. The favorable cation interaction with a polar surface, and more favorable anion interaction with a polar than with hydrocarbon surface, explains the observation (Fig. 6) that no Hofmeister salt stabilizes a 42% GC DNA duplex at high salt concentration, resulting in the shifted null point of the DNA Hofmeister series in Fig. 1.

To predict or interpret Hofmeister salt effects on processes involving charged biopolymers in an analogous manner to that done for urea and GB, Hofmeister and Coulombic effects must be separated. At high salt concentration such a separation appears feasible, and is described for lac DBD and the 12 base pair DNA duplex from Fig. 6 in ref. 7. Once this has been accomplished using modeling or a simple functional form suggested by Poisson–Boltzmann calculations (as in ref. 7), the results of ASA calculations on the native and unfolded states can be used (as described above) to make a quantitative comparison of predicted *m*-values (using the α values for the functional groups of proteins; Table 1) with experimental "Hofmeister *m*-values". Predicted and experimental *m*-values agree well for lacDBD, and appear consistent for the DNA duplex. Work in progress to obtain α values for Hofmeister salt–DNA base interactions will provide a more complete comparison with the experimental *m*-values.

4. Comparison to other approaches. von Hippel and collaborators found that while Hofmeister salts interact unfavorably with a homogeneous hydrocarbon surface, vicinal methylene groups significantly affect the affinity of salts for amide model compounds.[72] Recently, Rembert *et al.*[73] used thermodynamic and proton

NMR experiments along with MD simulations to study the interactions of Hofmeister ions with the polypeptide $(VPGVG)_{120}$. They concluded that destabilizing anions like SCN^- and I^- bind to amide N and the adjacent α-carbon (but not the valine side chains), neutral anions like Cl^- bind weakly to this site and stabilizing anions like SO_4^{2-} are excluded from all peptide surface types. The result for SO_4^{2-} is consistent with the model compound analysis showing strong exclusion from both nonpolar and polar amide surfaces. As in the case of urea and GB effects, a current point of debate is therefore whether interactions of Hofmeister ions with protein functional groups and Hofmeister salt effects on protein processes are dominated by backbone interactions or whether the most important distinction is in how the salt ion or solute interacts with different types of surface (C, N, O), irrespective of their location on backbone or side chains.

Acknowledgements

We thank members of our laboratory, both current (Ben Knowles, Irina Shkel) and previous (Elizabeth Courtenay, Wentao Zhang, Dan Felitsky, Jiang Hong, Jonathan Cannon, Charles Anderson, Ruth Saecker, Rogerio Sassonia) and a dedicated group of undergraduates (Megan Kratz, Brad Nelms, Robert Erdmann, Timothy Wendorff, Dana Bellissimo, Andrew LaCroix, Nickolas Deines, Michelle Pollock, Joanne Tsarouha, Xiaoyi Qu, Roger Diehl, Michael Kerins, Tyler Weid, Hyo Keun Cha, Rachel Wong, Chau Phan, Emily Lingeman, Katie Bieda, Shaenah Maguire, Matt Sternke, Shane Zhou, Kayla Coppens, Yurun Zhang, Audrey Hartzler, Laura Vian, Evan Buechel, August Melcher) for their contributions to the research reviewed here. Research from our laboratory was supported primarily by the NIH (GM47022). We also thank Pavel Jungwirth, Paul Cremer and other members of the committee for organizing the *Faraday Discussion* at which this paper was presented.

References

1 R. L. Baldwin, *Biophys. J.*, 1996, **71**, 2056–2063.
2 P. H. von Hippel and T. Schleich, *Biol. Macromol.*, 1969, **2**, 417–574.
3 S. N. Timasheff, *Adv. Protein Chem.*, 1998, **51**, 351–432.
4 M. T. Record, W. Zhang and C. F. Anderson, *Adv. Protein Chem.*, 1998, **51**, 281–353.
5 F. Hofmeister, *Arch. Exp. Pathol. Pharmakol.*, 1888, **24**, 247–260.
6 P. H. von Hippel and T. Schleich, *Acc. Chem. Res.*, 1969, **2**, 257–265.
7 L. M. Pegram, T. Wendorff, R. Erdmann, I. Shkel, D. Bellisimo, D. J. Felitsky and M. T. Record, *Proc. Natl. Acad. Sci. U. S. A.*, 2010, **107**, 7716–7721.
8 E. J. Guinn, L. M. Pegram, M. W. Capp, M. N. Pollock and M. T. Record, *Proc. Natl. Acad. Sci. U. S. A.*, 2011.
9 E. S. Courtenay, M. W. Capp, R. M. Saecker and M. T. Record, *Proteins: Struct., Funct., Genet.*, 2000, **41**, 72–85.
10 J. A. Holbrook, M. W. Capp, R. M. Saecker and M. T. Record, *Biochemistry*, 1999, **38**, 8409–8422.
11 E. S. Courtenay, M. W. Capp, C. F. Anderson and M. T. Record, *Biochemistry*, 2000, **39**, 4455–4471.
12 M. W. Capp, L. M. Pegram, R. M. Saecker, M. Kratz, D. Riccardi, T. Wendorff and J. G. C. a. M. T. Record, *Biochemistry*, 2009, **48**, 10372–10379.
13 J. K. Myers, C. N. Pace and J. M. Scholtz, *Protein Sci.*, 1995, **4**, 2138–2148.
14 D. J. Felitsky, J. G. Cannon, M. W. Capp, J. Hong, A. W. V. Wynsberghe, C. F. Anderson and M. T. Record, *Biochemistry*, 2004, **43**, 14732–14743.
15 J. Hong, M. W. Capp, C. F. Anderson, R. M. Saecker, D. J. Felitsky, M. W. Anderson and M. T. Record, *Biochemistry*, 2004, **43**, 14744–14758.
16 W. A. Rees, T. D. Yager, J. Korte and P. H. von Hippel, *Biochemistry*, 1993, **32**, 137–144.
17 V. M. Shelton, T. R. Sosnick and T. Pan, *Biochemistry*, 1999, **38**, 16831–16839.
18 D. Lambert and D. E. Draper, *J. Mol. Biol.*, 2007, **370**, 993–1005.
19 L. Levine, J. A. Gordon and W. P. Jencks, *Biochemistry*, 1963, **2**, 168–175.
20 E. S. Courtenay, M. W. Capp and M. T. Record, *Protein Sci.*, 2001, **10**, 2485–2497.

21 D. B. Knowles, A. S. LaCroix, N. F. Deines, I. Shkel and M. T. Record, *Proc. Natl. Acad. Sci. U. S. A.*, 2011, **108**, 12699–12704.
22 R. M. Kramer, V. R. Shende, N. Motl, C. N. Pace and J. M. Scholtz, *Biophys. J.*, 2012, **102**, 1907–1915.
23 L. L.-Y. Lee and J. C. Lee, *Biochemistry*, 1987, **26**, 7813–7819.
24 S. C. H. and C. J. B., *Biochemistry*, 1999, **38**, 496–508.
25 R. Bhat and S. N. Timasheff, *Protein Sci.*, 1992, **1**, 1133–1143.
26 Y. Zhang and P. S. Cremer, *Proc. Natl. Acad. Sci. U. S. A.*, 2009, **106**, 15249–15253.
27 L. M. Gross and U. P. Strauss, in *Chemical Physics of Ionic Solutions*, ed. B. E. Conway and R. G. Barradas, Wiley, New York, 1966, pp. 361–389.
28 A. W. Omta, M. F. Kropman, S. Woutersen and H. J. Bakker, *Science*, 2003, **301**, 347–349.
29 W. Melander and C. Horvath, *Arch. Biochem. Biophys.*, 1977, **183**, 200–215.
30 M. T. Record and C. F. Anderson, *Biophys. J.*, 1995, **68**, 786–794.
31 L. M. Pegram and M. T. Record, *Proc. Natl. Acad. Sci. U. S. A.*, 2006, **103**, 14278–14281.
32 D. J. Felitsky and M. T. Record, *Biochemistry*, 2004, **43**, 9276–9288.
33 L. M. Pegram and M. T. Record, *J. Phys. Chem. B*, 2008, **112**, 9428–9436.
34 L. M. Pegram and M. T. Record, *Chem. Phys. Lett.*, 2008, **467**, 1–8.
35 C. Tanford, *J. Mol. Biol.*, 1969, **39**, 539–544.
36 J. A. Schellman, *Biopolymers*, 1994, **34**, 1015–1026.
37 J. A. Schellman, *Biophys. J.*, 2003, **85**, 108–125.
38 Y. Nozaki and C. Tanford, *J. Biol. Chem.*, 1963, **238**, 4074–4081.
39 M. Auton, L. M. F. Holthauzen and D. W. Bolen, *Proc. Natl. Acad. Sci. U. S. A.*, 2007, **104**, 15317–15322.
40 M. Auton, J. Rosgen, S. Mikhail, L. M. F. Holthauzen and D. W. Bolen, *Biophys. Chem.*, 2011, **159**, 90–99.
41 L. M. Pegram and M. T. Record, *J. Phys. Chem. B*, 2007, **111**, 5411–5417.
42 M. Setchenow, *Ann. Chim. Phys*, 1892, **25**, 226–270.
43 I. A. Shkel and M. T. Record, *Soft Matter*, 2012, **8**, 9345–9355.
44 S. Weerasinghe and P. E. Smith, *J. Phys. Chem. B*, 2003, **107**, 3891–3898.
45 V. Pierce, M. Kang, M. Aburi, S. Weerasinghe and P. E. Smith, *Cell Biochem. Biophys.*, 2007, **50**, 1–22.
46 R. A. Robinson and R. H. Stokes, *J. Phys. Chem.*, 1961, **65**, 1954–1958.
47 C. F. Anderson and M. T. Record, *Biophys. Chem.*, 2004, **112**, 165–175.
48 W. Zhang, M. W. Capp, J. P. Bond, A. C. F. and M. T. Record, *Biochemistry*, 1996, **35**, 10506–10516.
49 J. G. Cannon, C. F. Anderson and M. T. Record, *J. Phys. Chem. B*, 2007, **111**, 9675–9685.
50 S. M. Timasheff and H. Inoue, *Biochemistry*, 1968, **7**, 2501–2513.
51 L. M. Pegram and M. T. Record, *J. Phys. Chem. C*, 2009, **113**, 2171–2174.
52 S. Gopalakrishnan, P. Jungwirth, D. J. Tobias and H. C. Allen, *J. Phys. Chem. B*, 2005, **109**, 8861–8872.
53 M. Mucha, T. Frigato, L. M. Levering, H. C. Allen, D. J. Tobias, L. X. Dang and P. Jungwirth, *J. Phys. Chem. B*, 2009, **105**, 7617–7623.
54 S. K. Sinha, S. Chakraborty and S. Bandyopadhyay, *J. Phys. Chem. B*, 2008, **112**, 8203–8209.
55 D. Jeruzalmi and T. A. Steitz, *J. Mol. Biol.*, 1997, **274**, 748–756.
56 E. J. Guinn, J. Schwinefus, H. Cha and M. T. Record, *J. Am. Chem. Soc.*, 2012, in preparation.
57 D. Lambert and D. E. Draper, *Biochemistry*, 2012, **51**, 9014–9026.
58 S. Cayley and M. T. Record, *Biochemistry*, 2003, **42**, 12596–12609.
59 T. Arakawa and S. N. Timasheff, *Biophys. J.*, 1985, **47**, 411–414.
60 U. D. Priyakumar, C. Hyeon, D. Thirumalai and A. D. MacKerell, J. Am. Chem. Soc., pp. 17759–17761.
61 J. Hong, M. W. Capp, R. M. Saecker and M. T. Record, *Biochemistry*, 2005, **44**, 16896–16911.
62 C. G. Kalodimos, G. E. Folkers, R. Boelens and R. Kaptein, *Proc. Natl. Acad. Sci. U. S. A.*, 2001, **98**, 6039–6044.
63 C. G. Kalodimos, A. M. J. J. Bonvin, R. K. Salinas, R. Wechselberger, R. Boelens and R. Kaptein, *EMBO J.*, 2002, **21**, 2866–2876.
64 J. M. Scholtz, D. Barrick, E. J. York, J. M. Stewart and R. L. Baldwin, *Proc. Natl. Acad. Sci. U. S. A.*, 1995, **92**, 185–189.
65 M. Auton and D. W. Bolen, *Proc. Natl. Acad. Sci. U. S. A.*, 2005, **102**, 15065–15068.
66 L. Ma, L. Pegram, M. T. Record and Q. Cui, *Biochemistry*, 2010, **49**, 1954–1962.
67 D. R. Canchi and A. E. Garcia, *Biophys. J.*, 2011, **100**, 1526–1533.
68 D. Horinek and R. R. Netz, *J. Phys. Chem. A*, 2011, **115**, 6125–6136.

69 L. B. Sagle, Y. Zhang, V. A. Litosh, X. Chen, Y. Cho and P. S. Cremer, *J. Am. Chem. Soc.*, 2009, **131**, 9304–9310.
70 W. K. Lim, J. Rosgen and S. W. Englander, *Proc. Natl. Acad. Sci. U. S. A.*, 2009, **106**, 2595–2600.
71 P. K. Nandi and D. R. Robinson, *J. Am. Chem. Soc.*, 1972, **94**, 1299–1308.
72 A. Hamabata and P. H. von Hippel, *Biochemistry*, 1973, **12**, 1264–1271.
73 K. B. Rembert, J. Paterova, J. Heyda, C. Hilty, P. Jungwirth and P. S. Cremer, *J. Am. Chem. Soc.*, 2012, **134**, 10039–10046.
74 T. Peters, *All about Albumin: Biochemistry, Genetics, and Medical Applications*, Academic Press, Inc., San Diego, CA, 1996.
75 D. J. Felitsky and M. T. Record, *Biochemistry*, 2003, **42**, 2202–2217.
76 J. Estrada, P. Bernado, M. Blackledge and J. Sancho, *BMC Bioinformatics*, 2009, **10**, 104.

Adsorption kinetics of non-ionic surfactants in micellar solutions: effects of added charge

Ci Yan, Anna Angus-Smyth and Colin D. Bain*

Received 11th June 2012, Accepted 25th June 2012
DOI: 10.1039/c2fd20118f

The kinetics of adsorption of micellar solutions of non-ionic surfactants have been studied in an overflowing cylinder. The addition of small amounts (<10% of the total surfactant concentration) of ionic surfactants, CTAB and STS, to solutions of $C_{16}E_8$ causes a dramatic reduction in the rate of adsorption of the nonionic surfactant. The results are rationalised by a combination of monomer and micelle adsorption to the air–water interface. In the presence of trace ionic surfactants, only uncharged micelles adsorb. The adsorption kinetics are independent of the sign of the charge on the micelles, only on its magnitude. The influence of ionic surfactants on the adsorption rate is reversed by addition of millimolar concentrations of salt. Electrolyte screens the repulsions between micelles and the adsorbed monolayer and allows charged micelles to adsorb without first breaking down into monomers.

1 Introduction

Aqueous formulations in commercial products almost always contain surfactants for one purpose or another, for example as wetting agents, dispersants, foam stabilisers, detergents or biocides. Many practical applications of such formulations involve the rapid production of air–water interfaces, including sprays, coating flows, foam formation and printing. The evolution of the surface properties (such as surface tension, elasticity, composition and disjoining pressure) following the generation of an air–water surface depend on the transport of surfactants to the surface and their kinetics of adsorption at the surface.[1] If the surfactant concentration is below the critical micelle concentration (CMC), then the mass transport is determined by solution of the convection–diffusion equation for the monomeric surfactant. Under diffusion-controlled conditions, the surface is locally at equilibrium with the subsurface and the surface excess, Γ, is then found from the equilibrium equation of state, $\Gamma(c)$, and the sub-surface concentration. A complicating factor is that the bulk hydrodynamics and surfactant adsorption can be strongly coupled by Marangoni effects. When the surfactant concentration is above the CMC, which is normally the case in practical applications of non-ionic surfactants owing to their low CMCs, there is the additional complexity of micelle transport and of exchange of material between monomers and micelles. In this paper, we study the effect of charge on adsorption in systems with very low CMCs where the surfactant is overwhelmingly present in micellar form. From these results, we derive insights into the mechanism of adsorption in micellar surfactant solutions.

The experimental platform we use is the overflowing cylinder (OFC), shown in Fig. 1. The use of the OFC in the study of adsorption kinetics has been reviewed elsewhere,[2] but the most important feature is that it generates a steady-state expanding liquid surface that is amenable to study by a wide range of surface sensitive

Department of Chemistry, Durham University, Durham, DH1 3LE, U. K. E-mail: c.d.bain@durham.ac.uk; Fax: +44 191 3342051; Tel: +44 191 3342138

Fig. 1 Photograph and schematic diagram of an overflowing cylinder.

techniques. The hydrodynamics are characterised by the surface expansion rate $\theta = r^{-1}\partial(rv_r)/\partial r$, where r is the radial position and v_r the radial surface velocity. In the presence of surfactants, Marangoni effects serve both to increase v_r and to make θ almost invariant across the surface of the OFC.[11] For a system containing a single surface-active component with a diffusion coefficient, D, solution of the steady-state convective diffusion equation yields

$$\Gamma = \sqrt{\frac{2D}{\pi\theta}}(c_b - c_s) \tag{1}$$

where c_b and c_s are the bulk and subsurface concentrations, respectively.[12] The equivalent surface age for time-dependent adsorption to a stagnant surface is given approximately by $(2\theta)^{-1}$.[2] Thus for a typical value of $\theta = 5\ s^{-1}$, the surface age is 0.1 s. In micellar systems, the convection-diffusion equation is also amenable to analytical solution provided that the micelles and monomers are in rapid equilibrium.[3] In this case, a micelle-free zone of thickness h develops near the surface. At distances from the surfaces, $z > h$, the monomer concentration is a constant and only micelles diffuse. Balancing micellar and monomer fluxes at $z = h$ and the mass balance at the free liquid surface then yields the following expression for Γ

$$\Gamma = \sqrt{\frac{2D}{\pi\theta}}(c_{cmc} - c_s)/\mathrm{erf}\left(h\sqrt{\frac{\theta}{2D}}\right) \tag{2}$$

where h is found from

$$\frac{c_b - c_{cmc}}{c_{cmc} - c_s} = \sqrt{\frac{D}{D_m}}\,\frac{\exp\left(\frac{-\theta h^2}{2D}\right)\mathrm{erfc}\left(h\sqrt{\frac{\theta}{2D_m}}\right)}{\exp\left(\frac{-\theta h^2}{2D_m}\right)\mathrm{erf}\left(h\sqrt{\frac{\theta}{2D}}\right)} \tag{3}$$

and D_m is the micellar diffusion coefficient. As c_s increases, h decreases gradually to zero. For higher surfactant concentrations, micelles persist right up to the surface and eqn (1) applies with D replaced by D_m.

The surfactants we use are the alkyl oligoethylene glycols ($CH_3(CH_2)_{n-1}(OC_2H_4)_mOH$), commonly abbreviated C_nE_m. The CMC of $C_{10}E_8$ is 1 mM, decreasing by approximately one order of magnitude for each increase of 2 in n. The CMC is much less sensitive to m, falling slowly with decreasing m. For $m = 8$, the surfactants form spherical or ellipsoidal micelles at room temperature. As m decreases, the micelles become rod-like and eventually form lamellar phases.

In a preliminary communication,[4] we studied the adsorption of $C_{14}E_8$ at the surface of a liquid jet with a surface age in the range of 2–50 ms. We observed that $C_{14}E_8$ adsorbed at a rate close to the diffusion-controlled limit. The conventional models for adsorption in micellar solutions assume that only monomers can adsorb to a surface: the role of micelles is solely to replenish monomer that has been depleted by adsorption.[5] As the expected lifetime of $C_{14}E_8$ micelles is orders of magnitude longer than the surface age in the liquid jet,[6] we inferred that the micelles adsorb directly to the interface without first breaking down into monomers. Subsequently, we have analysed theoretically the mechanism by which micelles break down into monomers.[7] The established model for equilibration in micellar solutions assumes that growth and dissolution of micelles occurs by sequential monomer addition and loss: the resulting set of coupled differential equations are known as the Becker–Döring equations.[8] Aniansson and Wall obtained an elegant solution to these equations for small perturbations from equilibrium, identifying two well-separated relaxation times.[9] The fast relaxation is associated with the exchange of monomers between micelles and the bulk solution and is assumed to be close to diffusion-controlled (the mean lifetime of a surfactant monomer in a micelle, $\tau_1/\mu s \approx 0.1$ mM/CMC for C_nE_8). The slow relaxation time, τ_2, is associated with the complete breakdown of micelles into monomers and occurs on much longer timescales (typically ms to seconds or longer). We have shown that τ_2 is a strong function of the size of the perturbation, decreasing as the monomer concentration decreases.[7] For a monomer concentration near zero, $\tau_2 \approx N\tau_1$, where N is the mean aggregation number of the micelles. Near a nascent surface the monomer concentration may be far below the CMC, in which case the micelle lifetime may be very much less than that measured by perturbation methods. Furthermore, the relaxation times predicted by the Becker–Döring mechanism are unphysically long for efficient surfactants such as $C_{14}E_8$.[16] Consequently, we have proposed an alternative mechanism in which micelle breakdown occurs through the fusion of two micelles followed by sequential monomer loss.[10] In this paper, we extend our original study[4] and examine whether our growing understanding of micellar relaxation offers alternative explanations of the experimental data, which do not involve adsorption of micelles. The hydrodynamics of the OFC are easier to model than a liquid jet and thus allow a more quantitative description of experimental data. The surface age is also longer in an OFC than a jet and consequently we have focused primarily on a surfactant with a lower CMC (and hence slower kinetics): $C_{16}E_8$. To confirm our proposed adsorption mechanism (or otherwise) we have compared data on $C_{16}E_8$ in the OFC with other members of the C_nE_m family. As in our preliminary study,[4] selectively adding charge to the non-ionic micelles enables us to discriminate among possible mechanisms.

2. Experimental

2.1 Overflowing cylinder

The OFC has been described in detail elsewhere.[11,12] It consists of two concentric cylinders, with an inner diameter of 8 cm. Liquid flows upwards in the inner cylinder at a flow rate of 18 $cm^3 s^{-1}$, passing through a resistance plate and flow straightener to ensure plug-flow well below the surface. The liquid overflows the rim of the inner cylinder and is collected in the outer cylinder before being recycled by a pump. The inner cylinder is gravity-fed to maintain a constant flow rate and to decouple the OFC from pump vibrations. The total volume of solution required is 1.5 dm^3. Solutions of the non-ionic surfactants were prepared in ultra-high quality water (Millipore, Milli-Q). Ellipsometry isotherms were obtained by sequential dilution of the initial concentration, in a series of overlapping concentration intervals. Ionic surfactants and salt were added dropwise into the OFC from concentrated stock solutions prepared with the non-ionic surfactant at its bulk concentration so that the concentration of non-ionic did not vary during addition of the ionic surfactant or salt.

Following each addition, readings were allowed to stabilise before being recorded. Experiments were conducted at 298 K.

2.2 Laser Doppler velocimetry (LDV)

The surface expansion rate was measured by LDV, with scattering from a sparse suspension of 2-μm titania particles. Experimental details may be found elsewhere.[11] The surface velocity, v_r, was measured as a function of the position, x, along a line through the axis of the OFC. The velocity was then fitted to a function of the form

$$v_r(x) = a_1(x - \delta) + a_3(x - \delta)^3. \tag{4}$$

Near the centre of the OFC, $\theta = 2a_1$.

2.3 Ellipsometry

The coefficient of ellipticity, $\bar{\rho}$, was measured at the centre of the OFC on a phase-modulation ellipsometer (Beaglehole Instruments, Wellington, NZ) equipped with a 2-mW HeNe laser. Data were acquired every 4 s and averaged over 5 min. The random errors in $\bar{\rho}$ are $<1 \times 10^{-5}$ and lie within the symbols in the figures. The accuracy of $\bar{\rho}$ is limited by alignment and calibration errors: repeated measurements of the same surfactant system over a period of years fall within ±5% of the mean value. Experimentally we have found a good linear relationship between $\bar{\rho}$ and the surface excess, Γ, for a number of a simple surfactant systems (data for C_8E_4OMe may be found in ref. 13).[14,15] Consequently, we convert values of $\bar{\rho}$ into values of Γ using a linear interpolation between the value of $\bar{\rho}$ for pure water in the OFC (typically 0.35×10^{-3}) and for the surfactant at its limiting surface excess (measured on a stagnant solution above the CMC). For $C_{16}E_8$, we use values of $\bar{\rho} = -2.45 \times 10^{-3}$ at $\Gamma_{max} = 3.3 \times 10^{-6}$ mol m^{-2}.[16] This method probably underestimates the surface excess at low surfactant coverages, for reasons discussed in detail elsewhere;[17] for the purposes of this paper, however, a linear interpolation is sufficient. The accuracy of values of Γ derived from ellipsometry is often limited by the accuracy with which the limiting surface excess at the cmc is known from other techniques. A direct calculation of Γ from $\bar{\rho}$ using a two-layer model of the interface is accurate to around 5% of a monolayer. When converting $\bar{\rho}$ into Γ for mixed monolayers we have assumed that all surfactants make the same contribution to the ellipticity per molecule: this approximation does not introduce significant errors here.

2.4 Neutron reflectometry

Neutron reflectometry (NR) was performed on the OFC on the FIGARO reflectometer at the Institut Laue Langevin, France.[18] Details of the experimental design and data analysis may be found elsewhere.[19] Briefly, a single isotopic contrast was employed, with protonated $C_{16}E_8$ at a concentration of 1.86 mM in null reflecting water. Chain-deuterated CTAB was then added to a final concentration of 0.11mM, amounting to 6% of the $C_{16}E_8$ concentration. The scattering lengths of $C_{16}E_8$ and dCTAB are 18 fm and 329 fm, respectively, and consequently the neutrons are scattered predominantly by the dCTAB. Given the extremely weak scattering expected, a reflectivity profile was only acquired at a single angle of incidence of 0.624° with the high flux settings of FIGARO at a constant $\Delta\lambda/\lambda = 9.8\%$ over a wavelength range of 2–20 Å. The NR profiles were reduced over a Q range of 0.008–0.045 Å^{-1}. Data from monolayers were recorded for 3.5 h and for nrw for 2h. Due to the very low scattering length density (sld) of the fractional monolayers, the background needs to be carefully removed. A smooth curve was fitted through the nrw profile and subtracted from the surfactant profiles. The specular reflectivity profiles thus obtained were fitted in MOTOFIT[20] using a simple one-layer model with a thickness of 14 Å.

2.5 Materials

Non-ionic surfactants (Fluka and Nikko) were used as received. Solutions were warmed gently to aid dissolution when necessary. Hexadecyltrimethylammonium bromide (CTAB, Sigma-Aldrich) was recrystallized 3 times from acetone/ethanol. d_{42}-CTAB was supplied by the Oxford deuteration facility and used as received. Sodium tetradecyl sulphate (STS, Wako Chemicals) was recrystallized three times from ethanol. NaBr (Fluka, Ultra) was used as received.

3. Results and discussion

Before exploring the behaviour of $C_{16}E_8$ in the OFC, it is instructive first to look at the behaviour of a surfactant with a much higher CMC. $C_{12}E_6$ has a CMC of 0.073 mM,[21] and exists as a mixture of globular micelles and short rods at room temperature.[22] Fig. 2 shows the dynamic surface excess of $C_{12}E_6$ as a function of bulk surfactant concentration, $c_{C_{12}E_6}$. The dashed lines indicates the limiting surface coverage, Γ_{max}, for $C_{12}E_6$. The arrow denotes the CMC. The solid line indicates the surface excess for diffusion-controlled adsorption with fast micelle–monomer equilibration, calculated with eqn (1) below the CMC and eqn (2) and (3) above the CMC ($D_{mon} = 2.9 \times 10^{-10}$ m^{2}s^{-1} and $D_{mic} = 5.0 \times 10^{-11}$ m^{2}s^{-1}, determined by pulsed field gradient (PFG) NMR). Except very close to Γ_{max}, the experimental data are close to the calculated curve; the small deviations at lower concentrations are probably due to a systematic error in the conversion of $\bar{\rho}$ into Γ rather than any kinetic limitations. The behaviour near Γ_{max} will be discussed elsewhere; the main point here is that over almost the whole concentration range, the experimental data conform to a simple diffusion-controlled model with no kinetic barriers either for micelle break-down or for adsorption to the surface.

Turning our attention now to the $C_{16}E_8$, Fig. 3 shows the raw ellipsometry data from the flowing surface of the OFC, as a function of the bulk concentration, $c_{C_{16}E_8}$. The CMC of $C_{16}E_8$ is around 1 μM, so all the measurements are taken at concentrations well above the CMC. Fig. 4 shows the surface expansion rate, θ, measured by laser Doppler velocimetry (LDV). Eqn (2) and (3) can be used to calculate the dynamic surface excess for the same diffusion-controlled, fast micelle break-down model as in Fig. 2. Since h is so small, however, it is easier just to use eqn (1)

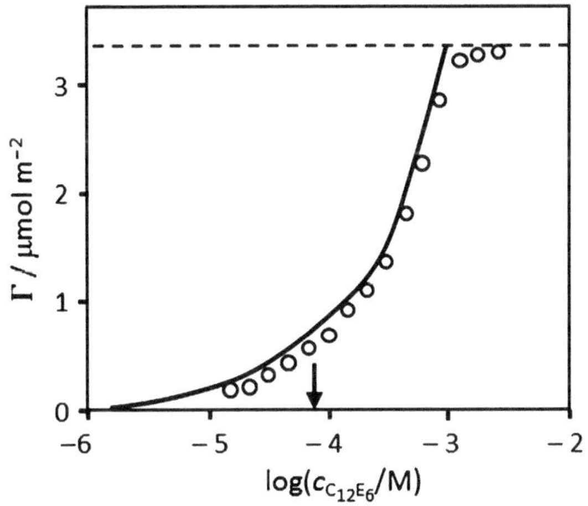

Fig. 2 Dynamic surface excess of $C_{12}E_6$ in an OFC. The dashed line marks the maximum surface excess observed in equilibrium solutions above the CMC. The CMC is marked with an arrow. The solid line is a theoretical prediction for diffusion-controlled adsorption with fast micelle kinetics.

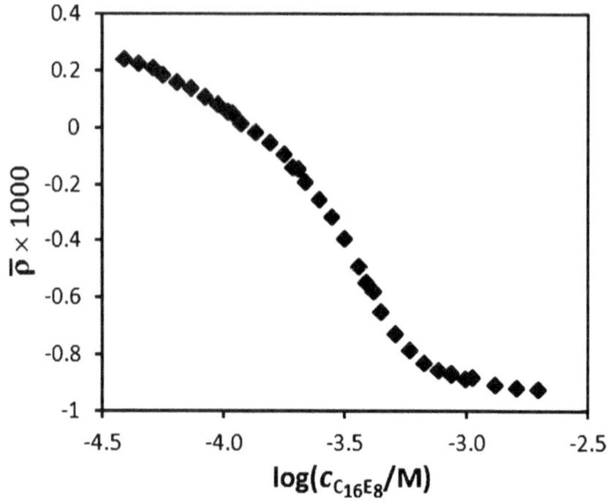

Fig. 3 Coefficient of ellipticity of aqueous solutions of $C_{16}E_8$ in an OFC. Pure water had a value of $\bar{\rho} = 0.35 \times 10^{-3}$ and at the limiting coverage of $C_{16}E_8$, $\bar{\rho} = -2.45 \times 10^{-3}$. $\text{Log}(\text{CMC}_{C_{16}E_8}/\text{M}) = -6$).

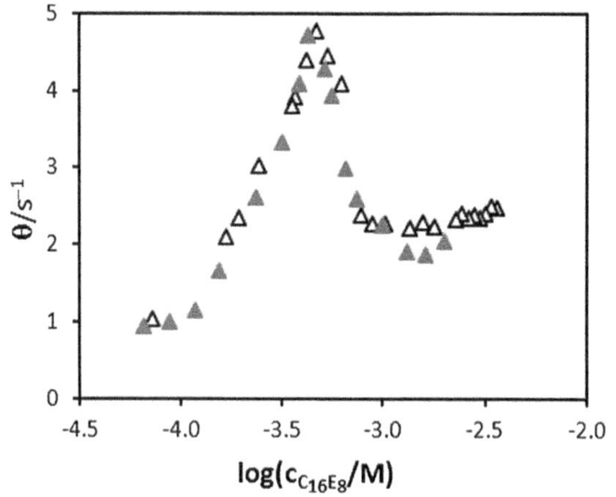

Fig. 4 Surface expansion rate in the OFC as a function of concentration of $C_{16}E_8$. The two different symbols represent measurements made by two researchers with different batches of surfactant and illustrate the repeatability of the data.

with D replaced by D_{mic} ($= 4.2 \times 10^{-11}$ m^2s^{-1}, measured by PFG-NMR). In Fig. 5, the ellipsometry data from Fig. 3 are converted to surface excess and replotted on a linear scale; the theoretical curve derived from eqn (1) and the LDV data in Fig. 4 is shown as a solid line in Fig. 5.

The contrast between the adsorption behaviour of $C_{16}E_8$ and $C_{12}E_6$ is marked. At concentrations up to 0.5 mM, the surface excess of $C_{16}E_8$ is in good agreement with the diffusion-controlled, fast micelle break-down model. At a surface coverage of around 40% of a monolayer, the surface coverage abruptly flattens out. The surface expansion rate also drops sharply, which is an indication that the surface excess (and hence surface tension) is much less sensitive to the subsurface surfactant concentration. At higher surfactant concentrations, the surface excess increases very slowly reaching 45% of a monolayer at $c_b = 2$ mM. We can postulate a number of possible explanations for the observed behaviour.

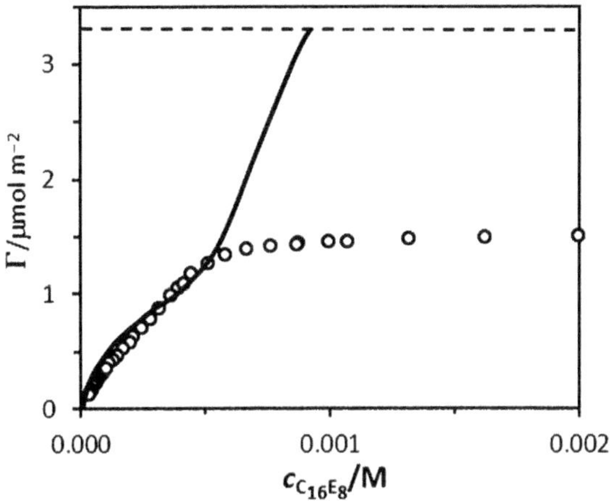

Fig. 5 Dynamic surface excess of $C_{16}E_8$ in the OFC. The dashed line marks the limiting surface excess, Γ_{max}. The solid line is the theoretical surface excess for diffusion-controlled adsorption with fast micelle kinetics.

(i) The mechanism changes from diffusion-controlled adsorption of monomers to adsorption limited kinetically by the release of monomers from micelles. If micelles break down by the Becker–Döring mechanism then the micelle break-down rate is a strong function of monomer concentration.[7] As Γ increases, so does the sub-surface concentration of monomers, c_s and hence the micelle break-down rate decreases. The adsorption isotherms of non-ionic surfactants are, however, rather shallow[23] and a surface that was only 40% covered with surfactant would be in equilibrium with a monomer concentration at least an order of magnitude lower than the cmc. Simulations of micelle breakdown show that the rate constant is insensitive to the monomer concentration below 0.5 × CMC (and very sensitive to monomer concentration above that value).[16] Consequently, is unlikely that the rate of micelle break-down in the subsurface region changes abruptly at 40% monolayer coverage.

(ii) The mechanism changes from diffusion-controlled adsorption of monomers to adsorption limited by kinetic barriers between the subsurface and the monolayer. As the surface coverage increases, so do the steric repulsions between the adsorbed monolayer and the poly(oxyethylene) head groups of surfactant monomers in solution. A previous study of the adsorption kinetics of a related surfactant, $C_{12}E_8$, at a stagnant interface concluded that the mechanism changed from diffusion-controlled to a mixed diffusion–kinetic regime as the concentration of surfactant increased.[24] If adsorption were kinetically limited, as the bulk surfactant concentration increased, the sub-surface concentration would rise above its equilibrium value until it attained the CMC. At this point the adsorption flux would be fixed at $k_{a,mon}(\Gamma) \times$ CMC, where $k_{a,mon}$ is the adsorption rate constant for monomers, and a further increase in the bulk surfactant concentration would have no effect on the adsorption rate. Experimentally, we observe a levelling out in the surface excess (Fig. 5), but Γ does still slowly increase with increasing c_b.

(iii) The diffusion-controlled adsorption at low surfactant concentrations does not proceed *via* adsorption of monomers but *via* adsorption of micelles. The micelles do not remain intact at the surface but are rapidly ripped apart by surface stresses: an alternative semantics would be to say that the surface catalyses the disintegration of the micelle. As the surface coverage of $C_{16}E_8$ increases, the steric repulsions between the micelles and the adsorbed monolayer increase and the rate constant for adsorption rapidly decreases. The rate of adsorption is then $k_{a,mic}(\Gamma) \times c_s$, where the subsurface concentration, c_s is given by eqn (1) with $D = D_{mic}$. Since the adsorption flux is balanced by $\Gamma\theta$ and c_s increases rapidly with c_b at concentrations above 0.5 mM, $k_{a,mic}(\Gamma)$ must decrease with increasing c_b and hence Γ must increase, as observed.

We argued previously that micelle break-down is not sufficiently fast to supply monomer to the surface on the OFC timescales, as required for mechanism (ii).[4] A micellar lifetime of 4 s for $C_{12}E_8$ has been reported[25] and since the lifetime scales inversely with the CMC,[26] the lifetime of micelles of $C_{16}E_8$ should be at least two orders of magnitude longer. We now believe that the rate of micelle break-down is a strong function of the size of the perturbation and hence it may not be safe to assume that the rate of relaxation to equilibrium can be used to predict micelle kinetics far from equilibrium.[7] At concentrations above 1 mM, $\Gamma\theta$ increases with c_b and hence so must $k_{a,mon}(\Gamma) \times$ CMC in mechanism (ii), which is inconsistent with the assumption that $k_{a,mon}(\Gamma)$ decreases with increasing Γ. Nevertheless, it is desirable to have a more categorical test to distinguish between adsorption of monomers and adsorption of micelles.

To discriminate between these two mechanisms we have measured the effect of charge on the adsorption kinetics. If small amounts of an ionic surfactant, such as the cationic surfactant hexadecyl trimethylammonium bromide (CTAB) or the anionic surfactant sodium tetradecyl sulphate (STS), are added to a micellar solution of a $C_{16}E_8$, the ionic surfactants partition between solution and the non-ionic micelles. The interaction parameter is favourable, since the non-ionic surfactants screen the electrostatic repulsions between the ionic headgroups and the presence of the small ionic headgroups increases the area per E_8 chain at the surface of the micelle and hence reduces steric repulsions between the E_8 chains. The added concentration of the ionic surfactant as a fraction of the total surfactant concentration is <0.1. In the limit that $c_{ionic} \ll c_{C_{16}E_8}$, the fraction of the ionic surfactant present in the micelles is given by

$$\left(\frac{c_{mic}}{c_{mic}+c_{mon}}\right)_{ionic} = \frac{e^{-\beta}\left(c_{C_{16}E_8}/\mathrm{CMC}_{ionic}\right)}{1+e^{-\beta}\left(c_{C_{16}E_8}/\mathrm{CMC}_{ionic}\right)}. \tag{5}$$

Carnero Ruiz and Aguiar measured an interaction parameter $\beta = -3$ for mixed micelles of $C_{12}E_8$ with the cationic surfactant TTAB (tetradecyl trimethylammonium bromide).[27] Assuming the same value of β for sixteen-carbon chains and a $C_{16}E_8$ concentration of 0.5 mM, we find that >90% of the ionic surfactant CTAB (CMC = 0.92 mM) resides in the micelles. For STS, the CMC is higher (CMC = 2.2 mM), but the interaction between poly(oxyethylene) groups and anionic head groups is more favourable than with cationic head groups,[28] so the STS is also expected to partition strongly into the micelles. Data presented below support this assumption. At low mole fractions in the micelle the surfactants are almost fully dissociated.[29] Consequently, adding a small amount of an ionic surfactant is a simple way of controlling the charge on the $C_{16}E_8$ micelles.

We now consider the effect of charge on the three possible mechanisms described above. If adsorption were to proceed *via* adsorption of monomers of $C_{16}E_8$ (mechanisms (i) and (ii)), then the adsorption kinetics would be unaffected by the presence of trace amounts of an ionic surfactant in the monolayer and in solution. The mutual diffusion coefficient of mixed micelles, D_{mic}, is, however, enhanced by migration fields so mass transport to the sub-surface layer would increase.[30] [Migration fields arise because the counterion diffuses faster than the surfactant in the micelles. To prevent the build-up of space charge, an electric field is established in the solution that enhances the movement of the micelles towards the surface and inhibits the movement of the counterions so that the overall transport rate is the same for both charges.] In mechanism (i), the flux to the surface would increase at all concentrations, while in mechanism (ii) it would increase until the sub-surface concentration of monomers attained the CMC. Conversely, if the micelles adsorb directly to the surface without first breaking down into monomers (mechanism (iii)), they will experience an electrostatic barrier arising from the presence of ionic surfactants in both the micelles and the adsorbed monolayer. This repulsion will deplete the sub-surface region of micelles and reduce the flux of surfactant to the surface.

Fig. 6 plots the coefficient of ellipticity of an 0.50 mM $C_{16}E_8$ solution as a function of the amount of added CTAB (expressed as a percentage of the non-ionic surfactant) together with the corresponding surface expansion rate. Fig. 6 shows that even adding 0.5% of CTAB has a measurable effect on the surface excess and that adding 5% of CTAB reduces the adsorbed amount by a factor of 4. Concomitantly, the surface expansion rate also decreases by a factor of 3, so the mass transport of surfactant to the surface of the OFC falls by an order of magnitude. These data provide strong support for mechanism (iii), namely that micelles adsorb without first breaking down into monomers.

Above 5% of CTAB, the ellipticity levels out at a value corresponding to about 10% of a monolayer of surfactant. The constancy of this value suggests that the residual adsorption arises from a different mechanism. The monomer alone at its CMC can only supply ~0.01 μmol m^{-2} of surfactant (from eqn (1)) at $\theta = 1.5\ s^{-1}$, and so cannot be responsible for the observed surface excess of 0.3 μmol m^{-2}. One plausible explanation is that the residual surfactant arises from the fast relaxation process (characterised by a timescale τ_1) in which each micelle sheds a fraction of its monomers without completely breaking down. To explain our experiments, each micelle would lose 10–15% of its monomers, which would then adsorb individually to the air–water interface. Such an amount is consistent with numerical simulations of micelle relaxation following a sudden dilution.[7]

It might seem that without a knowledge of the composition of the monolayer, and hence its surface charge, we cannot hope to explain the data in Fig. 6 quantitatively. It transpires, however, that a grossly simplified model provides useful insight into the adsorption kinetics. We first calculate the sub-surface concentration of micelles from eqn (1), neglecting any migration fields. We then assume that the local surfactant composition is equal to the bulk composition (that is, we neglect any preferential adsorption of one component) and that the CTAB molecules are distributed among the micelles according to the Poisson distribution (see Fig. 7), on the grounds that the average separation of the charges at the surface of the micelle exceeds the Bjerrum length in water. If the monolayer composition were to match the bulk composition, then the surface potential calculated from the Gouy-Chapman equation

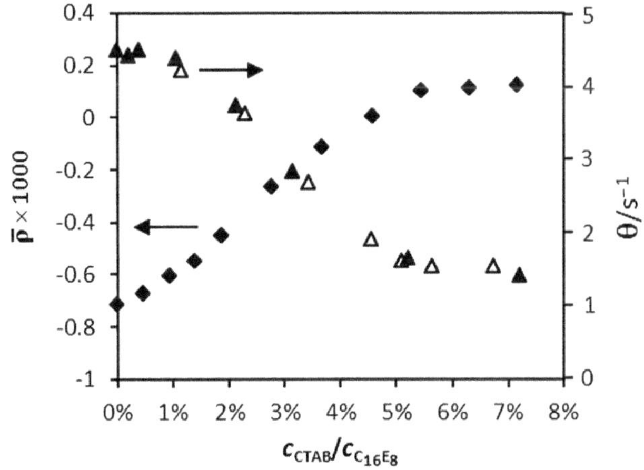

Fig. 6 Effect of addition of CTAB to a 0.50 mM $C_{16}E_8$ solution in an OFC. Diamonds: coefficient of ellipticity (on left axis). Triangles: surface expansion rate (on right axis).

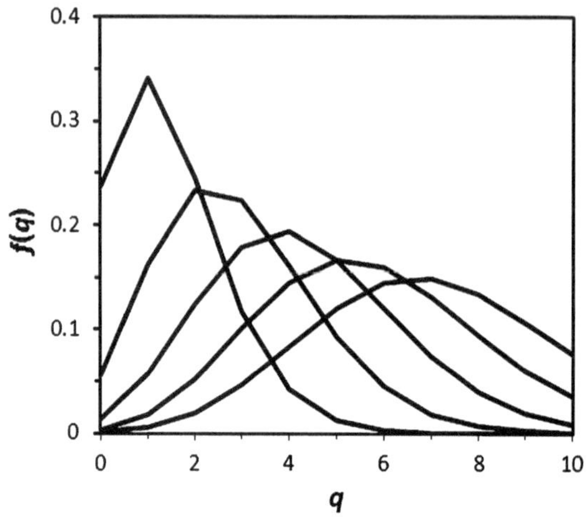

Fig. 7 Distribution of charge, q, on micelles of $C_{16}E_8$ for concentrations of CTAB of 1, 2, 3, 4 and 5% (from left to right) calculated from the Poisson distribution. Mean aggregation number of micelles, $N = 144$.

would lie in the range 80–110 mV for all concentrations of CTAB. Consequently, the electrostatic barrier to adsorption would be 3–4 k_BT for singly charged micelles and a multiple of this value for micelles containing more than one molecule of ionic surfactant. We therefore assume that only uncharged micelles can adsorb (thus obviating the need for a detailed knowledge of the electrostatics). Furthermore, we will assume that the rate constant for adsorption, $k_{a,mic}$ is not a function of Γ. While this assumption is unwarranted, $k_{a,mic}$ is really an effective rate constant that takes account also of how fast CTAB is exchanged between micelles to maintain a Poisson distribution of charges. Finally, we assume that the 'monomer contribution' to Γ is independent of c_{CTAB}, at 10% of a monolayer. The advantage of this simplified model is that there is only a single fitting parameter, $k_{a,mic}$ (but see comment in caption to Fig. 8).

Fig. 8 shows the calculated flux to the surface, $\Gamma\theta$, with the best fit value of $k_{a,mic} = 4 \times 10^{-5}$ ms^{-1}. This simple model captures the observed behaviour well. Fig. 8 also shows the adsorption flux from all micelles in the subsurface region, calculated on the assumption that the surface composition matches the bulk. The contribution of charged micelles is seen to be negligible. We provide evidence later that the surface is enriched in the ionic surfactant, which will increase the surface potential and make adsorption of charged micelles even less favourable.

If only uncharged micelles actually adsorb, then we would expect the adsorption kinetics to be independent of the sign of the charge, of the details of the interactions between the ionic and non-ionic surfactant (which influence the composition of the monolayer) or of counterion binding. To test this prediction, we repeated the experiment in Fig. 6 with the anionic surfactant, STS, in place of the cationic surfactant, CTAB. As the ellipsometry results in Fig. 9 show, the influence of the two ionic surfactants on the adsorption kinetics of $C_{16}E_8$ are virtually indistinguishable. Fig. 9 provides indirect confirmation that virtually all the ionic surfactant is present in the micelles; otherwise the differing CMCs and interaction parameters of CTAB and STS would lead to different partitioning between micelles and free surfactant and hence a different concentration of uncharged micelles.

If the decrease in adsorption of $C_{16}E_8$ with added ionic surfactant is due to electrostatic repulsion of charged micelles, then addition of salt should screen the repulsions and restore the adsorption of the non-ionic surfactant. Fig. 10 shows the effect of adding NaBr to the two solutions of highest concentration in Fig. 9.

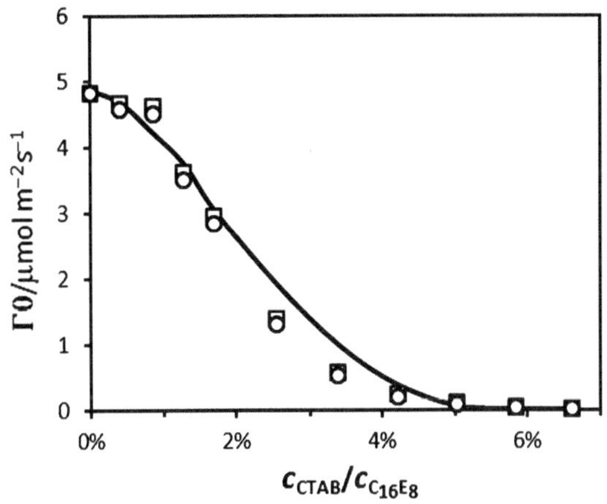

Fig. 8 Calculated flux to the surface with $k_{a,mic} = 4 \times 10^{-5}$ ms^{-1} if only uncharged micelles adsorb (circles), for mixturcs of CTAB and $C_{16}E_8$ (c = 0.5 mM). The micelle composition has been corrected for free CTAB (8% of the total CTAB from eqn (5)). Black line: experimental flux from ellipsometry and LDV. Squares: total flux from micelles of all charges if monolayer has same composition as bulk. For pure $C_{16}E_8$, the value of c_s from eqn (1) is very sensitive to the values of Γ_{max} and D_{mic} since adsorption is close to diffusion-controlled. For the 'best estimate' values of Γ_{max} and D_{mic}, c_s is slightly negative. To obtain a well-behaved function near $c_{CTAB}/c_{C_{16}E_8} = 0$, a scaling factor (near unity) was applied to the calculated surface excess; the value used for the fit above was 0.945.

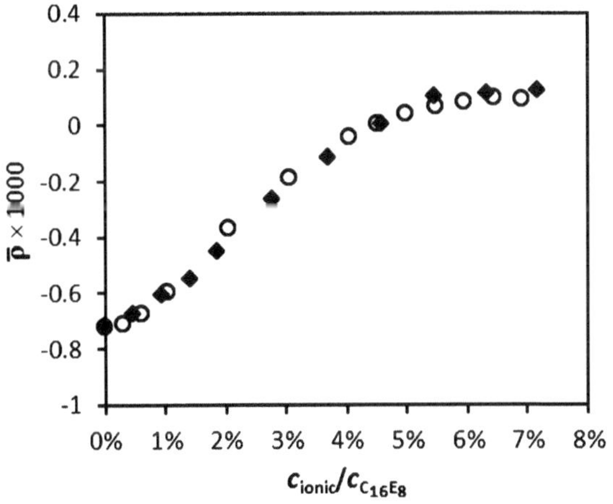

Fig. 9 Comparison of the effect of added STS (circles) and CTAB (diamonds) on the adsorption of $C_{16}E_8$ at a concentration of 0.5 mM.

Concentrations of NaBr <0.1 mM have no effect on surfactant adsorption, but higher salt concentrations switch adsorption back on. 10 mM salt is sufficient to return the surface excess to the value of the pure surfactant. The two ionic surfactants, CTAB and STS no longer behave the same since the adsorption rate for charged micelles depends sensitively on the electrostatics at the surface. Na^+ ions appear to be more effective at screening the repulsions in the presence of STS than Br^- ions are at screening repulsions with CTAB.

We have not yet explored the influence of different counterions on the adsorption kinetics of the same ionic amphiphile. We would anticipate that the nature of the

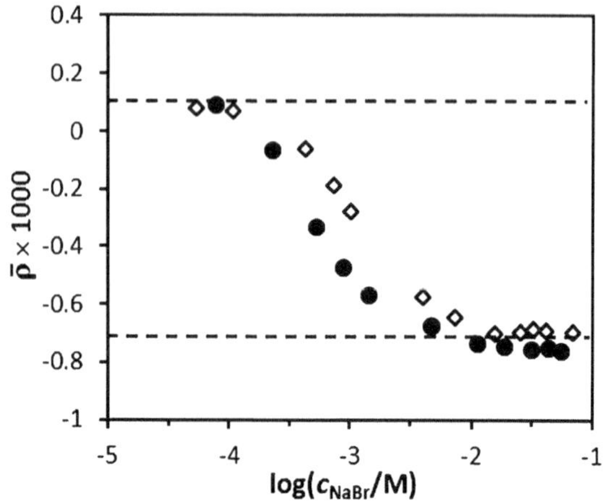

Fig. 10 Effect of adding electrolyte (NaBr) to an 0.50 mM solution of $C_{16}E_8$ containing 7% of either CTAB (diamonds) or STS (circles). The lower dotted line marks the coefficient of ellipticity of 0.50 mM $C_{16}E_8$ without added ionic surfactant or NaBr. The upper dotted line indicates the value with ionic surfactant but without NaBr.

counterion would have no effect in the absence of added electrolyte, unless there were ion pair formation with the ionic headgroups at the surface of the micelle – which would then behave like an uncharged micelle. In the presence of added electrolyte, stronger counterion binding both reduces the surface potential and decreases the effective charge on the micelles. Consequently, we might expect the efficiency of salts in turning the micellar adsorption back on would follow their efficiency in reducing the CMC. Differences may arise since counterion effects on CMCs are collective interactions with many charged headgroups, while in the systems studied here the charges are isolated and so, in principle, ion effects should be easier to model.

Ellipsometry and LDV do not provide direct information on the composition of the monolayer – chemically sensitive techniques are required. To estimate the composition, we employed neutron reflectometry (NR) using chain-deuterated CTAB (dCTAB) in null-reflecting water. Since $C_{16}E_8$ has a negligible scattering length, NR is sensitive only to the adsorbed dCTAB. Once the surface excess of dCTAB is known, the surface excess of $C_{16}E_8$ can be estimated from the coefficient of ellipticity. To improve the chances of success, we increased the concentration of $C_{16}E_8$. Ellipsometry measurements showed that, upon addition of CTAB, higher concentrations of $C_{16}E_8$ behaved in a way exactly analogous to the 0.5 mM solution shown in Fig. 6, except that the ellipticity is lower (*i.e.* more negative) at all concentrations. For example, for 1.74 mM $C_{16}E_8$, the plateau in $\bar{\rho}$ corresponds to around 13% of a monolayer ($\bar{\rho} = +0.01 \times 10^{-3}$). The specular neutron reflection experiment on the OFC was conducted with 1.86 mM $C_{16}E_8$ containing 6% dCTAB. The reflection of neutrons by the dCTAB was only above the background level from nrw at low values of $Q < 0.02$ Å^{-1}. From analysing the specular reflectivity profile, we estimate that the monolayer contained 30–40% CTAB. Hence the monolayer is strongly enriched in the ionic surfactant compared to the bulk (6% CTAB). The most likely explanation is that the exchange of CTAB between the bulk solution and the micelle is much faster than exchange of $C_{16}E_8$. We can estimate the exit rates of monomers from the micelle if we assume the entry rates are diffusion-limited. The exit rates are then proportional to the monomer concentrations in equilibrium with the mixed micelle, which is higher for CTAB (~3 μM) than for $C_{16}E_8$ (~1 μM) under the experimental conditions. CTAB molecules released from the micelles can then adsorb to the surface, depleting the monomer concentration of CTAB and favouring the release of further CTAB molecules from the micelles. Enrichment of the monolayer in CTAB has two consequences: first, the surface potential is increased and therefore the electrostatic repulsion experienced by charged micelles is higher; second, the micelles in the subsurface region are depleted of CTAB and hence there is a larger number of uncharged micelles with the potential to adsorb.

The relaxation kinetics of micelles are very sensitive to the length of the alkyl chain. The fast (τ_1) relaxation time is inversely proportional to the CMC, which increases by an order of magnitude for each C_2H_4 unit that is added to the hydrocarbon chain. The slow (τ_2) relaxation time is expected to be even more sensitive to chain length.[9] On the other hand, the monomer and micelle diffusion coefficients are rather insensitive to chain length, so long as the micelles remain globular. For the series, $C_{12}E_8$, $C_{14}E_8$ and $C_{16}E_8$, the mean aggregation numbers in dilute solution obtained by small angle neutron scattering are approximately 80, 110 and 145, respectively,[16] indicative of globular micelles. The interplay between diffusion and kinetic control, and between monomer and micelle adsorption, can therefore be explored through variation of the alkyl chain length.

Fig. 11 shows the effect of adding CTAB to solutions of C_nE_8 ($n = 12$, 14 and 16) at concentrations around 0.5 mM. The effect of CTAB on the adsorption kinetics of $C_{14}E_8$ is qualitatively similar to that of $C_{16}E_8$: the adsorbed amount decreases up to 5% CTAB and then levels out at a constant value. The obvious difference is that the surface excess at high c_{CTAB} is four times larger for the smaller surfactant. While the plateau in adsorption for $C_{16}E_8$ was consistent with each micelle shedding 10–15% of its monomers while still remaining intact, the surface excess for $C_{14}E_8$ is too high to be accounted for by this mechanism, implying that at least a portion of the micelles are breaking down completely into monomers. In our earlier study on $C_{14}E_8$ in a liquid jet, where the timescale of the experiment is an order of magnitude faster, the addition of 5% CTAB to 1 mM $C_{14}E_8$ reduced the adsorbed amount by three-quarters. This comparison suggests that micelle break-down is fast enough to influence the adsorption kinetics on the OFC timescale but not on the timescale of the jet.

The effect of CTAB on $C_{12}E_8$ adsorption is quite different. In the absence of CTAB, $C_{12}E_8$ adsorbs at a diffusion-controlled rate with fast micelle break-down. Addition of CTAB increases the surface excess until Γ reaches a plateau above 5% CTAB. This result is qualitatively consistent with an enhancement of the micellar diffusion coefficient from migration fields.

The different kinetic behaviour of $C_{12}E_8$ is confirmed by the addition of salt (Fig. 12). The surface excess of $C_{14}E_8$ recovers to the value of the pure non-ionic

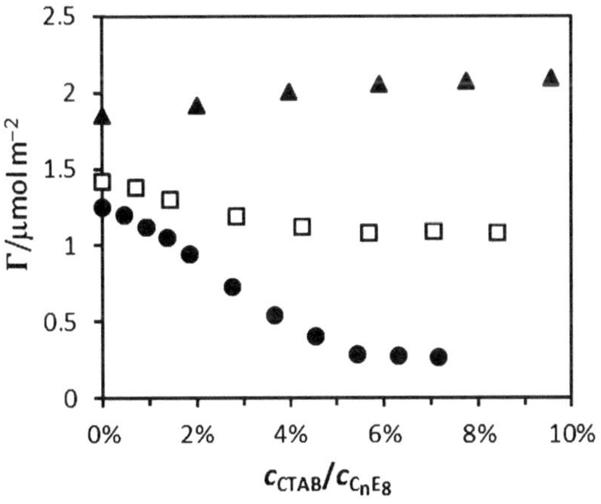

Fig. 11 Effect of addition of CTAB on the surface excess of non-ionic surfactants C_nE_8 in the OFC. Diamonds: $C_{16}E_8$ at 0.50 mM; squares: $C_{14}E_8$ at 0.54 mM; triangles: $C_{12}E_8$ at 0.54 mM.

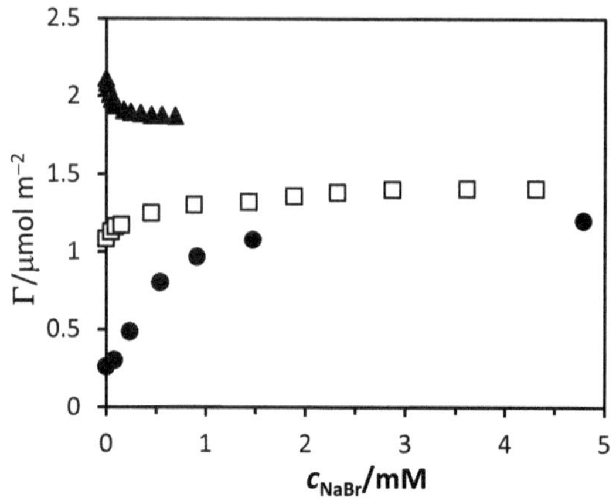

Fig. 12 Effect of addition of NaBr on the surface excess of non-ionic surfactants C_nE_8 in the OFC, in the presence of CTAB. Diamonds: $C_{16}E_8$ at 0.50 mM with 7% CTAB; squares: $C_{14}E_8$ at 0.54 mM with 8% CTAB; triangles: $C_{12}E_8$ at 0.54 mM with 10% CTAB. For ease of comparison with Fig. 11, the ellipticity has been converted to surface excess on the assumption that a CTAB molecule makes the same contribution to $\bar{\rho}$ as a C_nE_8 molecule.

surfactant with addition of 3 mM NaBr, which is similar to the behaviour of $C_{16}E_8$. For $C_{12}E_8$ the increased adsorption is due to migration fields, which are screened when the added electrolyte is in large excess over the counterions. In this experiment, the counterion concentration in the absence of NaBr is 0.05 mM. We observe that a ten-fold excess (0.5 mM) of NaBr is sufficient to remove completely the effect of migration fields on the adsorption kinetics.

Finally, we briefly explore the effect of the shape of the micelle on the adsorption mechanism. $C_{16}E_6$ forms cylindrical micelles whose length depends on temperature and concentration.[31] For example, from SANS experiments on micelles aligned by a shear flow in a Couette cell, Cummins *et al.* determined a micelle length of 400 nm for 20 mM $C_{16}E_6$ at 28 °C, corresponding to an aggregation number $N \sim 10^4$. Fig. 13 shows the ellipticity of an 0.56 mM solution of $C_{16}E_6$ in the OFC as a function of CTAB concentration. For the pure surfactant, less $C_{16}E_6$ adsorbs than for $C_{16}E_8$

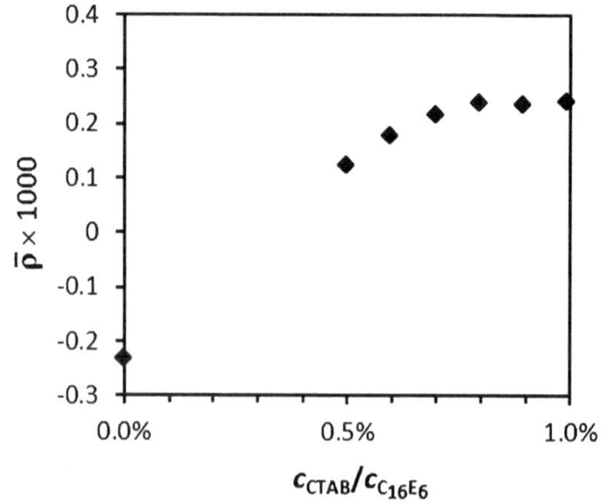

Fig. 13 Effect of addition of CTAB on the coefficient of ellipticity of a 0.56 mM solution of $C_{16}E_6$ in the OFC. The experiments were conducted at 298 K, which is just below the reported Krafft point of $C_{16}E_6$ (301 K)[31] but the solutions remained stable at this high dilution.

at the same concentration because cylindrical micelles diffuse more slowly than globular ones. The adsorption rate of $C_{16}E_6$ is much more sensitive to addition of CTAB: only 0.8% of CTAB (<5 μM) was needed to reduce the ellipticity (and hence surface excess) to its plateau value. This observation is consistent with the hypothesis that only uncharged micelles adsorb: the higher the aggregation number the less CTAB is needed to ensure that virtually all the micelles are charged. From the amount of CTAB needed to turn off micellar adsorption, we can make a rough estimate of the aggregation number of the $C_{16}E_6$ micelles of 10^3, which is an order of magnitude smaller than that measured directly by SANS. We note, however, that our solutions are much more dilute and the length of rods is expected statistically to decrease with decreasing concentration. When NaBr was added to the 1% CTAB mixture, 1 mM of salt was sufficient to return $\bar{\rho}$ to its original value in the absence of CTAB.

One puzzling feature of Fig. 13 is the very low amount of surfactant adsorbed once direct micellar adsorption has been inhibited. In a one-dimensional aggregation model, the chemical potential of a monomer in a micelle is very weakly dependent on the length of the rod.[32] Consequently, if the monomer concentration is reduced by adsorption to the surface, one would expect that the micelles would readily shed most of their monomers by sequential monomer loss. Hence the adsorption rate should be close to the diffusion-controlled limit even without direct micellar adsorption or complete micelle break-down. The experimental evidence would therefore imply that the rate of loss of monomers from the $C_{16}E_6$ micelles is much less than would be expected if exchange between monomers and micelles occurred at a diffusion-controlled rate. Cummins *et al.* noted, however, that the polydispersity of the $C_{16}E_6$ micelles was much smaller than expected for a 1-D aggregation model, suggesting that there is deeper minimum in the chemical potential at the size of the most favourable aggregate. Such a minimum would restrict the number of monomers that could be shed sequentially by a rod-like micelle. Shchekin and coworkers have recently considered the kinetics of exchange between rod-like and globular micelles.[33]

4. Conclusion

The overflowing cylinder provides a convenient platform for studying the adsorption kinetics of surfactants at the air–water interface. The flowing surface of the OFC is extremely stable, permitting the study of the interface by a range of experimental techniques. It is also at steady-state, so experiments can be run over many hours if required. The mass balance at the surface requires that the removal of surfactant by surface expansion, $\Gamma\theta$, is equal to the flux of surfactant to the surface from the bulk solution. Measurement of the surface excess, Γ, by ellipsometry and the surface expansion rate, θ, by laser Doppler scattering thus allows us to characterise the adsorption kinetics as a function of the surface excess. In this work, we have used the OFC to explore the kinetics of adsorption of non-ionic surfactants with very low CMCs such that the overwhelming majority of the surfactant in solution is in micellar form. Consequently, micellar processes have a dominant role in determining the adsorption kinetics. The focus of our study has been $C_{16}E_8$. We find that up to 40% of a monolayer coverage, the adsorption is diffusion-controlled on the OFC timescale (0.1–1 s). Above that coverage, the surface excess increases only very slowly with bulk surfactant concentration. We interpret this behaviour in terms of a direct adsorption flux of micelles which do not first break down into monomers. Once the surfactants on the surface reach a certain coverage, steric repulsions between the polyethyleneglycol chains of the monolayer and the micelles prevent the micelles from approaching close enough to the surface for the interfacial stresses to rip the micelles apart.

To support this hypothesis we have doped the micelles with small amounts of the cationic surfactant, CTAB, or the anionic surfactant, STS. We find that either surfactant substantially reduces Γ at trace levels (<5% of the total surfactant concentration). We ascribe this effect to electrostatic repulsions between the charged monolayer and the charged micelles and find that a simple model in which only uncharged micelles can adsorb provides qualitative agreement with all our observations. Addition of millimolar quantities of NaBr to the $C_{16}E_8$/ionic surfactant mixture screens the repulsions between micelles and monolayer and causes Γ to recover to the value of the pure surfactant. The adsorption behaviour of the shorter chain surfactants, $C_{12}E_8$ and $C_{14}E_8$ are consistent with a model in which both micelles and monomers can adsorb, with the relative contributions depending on the CMC and the kinetics of release of monomers from the surfactant (through either the fast or slow relaxation mechanisms). On the shorter timescale of liquid jets, the micellar route is more important. Further support for our model is provided by measurements on rod-like micelles, which have a higher aggregation number than globular micelles: a much lower concentration of ionic surfactant is needed to inhibit the adsorption of $C_{16}E_6$ than $C_{16}E_8$.

This study has used just two different ionic surfactants: CTAB and STS. The identical effect of these two surfactants on the adsorption kinetics of $C_{16}E_8$ suggests that specific ion effects will not be important in the adsorption kinetics of micelles in salt-free solutions, except to the extent that they influence the thermodynamics of partitioning of the ionic surfactant between monomers and micelles. When electrolyte was added to the system to screen the electrostatic repulsions between monolayer and micelles, the two ionic surfactants showed different behaviour. The highly dilute solution of ionic surfactants within the non-ionic micelles and the low charge density in the adsorbed monolayer mean that any specific interactions would occur between isolated anions and cations. Future work will explore the influence of different cations and anions on the electrostatic barriers to adsorption of micelles at surfaces.

Acknowledgements

We thank Dr N. Péron for modelling the data in Fig. 2, J. K. Thye for recrystallising the STS and Dr D. M. Colegate for useful discussions.

References

1 S. S. Dukhin, G. Kretzschmar and R. Miller, *Dynamics of Adsorption at Liquid Interfaces*, Elsevier, Amsterdam, 1995.
2 C. D. Bain, *Adv. Colloid Interface Sci.*, 2008, **144**, 4.
3 D. S. Valkovska, G. Shearman, C. D. Bain, R. C. Darton and J. Eastoe, *Langmuir*, 2004, **20**, 4436.
4 D. M. Colegate and C. D. Bain, *Phys. Rev. Lett.*, 2005, **95**, 198302.
5 B. A. Noskov, *Adv. Colloid Interface Sci.*, 2002, **95**, 237.
6 A. Patist, J. R. Kanicky, P. K. Shukla and D. O. Shah, *J. Colloid Interface Sci.*, 2002, **245**, 1.
7 I. M. Griffiths, C. D. Bain, C. J. W. Breward, D. M. Colegate, P. D. Howell and S. L. Waters, *J. Colloid Interface Sci.*, 2011, **360**, 662; I. M. Griffiths, C. D. Bain, C. J. W. Breward, S. J. Chapman, P. D. Howell and S. L. Waters, *SIAM J. Appl. Math.*, 2012, **72**, 201.
8 R. Becker and W. Döring, *Ann. Phys. (Leipzig)*, 1935, **24**, 719.
9 E. A. G. Aniansson, S. N. Wall, M. Almgren, H. Hoffmann, I. Kielmann, W. Ulbricht, R. Zana, J. Lang and C. Tondre, *J. Phys. Chem.*, 1976, **80**, 905.
10 I. M. Griffiths and C. D. Bain, unpublished results.
11 S. Manning-Benson, C. D. Bain and R. C. Darton, *J. Colloid Interface Sci.*, 1997, **189**, 109.
12 C. D. Bain, S. Manning-Benson and R. C. Darton, *J. Colloid Interface Sci.*, 2000, **229**, 247.
13 D. Valkovska, K. M. Wilkinson, R. A. Campbell, C. D. Bain, R. Wat and J. Eastoe, *Langmuir*, 2003, **19**, 5960.
14 S. Manning-Benson, S. R. W. Parker, C. D. Bain and J. Penfold, *Langmuir*, 1998, **14**, 990.
15 T. Battal, G. C. Shearman, D. Valkovska, C. D. Bain, R. C. Darton and J. Eastoe, *Langmuir*, 2003, **19**, 1244.
16 D. M. Colegate, PhD thesis, Durham University, U.K., 2009.
17 A. Angus-Smyth, R. A. Campbell and C. D. Bain, *Langmuir*, 2012, **28**, 12479.

18 R. A. Campbell, H. P. Wacklin, I. Sutton, R. Cubitt and G. Fragneto, *Eur. Phys. J. Plus*, 2011, **126**, 107.
19 S. Manning-Benson, S. R. W. Parker, C. D. Bain, R. C. Darton and J. Penfold, *Langmuir*, 1998, **14**, 990.
20 http://motofit.sourceforge.net/wiki/index.php/Main_Page.
21 S. R. Patil, N. Buchavzov, E. Carey and C. Stubenrauch, *Soft Matter*, 2008, **4**, 840.
22 P. Bäverbäck, C. L. P. Oliveira, V. M. Garamus, I. Varga, P. M. Claesson and J. S. Pedersen, *Langmuir*, 2009, **25**, 7296.
23 J. R. Lu, Z. X. Li, R. K. Thomas, E. J. Staples, I. Tucker and J. Penfold, *J. Phys. Chem.*, 1993, **97**, 8012.
24 S.-Y. Lin, R.-Y. Tsay, L.-W. Lin and S.-I. Chen, *Langmuir*, 1996, **12**, 6530.
25 A. Patist, J. R. Kanicky, P. K. Shukla and D. O. Shah, *J. Colloid Interface Sci.*, 2002, **245**, 1.
26 E. A. G. Aniansson and S. N. Wall, *J. Phys. Chem.*, 1974, **78**, 1024.
27 C. Carnero Ruiz and J. Aguiar, *Colloids Surf., A*, 2003, **224**, 221.
28 E. D. Goddard, *Colloids Surf.*, 1986, **19**, 255.
29 Y. Nogami, H. Watanabe, H. Ohtaka-Saiki and T. Tominaga, *Colloids Surf., A*, 2000, **169**, 227.
30 Y. Nogami, M. Iwata and T. Tominaga, *J. Mol. Liq.*, 2005, **119**, 83.
31 P. G. Cummins, E. Staples, J. Penfold and R. K. Heenan, *Langmuir*, 1989, **5**, 1195.
32 J. Israelachvili, *Intermolecular and Surface Forces*, 2nd edn, Academic, London, 1991.
33 A. K. Shchekin, M. S. Kshevetskiy and O. S. Pelevina, *Colloid J.*, 2011, **73**, 406.

Adsorption of solutes at liquid–vapor interfaces: insights from lattice gas models

Suriyanarayanan Vaikuntanathan,[a] Patrick R. Shaffer[b] and Phillip L. Geissler*[ab]

Received 10th May 2012, Accepted 28th May 2012
DOI: 10.1039/c2fd20106b

The adsorption behavior of ions at liquid–vapor interfaces exhibits several unexpected yet generic features. In particular, energy and entropy are both minimum when the solute resides near the surface, for a variety of ions in a range of polar solvents, contrary to predictions of classical theories. Motivated by this generality, and by the simple physical ingredients implicated by computational studies, we have examined interfacial solvation in highly schematic models, which resolve only coarse fluctuations in solvent density and cohesive energy. Here we show that even such lattice gas models recapitulate surprising thermodynamic trends observed in detailed simulations and experiments. Attention is focused on the case of two dimensions, for which approximate energy and entropy profiles can be calculated analytically. Simulations and theoretical analysis of the lattice gas highlight the role of capillary wave-like fluctuations in mediating adsorption. They further point to ranges of temperature and solute–solvent interaction strength where surface propensity is expected to be strongest.

I. Introduction

Experiments and computer simulations have demonstrated that certain small ions preferentially adsorb at liquid–vapor interfaces,[1–5] contrary to expectations from classic theories of ion solvation.[6] Efforts to explain this behavior have focused primarily on accounting for effects of solute polarizability and the thermodynamic cost of excluding volume in bulk solution.[1,7–10] For small excluded volumes, this cost is primarily entropic and reflects constraints on the available arrangements of nearby solvent molecules.[11] From this perspective the adsorption of a non-polarizable ion is expected to be opposed energetically (through loss of dielectric polarization energy) and favored entropically (through recovered freedom of local solvent arrangements). Very recent work has shown, however, that these thermodynamic driving forces generally follow the opposite trend, with energy and entropy both exhibiting minima when an ion resides near the interface.[5,12,13] In these studies we and others also revealed important driving forces governing ion adsorption that had been largely ignored in discussions of interfacial solvation.

The interface between a dense polar liquid and its coexisting vapor is a region of high energy density (relative to that of the bulk liquid), due to incomplete or strained coordination of solvent molecules. The liquid's boundary can nonetheless be quite soft, with substantial wavelike shape fluctuations even on molecular length scales. These generic facts appear from simulations to have important consequences for the thermodynamics of ion adsorption. In particular, moving a solute from bulk solution to the interface effects more than reduced coordination of the ion (which

[a]*Chemical Sciences Division, Lawrence Berkeley National Lab, Berkeley, CA 94720*
[b]*Department of Chemistry, University of California, Berkeley, CA 94720*

is exclusively emphasized in classic theories). An adsorbed solute also occupies volume in the high-energy region, effectively returning solvent density to the bulk environment.[5,14,15] The consequence of this simple but often overlooked accounting is a substantially favorable energetic contribution. For ions of modest size it is sufficient to render adsorption energies negative.[5,12–15] Although electrostatic forces clearly figure prominently in this mechanism, their long spatial range is not manifested by the dominant energetic changes accompanying adsorption.[5] Instead, a spatially local approximation, which expresses the total energy in terms of the populations of different solvent regions (bulk, interface, and solute coordination zone),

$$U_{\text{local}} = \varepsilon_{\text{bulk}} n_{\text{bulk}} + \varepsilon_{\text{interface}} n_{\text{interface}} + \varepsilon_{\text{coord}} n_{\text{coord}}, \quad (1)$$

was found to be quite accurate for systems with non-polarizable ions.[5] Here, ε_γ is the energy per molecule of solvent in region γ, and n_γ is the corresponding number of molecules in that region.

Softness of the interface, also neglected by conventional theories, is implicated by these simulations to significantly influence the entropy of adsorption.[5] A solute that substantially attracts solvent, such as an ion in water, tends to suppress surface undulations when it resides near the interface. The entropic cost of this suppression can be considerable, opposing adsorption by many $k_B T$. Harmonic analysis of capillary wave-like fluctuations in the adsorbed and non-adsorbed states was found to roughly account for observed entropy profiles.[5]

The two physical ingredients speculated in Ref. 5 to be key for ion adsorption, namely surface shape fluctuations and heterogeneous energy density near the liquid's boundary, are so generic as to be described by the very simplest microscopic models of the liquid state. Motivated by this apparent generality, we examine in this paper the classic reduced model of liquid–vapor coexistence, suitably adapted to address questions of solvation. Specifically, we consider a two-dimensional lattice gas, with a solute that occupies a single lattice cell and attracts only its nearest neighbors. Presenting results of both theory and simulation, we demonstrate that the unexpected thermodynamic trends revealed by recent atomistic simulations can indeed be captured by such a simple model. The very limited microscopic detail in the case of a lattice gas allows unambiguous conclusions regarding the source of energy and entropy changes as the solute moves through the interface. It allows as well crude predictions for the varying propensity ions exhibit for liquid–vapor interfaces.

The rest of the paper is organized as follows. In Sec. II, we describe our lattice gas model for solvation. We present results of Monte Carlo simulations in Sec. III. In Sec. IV we develop a theory for the behavior of this model based on the so-called solid-on-solid simplification and compare its predictions with numerical results. We conclude in Sec. V with a discussion of specific implications for three-dimensional molecular systems.

II. Lattice gas model

We consider a two-dimensional $L \times L$ lattice gas system with a single solute, that occupies one lattice cell. Cells are indexed by a horizontal coordinate x ranging from 1 to L and a vertical coordinate y ranging from $-(L-1)/2$ to $(L-1)/2$†. The x-coordinate of the solute is constrained to $x = (L+1)/2$. The solute interacts only with the 4 nearest neighbor cells. It additionally excludes volume, so that a cell may not be simultaneously occupied by both solvent and solute. This system is described by the Hamiltonian

$$H = H_0 - \varepsilon_1 \sum_{i,j \in nn} n_i^s n_j, \quad (2)$$

† Without loss of generality, we assume that L is odd.

where

$$H_0 \equiv -\varepsilon_0 \sum_{i,j \in nn} n_i n_j - \mu \sum_i n_i \tag{3}$$

describes the pure solvent. The occupation variables n and n_i^s are binary: $n_i = 1$ if the ith lattice site is occupied by a solvent and zero otherwise, $n_i^s = 1$ if the ith lattice site is occupied by a solute and zero otherwise. $\sum_{i,j \in nn}$ denotes a sum over all nearest neighbor lattice pairs i, j. The coupling constant $\varepsilon_0 > 0$ describes attraction between neighboring solvent cells; ε_1 is the solvent–solute coupling constant.

For both theory and simulation we work with a grand canonical ensemble, in which the number of solvent cells can fluctuate subject to chemical potential μ. (The number of solute cells is fixed.) Conditions of liquid–vapor coexistence are imposed by setting $\mu = -2\varepsilon_0$, and applying boundary conditions that favor liquid in $y < 0$ and vapor in $y > 0$. Specifically, lattice cells with $y = -(L-1)/2$ are constrained to be occupied by solvent; those with $y = (L-1)/2$ are constrained to be unoccupied (*i.e.*, vapor). The mean interface position, or the Gibbs Dividing Surface (GDS), is fixed at $y = 0$ by constraining as well the state of lattice cells with $x = 1$ and $x = L$; those with $y < 0$ are occupied, and those with $y > 0$ are unoccupied.

We will examine values of ε_1 between 0 and 1. By symmetry (see Appendix A), a solute with $\varepsilon_1 = \varepsilon_0/2$ is equally likely to reside in a cell well within the liquid phase as in a cell well within the vapor phase. Judging by the free energy of transfer from liquid to vapor, one would therefore consider solutes with $\varepsilon_1 = \varepsilon_0/2$ to be solvophobic, and those with $\varepsilon_1 < \varepsilon_0/2$ to be solvophilic (although the latter may attract solvent significantly less strongly than adjacent solvent cells attract each other).

Using theory and simulation, we have determined the free energy F, average energy E, and the entropy S as functions of the height $\bar{h}$ of the solute (*i.e.*, its y-coordinate).

III. Monte Carlo simulations

We present results of Monte Carlo simulations at temperature $T = \varepsilon_0/3k_B$ (well below the critical temperature $T_c \approx \varepsilon_0/1.764k_B$) for a lattice with linear dimension $L = 30$. Trial moves included (i) changing the state of a cell from solvent to vapor, (ii) changing a cell from vapor to solvent, and (iii) translating the solute along the y axis from a lattice cell s to a cell s'. In the case of solute displacements, two different trial moves were used, distinguished by their treatment of solvent occupation in the cell s vacated by the solute. In one move, cell s adopts the previous state of s', *i.e.*, $n_s^{\text{trial}} = n_{s'}$. In another move, cell s adopts the complementary state of s', *i.e.*, $n_s^{\text{trial}} = 1 - n_{s'}$. In all cases detailed balance was ensured through appropriate Metropolis acceptance criteria, preserving the Boltzmann distribution $p(\{n_i,n_i^s\}) \propto e^{-\beta H}$, where $\beta = 1/k_B T$ is inverse temperature. We obtained estimates of $F(\bar{h})$ by computing the equilibrium probability $p(\bar{h}) \propto \exp[-\beta F(\bar{h})]$ that the solute resides at height $\bar{h}$. The average energy $E(\bar{h})$ was estimated directly from simulations, while estimates of the entropy $S(\bar{h})$ were calculated from $TS(\bar{h}) = E(\bar{h}) - F(\bar{h})$.

The profiles of $F(\bar{h})$, $E(\bar{h})$, and $S(\bar{h})$ obtained from simulations with $\varepsilon_1/\varepsilon_0 = \{0.3,0.4,0.5,0.6,0.7,0.8\}$, are plotted in Fig. 1. These results echo several nontrivial thermodynamic trends observed in molecular simulations. In particular, free energy profiles for $\varepsilon_1 = \{0.4\varepsilon_0,0.5\varepsilon_0,0.6\varepsilon_0,0.7\varepsilon_0\}$ exhibit minima near $\bar{h} = 0$. These solutes adsorb to the interface, *i.e.*, their density at $\bar{h} = 0$ is higher than in either bulk phase. Entropy minima at the same values of ε_1 indicate that this adsorption reduces the diversity of accessible solvent configurations. The driving force for surface propensity is instead energetic, as evidenced by strong minima in $E(\bar{h})$ near $\bar{h} = 0$. Interestingly, adsorption is strongest for the case $\varepsilon_1 = \varepsilon_0/2$, at the crossover between solvophobic and solvophilic regimes.

These trends were also observed in simulations with smaller and larger system sizes ($L = \{20,30,40,50,60,70\}$), and for various temperatures in the range

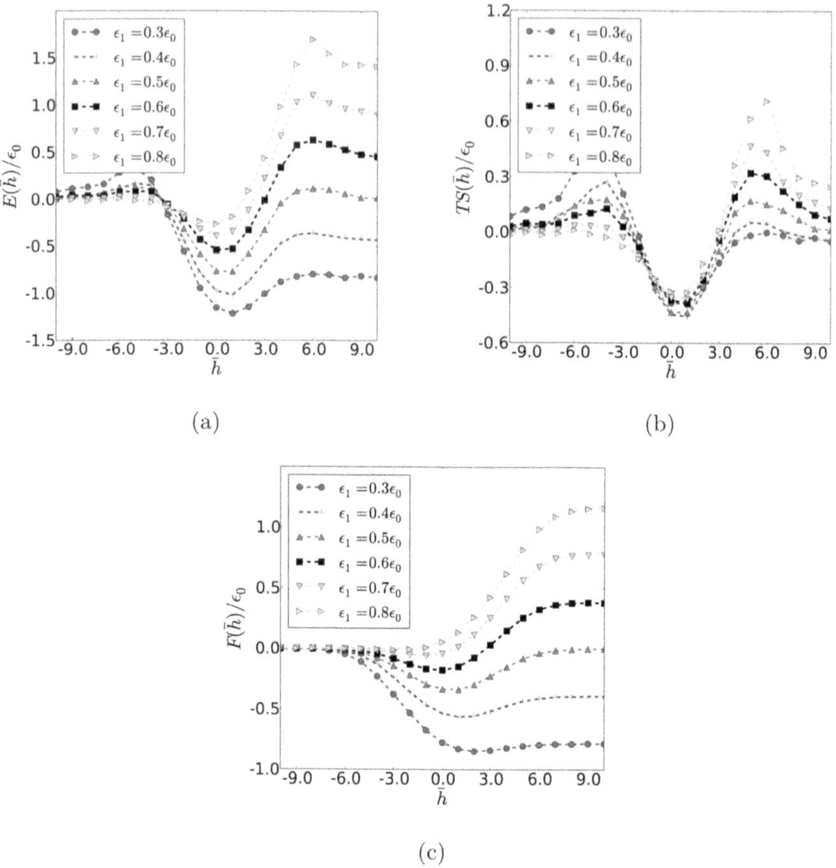

Fig. 1 Profiles of (a) energy $E(\bar{h})$, (b) entropy $S(\bar{h})$, and (c) free energy $F(\bar{h})$, as a function of the height of the solute, $\bar{h}$, for various values of ε_1 at $T = \varepsilon_0/3k_B$. The height of $\bar{h} = 0$ coincides with the GDS. These profiles were obtained from Monte Carlo simulations of the lattice gas.

$\varepsilon_0/5k_B \le T \le \varepsilon_0/2k_B$. In all cases, profiles of free energy, energy, and entropy exhibit minima when the solute is at the interface, and maximum adsorption occurs at $\varepsilon_1 = \varepsilon_0/2$.

IV. Theory

The 2-d lattice gas model of solvation described by eqn (2), within approximations that are well justified for $T \ll T_c$, is sufficiently simple to permit an analytical solution. This section presents our mathematical analysis, which draws heavily on results from theoretical studies of surface roughening in Ising-like models.[16–19] Our key assumptions are that the liquid–vapor interface is well defined and that the solute does not significantly alter fluctuations within either phase. More specifically, we restrict attention to the so-called "solid-on-solid" (SOS) limit,[20,21] in which: (1) all cells in the liquid phase are occupied by solvent (except the cell occupied by the solute when it is present), (2) all cells in the vapor phase are unoccupied (except the solute cell), and (3) the boundary between liquid and vapor is a single-valued function of x. With this neglect of bubbles, islands, and overhangs, a configuration can be specified through the heights of L columns representing the instantaneous domain of the liquid phase.

Under conditions of coexistence, the Hamiltonian describing the solvent is then given by[20,22]

$$\bar{H}_0 = \frac{\varepsilon_0}{2} \sum_{i,j \in nn} |h_i - h_j| , \tag{4}$$

where h_i denotes the height of the liquid in the ith column. Boundary conditions described in Sec. 2 fix the heights of the first and last columns, $h_1 = 0$ and $h_L = 0$. They also bound the range of column height fluctuations, $-L/2 < h_i < L/2$. As an added simplification, we work with the Restricted SOS (RSOS) model[17] where the height difference between neighboring columns is constrained to be at most 1, $|h_i - h_{i+1}| \leq 1$. A few allowed configurations of the RSOS model are illustrated in Fig. 2. Previous work has shown that fluctuations of the neat liquid–vapor interface of a two-dimensional lattice gas are well described by the RSOS model for temperatures $T \ll T_c$.[17,19] As in that work, the restriction on height fluctuations is necessary to derive simple closed form expressions for thermodynamic quantities of interest.

The liquid–vapor interface of the RSOS model supports shape fluctuations akin to capillary waves. It has been shown in particular that long-wavelength Fourier modes are Gaussian distributed to a good approximation. For a 2-d RSOS interface of length M, whose boundaries are fixed at $h_1 = p$ and $h_M = q$, the partition function is given by[17,19,23]

$$Z^{1,M}(p,q) \approx \frac{e^{2aM}}{\sqrt{4\pi aM}} \exp\left[\frac{-(p-q)^2}{4aM}\right] , \tag{5}$$

in the limit $Ma \gg 1$, where $a \equiv \exp(-\beta\varepsilon_0/2)$. This result is strictly valid only in the low temperature limit, where terms of order $O(a^2)$ can be neglected. Empirically, however, the thermodynamic properties we compute using eqn (5) agree well with those obtained from lattice gas simulations even at moderate temperatures.

To construct a theory for interfacial solvation from eqn (5), we first partition the ensemble of solvent fluctuations consistent with a given solute height $\bar{h}$. We classify configurations in which the interface lies entirely above the solute as belonging to a completely solvated class, denoted *in*. Immersion in this sense requires $h_s > \bar{h}$, where s is the solute's x-coordinate. Configurations in which the solute makes no contact with the liquid phase, $h_s < \bar{h} - 1$, we classify as *out*. In the remaining configurations, which we classify as *pinned*, the solute resides at the interface, $h_s = \bar{h} - 1$, in effect pinning local height fluctuations.

Defining $Z_{in}(\bar{h})$, $Z_{out}(\bar{h})$, and $Z_{pinned}(\bar{h})$ as the partition functions for these three subensembles, we can reconstruct the total partition function for a given solute height simply as

$$Z(\bar{h}) = Z_{in}(\bar{h}) + Z_{out}(\bar{h}) + Z_{pinned}(\bar{h}). \tag{6}$$

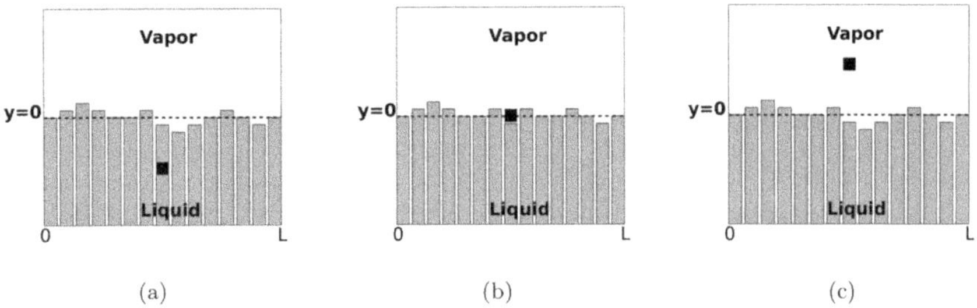

Fig. 2 Illustration of three configurational subensembles in the RSOS model. The fully solvated *in* state (a) features a liquid vapor interface entirely above the solute. In the *pinned* state subensemble (b), this interface directly contacts the solute. In the unsolvated state *out* state, (c), the interface lies below the solute.

Each of the three configurational classes corresponds to a constrained ensemble of pure liquid–vapor configurations, reweighted by solvent–solute interactions. We will calculate their partition functions in turn.

The *out* subensemble is simplest, since solvent and solute do not interact. The partition function for this case is given by

$$Z_{out}(\bar{h}) = \sum_{h_1=0\ ,h_L=0\ ,h_s<\bar{h}-1} \cdots \sum \mathrm{e}^{-\beta \bar{H}_0} = \sum_{h_s<\bar{h}-1} Z^{1,s}(0,h_s)Z^{s,L}(h_s,0), \tag{7}$$

where the notation $\sum \cdots \sum_{h_1=0,h_L=0,h_s<\bar{h}-1}$ implies a sum over all configurations of the RSOS system subject to the constraints $\{h_1 = 0,\ h_L = 0,\ h_s < \bar{h} - 1\}$. To obtain a closed-form expression, we convert sums to integrals in the limit $L \gg 1$, and set the solute's horizontal position at $s = (L + 1)/2$, yielding,

$$\begin{aligned} Z_{out}(\bar{h}) &\approx \frac{\mathrm{e}^{2aL}}{2\pi aL} \int_{-(L-1)/2}^{\bar{h}-1} dh_s \mathrm{e}^{-h_s^2/aL} \\ &\approx \frac{\mathrm{e}^{2aL}}{2\pi aL} \int_{-\infty}^{\bar{h}-1} dh_s \mathrm{e}^{-h_s^2/aL} = \frac{\mathrm{e}^{2aL}}{2\pi aL} \sqrt{\frac{\pi aL}{4}} \left[\mathrm{erf}\left(\frac{\bar{h}-1}{\sqrt{aL}} \right) + 1 \right] \end{aligned} \tag{8}$$

Computing the energy $\bar{H}_{in}$ of a fully solvated configuration (belonging to the *in* state) requires accounting for direct interactions between solvent and solute, as well as counting broken solvent–solvent bonds and the thermodynamic consequences of removing a solvent cell to accommodate the solute. The four solvent–solute bonds contribute an energy $-4\varepsilon_1$. Loss of 4 solvent–solvent interactions is partially offset by the reversible work $\mu = -2\varepsilon_0$ of returning one solvent cell to the bath, yielding a total energy

$$\bar{H}_{in} = \bar{H}_0 - 4(\varepsilon_1 - \varepsilon_0/2) \tag{9}$$

The corresponding partition function can therefore be written as

$$Z_{in}(\bar{h}) = \sum_{h_1=0\ ,h_L=0\ ,h_s>\bar{h}} \cdots \sum \mathrm{e}^{-\beta \bar{H}_0} \mathrm{e}^{\beta 4(\varepsilon_1-\varepsilon_0/2)} \tag{10}$$

Again, using eqn (5) in the limit $L \gg 1$, and rewriting the sums as integrals, we obtain

$$\begin{aligned} Z_{in}(\bar{h}) &\approx \mathrm{e}^{\beta 4(\varepsilon_1-\varepsilon_0/2)} \frac{\mathrm{e}^{2aL}}{2\pi aL} \int_{\bar{h}}^{(L-1)/2} dh_s \mathrm{e}^{-h_s^2/aL} \approx \mathrm{e}^{\beta 4(\varepsilon_1-\varepsilon_0/2)} \frac{\mathrm{e}^{2aL}}{2\pi aL} \int_{\bar{h}}^{\infty} dh_s \mathrm{e}^{-h_s^2/aL} \\ &= \mathrm{e}^{\beta 4(\varepsilon_1-\varepsilon_0/2)} \frac{\mathrm{e}^{2aL}}{2\pi aL} \sqrt{\frac{\pi aL}{4}} \mathrm{erfc}\left(\frac{\bar{h}}{\sqrt{aL}} \right) \end{aligned} \tag{11}$$

Statistics of the pinned state are complicated by variation in the solvent–solute interaction energy. Within the RSOS model, allowed heights of columns $s - 1$ and $s + 1$ are $\bar{h}$, $\bar{h} - 1$, and $\bar{h} - 2$. Only in the first case does the solute interact with an adjacent liquid column. Since it is not necessary to break solvent–solvent bonds or remove solvent cells in this case, the corresponding energy function can be written as

$$\bar{H}_{pinned} = \bar{H}_0 + \Delta H(h_{s-1}, h_{s+1}; \bar{h}) \tag{12}$$

where

$$\Delta H(h_{s-1},h_{s+1};\bar{h}) = -\varepsilon_1(1 + \delta_{h_{s-1},\bar{h}} + \delta_{h_{s+1},\bar{h}}). \tag{13}$$

Summation over the column heights $h_2, h_3, \ldots, h_{s-2}$ and $h_{s+2}, h_{s+3}, \ldots, h_{L-1}$ may now be performed using eqn (5):

$$Z_{pinned} = \sum_{h_{s-1}=\bar{h}-2}^{\bar{h}} \sum_{h_{s+1}=\bar{h}-2}^{\bar{h}} Z^{(1,s-1)}(0, h_{s-1}) \exp\left[-\beta \Delta H(h_{s-1}, h_{s+1}; \bar{h})\right]$$

$$\times \exp\left[\frac{-\beta\varepsilon_0}{2}(|h_{s-1} - (\bar{h}-1)| + |h_{s-1} - (\bar{h}-1)|)\right] Z^{(s+1,L)}(h_{s+1}, 0)\ . \qquad (14)$$

The constraint $h_s = \bar{h} - 1$ defining the pinned state, together with the locality of inter-column interactions, decouples fluctuations to the left and right of the solute. As a result, summations over h_{s-1} and h_{s+1} can be carried out independently; with the choice $s = (L + 1)/2$ they yield identical contributions in the $L \gg 1$ limit. We then obtain

$$Z_{pinned}(\bar{h}) \approx b^2 \frac{e^{2aL}}{2\pi a L} e^{-(\bar{h}-1)^2/aL}\ , \qquad (15)$$

where

$$b \equiv e^{-2a}[e^{\beta(3\varepsilon_1/2 - \varepsilon_0/2)} e^{(-2\bar{h}+1)/(2aL)} + e^{\beta(\varepsilon_1/2)} + e^{\beta((\varepsilon_1-\varepsilon_0)/2)} e^{(2\bar{h}-3)/(2aL)}], \qquad (16)$$

Using eqn (6), (8), (11), (15), we can readily obtain profiles of free energy $F(\bar{h}) = -\beta^{-1}\ln Z(\bar{h})$, energy $E(\bar{h}) = -(\partial \ln Z(\bar{h})/\partial\beta)_L$, and entropy $S(\bar{h}) = (E - F)/T$. These analytical results are plotted in Fig. 3 for a lattice of size $L = 30$, at temperature $T = \varepsilon_0/3.6k_B$, and for various values of $\varepsilon_1/\varepsilon_0$. They agree very well with simulation results. In particular, the depth of minima at $\bar{h} = 0$ are accurately predicted. In Fig. 4 we focus on these adsorption properties, $\Delta g_{ads} \equiv g(0) - g(-(L - 1)/2)$ for thermodynamic property g, for the specific case $\varepsilon_1 = \varepsilon_0/2$‡. Theoretical predictions are plotted as functions of temperature at $L = 30$ alongside results of Monte Carlo simulations. Agreement is very good for temperatures in the range $\varepsilon_0/6k_B < T < \varepsilon_0/3k_B$ for $L = 30$.

The failure of our theory at high temperature stems from neglect of contributions to Z from terms of order $O(a^2)$ and higher. At high temperature ruggedness of the interface can also render the RSOS model a poor proxy for the SOS model. At higher temperatures still, neglect of bulk inhomogeneities in the SOS model is itself not justified. At low temperatures a different prerequisite for the validity of eqn (5) is violated, namely, aL is no longer a large number. For the RSOS fluid with $L = 30$ columns, $aL \approx 1$ when $T \approx 0.147\varepsilon_0/k_B$. For temperatures around or lower than this value, height fluctuations in the RSOS fluid are no longer Gaussian to a good approximation.

Success of the SOS approximation in the temperature range $\varepsilon_0/6k_B < T < \varepsilon_0/3k_B$ reveals clearly the source of entropy variations observed in our lattice gas simulations. Absent density fluctuations in the liquid or vapor phase (which are neglected in the SOS description), fluctuations in interfacial shape provide the sole source of entropy in these schematic models. The minimum in $S(\bar{h})$ near $\bar{h} = 0$ thus directly indicates suppression of capillary fluctuations when the solute resides at the liquid–vapor interface. Similarly, local maxima of $S(\bar{h})$ just above and below $\bar{h} = 0$ signify enhancement of surface undulations when the solute pulls the interface a few lattice spacings away from its natural equilibrium position.

Interfacial pinning is strongest for a solute with coupling constant $\varepsilon_1 \approx \varepsilon_0/2$. In simulations this conclusion is suggested by the depth of entropy minima at $\bar{h} = 0$,

‡ Recall that the bulk values of F,E,S are equal in the liquid and vapor states when $\varepsilon_1 = \varepsilon_0/2$. Hence, the results would not have changed if the bulk vapor was chosen as the reference state.

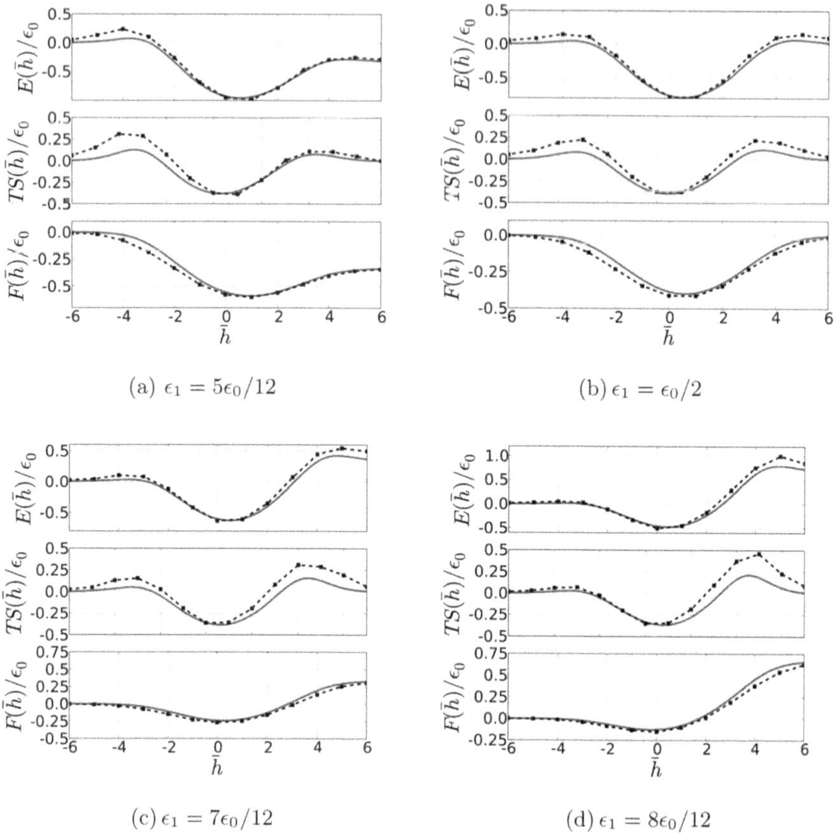

Fig. 3 Profiles of the energy, entropy, and the free energy as a function of $\bar{h}$ for various values of ε_1 at $T = \varepsilon_0/3.6k_B$. The solid lines were calculated from eqn (6), while the dashed lines with squares were obtained from simulations.

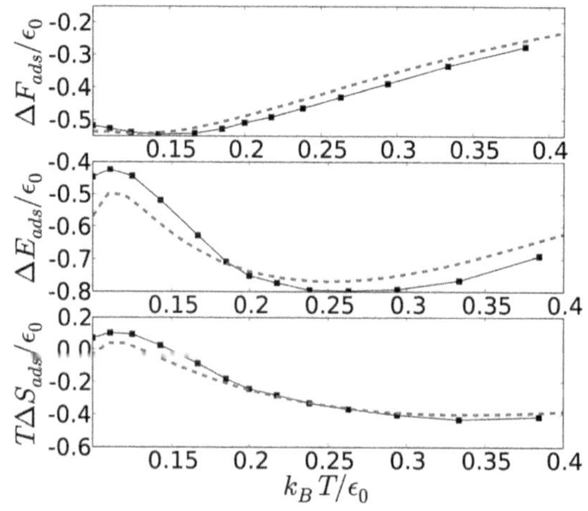

Fig. 4 Free energy, energy, entropy of adsorption as a function of T for $\varepsilon_1 = \varepsilon_0/2$, with $L = 30$. The dashed line is calculated from theory, while the solid line with circles is obtained from simulations.

which are most pronounced at intermediate coupling. It can be drawn more directly in our RSOS theory from the statistical weight $P_{pinned}(\bar{h}) = Z_{pinned}(\bar{h})/Z(\bar{h})$ of the pinned subensemble. For a given value of ε_1, $P_{pinned}(\bar{h})$ is always greatest at $\bar{h} = 0$, where pinning does not require deforming the interface. (See Fig. 5(a)) $P_{pinned}(0)$ should clearly increase with ε_1 in the strongly solvophobic regime, where weak attraction to solvent is insufficient to offset the entropic price of suppressing capillary-like waves. It is also clear that $P_{pinned}(0)$ should decrease with ε_1 in the strongly solvophilic regime, where extremely favorable solvation encourages the fluctuating interface to deform above a solute at $\bar{h} = 0$ (*i.e.*, to remain in the *in* state). Fig. 5(a) bears out the resulting expectation that pinning is most effective at intermediate ε_1, very close to $\varepsilon_1 = \varepsilon_0/2$.

Subensemble weights $P_\alpha(\bar{h}) \equiv Z_\alpha(\bar{h})/Z(\bar{h})$, where $\alpha = \{in,out,pinned\}$, also shed light on the nature of entropy maxima at values of $\bar{h}$ a few lattice spacings above and below the GDS. In the RSOS model the total entropy can be decomposed by subensemble:

$$S(\bar{h}) = \sum_\alpha S_\alpha(\bar{h})P_\alpha(\bar{h}) - k_B \sum_\alpha P_\alpha(\bar{h})\ln P_\alpha(\bar{h}), \quad (17)$$

where $S_\alpha = (E_\alpha - F_\alpha)/T$ characterizes fluctuations within each subensemble. The final term in eqn (17) is an entropy of mixing associated with fluctuations between *in*, *out*, and *pinned* states. Plots of $P_\alpha(\bar{h})$ in Fig. 5(a) show that *in* and *pinned* states become about equally probable near $\bar{h} = -2$, contributing a mixing entropy of roughly $k_B\ln 2$. This gain is partially offset by the intrinsically low entropy of the *pinned* state, and by decreasing S_{in} as the requirement $h_s > \bar{h}$ becomes a nonnegligible constraint on interfacial fluctuations in the solvated state. The residual increase in entropy is a signature of spontaneous switching from the pinned state to an unpinned state, and *vice versa*. Analogous switching between pinned and unsolvated states is responsible for the entropy maximum at $\bar{h} = +4$.

Energy can also be decomposed by subensemble:

$$E(\bar{h}) = \sum_\alpha E_\alpha(\bar{h})P_\alpha(\bar{h}), \quad (18)$$

where $E_\alpha(\bar{h})$ is the average energy in subensemble α when the solute is held at $\bar{h}$. Eqn (4), (9), (12), allow rough estimates of $E_\alpha(\bar{h})$ and thus simple predictions for energies of adsorption. For $\bar{h} \gg 0$ the unsolvated state dominates ($P_{out} = 1$), giving $E(\infty) \approx E_{out} \approx E_0$, where E_0 is the average energy of the solute-free system. For $\bar{h} \ll 0$ the

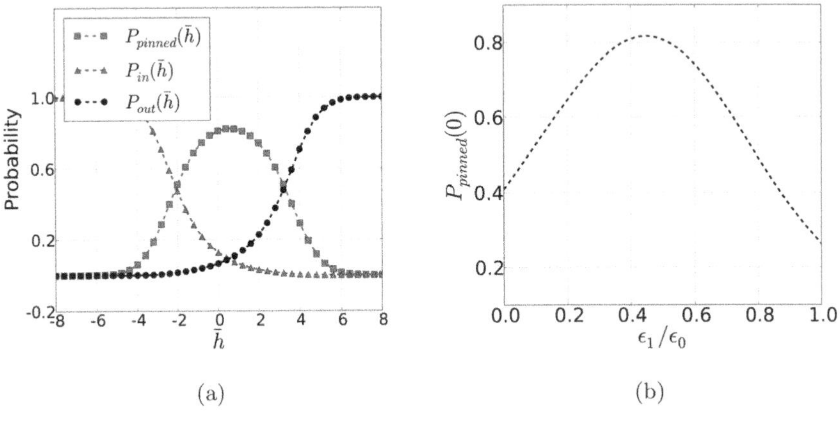

Fig. 5 (a) Probabilities of the *pinned*, *in*, and *out* states as a function of $\bar{h}$ for $T = \varepsilon_0/3.6k_B$, $\varepsilon_1 = \varepsilon_0/2$ and (b) probability of the *pinned* state at $\bar{h} = 0$ as a function of ε_1 for $T = \varepsilon_0/3.6k_B$. When the solute is near the GDS at $\bar{h} = 0$, most of the configurations are in the *pinned* state.

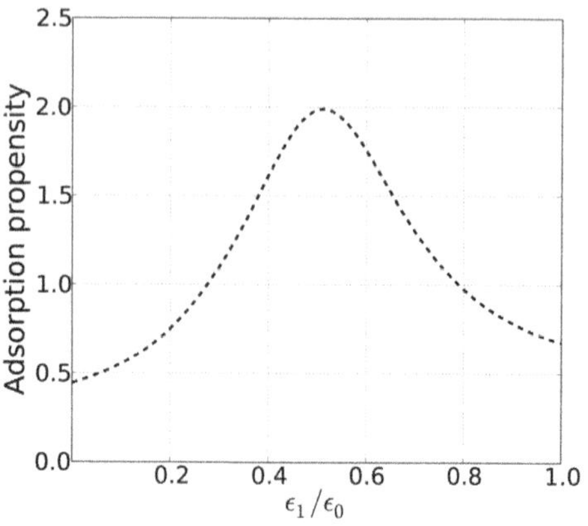

Fig. 6 Adsorption propensity defined as $\frac{Z(0)}{Z(-(L-1)/2)+Z((L-1)/2)}$ as a function of ε_1 for $T = \varepsilon_0/3.6k_B$. These estimates were calculated using eqn (6).

solvated state dominates ($P_{in} = 1$), giving $E(-\infty) \approx E_{in} \approx E_0 - 4(\varepsilon_1 - \varepsilon_0/2)$ from eqn (9). To the extent that pinning dominates at $\bar{h} = 0$, we have $E(0) \approx E_{pinned}(0) \approx E_0 + \Delta H(h_{s-1},h_{s+1};\bar{h})$. At temperatures low enough to safely set $h_{s-1} = h_{s+1} = \bar{h}$ for the pinned state, we have $E(0) \approx E_0 - 3\varepsilon_1 - \varepsilon_0$. Interfacial adsorption of the solute from the liquid phase therefore produces an energy change $E(0) - E(-\infty) \approx \varepsilon_1 - \varepsilon_0$, accounting roughly for replacing a solvent–solute bond with a solvent–solvent bond. For a modestly solvophilic solute $\varepsilon_0/2 < \varepsilon_1 < 1$, adsorption is energetically favorable, and the energetic driving force strengthens with decreasing ε_1. This force is stronger still in the solvophobic regime ($\varepsilon_1 < \varepsilon_0/2$), but these solutes reside primarily in the vapor phase. The strongest relevant energy of adsorption from the liquid phase is thus obtained for $\varepsilon_1 \approx \varepsilon_0/2$, for which $\Delta E_{ads} \approx -\varepsilon_0/2$.

These crude predictions are consistent at low temperatures (below $T = 0.15\varepsilon_0/k_B$ for $L = 30$) with results from our simulations and detailed theoretical analysis (see Fig. 4). At temperatures above $T = 0.15\varepsilon_0/k_B$ for a lattice of size $L = 30$, interfacial fluctuations modify ΔE_{ads} significantly. Reduced coordination at a rough interface enhances the energetic driving force for adsorption, effectively increasing the number of solvent–solute and solvent–solvent bonds exchanged when the solute moves to the surface (see Fig. 4). At the same time, however, the opposing entropic forces discussed above become more pronounced as temperature increases. It is not obvious from our simple estimates how this competition plays out, though the sign and magnitude of ΔS_{ads} ensures that net surface propensity decreases smoothly with increasing temperature (see Fig. 4). Regardless of temperature, adsorption is always strongest when $\varepsilon_1 = \varepsilon_0/2$ (see Fig. 6).

V. Conclusions

Among models that resolve microscopic fluctuations of the liquid–vapor interface, the two-dimensional lattice gas we have studied may be the simplest from which mechanisms of solute surface propensity might be learned. Any insight it offers into the behavior of real molecular systems is no doubt schematic. But the fact that it captures thermodynamic trends which elude celebrated theories of solvation encourages a detailed view of its conclusions in the context of more realistic molecular models.

Any such comparison requires assigning interaction parameters ε_0 and ε_1 to systems with much more complicated energetics. We do so through the local approximation of eqn (1), which for simulations accurately describes the potential energy of a non-polarizable ion in water at a level of detail akin to that of a lattice gas. For a three-dimensional lattice gas model of bulk solvent, the total energy could be calculated by attributing to each liquid cell half of the interactions with its 6 neighboring liquid cells, so that $\varepsilon_{\text{bulk}} = -3\varepsilon_0$ (neglecting solvent density inhomogeneities). Similarly, since solvent cells at an ideally flat liquid–vapor interface engage in one fewer interaction than in bulk, we estimate $\varepsilon_{\text{interface}} \approx -(5/2)\varepsilon_0$. From the value of $\varepsilon_{\text{interface}} - \varepsilon_{\text{bulk}} = 6.52$ kJ mol^{-1} computed from atomistic simulations of water,[5] we thus infer a coarse-grained solvent–solvent attraction strength $\varepsilon_0 \approx$ 13.04 kJ mol^{-1}. Within the local approximation, solvent cells adjacent to the solute would be assigned half of their solvent–solvent interactions as well as a single solute bond. For a solute well within bulk solution, we then have $\varepsilon_{\text{coord}} = -\varepsilon_1 - (5/2)\varepsilon_0$ in the absence of solvent density inhomogeneitics (*i.e.*, within the SOS approximation). From the value $\varepsilon_{\text{coord}} - \varepsilon_{\text{bulk}} = -4.44$ kJ mol^{-1} determined for a fractionally charged non-polarizable iodine ion in water, $I^{-0.8}$, by simulation[5] we finally obtain $\varepsilon_1 =$ 10.96 kJ mol^{-1}, or $\varepsilon_1/\varepsilon_0 \approx 0.84$.

From a lattice gas perspective, the simulated solute $I^{-0.8}$(aq) therefore falls within the hydrophilic regime $\varepsilon_1/\varepsilon_0 > 1/2$. Further reducing the magnitude of the solute's charge should certainly weaken solute–solvent attractions. Decreasing ε_1 by a small amount in this way should bring the solute closer to conditions of maximum adsorption, $\varepsilon_1 = \varepsilon_0/2$, where the scales of adsorption energy and entropy are both largest. Based on results we have presented, we thus expect that changing the ion's charge from $-0.8e$ to $-0.75e$ would deepen the interfacial minima of energy and entropy profiles. Using the same molecular simulation techniques described in Ref. 5, we find this prediction to indeed hold true.

In drawing this comparison we have assumed that basic interfacial solvation properties of the lattice gas are insensitive to dimensionality *d*. Although scaling behavior of a lattice gas near the critical point is famously sensitive to *d*, there is good reason to expect that the essence of our conclusions regarding solvation will be unchanged when moving from $d = 2$ to $d = 3$. The liquid–vapor interface of a three-dimsional lattice gas similarly supports capillary wave-like fluctuations for temperatures in the range $T_R < T < T_c$, where T_R denotes the so-called roughening transition temperature.[24] Furthermore, these surface undulations are Gaussian distributed to a good approximation at long wavelengths,[16,25] just as in eqn (5). Partition functions of the *in*, *out*, and *pinned* states should therefore have similar forms in $d = 3$. In qualitative terms, only the scale of height fluctuations governing Z_{in}, Z_{out}, and Z_{pinned} will change, growing with system size only as $\sqrt{\ln L}$ rather than $\sqrt{L}$.[22] Preliminary simulation results for the $d = 3$ case bear out these expectations.

We conclude that lattice gas models of interfacial solvation are a promising starting point for scrutinizing essential effects of surface roughness and energetic heterogeneity on solute adsorption. For the especially interesting case of charged solutes in highly polar solvents, a more realistic treatment of electrostatics is clearly in order. Given the success of lattice models for dielectric response,[26] this improvement appears quite feasible. An ability to treat solutes of varying size and geometry would also greatly aid connections with experiment. These developments are underway.

Appendix A: appendix

Here we present arguments underlying the assertion that a solute with coupling constant $\varepsilon_1 = \varepsilon_0/2$ is equally likely to reside in either bulk phase. Consider a pair of configurations $\{n_i, n_i^s\}$ and $\{n_i', n_i^s\}$ related by a global transformation that swaps liquid and vapor phases. Specifically, the configuration $\{n_i', n_i^s\}$ is obtained from $\{n_i, n_i^s\}$ by changing the solvent occupation state of all lattice cells not occupied by the solute, $n_i' = (1 - n_i^s)(1 - n_i)$. Repeating this transformation, $n_i = (1 - n_i^s)(1 - n_i')$, we recover the original configuration $\{n_i, n_i^s\}$. In simple terms, if the solute resides in the liquid phase in $\{n_i, n_i^s\}$, then it resides in the vapor phase in $\{n_i', n_i^s\}$, and *vice versa*.

With an accompanying transformation of the solute–solvent coupling ε_1, the energies of configurations $\{n_i, n_i^s\}$ and $\{n_i', n_i^s\}$ can be directly related. Defining $E(\{n_i, n_i^s\}; \{\varepsilon_0, \varepsilon_1\})$ as the energy of configuration $\{n_i, n_i^s\}$ with solvent–solvent coupling ε_0, and solvent–solute coupling ε_1, we have

$$\begin{aligned} E(\{n_i', n_i^s\}; \{\varepsilon_0, \varepsilon_1\}) &= -\varepsilon_0 \sum_{i,j \in nn} n_i' n_j' - \mu \sum_i n_i' - \varepsilon_1 \sum_{i,j \in nn} n_i^s n_j' \\ &= -\varepsilon_0 \sum_{i,j \in nn} n_i n_j - \mu \sum_i n_i - (\varepsilon_0 - \varepsilon_1) \sum_{i,j \in nn} n_i^s n_j + 4(\varepsilon_1 - \varepsilon_0/2) \\ &= E(\{n_i, n_i^s\}; \{\varepsilon_0, \varepsilon_0 - \varepsilon_1\}) + 4(\varepsilon_1 - \varepsilon_0/2). \end{aligned} \tag{A1}$$

When $\varepsilon_1 = \varepsilon_0/2$, the energy of the two configurations are the same. Thus for every configuration in which the solute is in the liquid phase, there exists a configuration with the same statistical weight in which the solute is in the vapor phase and *vice versa*. Hence solutes with $\varepsilon_1 = \varepsilon_0/2$ exhibit equal preference for the liquid and vapor phases.

Acknowledgements

Kelsey Schuster performed preliminary simulations on the lattice gas model. We gratefully acknowledge useful discussions with Yan Levin and David Limmer. This project was supported by the US Department of Energy, Office of Basic Energy Sciences, through the Chemical Sciences Division (CSD) of the Lawrence Berkeley National Laboratory (LBNL), under Contract DE-AC02-05CH11231.

References

1 P. Jungwirth and D. J. Tobias, *Chem. Rev.*, 2006, **106**, 1259–1281.
2 P. Jungwirth and D. Tobias, *J. Phys. Chem. B*, 2002, **106**, 6361–6373.
3 P. Petersen and R. Saykally, *Chem. Phys. Lett.*, 2004, **397**, 51–55.
4 J. Noah-Vanhoucke and P. Geissler, *Proc. Natl. Acad. Sci. U. S. A.*, 2009, **106**, 15125–15130.
5 D. Otten, P. Shaffer, P. Geissler and R. Saykally, *Proc. Natl. Acad. Sci. U. S. A.*, 2012, **109**, 701–705.
6 L. Onsager and N. Samaras, *J. Chem. Phys.*, 1934, **2**, 528.
7 G. Archontis and E. Leontidis, *Chem. Phys. Lett.*, 2006, **420**, 199–203.
8 Y. Levin, *Phys. Rev. Lett.*, 2009, **102**, 147803.
9 Y. Levin, A. P. D. Santos and A. Diehl, *Phys. Rev. Lett.*, 2009, **103**, 257802.
10 V. Markin and A. Volkov, *J. Phys. Chem. B*, 2002, **106**, 11810–11817.
11 D. M. Huang and D. Chandler, *J. Phys. Chem. B*, 2002, **106**, 2047–2053.
12 C. Caleman, J. Hub, P. van Maaren and D. van der Spoel, *Proc. Natl. Acad. Sci. U. S. A.*, 2011, **108**, 6838.
13 S. Iuchi, H. Chen, F. Paesani and G. A. Voth, *J. Phys. Chem. B*, 2009, **113**, 4017–4030.
14 S. Stuart and B. Berne, *J. Phys. Chem.*, 1996, **100**, 11934–11943.
15 S. Vaitheeswaran and D. Thirumalai, *J. Am. Chem. Soc.*, 2006, **128**, 13490–13496.
16 J. Weeks, *J. Chem. Phys.*, 1977, **67**, 3106.
17 S. Chui and J. Weeks, *Phys. Rev. B*, 1981, **23**, 2438.
18 D. B. Abraham, *Phys. Rev. Lett.*, 1980, **44**, 1165–1168.
19 M. Fisher, *J. Stat. Phys.*, 1984, **34**, 667–729.
20 D. Nelson, T. Piran and S. Weinberg, *Statistical Mechanics of Membranes and Surfaces*, World Scientific Pub., 2004.
21 H. N. V. Temperley, *Math. Proc. Cambridge Philos. Soc.*, 1952, 683–697.
22 S. Safran, *Statistical Thermodynamics of Surfaces, Interfaces, and Membranes*, Westview Press, 2003.
23 M. Kardar, *Statistical Physics of Fields*, Cambridge University Press, 2007.
24 J. Weeks, G. Gilmer and H. Leamy, *Phys. Rev. Lett.*, 1973, **31**, 549–551.
25 S. Vaikuntanathan, P. Shaffer and P. L. Geissler, Unpublished, 2012.
26 X. Song and D. Chandler, *J. Chem. Phys.*, 1998, **108**, 2594–2600.

Surface and interfacial tensions of Hofmeister electrolytes

Alexandre P. dos Santos and Yan Levin*

Received 9th April 2012, Accepted 25th April 2012
DOI: 10.1039/c2fd20067h

We present a theory that is able to account quantitatively for the surface and interfacial tensions of different electrolyte solutions. It is found that near the interface, ions can be separated into two classes: the kosmotropes and the chaotropes. While the kosmotropes remain hydrated near the interface and are repelled from it, the chaotropes loose their hydration sheath and become adsorbed to the surface. The anionic adsorption is strongly correlated with the Jones–Dole viscosity *B*-coefficient. Both hydration and polarizability must be taken into account to obtain a quantitative agreement with the experiments. To calculate the excess interfacial tension of the oil–electrolyte interface, the dispersion interactions must also be included. The theory can also be used to calculate the surface and the interfacial tensions of acid solutions, predicting a strong surface adsorption of hydronium ion.

Introduction

Understanding the behavior of ions at the air–water and oil–water interfaces should help shed light on how the ions interact with proteins and colloidal particles. Over a hundred years ago Hofmeister organized various electrolytes according to their ability to salt-out protein solutions. The sequences of anions and cations, which now bear his name, have been also observed in the fields of science as diverse as biophysics, biochemistry, electro-chemistry and colloidal science.[1–5]

Although the bulk thermodynamics of electrolyte solutions is fairly well understood,[6] we still know little about ionic behavior when the translational symmetry is broken.[7–11] The surface and the interfacial tensions provide us with an indirect indication of ionic distribution near an interface. A long time ago Heydweiller[12] observed that addition of salt increases the surface tension of the air–water interface. Heydweiller also noted that the effect of different electrolytes followed the sequence found by Hofmeister some years earlier. The increase of the surface tension by electrolytes was soon attributed to the ionic depletion from the interfacial region.[7] Wagner[8] and Onsager and Samaras[9] suggested that this was a consequence of the charge induced on the dielectric interface separating water from air. The theory was able to quantitatively account for the surface tension of sodium chloride solution at very large dilutions, but failed for larger concentrations of electrolyte. Inclusion of ionic hydration into the theory by Levin and Flores-Mena[10] extended its validity up to 1 M concentration. However when the same theory was applied to study the surface tension of sodium iodide solutions it was found that it predicted a qualitatively incorrect behavior—the surface tension of NaI was found to be larger than of NaCl, contrary to the experiments. The origin of this discrepancy was not clear. The fundamental insights, however, appeared soon after in the form of polarizable force fields simulations[13–19] and

Instituto de Física, Universidade Federal do Rio Grande do Sul, Caixa Postal 15051, CEP 91501-970 Porto Alegre, RS, Brazil. E-mail: levin@if.ufrgs.br

experiments.[20–22] The new simulations and experiments demonstrated that it was possible for large halogen anions to become adsorbed to the interface. The physical mechanism of this adsorption, however, remained unclear. Boström *et al.*[11] suggested that the dispersion (van der Waals) interactions, neglected within the Wagner–Onsager–Samaras (WOS) theory, were responsible for the ionic adsorption. This interesting suggestion, however, contradicts both experiments and simulations. Dispersion forces are proportional to the ionic polarizability. Therefore, strong dispersion interactions between ions and water should favor bulk solvation. Since anions are much more polarizable than cations, the dispersion interactions should want to keep these ions in the bulk, away from the interface. This means that a theory based on dispersion forces will predict that weakly polarizable cation might be adsorbed at the interface, contrary to what was found in experiments and simulations.

A different theory was recently advanced by Levin *et al.*[23,24] These authors argued that the driving force behind the adsorption of highly polarizable anions was due to the hydrophobic effect. To solvate an ion, a cavity must be created. The cavity perturbs the hydrogen bond network of water molecules, resulting in a free energy cost. Clearly if the ion moves towards the interface, the cavitational energy will diminish. There is, however, an electrostatic self-energy penalty of exposing the ionic charge to the low dielectric air environment. For hard, non-polarizable, ions of the WOS theory the self-energy penalty completely overwhelms the gain in the hydrophobic free energy, forcing these ions to remain in the bulk. The situation is very different for large polarizable anions. When such ions move towards the interface, their electronic charge distribution shifts so that it remains mostly hydrated. This drastically diminishes the electrostatic self-energy penalty of having such ions located at the interface. A careful calculation shows that for highly polarizable ions the electrostatic self-energy penalty becomes comparable to the gain in the hydrophobic free energy resulting from moving an ion from the bulk to the surface.[23]

In this paper we will show how the ideas presented above can be used to calculate the surface and the interfacial tensions of electrolytes and acid solutions, as well as their electrostatic potential difference across the dielectric interface.

The drop model

To perform the electrostatic calculations it is convenient to consider an electrolyte solution inside a spherical water drop of radius R.[24,25] Here we will choose R = 300 Å, which is sufficiently large to avoid all finite size effects so that the excess surface tension calculated inside the drop will be the same as the surface tension of an extended thermodynamic interface. Outside the drop there is a low dielectric medium (air or oil), so that the interface at $r = R$ corresponds to the Gibbs dividing surface (GDS). N salt or acid "molecules" are dissociated inside the drop, resulting in N cations and N anions of charge $+q$ and $-q$, respectively. In the case of divalent anions, for each anion of charge $-2q$ there will be 2 cations of charge $+q$. The water and the external medium will be treated as dielectrics of permittivities ε_w and ε_o, respectively. The Bjerrum length is defined as $\lambda_B = \beta q^2/\varepsilon_w$.

The interfacial tensions are calculated by integrating the Gibbs adsorption isotherm equation,

$$d\gamma = -\Gamma_+ d\mu_+ - \Gamma_- d\mu_-, \tag{1}$$

where $\Gamma_\pm = [N - V\rho_\pm(0)]/S$ are the ionic excess due to interface, $\mu_\pm$ are the ionic chemical potentials, $\rho_\pm(0)$ are the ionic bulk concentrations, and S and V are the surface and the volume of the drop, respectively. The bulk concentrations are obtained from the numerical solution of the modified Poisson–Boltzmann (PB) equation:

$$\nabla^2 \phi(r) = -\frac{4\pi q}{\varepsilon_w}\left[\rho_+(r) - \rho_-(r)\right],$$

$$\rho_\pm(r) = A_\pm e^{\left[\mp\beta q\phi(r) - \beta U_\pm(r)\right]},$$

$$A_\pm = N\Big/\left[4\pi\int_0^{r_{\max}} \mathrm{d}r\, r^2 e^{\left[\mp\beta q\phi(r) - \beta U_\pm(r)\right]}\right], \tag{2}$$

where $\phi(r)$ is the electrostatic potential, $\rho_\pm(r)$ are the ionic concentrations, and $r_{\max}$ is the maximum ionic distance from the center of the drop. For chaotropes $r_{\max} = R + a$ and for kosmotropes $r_{\max} = R - a$. The ion-interface interaction potentials, $U_\pm(r)$, will be discussed in the following sections. The chemical potentials inside the drop are constant and within the PB approximation are given by $\beta\mu_\pm = \log(\Lambda_\pm^3\rho_\pm(0))$, where $\Lambda_\pm$ are the thermal de Broglie wavelengths.

Air–water interface

When an ion moves close to the dielectric interface, there are two effects: (1) the interface becomes polarized; and (2) there is a loss of solvation free energy arising from the imperfect screening of the ionic electric field by the rest of electrolyte. Both of these effects lead to a repulsive force from the interface. The work necessary to bring an ion from the bulk to a distance z from the GDS is found to be[10,26]

$$\beta U_i(z) = \beta W \frac{a}{z} e^{-2\kappa(z-a)}, \tag{3}$$

where $\kappa = \sqrt{8\pi\lambda_B\rho(0)_\pm}$ is the inverse Debye length and a is the ionic radius. The contact value, W, is calculated by solving the Poisson equation with the appropriate boundary conditions,[10]

$$\beta W = \frac{\lambda_B}{2}\int_0^\infty \mathrm{d}p \frac{p\left[s\cosh(pa) - p\sinh(pa)\right]}{s\left[s\cosh(pa) + p\sinh(pa)\right]}, \tag{4}$$

where

$$s = \sqrt{\kappa^2 + p^2}\cdot$$

For polarizable ions, Levin[23] calculated the variation in the electrostatic self-energy as an ion crosses the dielectric interface. The ion was modeled as an imperfect conducting sphere of relative polarizability $\alpha = \gamma_i/a^3$, where γ_i is the absolute ionic polarizability measured in $\mathring{A}^3$. The electrostatic self-energy of an ion whose center is at distance $-a < z < a$ from the GDS is found to be

$$\beta U_p(z) = \frac{\lambda_B}{2a}\left[\frac{\pi x^2}{\theta(z)} + \frac{\pi[1-x]^2\varepsilon_w}{[\pi - \theta(z)]\varepsilon_o}\right] + g\left[x - \frac{1-\cos[\theta(z)]}{2}\right]^2, \tag{5}$$

where $\theta(z) = \arccos[-z/a]$ and $g = (1 - \alpha)/\alpha$. The fraction of charge that remains hydrated x, is obtained by minimizing eqn (5),

$$x(z) = \left[\frac{\lambda_B\pi\varepsilon_w}{a\varepsilon_o[\pi - \theta(z)]} + g\left[1 - cos[\theta(z)]\right]\right]\Big/\left[\frac{\lambda_B\pi}{a\theta(z)} + \frac{\lambda_B\pi\varepsilon_w}{a\varepsilon_o[\pi - \theta(z)]} + 2g\right]. \tag{6}$$

To solvate an ion in water requires creation of a cavity. For small cavities this hydrophobic free energy scales with the volume of the void.[27] When the ion crosses the GDS, the perturbation to the hydrogen bond network diminishes, resulting in a thermodynamic force that drives the ion towards the air–water interface. The cavitational potential energy is found to be

$$\beta U_c(z) = \begin{cases} \nu a^3 \text{ for } z \geq a, \\ \frac{1}{4}\nu a^3\left(\frac{z}{a}+1\right)^2\left(2-\frac{z}{a}\right) \text{ for } -a<z<a, \end{cases} \tag{7}$$

where $\nu \approx 0.3/\text{Å}^3$ is obtained using the bulk simulations.[27]

Physical chemists have known for a long time that ions come in two categories: structure-makers (kosmotropes) and structure-breakers (chaotropes). The separation into these two classes is often based on the Jones–Dole (JD) viscosity *B*-coefficient[28] which also correlates well with the ionic enthalpy of hydration.[29] In 1929 Jones and Dole observed that the relative viscosity η_r of an electrolyte solution is very well fitted by a simple formula $\eta_r = 1 + A\sqrt{c} + Bc$, where c is the bulk concentration of electrolyte. The square root term is universal and can be calculated using the Debye–Hückel–Onsager theory. On the other hand, the linear term in c is electrolyte specific. Ions with a positive *B* coefficient (kosmotropes) are supposed to organize water making it "more" viscous, while the ions with negative *B* are supposed to make water more disordered and "less" viscous. To what extent this physical picture is realistic is not clear, and recent experiments suggest that the action of ions on water molecules is much more local, not extending much beyond the first hydration shell.[30] This also agrees well with the theory of surface tensions of electrolyte solutions that will be presented in the present paper. We find that kosmotropic ions remain strongly hydrated near the interface and are repelled from it, while the chaotropic ions loose their hydration sheath and as the result of their large polarizability become adsorbed to the interface.

In the case of halides, the separations into kosmotropes and the chaotropes correlates well with the ionic size. Heavy halogen anions produce a weak electric field which is not sufficient to bind strongly the adjacent water molecules and prevent them from dissociating when a chaotropic ion approaches the interface. This is the case for I^- and Br^-, whose JD viscosity *B*-coefficients are −0.073 and −0.033, respectively. On the other hand, the small fluoride anion has large positive *B*-coefficient equal to 0.107. This ion, should therefore, remain strongly hydrated near a hydrophobic interface. The chloride ion, with a *B*-coefficient close to zero, −0.005, is on the borderline between the two classes.

The total interaction potential for strongly hydrated kosmotropes is dominated by the charge–image interaction, $U_\pm(z) = U_i(z)$ and the hard core repulsion (at one hydrated radius) from the GDS. On the other hand, large chaotropic anions, such as iodide and bromide, are able to cross the GDS with a relatively small electrostatic self-energy penalty, gaining the hydrophobic cavitational free energy. For such chaotropic anions the total ion–interface interaction potential is $U_-(z) = U_i(z) + U_p(z) + U_c(z)$. The radius[31] of I^- is $a = 2.26$ Å and its relative polarizability[32] is $\alpha = 0.64$; for Br^-, $a = 2.05$ Å and $\alpha = 0.59$.

The partially hydrated radius of the sodium cation Na^+ is the only free adjustable parameter of the theory. It is obtained by fitting the surface tension of the NaI solution. The calculation is performed by first numerically solving the PB equation to obtain the bulk concentration of electrolyte $\rho_\pm(0)$ at the center of the drop. Then, the Gibbs adsorption isotherm eqn (1) is integrated numerically to calculate the excess surface tension of the electrolyte solution. We find that $a = 2.5$ Å for Na^+ gives an excellent fit to the experimental data,[24] Fig. 1. The same radius of Na^+ is then used to calculate the excess surface tensions of other electrolyte solutions. Note that while Br^- is a chaotrope, both F^- and Cl^- are kosmotropes. Furthermore, for F^- the JD viscosity *B*-coefficient is large and positive while for Cl^- it is almost zero. Therefore near the interface, F^- will remain fully hydrated with the effective radius[33] $a = 3.52$ Å, while Cl^- is very weakly hydrated, with the radius $a = 2$ Å close

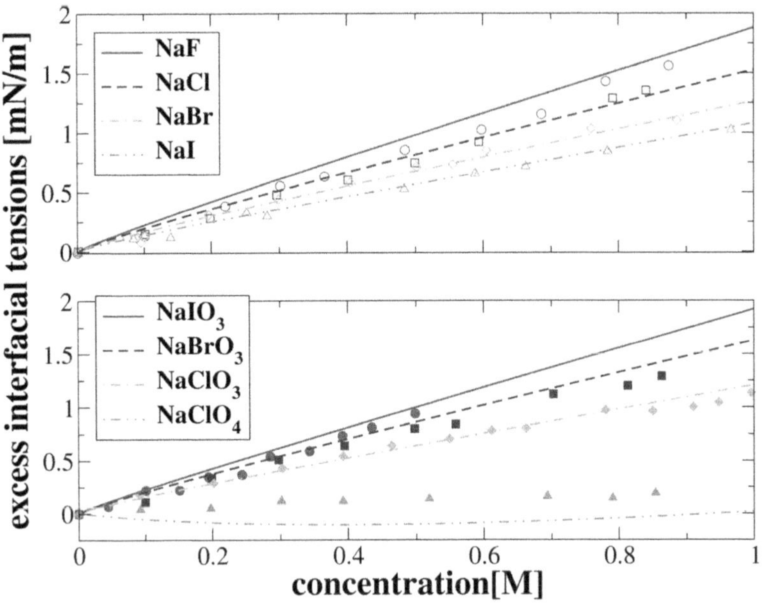

Fig. 1 Excess interfacial tensions for various salts at the electrolyte–air interface. The open circles, squares, diamonds and triangles represent experimental data[34–36] for NaF, NaCl, NaBr and NaI, respectively. The full circles, squares, diamonds and triangles represent experimental data[36] for $NaIO_3$, $NaBrO_3$, $NaClO_3$ and $NaClO_4$, respectively. The lines represent the present theory.

to its crystallographic size. The calculated excess surface tension for all halide salts are in excellent agreement with the experimental measurements, Fig. 1.

At the moment there is no general theory for ionic hydration. For halide anions we saw that there was a very good correlation between the size of the ion, its JD viscosity *B*-coefficient, and the hydration characteristics near the air–water interface. One might hope that such correlations will also persist for more complicated anions as well. This, however, is not the case. For example, iodate, IO_3^-, is a very large anion, yet its JD viscosity *B*-coefficient is similar to that of fluoride. Indeed, calculating the surface tension of $NaIO_3$ solution, we find that IO_3^- must be treated as a strongly kosmotropic anion. This is also consistent with the recent *ab initio* simulations of Baer *et al.*[37] Although the correlation between the ionic size and the ionic hydration is lost for more complicated oxy-anions, the correlation between the JD viscosity *B*-coefficient and the interfacial hydration seems to persist. This correlation can, therefore, be used to distinguish between the kosmotropes and the chaotropes in the case of the more complex anions.[26]

We now consider salts with oxy-anions. Although oxy-anions are not spherical, their effective radii are well described by an empirical formula based on experimentally measured entropies of hydration,[38] $a = \frac{n_{\text{oxy}}}{4}(d + 1.4/\text{Å})$,where n_{oxy} is the number of oxygens in the anion and d is the halogen–oxygen covalent bond length in the corresponding salt crystal.[33] We first consider $NaIO_3$ solution. The partially hydrated radius of sodium is the same as before, $a = 2.5$ Å. The JD viscosity *B*-coefficient[39] of IO_3^- is large and positive, 0.14. This means that iodate will remain fully hydrated near the interface, keeping its bulk hydration radius[33] $a = 3.74$ Å. The ion-interface potential of IO_3^- is then $U_-(z) = U_i(z)$, with a hardcore repulsion at $z = 3.74$ Å from the GDS. Calculating the excess surface tension for $NaIO_3$, we find a good agreement with experiment, Fig. 1. The ion BrO_3^- has the *B*-coefficient[39] near zero, 0.009, similar to Cl^- so, once again, we will treat these ion as a kosmotrope with a partially hydrated radius equal to its bare size $a = 2.41$ Å. The result of this calculation is shown in Fig. 1. The ions ClO_3^- and ClO_4^- have negative B coefficients,[39] −0.022 and −0.058, respectively, and are chaotropes. The total ion–interface interaction potential for these ions is $U_-(z) = U_i(z) + U_p(z) + U_c(z)$, with bare ionic radii,[38] $a = 2.16$ and $a = 2.83$ Å; and relative polarizabilities,[26,32] $\alpha = 0.52$ and $\alpha = 0.24$, respectively. The calculated excess interfacial tension for $NaClO_3$ agrees very well with the experimental data, Fig. 1. The agreement is not very good for sodium perchlorate, Fig. 1. This ion is very big, so that a small error in its effective radius calculated using the empirical formula presented above leads to a large error in its cavitational free energy—cavitational energy scales with the cube of the radius—resulting in an incorrect estimate of adsorption. The theory also allows us to estimate the electrostatic potential difference across the interface,[26] $\phi(\infty) - \phi(0)$, which is reported in Table 1 for various salts at 1 M concentration. The results are in a qualitative agreement with the experimental measurements of Frumkin[40] and Jarvis and Scheiman.[41] Furthermore, if the ions are arranged in the order of increasing surface potential, one finds precisely the celebrated Hofmeister series.[26]

Acids solutions

While most salts tend to increase the surface tension of the air–water interface, most acids do precisely the opposite. It is well known that proton H^+ interacts strongly with the water molecules,[43–45] forming complexes such as H_3O^+ and $H_2O_5^+$. The trigonal pyramidal structure of hydronium[46,47] favors strong adsorption at the water–air interface, with the oxygen pointing towards the air.[17,46–49] For many acids the protonation of the interface is so strong as to result in a negative excess surface tension. To take this into account[50] we add an additional adsorption potential for H^+,

$$\beta U_h(z) = \begin{cases} 0 \text{ for } z \geq 1.97\text{Å}, \\ -3.05 \text{ for } z < 1.97\text{Å}. \end{cases} \quad (8)$$

where the value −3.05 was adjusted in order to obtain the correct excess interfacial tension for the hydrochloric acid, Fig. 2. The range of this potential is taken to be 1.97 Å, the length of the hydrogen bond. The anions are treated as before—classified as kosmotropes or chaotropes[50]—while the proton interacts with the interface through the potential $U_+(z) = U_i(z) + U_h(z)$. In the image part of this potential the radius of a proton is set to zero.

Table 1 Surface potentials difference at 1 M for various salts

Salts	Calculated/mV	Frumkin/mV40,42	Jarvis *et al.*/mV41
NaF	4.7	—	—
NaCl	−2.1	−1	≈ −1
NaBr	−9.4	—	≈ −5
NaI	−14.3	−39	≈ −21
$NaIO_3$	5	—	—
$NaBrO_3$	−0.12	—	—
$NaNO_3$	−8.27	−17	≈ −8
$NaClO_3$	−11.02	−41	—
$NaClO_4$	−31.1	−57	—
Na_2CO_3	10.54	3	≈ 6
Na_2SO_4	10.17	3	≈ 35

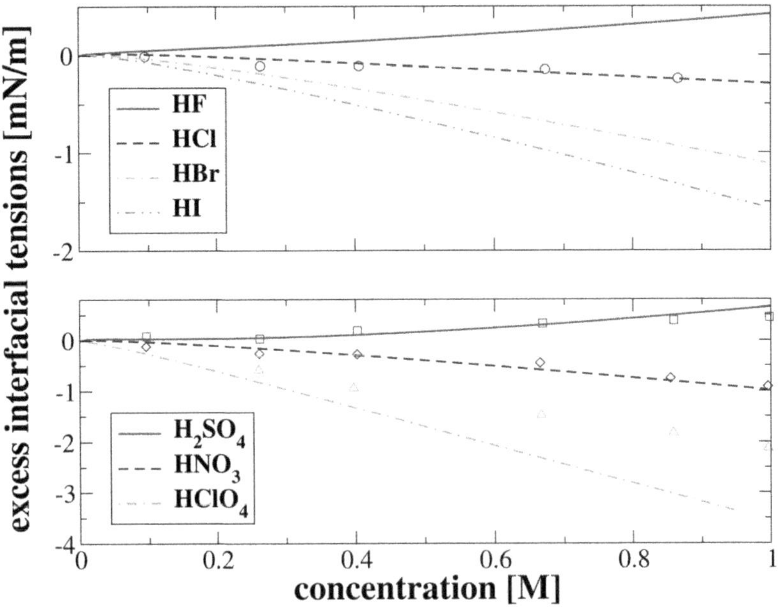

Fig. 2 Excess interfacial tensions for various acids at the electrolyte–air interface. The open circles, squares, diamonds and triangles represent experimental data[51] for HCl, H_2SO_4, HNO_3 and $HClO_4$, respectively. The lines represent the present theory.

In Fig. 2 the excess interfacial tensions for various acids are plotted. The agreement with the experimental data is very good for H_2SO_4 and HNO_3. As for sodium perchlorate, $HClO_4$ also shows a significant deviation from the experimental data, indicating again that our estimate of the effective radius of ClO_4^- is too large.

To calculate the electrostatic potential difference across the interface we integrate the PB equation. This, however, will not account for the reorientation of the interfacial water molecules when hydronium complex is formed, resulting in a dipole layer with a corresponding potential drop.[52] As was discussed earlier H_3O^+ prefers to orient itself with the hydrogens pointing towards the bulk water. The number of hydroniums formed can be estimated from the proton adsorption, $N_+ = [N - V\rho_+(0)]$, where $\rho_+(0)$ is the bulk concentration, given by eqn (2). The internal electric field inside the dipole layer can be estimated to be $E = 4\pi pN_+/\varepsilon_0 Sd$, where $p = 0.3854q$ Å the water dipole moment, d is the dipole length, and S is the interfacial area. Integrating this field across the interface, we find the contribution of the surface hydroniums to the overall potential difference across the interface to be $\Delta\phi_w = -69604.5\Gamma_+$, in mV. Summing this with the contribution arising from the PB equation, we obtain the overall potential drop across the air–water interface,

Table 2 Electrostatic potential differences across the water–air interface for various acids (including the contributions from PB equation and the oriented water dipoles)

Acids	calculated/mV	Frumkin/mV40
HF	85.5	−71
HCl	1.24	−23
HNO_3	−84.4	−48
HBr	−95	−34
HI	−144.8	−61
$HClO_4$	−412	−82

reported in Table 2. The theoretical results are also compared with the data of Frumkin.[40] In view of the roughness of the estimates presented above the qualitative agreement between the theory and experiment is quite reasonable. The value −71mV for HF reported in ref. 40 most like has as a wrong sign, since it falls completely outside the general trend.

Electrolyte–oil interface

The good agreement between the theory and experiments suggests that the physical picture behind the mechanism of the ion–interface interaction presented above is more-or-less correct. In particular we see that the ions near the air–water interface must be divided into two classes: kosmotropes and chaotropes.[26] While the kosmotropes remain hydrated near the interface the chaotropes lose their hydration shell and, as the result of the hydrophobic cavitational forces and high polarizability, become partially adsorbed to the interface. In their recent *ab initio* simulations, Baer and Mundy[53] have calculated the potential of mean force for iodide near the air–water interface, finding an almost perfect agreement with the theory presented above.[23] This suggests that the dispersion (van der Waals) interactions do not play a significant role at the air–water interface. To see why this might be the case, let us first consider a kosmotropic ion. Near the interface, such an ion remains fully hydrated, interacting with almost the same number of water molecules as in the bulk, so that the dispersion contribution to its total free energy of solvation should not be affected by the presence of the interface.

For chaotropic ions, the absence of dispersion interactions is not so easily understood. It is possible, however, to make the following argument: most of the ionic charge of a chaotropic anion as it crosses the GDS concentrates in water, resulting in a large electric field.[23] This strong field attracts water molecules so that it is possible for a chaotropic ion to interact dispersively with the same number of water molecules as it did in the bulk. To see if this argument is consistent, we will now study the effect of electrolyte on the interfacial tension of the oil–water interface.

Similar to what happens at the air–water interface, the kosmotropic ions near the oil–water interface will feel the ion–image interaction and the hardcore repulsion from the GDS. Oil, like air, has low dielectric constant, $\varepsilon_o \approx 2$, so that the ion-image and the polarization potentials, eqn (3) and (5), will remain the same as at the air–water interface. The chaotropic ions are driven towards the interface by their cavitational potential. When part of the ion penetrates into oil, there is also a cavitation energy penalty from the oil side. The cavitational energy, is mostly entropic—related to the number of water/oil molecules excluded from the cavity produced by the ion. The molecular weight of oil (dodecane used in the experiments) is 10 times higher, while its mass density is the same as that of water. This means that the number of excluded molecules in the ion cavity, and consequently the cost of cavitational energy, in oil will be about 10 times smaller than in water, and can be safely neglected. Therefore, the cavitational potential of a chaotrope at the water–oil interface will remain the same as at the air–water interface, eqn (7).

Since at the air–water interface there is no dispersive contribution to the adsorption potential, the dispersion interaction with the oil–water interface should be proportional to the ionic polarizability and the ionic volume exposed to oil. We, therefore, suggest the following simple phenomenological expression[54]

$$U_{\rm d}(z) = \begin{cases} 0 \ \text{for} \ z \geq a, \\ A_{\rm eff}\alpha\left[1 - \dfrac{(z/a+1)^2(2-z/a)}{4}\right] \text{for} \ -a < z < a, \end{cases} \tag{9}$$

where $A_{\rm eff}$ is the effective Hamaker constant, $A_{\rm eff} \approx A^{v}_{\rm mw} - A^{v}_{\rm mo}$, where $A^{v}_{\rm mw}$ and $A^{v}_{\rm mo}$ are the metal–water and metal–oil (dodecane) Hamaker constants in vacuum.[54] The metal constants are used since the ionic polarizability is already included in eqn (9). Using the tabulated values of the Hamaker constants, from ref. 55, we obtain $A_{\text{eff}} \approx -4k_BT$. Note that this is only a rough estimate of the strength of the dispersion interaction. In practice, we will adjust the value of A_{eff} to obtain the measured interfacial tension of the KI solution.

The excess interfacial tensions will be calculated as before. We will solve the modified PB equation eqn (2) inside the drop with potentials: $U_+(z) = U_i(z)$ for K^+, $U_-(z) = U_i(z)$ for kosmotropes, and $U_-(z) = U_i(z) + U_p(z) + U_c(z) + U_d(z)$ for chaotropes. From this solution we will calculate the ionic adsorption and, integrating the Gibbs adsorption isotherm eqn (1), will obtain the interfacial tensions. All the parameters used for the kosmotropes and the chaotropes are the same as in the previous sections. The hydrated radius of the K^+ is adjusted to obtain the experimentally measured surface tension of KCl solution, Fig. 3. We find that the potassium ion is partially hydrated with radius of $a = 2$ Å. This radius will be used for all the potassium salts. To obtain the effective Hamaker constant for chaotropic ions, we study the KI solution, Fig. 3. Fitting the experimental data we obtain $A_{\text{eff}} = -4.4k_BT$, which is in excellent agreement with our theoretical estimate, suggesting that our physical picture about the role of dispersion interactions at the air–water and oil–water interfaces is correct. This Hamaker constant will be used for all the chaotropic anions. In Fig. 3, we present the calculated interfacial tensions for various potassium salts. The only additional experimental data available to us is for KBr, which agrees very well with the predictions of the present theory.

Acids–oil interface

We will now explore the effect of acids on the interfacial tension of the water–oil interface. It was shown previously that the hydronium ion H_3O^+ has a particular preference for the interfacial solvation. This happens because the hydrogens of the

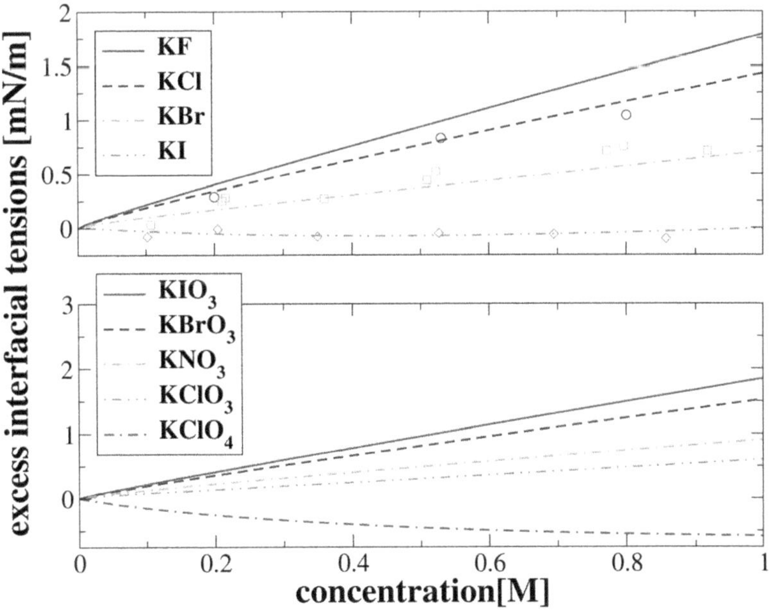

Fig. 3 Excess interfacial tensions for various electrolyte solutions. The open circles, squares and diamonds represent experimental data[56] for KCl, KBr and KI, respectively. The lines are calculated using the present theory.

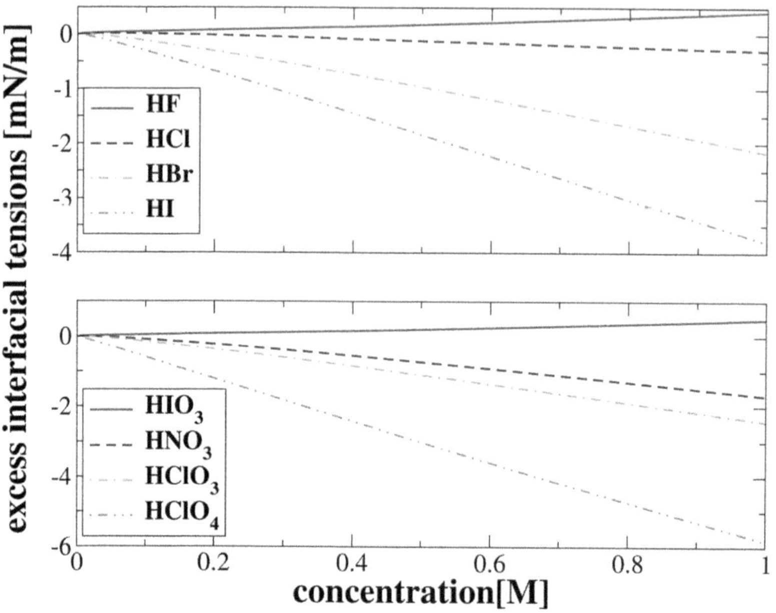

Fig. 4 Excess interfacial tensions for different acids. The lines are calculated using the present theory.

hydronium ion are very good hydrogen bond donors, while the oxygen is a bad receptor.[46] This leads to a preferential orientation of the hydronium ion at the water–air interface, with the hydrogens pointing into the aqueous environment and the oxygen sticking out. Calculations of solvation free energy confirm this interfacial behavior.[57] Here we will suppose that this basic picture persists for hydronium ion at the water–oil interface as well. Since at the moment there are no experimental measurements of the interfacial tension of acid solutions which could be used to re-parameterize our model, we will use the same adsorption energy of proton as at the air–water interface,[50] $-3.05\ k_BT$. The dispersion interaction and the cavitational potential are also the same as used in the previous sections. Integrating the modified PB equation and the Gibbs adsorption isotherm, we obtain the ionic adsorption and the excess interfacial tensions of different acids. In Fig. 4 we present our results. A significant decrease in the pure water–oil interfacial tension is observed for acids containing chaotropic anions. Unfortunately at the moment there is no experimental data available to test the predictions of the present theory.

The theory can also be used to estimate the electrostatic potential across the acid–oil interface. The calculation is analogous to the one performed for the acid–air interface.[50] The results are presented in Table 3. It is very probable that the

Table 3 Electrostatic potential differences across the oil–water interface for various acids (including the contributions from PB equation and the oriented water dipoles)

Acids	Calculated/mV
HNO_3	−154.53
HBr	−196.67
$HClO_3$	−218.36
HI	−320.65
$HClO_4$	−453.78

calculated potential differences are too large, since the theory is not fully self-consistent. Nevertheless the results provide us with a magnitude of the electrostatic potential difference that can be expected across the water–oil interface for different acids.

Conclusions

We have presented a general theory which allows us to calculate the surface and the interfacial tensions of electrolyte solutions. The theory provides a very interesting picture of ionic specificity. We find that near the air–water interface or a general hydrophobic surface, ions can be divided into two classes: kosmotropes and chaotropes.[5,26] Near the interface kosmotropes remain hydrated and are repelled from the GDS. On the other hand, chaotropes lose their hydration sheath and, as a result of large polarizability, can become adsorbed to the hydrophobic interface. The theory also shows that the hydronium ion has a strong preference for interfacial solvation.[50]

It is believed that the surface water molecules are preferentially oriented with the hydrogens sticking out towards the air. To account for the measured electrostatic potential difference of acid solutions, we find that hydronium must orient itself opposite to the surface water, with its hydrogens pointing toward the bulk.[52]

Beyond a qualitative picture, the theory presented in this paper allows us to make quantitative predictions about the surface and the interfacial tensions of electrolyte and acid solutions as well as to calculate the electrostatic potential difference across the interface. The theory can also be extended to calculate quantitatively the critical coagulation concentrations of hydrophobic colloidal suspensions, providing a new insight into the physical mechanisms responsible for the ionic specificity.[5]

Although the theory helps us to understand the physics behind the Hofmeister series, there are still a number of issues that must be explored. One of them is the role of the electrostatic surface potential of water. The dielectric continuum theory presented in this paper completely ignores the surface potential of water, suggesting that it should be negligibly small. This is justified *a posteriori* by the good agreement between the theory and the experimental measurements of the surface and the interfacial tensions of acids and electrolytes. On the other hand, the classical point charge water models[58–60] predict that across the air–water interface there should be an electrostatic potential drop of approximately −600 mV. If such surface potential really exists, it would completely change the electrostatics of ionic solvation, favoring a much stronger adsorption of the chaotropic anions than is predicted by the present theory. This, however, cannot be true since a stronger adsorption would result in an erroneous surface tensions of electrolyte solutions! In fact, it is now believed that the polarizable force field simulations, which have stimulated the development of the present theory, lead to too much adsorption of the chaotropic anions. We speculate that the reason for this excessive adsorption is precisely the artificial surface potential of the point charge water models.

If the existing classical water models cannot account for the surface properties of water, what about the full quantum mechanical calculations? In fact, the recent *ab initio* simulations[61–63] show that the surface potential of water is not −600 mV, but is +3000 mV. Note the difference in sign and the magnitude of this potential! The +3 V electrostatic potential difference across the air–water interface has been measured by the high energy electron holography.[58] Nevertheless, the authors of ref. 58, argue that this huge potential is irrelevant for electrochemistry. They suggest that the quantum electrostatic potential must be coarse grained on the scale of the ion. If this is done properly, they argue, the surface potential of water felt by an ion such as I^- will drop to a few mV and can be safely ignored. This is possibly the reason why the *ab initio* potential of mean force for I^- agrees so well[53] with the present dielectric continuum theory, which completely neglects the electrostatic surface potential of water. More work, however, still is necessary to fully elucidate this point.

Acknowledgements

This work was partially supported by the CNPq, FAPERGS, INCT-FCx, and by the US-AFOSR under the grant FA9550-09-1-0283.

References

1 R. Komsa-Penkova, R. Koynova, G. Kostov and B. G. Tenchov, *Biochim. Biophys. Acta, Protein Struct. Mol. Enzymol.*, 1996, **1297**, 171.
2 N. Jiang, P. X. Li, Y. L. Wang, J. B. Wang, H. K. Yan and R. K. Thomas, *J. Phys. Chem. B*, 2004, **108**, 15385.
3 L. M. Pegram, T. Wendorff, R. Erdmann, I. Shkel, D. Bellissimo, D. J. Felitsky and M. T. Record, *Proc. Natl. Acad. Sci. U. S. A.*, 2010, **107**, 7716.
4 J. M. Peula-Garcia, J. L. Ortega-Vinuesa and D. Bastos-Gonzalez, *J. Phys. Chem. C*, 2010, **114**, 11133.
5 A. P. dos Santos and Y. Levin, *Phys. Rev. Lett.*, 2011, **106**, 167801.
6 P. W. Debye and E. Hückel, *Phys. Z.*, 1923, **24**, 185.
7 I. Langmuir, *J. Am. Chem. Soc.*, 1917, **39**, 1848.
8 C. Wagner, *Phys. Z.*, 1924, **25**, 474.
9 L. Onsager and N. N. T. Samaras, *J. Chem. Phys.*, 1934, **2**, 528.
10 Y. Levin and J. E. Flores-Mena, *Europhys. Lett.*, 2001, **56**, 187.
11 M. Boström, D. R. M. Williams and B. W. Ninham, *Langmuir*, 2001, **17**, 4475.
12 A. Heydweiller, *Ann. Phys. (Leipzig)*, 1910, **33**, 145.
13 L. Perera and M. L. Berkowitz, *J. Chem. Phys.*, 1991, **95**, 1954.
14 L. X. Dang and D. E. Smith, *J. Chem. Phys.*, 1993, **99**, 6950.
15 S. J. Stuart and B. J. Berne, *J. Phys. Chem. A*, 1999, **103**, 10300.
16 P. Jungwirth and D. J. Tobias, *J. Phys. Chem. B*, 2002, **106**, 6361.
17 P. Jungwirth and D. J. Tobias, *Chem. Rev.*, 2006, **106**, 1259.
18 M. A. Brown, R. D'Auria, I. F. W. Kuo, M. J. Krisch, D. E. Starr, H. Bluhm, D. J. Tobias and J. C. Hemminger, *Phys. Chem. Chem. Phys.*, 2008, **10**, 4778.
19 D. Horinek, A. Herz, L. Vrbka, F. Sedlmeier, S. I. Mamatkulov and R. R. Netz, *Chem. Phys. Lett.*, 2009, **479**, 173.
20 G. Markovich, S. Pollack, R. Giniger and O. Cheshnovski, *J. Chem. Phys.*, 1991, **95**, 9416.
21 B. Garrett, *Science*, 2004, **303**, 1146.
22 S. Ghosal, J. Hemminger, H. Bluhm, B. Mun, E. L. D. Hebenstreit, G. Ketteler, D. F. Ogletree, F. G. Requejo and M. Salmeron, *Science*, 2005, **307**, 563.
23 Y. Levin, *Phys. Rev. Lett.*, 2009, **102**, 147803.
24 Y. Levin, A. P. dos Santos and A. Diehl, *Phys. Rev. Lett.*, 2009, **103**, 257802.
25 C. H. Ho, H. K. Tsao and Y. J. Sheng, *J. Chem. Phys.*, 2003, **119**, 2369.
26 A. P. dos Santos, A. Diehl and Y. Levin, *Langmuir*, 2010, **26**, 10778.
27 S. Rajamani, T. M. Truskett and S. Garde, *Proc. Natl. Acad. Sci. U. S. A.*, 2005, **102**, 9475.
28 Y. Marcus, *Chem. Rev.*, 2009, **109**, 1346.
29 K. D. Collins, *Biophys. J.*, 1997, **72**, 65.
30 A. W. Omta, M. F. Kropman, S. Woutersen and H. J. Bakker, *Science*, 2003, **301**, 347.
31 W. M. Latimer, K. S. Pitzer and C. M. Slansky, *J. Chem. Phys.*, 1939, **7**, 108.
32 N. C. Pyper, C. G. Pike and P. P. Edwards, *Mol. Phys.*, 1992, **76**, 353.
33 E. R. Nightingale Jr, *J. Phys. Chem.*, 1959, **63**, 1381.
34 N. Matubayasi, K. Tsunemoto, I. Sato, R. Akizuki, T. Morishita, A. Matuzawa and Y. Natsukari, *J. Colloid Interface Sci.*, 2001, **243**, 444.
35 N. Matubayasi, H. Matsuo, K. Yamamoto, S. Yamaguchi and A. Matuzawa, *J. Colloid Interface Sci.*, 1999, **209**, 398.
36 N. Matubayasi, (unpublished).
37 M. D. Baer, V. T. Pham, J. L. Fulton, G. K. Schenter, M. Balasubramanian and C. J. Mundy, *J. Phys. Chem. Lett.*, 2011, **2**, 2650.
38 A. M. Couture and K. L. Laidler, *Can. J. Chem.*, 1957, **35**, 202.
39 H. D. B. Jenkins and Y. Marcus, *Chem. Rev.*, 1995, **95**, 2695.
40 A. Frumkin, *Z. Phys. Chem.*, 1924, **109**, 34.
41 N. L. Jarvis and M. A. Scheiman, *J. Phys. Chem.*, 1968, **72**, 74.
42 J. E. B. Randles, *Advances in Electrochemistry and Electrochemical Engineering*, Interscience, New York, 1963, vol. 3, p. 1.
43 M. Eigen, *Angew. Chem., Int. Ed. Engl.*, 1964, **3**, 1.
44 G. Zundel, *Adv. Chem. Phys.*, 2000, **111**, 1.
45 D. Marx, M. E. Tuckerman, J. Hutter and M. Parrinello, *Nature*, 1999, **397**, 601.

46 M. K. Petersen, S. S. Iyengar, T. J. F. Day and G. A. Voth, *J. Phys. Chem. B*, 2004, **108**, 14804.
47 M. Mucha, T. Frigato, L. M. Levering, H. C. Allen, D. J. Tobias, L. X. Dang and P. Jungwirth, *J. Phys. Chem. B*, 2005, **109**, 7617.
48 L. M. Levering, M. R. Sierra-Hernández and H. C. Allen, *J. Phys. Chem. C*, 2007, **111**, 8814.
49 S. S. Iyengar, T. J. F. Day and G. A. Voth, *Int. J. Mass Spectrom.*, 2005, **241**, 197.
50 A. P. dos Santos and Y. Levin, *J. Chem. Phys.*, 2010, **133**, 154107.
51 P. K. Weissenborn and R. J. Pugh, *J. Colloid Interface Sci.*, 1996, **184**, 550.
52 J. E. B. Randles and D. J. Schiffrin, *Trans. Faraday Soc.*, 1966, **62**, 2403.
53 M. D. Baer and C. J. Mundy, *J. Phys. Chem. Lett.*, 2011, **2**, 1088.
54 A. P. dos Santos and Y. Levin, *Langmuir*, 2012, **28**, 1304.
55 D. B. Hough and L. R. White, *Adv. Colloid Interface Sci.*, 1980, **14**, 3.
56 R. Aveyard and S. M. Saleem, *J. Chem. Soc., Faraday Trans. 1*, 1976, **72**, 1609.
57 L. X. Dang, *J. Chem. Phys.*, 2003, **119**, 6351.
58 S. M. Kathmann, I. F. W. Kuo, C. J. Mundy and G. K. Schenter, *J. Phys. Chem. B*, 2011, **115**, 4369.
59 W. L. Jorgensen, J. Chandrasekhar, J. D. Madura, R. W. Impey and M. L. Klein, *J. Chem. Phys.*, 1983, **79**, 926.
60 L. X. Dang and T. M. Chang, *J. Chem. Phys.*, 1997, **106**, 8149.
61 S. M. Kathmann, I. F. W. Kuo and C. J. Mundy, *J. Am. Chem. Soc.*, 2008, **130**, 16556.
62 S. M. Kathmann, I. F. W. Kuo and C. J. Mundy, *J. Am. Chem. Soc.*, 2009, **131**, 17522.
63 K. J. Leung, *J. Phys. Chem. Lett.*, 2010, **1**, 496.

An *ab initio* approach to understanding the specific ion effect

Marcel D. Baer and Christopher J. Mundy*

Received 1st June 2012, Accepted 18th June 2012
DOI: 10.1039/c2fd20113e

Although there is a consensus that large, polarizable anions will adsorb to the air–water interface, the precise interactions that give rise to surface enhancement are still being debated. Previously, we have demonstrated that there is a significant dependence on the choice of molecular interaction potential for I^- adsorption at the air–water interface. Specifically, density functional theory (DFT) based interaction potentials lead to significantly less adsorption than empirical interaction potentials that include polarization. We have also demonstrated that DFT based interaction potentials can accurately capture the structure of the first solvation shell of both simple and polyatomic anions as compared to multi-edge X-ray absorption fine structure (XAFS) experiments. We utilize DFT to examine the local hydration structure of SCN^- and IO_3^- that exhibit both chaotropic and kosmotropic characteristics. We compare and contrast the solvation structure of the polyatomic anions with the series of halide anions. A picture emerges where we can correlate local solvation structure to the ions' position in the Hofmeister series.

1 Introduction

Recent developments in atmospheric chemistry,[1–7] have inspired physical chemists to revisit the subtleties of liquid–air and liquid–liquid interfaces. Here, interest is largely focused on chemical reactions that may be enhanced in the presence of ions or other organic material in the vicinity of an interface. However, interest regarding ion interactions with interfaces goes back hundreds of years. The very nature of the stability of proteins themselves are subject to how ions interact with hydrophobic surfaces.[8] The aforementioned phenomena regarding proteins has long been studied and a classification of cations and anions and their effect on the protein stability, known as the Hofmeister series, has been a cornerstone in our understanding of biology.[8,9] The fact that this empirical ordering persists in seemingly unrelated phenomena such as the surface tension of electrolytes and colloid stability has captured the attention of physical chemists seeking a molecular understanding of the Hofmeister series.[8,10–13] To this end, researchers looked to simpler systems where the specific ion effect was present. The air–water interface, although not simple, is now regarded as a testing ground for many theoretical[2,14–37] and experimental[11,12,29,38–57] studies on the specific ion effect.

As alluded to above, the first clues to the behavior of ions near the water–air interface came from measurements of the excess surface tension of electrolyte solutions.[58] In particular it was observed that halogen salts lead to increased surface tension of the air–water interface. For this to be the case, thermodynamics requires that ions must be depleted from the interfacial region relative to the bulk concentration.[59] On the other hand, electrochemical measurements of Frumkin showed that the

Pacific Northwest National Laboratory, Richland, Washington, 99352, USA. E-mail: chris.mundy@pnnl.gov

electrolyte solutions possess an electrostatic potential gradient across their surface.[60] Frumkin's work clearly indicated that cations and anions can behave very differently near the surface – with anions penetrating much farther into air than cations.

Not surprisingly, the long held belief was that the key to understanding the Hofmeister series of the specific ion effect was to be found in how the ion effected the long range structure of liquid water.[61,62] The classification of ions into kosmotropes (structure makers) and chaotropes (structure breakers) became a moniker for understanding complex phenomena that emerged when ions interacted with hydrophobic interfaces.[13] Recently, the ideas of the structure-maker/breaker concept were experimentally put to the test with a result that the perturbation to water's structure appeared to be confined to the first solvation shell of the ion.[62–64] Other clues to the local nature of the Hofmeister series have come through a deeper theoretical analysis.

A theoretical understanding of the surface tension of electrolytes dates back to the work of Onsagar and Samaras (OS) in which was derived a limiting law for electrolytes.[59] The OS theory worked very well for describing the surface tension of symmetric electrolytes at low concentrations. Serious deviations from experiment were already seen at concentrations approaching 1 M. A recent extension of the OS theory to higher concentrations (up to 1 M) relied on the description of the hydration radius of the ion. Thus, by inclusion of the finite ion size within a dielectric continuum theory, good agreement with the experimentally determined surface tensions for strongly hydrating salts (*e.g.* LiCl, NaCl, KCl) was obtained.[65] The local nature of the specific ion effect was also studied recently with molecular simulation.[66] In this study it was found that indeed, the local water arrangement around the ion is responsible for the majority of the energetic driving force of ion adsorption. Although the two theoretical studies differ, the emerging concept of describing local hydration of an ion either modeled as a radius in a dielectric or by the explicit accounting of solvating water molecules is an important distinction from previous studies.

The conventional wisdom, with regard to the Hofmeister series is shifting from implicating water's long-range structure to its local hydration structure around the ion. Understanding local effects is amenable to the detailed study of ion solvation using a variety of interaction potentials including *ab initio* density functional theory (DFT) based methods. In fact, it has long been known that the local structure around ions utilizing DFT can be different than that obtained with empirical potentials. Despite the known shortcomings of DFT, it has been shown that local solvation structure around ions is, in general, a more faithful representation of reality as determined by modern extended X-ray adsorption fine structure (EXAFS) measurements.[67–69]

Here, we examine the local structure of a variety of anions from the Hofmeister series using DFT interaction potentials. The hydration structure will thus result from an explicit treatment of valence electrons, allowing internal charge transfer that responds to the local solvent fluctuations. It is a serious challenge to account for these types of effects using an empirical molecular interaction potential. We will first motivate the use of DFT methods by comparing the hydration structure of "simple" halide anions under bulk solvation. In line with our previous work, we will relate the surface propensity of the halide series with the local solvation structure of the anion. Herein, we will extend our studies to molecular anions. The polyatomic ions studied here are chosen to be on the opposite sides of the Hofmeister series even though they have similar polarizabilities (*e.g.* between 6 and 8 Å^3).[70] To this end, we examine the solvation of thiocyanate (SCN^-) and iodate (IO_3^-) under both bulk and interfacial conditions. The overarching aim of this study will be to rationalize the surface propensity of anions based on the principles of local solvation.

2 Methods

We explored the hydration structure about SCN^-, IO_3^- and the halide anions (F^- to I^-) using molecular dynamics simulations employing DFT-based methods under bulk and interfacial solvation. The details for the bulk simulations of IO_3^- and I^- are given in ref. 71 and ref. 68 & 72 respectively. The simulations for F^-,Cl^- and Br^- are performed using the exact protocol as described for the I^- in ref. 68. All simulations presented herein utilize the CP2K (http://cp2k.org) simulation suite containing the *QuickStep* module for the DFT calculations.[73] The molecular dynamics simulations utilized a density cutoff of 280 Ry in conjunction with a double zeta basis set that has been optimized for the condensed phase.[74] Only the valence electrons were considered explicitly, and the core-electron states were represented *via* Goedecker–Teter–Hutter pseudopotentials.[75]

All of the simulations were performed in the canonical ensemble at 300 K using a Nosé–Hoover thermostat attached to every degree of freedom to ensure proper thermalization.[76] The Becke exchange[77] and correlation due to Lee, Yang and Parr[78] (BLYP) is utilized in addition to the dispersion correction put forth by Grimme[79] with a 40 Å cut-off. Each simulation was carried out for at least 20 ps duration, using a time-step of 0.5 fs. The bulk system contained 96 water molecules and a single anion within a supercell of 14.4 Å × 14.4 Å × 14.4 Å following the approach of earlier studies. This system size has proven to provide a good description of the experimentally determined solvation shell of low concentration electrolytes.[68,80,81]

The free energy to transfer a halide anion from the interior to the surface in a water cluster of a single anion and 27 water molecules is calculated following the same protocol as ref. 72. The distance r of the halide anion with respect to the center of mass of the water molecules was restrained to different distances r_0 in the range from 0.0 to 4.0 Å in 0.25 Å increments. The restraining potential was harmonic $\mathcal{V}(r) = 0.5k(r - r_0)^2$ with a force constant k of 444.8168 kcal mol^{-1} Å^{-2}. Each window was equilibrated for 4 ps, and an additional 12 ps were used to estimate the underlying free-energy using the weighted histogram analysis method (WHAM).[82,83] The probability is given by $P(r) = 4\pi r^2 \exp(-\beta A(r))$. Here, $\beta = 1/k_BT$, where k_B is Boltzmanns constant, T is temperature, and A is the potential of mean force (PMF).[84]

The extended interfacial system contained 215 water molecules and a single anion within a supercell of 15 Å × 15 Å × 71.44 Å. This choice of system size has been shown to produce a stable bulk liquid in the center of the slab.[29,71,85–87] It is imperative that we reproduce the correct bulk solvation of the iodide anion in the center of the slab, and therefore recover the correct bulk density for liquid water at 300 K in the interior of our slab. Interfacial studies of both IO_3^- and SCN^- were performed by constraining the anion either by an external potential or by constraining its orientation at the surface in distinct starting orientations. Chosen locations of the interface were determined by the distance we found for the free energy minimum in our previous DFT study of I^-.[72] Dipole moments for the halide anions are calculated using maximally localized Wannier centers.[88–90] Those for the polyatomic anions (SCN^- and IO_3^-) are calculated using a Bader analysis[91] to determine the molecular electron density and integrating the resulting charge density with respect to the center of mass.[92] Calculating molecular dipole moments in the condensed phase is ambiguous and different approaches will lead to slightly different results, but similar trends.[93]

3 Results and discussion

First we will review and discuss the bulk solvation studies of the halide anions (F^- to I^-) to establish a baseline for more complex polyatomic anions. As stated previously, our goal is to connect local solvation to surface-propensity of anions at the air–water interface.

Fig. 1 (a) and (b) show the radial distribution function (RDF) around the halide anion with respect to the water hydrogen and oxygen atoms under bulk solvation conditions. The RDFs compare well with recent DFT literature values (see Table 1).[94–99] The largest discrepancy is seen in the coordination shell around the

F⁻. Our simulations predict a tetrahedral structure, with four hydrogen bonded water molecules that do not exchange on the time scale of our simulations. This is evidenced by the drop to zero in the F^- to O RDF and was also observed in the interfacial study with the uncorrected BLYP in a slab geometry.[86] This is in contrast with older simulations by Heuft and Meijer which were run at a density of 1 g cm^{-3} using the uncorrected BLYP that corresponds to a high pressure liquid.[95,100] This density effect is known for ions with a tight solvation shell.[67] Thus, we believe that it is important to use this protocol to simulate at the correct state point of interest with DFT.[79,80,100]

A novel analysis is presented here to estimate solvated ion radii for direct comparison to literature values used in recent dielectric continuum theory (DCT) studies.[101,102] In a recent DCT study, the parameterization of an ion's radius was used to deduce the free energy of adsorption for both simple and complex anions. The good agreement of the potential of mean force obtained with DCT compared with our DFT study of I^- at the air–water interface prompted us to investigate whether or not other outcomes of our *ab initio* investigation could be verified with the assumptions of DCT. To this end, we use the charge density to infer the size of the cavity occupied by an ion. This can be done by sitting at the center of the ion and integrating the charge density, $4\pi\int_{r_{\text{center}}}^{r_{q=\pm1}} r^2\rho(r)\mathrm{d}r$, sampled over all configurations, until a value of ±1 (for the monovalent cations and anions) is obtained. The value, $r_{q=\pm1}$, of the cavity radius, can be interpreted as the average ionic radius, beyond which the effect of the dielectric of the liquid is felt. For the halide series,

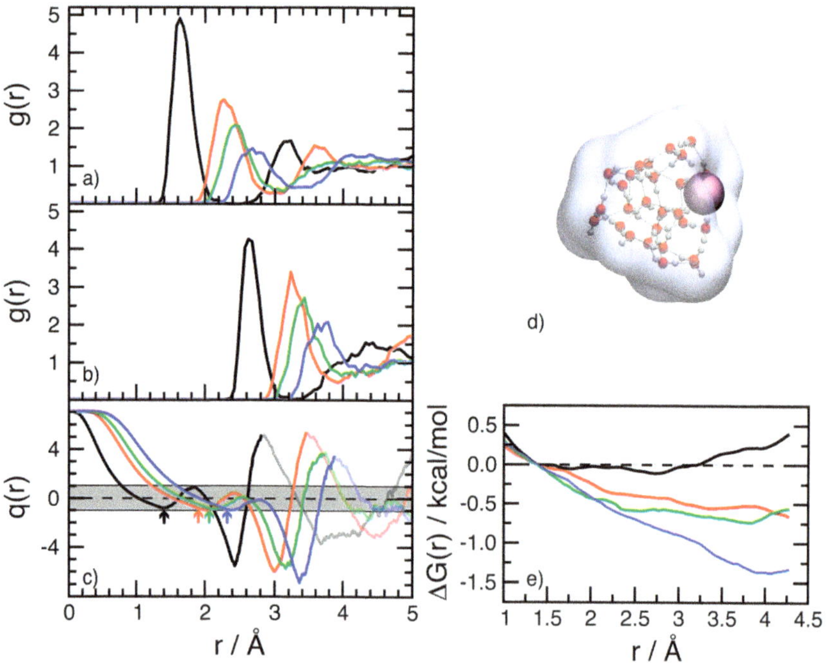

Fig. 1 (a) Radial distribution functions (RDF) of the halide to hydrogen, (b) halide to oxygen, (c) integrated radial charge-density for the halide anions F^- (black), Cl^- (red), Br^- (green) and I^- (blue). The radial integration uses the formula $4\pi\int_{r_{\text{center}}}^{r_{q=\pm1}} r^2\rho(r)\mathrm{d}r$ as described in the text. Because of the use of pseudopotentials, the core charge integrates to $+7e$ at short distances. Once one starts to sample the valence charge density, the integral drops to values near $-1e$ which we define as an effective radius for the anion. The grey scaled region highlights the range of charge from $-1e$ to $+1e$, where the dashed line is zero. Arrows denote the estimated ionic radii. Panel (e) is the potential of mean force for halide anions in small water clusters [see snap shot in panel (d)] as described in the Methods section.

Table 1 Structural and molecular parameters from cluster and bulk simulations. Here μ denotes the magnitude of the dipole moment in Debye in the bulk (b) and cluster (c). The cluster dipole moments were computed 4 Å from the cluster center. R denotes the radii in Å of the ion computed as described in the text and shown in Fig.1 and $R_{\mathrm{Lit}}^{\mathrm{ion}}$ are accepted values taken from ref. 103. CN denotes the coordination number of the first solvation shell. The geometric radius, R_{geo}, is defined as the position of the first minimum of the halide-oxygen radial distribution function minus half the position of the first oxygen–oxygen peak for water.[72] With the exception of the strongly hydrated F^- ion, good agreement between R_{geo} and R is found

	$\langle\mu_b\rangle$	$\langle\mu_c\rangle$	R	R_{geo}	$R_{\mathrm{Lit}}^{\mathrm{ion}}$	CN
F^-	0.43 ± 0.18	0.89 ± 0.14	1.41	1.25	1.36	4.0
Cl^-	0.62 ± 0.27	1.61 ± 0.20	1.91	1.94	1.81	6.2
Br^-	0.84 ± 0.36	1.95 ± 0.17	2.05	2.05	1.95	6.5
I^-	1.31 ± 0.41	2.13 ± 0.21	2.26	2.27	2.16	6.9

the running integrals of the charge density are shown in Fig. 1c), and the corresponding effective cavity radii are tabulated in Table 1. It can be seen that the radii determined by this simple method are in good agreement with other estimates, including the Latimer radii employed in dielectric continuum theory.[101,102] Although this does not guarantee that DFT will produce the correct free energies of solvation for the halides, it suggests that the electron density is giving rise to the correct *local structure*, as determined by XAFS[68] and yields a cavity size consistent with experimental estimates based on *thermodynamic* properties.

Here we want to examine the assumption that ions with small, strongly hydrated anions will be repelled from the interface, whereas large, polarizable anions which are weakly hydrated are more likely to be free energetically favorable at the air–water interface. This concept has been exploited in successful dielectric continuum theories and molecular simulations using point charge models to elucidate the driving force of ions to interfaces.

Because of the computational expense of using *ab initio* methods in the extended interface, here we present results for the full halide series in small water droplets at 150 K. The technical details of these calculations have been discussed in a previous study on I^-.[72] We present the results for the halide series in Fig. 1e). Overall, the results demonstrate that the F^- is repelled from the interface and I^- is adsorbed consistent with the conventional picture of aqueous solvation. However, Cl^- and Br^- show very "flat" to slightly adsorbing PMF for this cluster study. This result can also be rationalized with the differences in the aqueous solvation as discussed above. Specifically, Cl^- and Br^- are very similar in terms of size (see Table 1), coordination number and induced dipole moment. Whether this similarity between Cl^- and Br^- persists to the extended interfaces is yet to be determined. It suffices to say F^- and I^- show good qualitative agreement with results using the extended interface.[72,86] However, in the case of I^- the extended interface free energy minimum is less pronounced than the corresponding cluster study. It should be remembered that the cluster studies were done at 150 K and that direct comparison to the 300 K results is difficult to make.

Now that we have highlighted the correlation between hydration structure and the Hofmeister series for halide ions, we turn our attention to the polyatomic anions. These pose a more serious computation challenge for many reasons. First, and foremost, the lack of high symmetry requires additional sampling in the extended interface geometry. Second, the detailed representation of the interaction potential is likely a challenge due to the complex electronic response due to the resonance structure of SCN^-.[81,87,104] We remind the reader that the text book view of the charge distribution of SCN^- is indeed a resonance structure between two Lewis structures.

One Lewis structure localizes the excess electron on the sulfur ($^{-}S–C\equiv N$), whereas the other locates the electron on the nitrogen ($S=C=N^{-}$) atom. Thus, although both ends of the SCN^{-} molecule can accept hydrogen bonds, the chemical identities of each end will be very different.

Experimentally, the solvation structure of SCN^{-} has been studied using X-ray scattering,[105,106] which lead to an "egg-shell model" for the asymmetric solvation of the SCN^{-}. Our DFT results are presented in Fig. 2. Here we obtain very good agreement with the experimental estimates of the first peak of the radial distribution function of each atom in SCN^{-} with oxygen and hydrogen.[105,106] We refer the reader to Table 2 for the estimated coordination numbers, peak positions, and hydrogen bonding properties. The right panel of Fig. 2 gives the precise spatial distribution of oxygen and hydrogen density around the anion under bulk solvation. We also find the general picture put forth by the egg-shell is corroborated by our study where we find three strong hydrogen bonds around the N, no hydrogen bonds around the C, and approximately 2–3 hydrogen bonds around the S.[105]

Because the experimental estimates are fitted using a reduced model of SCN^{-} based on a point-charge model of this complicated resonance structure, some of the assignments may be ambiguous. Our study suggests a more directed structure of three strong hydrogen bonded molecules towards the N with an average O–N–C angle of 120 degrees. We also see two to three weaker hydrogen bonds toward the S with an angle of roughly 90 degrees to the SCN^{-} molecular axis. This is depicted schematically in Fig. 3. Here we denote "weak" and "strong" hydrogen bonds in analogy with the halide RDFs in Fig. 1. We see that the RDF around N are similar to those of F^{-} which is known to have a strong hydration shell. Whereas the RDF around S resembles the aqueous structure of I^{-} where hydrogen bonding is weaker. The analogy can be taken a step further where the hydrogen bond

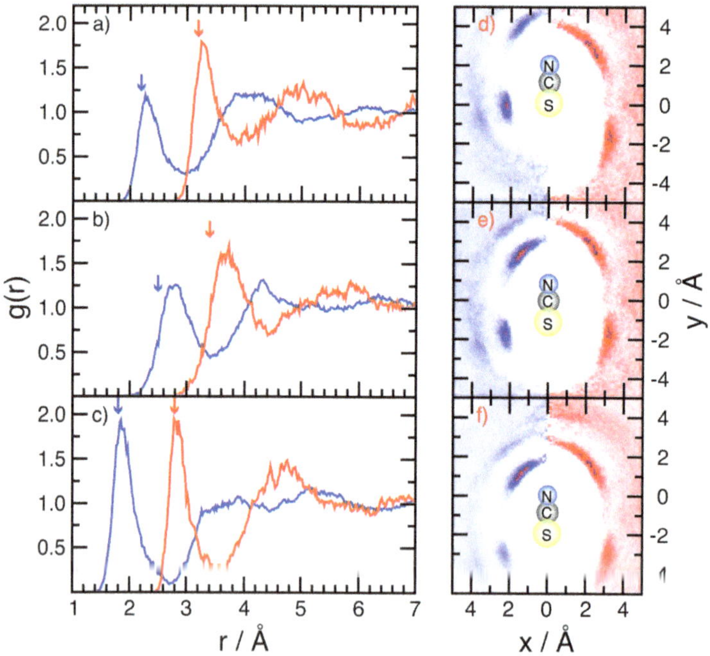

Fig. 2 Left panel: RDF of water oxygen (red) and hydrogen (blue) for (a) S, (b) C, and (c) N. Arrows mark the experimentally determined estimates for the first maxima.[105] Right panel: radial angular distribution function of oxygen (red) and hydrogen (blue) around (d) the S along the C–S bond, (e) C along the N–S bond, and (f) N along the C–N bond.

Table 2 Stuctural parameters for the bulk sovlation of SCN^-. $r_{max(min)}$ denotes the maximum (minimum) of the oxygen RDF used to compute the average number of water molecules, $\langle n_{tot} \rangle$. $\langle n_{hb} \rangle$ denotes the average number of hydrogen bonds to the atom center and α_{hb} is the C–heavy atom–O angle (in degrees), where the heavy atom is either S or N

Atom	r_{max}/Å	r_{min}/Å	$\langle n_{tot} \rangle$	$\langle n_{hb} \rangle$	α_{hb}
S	3.4	4.0	5.2	2–3	90
C	3.8	4.4	7.9	0	—
N	2.8	3.6	3.2	3	120

structure around S resembles that of a cavity found in a similar study for an over polarized model of I^-.[107]

Having analysed the detailed solvation structure of SCN^- under bulk aqueous conditions, it is now of interest to have a direct comparison of the salient features of the hydration shell with two different monovalent anions that are roughly the same size based on the oxygen-anion RDF, and similar polarizabilities.[70] Fig. 3 is a schematic of IO_3^-, I^-, and SCN^-. Previous studies comparing DFT directly to experimental X-ray absorption fine structure spectroscopy (XAFS) studies for both I^- and IO_3^- have revealed the details of their first solvation shell under bulk aqueous solvation. IO_3^- contains a region around the halide that is strongly hydrated as a cation.[80] The local region adjoining the iodine atom is sufficiently electropositive that three hydrating water molecules are oriented with their oxygen atoms directly interacting with the iodine atom, as schematically depicted in Fig. 3. There are an additional 2–3 water molecules that are hydrogen bonded to each oxygen of IO_3^- as is expected for an anion. [80] For I^- we found a flexible solvation shell containing roughly six hydrogen bonded water molecules in the first solvation shell (see Fig. 3).[68]

Given our schematic view of the three different anions shown in Fig. 3 we can speculate on the nature of their surface propensities based on the relative comparison of the anions under bulk solvation. Through the DFT simulations of the extended aqueous air–water interface, it was determined that I^- is weakly stabilized at the air–water interface in good agreement with what is expected through fitting the potential of mean force of iodide adsorption to surface tension data.[34,101,102]

There are additional empirical measures of hydration and cavitation that could also be used to infer surface propensity. The Jones–Dole viscosity *B* coefficient can be directly related to the viscosity of dilute electrolytes.[101,102] A positive *B* reflects a strong hydrated ion, whereas a negative value suggests a cavity-forming ion. The

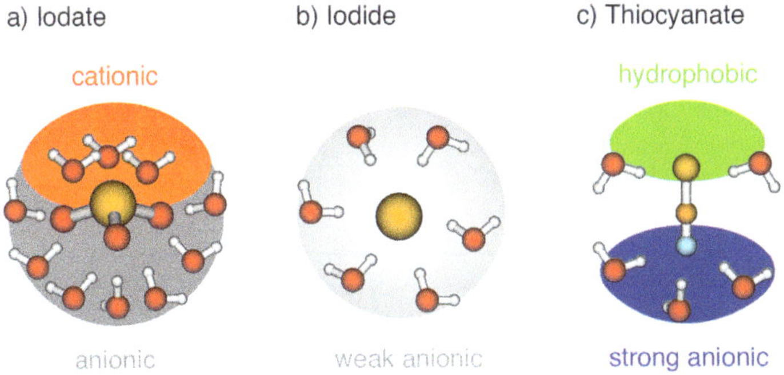

Fig. 3 Schematic of the solvation structure of iodate, iodide, and thiocyanate. The sphere size represents the similar radius of the complete first solvation shell. The salient classification of each anion is given in the Figure based on our analysis as outlined in the main text.

B coefficients based on viscosity are 0.140, −0.073, and −0.022 for IO_3^-, I^-, and SCN^- respectively.[61] It is clear that these values can be interpreted through observations made in the bulk solvation studies. Interestingly, for SCN^- the *B* coefficient is more positive than that of I^- suggesting that SCN^- is more strongly hydrated. Although this is reflected in the bulk solvation studies, how this translates into surface propensity is still not clear. Through our understanding of the Hofmeister series we would be compelled to believe that SCN^- would be more strongly adsorbed at the air–water interface based on surface tension data.[70] We now need to focus our attention on the interfacial solvation structure of SCN^- to discern whether the bulk structure persists to the interface as is the case for I^-.[66,72]

The surface activity of SCN^- has been studied with surface sensitive techniques such as sum frequency generation (SFG) and second harmonic generation (SHG) methods.[49,50,66,108] The SFG study estimates that there is no large pertubation of the asymmetric SCN^- vibration relative to bulk solvation, suggesting no significant difference in the bulk *versus* interfacial solvation. They also conclude that the preferential orientation of the asymmetric stretch of SCN^- with respect to the surface normal is roughly 45 degrees. This is in contrast to a recent empirical model that was used to reproduce the mechanism of surface adsorption as investigated by the SHG experiments.[66,108] Although the reduced models are able to capture the measured free-energy of adsorption as well as the measured enthalpic and entropic contributions, the orientation of the SCN^- anion is not consistent with the complementary SFG study. It was found that the orientation of the SCN^- lays perpendicular to the surface normal.[108]

To examine the orientation with respect to the surface normal of SCN^- using DFT, we place the anion at the surface in three distinct starting orientations as shown in Fig. 4. Our results indicate that regardless of the initial orientation of the anion, the equilibrium angle converges to about 60 degrees with a fluctuation of ±15 degrees (see Fig. 4). This is in good agreement with the SFG study.[50] It must be remembered that the N–S bond vector used in this study is only an approximation to the transition dipole moment of the asymmetric stretch that is referenced in the measurement. Nevertheless, our results are distinct from the simple empirical model.[108,109]

The SFG experiment indicates that the vibrational signature of the SCN^- is not changing dramatically between bulk and interfacial solvation.[50] We can examine this quantitatively through our DFT simulations. To this end, we analyse the hydration of SCN^- in the vicinity of the air–water interface using trajectories that are shown in Fig. 4 after suitable equilibration of 12 ps. The comparison of bulk *versus* interfacial solvation for SCN^- is shown in Fig. 5. It is remarkable that there are no

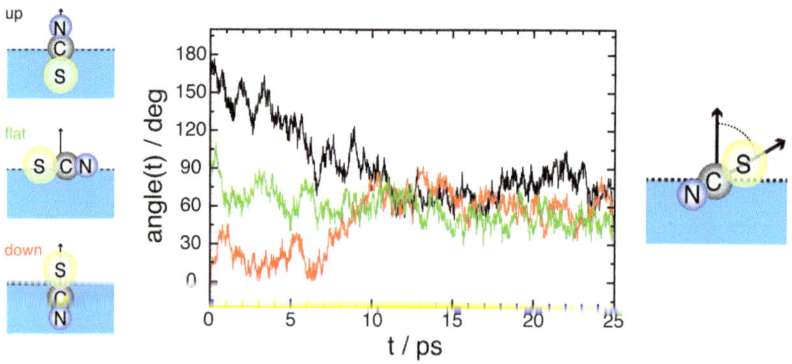

Fig. 4 Time evolution of the angle between the N–S bond with respect to the surface normal. See the left panel for a schematic definition. The middle panel shows the actual evolution of this angle with respect to time using *ab initio* molecular dynamics as discussed in the text. The right panel denotes the angle we measure and the schematic average orientation that is in good agreement with the experimental value.[49,50]

significant differences in the hydration structure, in excellent agreement with the conjectures of the SFG experiments.[50] The only difference is the number of hydrogen bonds toward the hydrophobic sulfur at the interface, namely at the interface an average we observe 1–2 hydrogen bonds which is one less than under bulk solvation (see Fig. 1).

We also observe that the analogies of SCN^- interfacial solvation with I^- persist. For I^-, we found that there is no overall change in the solvation structure in terms of hydrating water molecules approaching from bulk to interface.[72] The flexible solvation shell was able to rearrange around the stronger induced dipole moment as I^- approached the interface. Thus, we speculate that the mechanism of interfacial adsorption will have a strong model dependence based on the ability of the anion to adapt its charge density in accordance with the broken symmetry. Our studies of SCN^- suggest that there is a flexible solvation shell associated with the S end that will behave in a similar fashion to a weakly solvated anion (see Fig. 3). The strongly hydrated N end of the SCN^- molecule is thus unlikely to penetrate the surface and loose its solvation structure. We believe this is the origin of the preferential orientation of the SCN^- molecule, which can be viewed as an amphiphile.

It is now of interest to explore the solvation structure of a polyatomic anion that is strongly hydrated under bulk conditions. Fig. 3 schematically represents the bulk hydration structure of IO_3^-. Through molecular simulation using the DFT interaction potential it was found that the iodine atom is strongly hydrated by three water molecules in excellent agreement with EXAFS studies.[80] Moreover, DCT studies of the surface tension of polyoxyanions needed to treat IO_3^- as a strongly hydrated anion with a radius of 3.5 Å in order to reproduce the concentration dependent surface tension.[30] Treating IO_3^- as strongly hydrating, as was empirically determined through the Jones–Dole *B* coefficient from viscosity measurements, can be rationalized through the aforementioned molecular simulation and experimental studies.[80] We now turn to the examination of the interfacial hydration structure of IO_3^- to see if there are any dramatic differences with respect to its bulk solvation structure. From Fig. 6 it is clear that the interfacial solvation is very different from its bulk structure. Specifically, one can see the loss of three strongly hydrating water molecules as IO_3^- approaches the interface. This loss of strongly hydrating water molecules should lead to a high enthalpic penalty for IO_3^- to be at the interface. This is in contrast to both SCN^- and I^- where there seems to be more ordering in the solvation structure, but no loss of solvating water molecules.

4 Conclusion

We have presented a study on the complex solvation structure of both monatomic and polyatomic anions utilizing DFT interaction potentials. One of our aims was

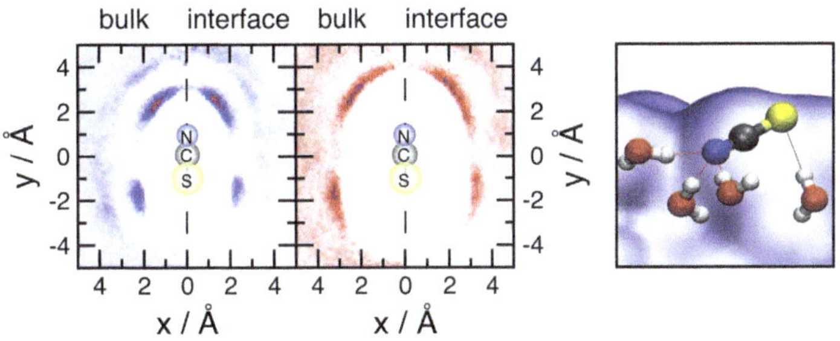

Fig. 5 Comparison of the spatial distribution of water oxygen (red) and hydrogen (blue) around the SCN^- anion under bulk and interfacial solvation. The density maps are centered around the C atom. The right shows a representative snapshot of interfacial solvation with hydrogen bonds highlighted with dashed lines.

to determine the local solvation structure and its correlation with where the ion appears in the Hofmeister series. To this end, we chose to study ions of similar size and polarizability, that span the full range of the Hofmeister series. IO_3^- is considered a strong kosmotrope, and both I^- and SCN^- are chaotropes. Our analysis of the solvation structure of the halide series allows us to characterize the solvation shells of IO_3^- and SCN^- as either weakly or strongly hydrating solvation based on comparisons of the relevant RDFs. An interesting outcome, though not surprising, is that one cannot view the polyatomic anions studied here as a symmetric spherical anion. In fact, we argue that the complex solvation structure that is determined for IO_3^- and SCN^- contribute to its position in the Hofmeister series. It is important to note that our solvation structure for all the ions presented is in good agreement with the available experimental data.[68,71,105,106] We thus provide a picture where indeed local solvation structure is an important aspect of understanding the specific ion effect in line with ideas put forth in a recent review.[13]

It is of interest to comment on the role of the local electronic structure, determined by the dipole moment, as a driving force for ion adsorption. As stated earlier, all three anions in this study have similar polarizablities. This results in dipole moments under bulk solvation for IO_3^-, SCN^-, and I^- that are 1.47 ± 0.33, 3.08 ± 0.64, and 1.1 $\pm$ 0.4 respectively. Under interfacial solvation, the dipole moments change to 2.43 ± 0.28, 3.35 ± 0.43, and 1.6 ± 0.36. These values already reflect the concomitant changes seen in the solvation structure, where the SCN^- shows the smallest deviation from its bulk structure under interfacial solvation (see Fig. 5). Interestingly, it is the IO_3^- that shows the largest increase in dipole moment from bulk to interface. Given how the ions reside in the Hofmeister series, if the dipole moment increase correlated with ion stabilization at the interface, we would expect a different ordering than what is observed. Although the principles of hydration and polarizabilty are not easily separated, from this study we can speculate that it is the flexibility of the first hydration shell rather than the dipole moment (or polarizability) that best correlates with the specific ion effect.

It is clear that the principles of ion solvation are still being worked out by the physical chemistry community. We believe that pursuing a reduced representation of the water–ion interaction will be necessary to form a complete picture of the specific ion effect. It is also clear that the aforementioned simplified descriptions of the interactions (*e.g.* point charge models, DCT, *etc.*) have found great success connecting the molecular picture with observed thermodynamic quantities.[30,34,66,101,102,110] Despite the aforementioned successes there are still considerable differences in the description of the reduced representation.[111] The present study suggests that the principles of anion adsorption can be related to a local picture of solvation. Future research will be focused on how one can effectively use the accurate information afforded by *ab initio* studies of solvation structure to generate a

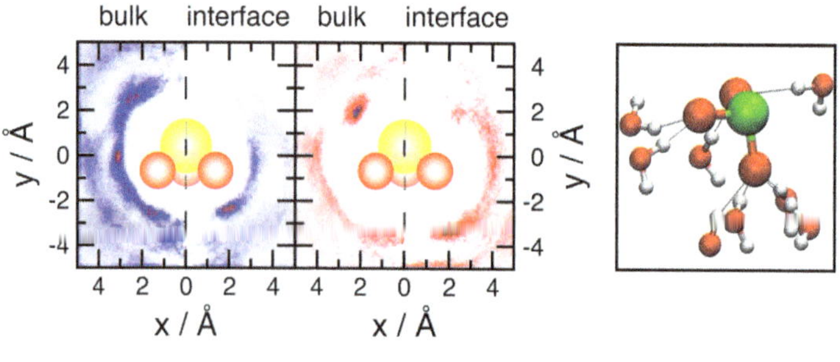

Fig. 6 Comparison of the spatial distribution of water oxygen (red) and hydrogen (blue) around the IO_3^- anion under bulk and interfacial solvation. The density maps are centered around the I atom. The right shows a representative snapshot of interfacial solvation with hydrogen bonds highlighted with dashed lines.

reduced representation where one can efficiently calculate the relevant thermodynamic quantities as a function of electrolyte concentration.

Acknowledgements

We acknowledge Abe Stern and Shawn Kathmann for a careful reading of this manuscript. This work was supported by the U.S. Department of Energy's (DOE) Office of Basic Energy Sciences, Division of Chemical Sciences, Geosciences and Biosciences. Pacific Northwest National Laboratory (PNNL) is operated for the Department of Energy by Battelle. The SCN^- calculations presented herein were performed with PNNL Institutional Computing. MDB is grateful for the support of the Linus Pauling Distinguished Postdoctoral Fellowship Program at PNNL. Parts of the research described in this paper was conducted under the Laboratory Directed Research and Development Program at PNNL. The surface calculations for IO_3^- were carried out under an INCITE 2008–2012 award, using the resources of the Oak Ridge Leadership Computing Facility at the Oak Ridge National Laboratory, which is supported by the Office of Science of the U.S. Department of Energy under Contract No. DE-AC05-00OR22725. Cluster calculations were performed using the CRAY XT5 at NERSC at Lawrence Berkeley National Laboratory.

References

1 J. H. Hu, Q. Shi, P. Davidovits, D. R. Worsnop, M. S. Zahniser and C. E. Kolb, *J. Phys. Chem.*, 1995, **99**, 8768–8776.
2 E. M. Knipping, M. J. Lakin, K. L. Foster, P. Jungwirth, D. J. Tobias, B. J. Gerber, D. Dabdub and B. J. Finlayson-Pitts, *Science*, 2000, **288**, 301–306.
3 S. W. Hunt, M. Roeselova, W. Wang, L. M. Wingen, E. M. Knipping, D. J. Tobias, D. Dabdub and B. J. Finlayson-Pitts, *J. Phys. Chem. A*, 2004, **108**, 11559–11572.
4 A. Laskin, H. Wang, W. H. Robertson, J. P. Cowin, M. J. Ezell and B. J. Finlayson-Pitts, *J. Phys. Chem. A*, 2006, **110**, 10619–10627.
5 D. Clifford and D. J. Donaldson, *J. Phys. Chem. A*, 2007, **111**, 9809–9814.
6 L. M. Wingen, A. C. Moskun, S. N. Johnson, J. L. Thomas, M. Roeselova, D. J. Tobias, M. T. Kleinman and B. J. Finlayson-Pitts, *Phys. Chem. Chem. Phys.*, 2008, **10**, 5668–5677.
7 N. K. Richards, L. M. Wingen, K. M. Callahan, N. Nishino, M. T. Kleinman, D. J. Tobias and B. J. Finlayson-Pitts, *J. Phys. Chem. A*, 2011, **115**, 5810–5821.
8 W. Kunz, J. Henle and B. W. Ninham, *Curr. Opin. Colloid Interface Sci.*, 2004, **9**, 19–37.
9 P. Lo Nostro and B. W. Ninham, *Chem. Rev.*, 2012, **112**, 2286–2322.
10 K. D. Collins and M. W. Washabaugh, *Q. Rev. Biophys.*, 1985, **18**, 323–422.
11 L. M. Pegram and M. T. Record Jr, *Proc. Natl. Acad. Sci. U. S. A.*, 2006, **103**, 14278–14281.
12 L. M. Pegram and M. T. Record Jr, *J. Phys. Chem. B*, 2007, **111**, 5411–5417.
13 D. J. Tobias and J. C. Hemminger, *Science*, 2008, **319**, 1197–1198.
14 J. Caldwell, L. X. Dang and P. A. Kollman, *J. Am. Chem. Soc.*, 1990, **112**, 9144–9147.
15 L. Perera and M. L. Berkowitz, *J. Chem. Phys.*, 1991, **95**, 1954–1963.
16 L. Perera and M. L. Berkowitz, *J. Chem. Phys.*, 1992, **96**, 8288–8294.
17 L. Perera and M. Berkowitz, *J. Chem. Phys.*, 1993, **99**, 4222–4224.
18 L. X. Dang and D. E. Smith, *J. Chem. Phys.*, 1993, **99**, 6950–6956.
19 S. J. Stuart and B. J. Berne, *J. Phys. Chem.*, 1996, **100**, 11934–11943.
20 P. Jungwirth and D. J. Tobias, *J. Phys. Chem. B*, 2002, **106**, 6361–6373.
21 L. X. Dang, *J. Phys. Chem. B*, 2002, **106**, 10388–10394.
22 L. X. Dang and T.-M. Chang, *J. Phys. Chem. B*, 2002, **106**, 235–238.
23 L. Vrbka, M. Mucha, B. Minofar, P. Jungwirth, E. C. Brown and D. J. Tobias, *Curr. Opin. Colloid Interface Sci.*, 2004, **9**, 67–73.
24 D. H. Herce, L. Perera, T. A. Darden and C. Sagui, *J. Chem. Phys.*, 2005, **122**, 024513.
25 P. Jungwirth and D. J. Tobias, *Chem. Rev.*, 2006, **106**, 1259–1281.
26 G. Archontis and E. Leontidis, *Chem. Phys. Lett.*, 2006, **420**, 199–203.
27 T. Ishiyama and A. Morita, *J. Phys. Chem. C*, 2007, **111**, 721–737.
28 T. Ishiyama and A. Morita, *J. Phys. Chem. C*, 2007, **111**, 738–748.
29 M. D. Baer, I.-F. W. Kuo, H. Bluhm and S. Ghosal, *J. Phys. Chem. B*, 2009, **113**, 15843–15850.
30 A. P. dos Santos, A. Diehl and Y. Levin, *Langmuir*, 2010, **26**, 10778–10783.

31 B. L. Eggimann and J. I. Siepmann, *J. Phys. Chem. C*, 2008, **112**, 210–218.
32 G. L. Warren and S. Patel, *J. Phys. Chem. C*, 2008, **112**, 7455–7467.
33 G. L. Warren and S. Patel, *J. Phys. Chem. B*, 2008, **112**, 11679–11693.
34 D. Horinek, A. Herz, L. Vrbka, F. Sedlmeier, S. I. Mamatkulov and R. R. Netz, *Chem. Phys. Lett.*, 2009, **479**, 173–183.
35 J. Noah-Vanhoucke and P. L. Geissler, *Proc. Natl. Acad. Sci. U. S. A.*, 2009, **106**, 15125–15130.
36 C. Caleman, J. S. Hub, P. J. van Maaren and D. van der Spoel, *Proc. Natl. Acad. Sci. U. S. A.*, 2011, **108**, 6838–6842.
37 T. Yagasaki, S. Saito and I. Ohmine, *J. Phys. Chem. A*, 2010, **114**, 12573–12584.
38 J. Cheng, C. D. Vecitis, M. R. Hoffmann and A. J. Colussi, *J. Phys. Chem. B*, 2006, **110**, 25598–25602.
39 R. M. Onorato, D. E. Otten and R. J. Saykally, *J. Phys. Chem. C*, 2010, **114**, 13746–13751.
40 Y. Hiranuma, K. Kaniwa, M. Shoji and F. Mafune, *J. Phys. Chem. A*, 2011, **115**, 8493–8497.
41 S. Ghosal, J. C. Hemminger, H. Bluhm, B. S. Mun, E. L. D. Hebenstreit, G. Ketteler, D. F. Ogletree, F. G. Requejo and M. Salmeron, *Science*, 2005, **307**, 563–566.
42 S. Ghosal, M. A. Brown, H. Bluhm, M. J. Krisch, M. Salmeron, P. Jungwirth and J. C. Hemminger, *J. Phys. Chem. A*, 2008, **112**, 12378–12384.
43 M. J. Krisch, R. D'Auria, M. A. Brown, D. J. Tobias, J. C. Hemminger, M. Ammann, D. E. Starr and H. Bluhm, *J. Phys. Chem. C*, 2007, **111**, 13497–13509.
44 D. F. Liu, G. Ma, L. M. Levering and H. C. Allen, *J. Phys. Chem. B*, 2004, **108**, 2252–2260.
45 V. Ostroverkhov, G. A. Waychunas and Y. R. Shen, *Phys. Rev. Lett.*, 2005, **94**, 046102.
46 C. S. Tian and Y. R. Shen, *Chem. Phys. Lett.*, 2009, **470**, 1–6.
47 H. T. Bian, R. R. Feng, Y. Y. Xu, Y. Guo and H. F. Wang, *Phys. Chem. Chem. Phys.*, 2008, **10**, 4920–4931.
48 N. Ji, V. Ostroverkhov, C. Y. Chen and Y. R. Shen, *J. Am. Chem. Soc.*, 2007, **129**, 10056–10057.
49 P. Viswanath and H. Motschmann, *J. Phys. Chem. C*, 2007, **111**, 4484–4486.
50 P. Viswanath and H. Motschmann, *J. Phys. Chem. C*, 2008, **112**, 2099–2103.
51 X. Chen, T. Yang, S. Kataoka and P. S. Cremer, *J. Am. Chem. Soc.*, 2007, **129**, 12272–12279.
52 P. B. Petersen and R. J. Saykally, *J. Am. Chem. Soc.*, 2005, **127**, 15446–15452.
53 P. B. Petersen and R. J. Saykally, *Chem. Phys. Lett.*, 2004, **397**, 51–55.
54 P. B. Petersen and R. J. Saykally, *Chem. Phys. Lett.*, 2008, **458**, 255–261.
55 D. E. Otten, P. B. Petersen and R. J. Saykally, *Chem. Phys. Lett.*, 2007, **449**, 261–265.
56 P. B. Petersen and R. J. Saykally, *J. Phys. Chem. B*, 2006, **110**, 14060–14073.
57 P. B. Petersen and R. J. Saykally, *J. Phys. Chem. B*, 2005, **109**, 7976–7980.
58 A. Heydweiller, *Ann. Phys.*, 1910, **338**, 145–185.
59 L. Onsager and N. N. T. Samaras, *J. Chem. Phys.*, 1934, **2**, 528–537.
60 A. Frumkin, *Z. Phys. Chem.*, 1924, **109**, 34–48.
61 Y. Marcus, *Chem. Rev.*, 2009, **109**, 1346–1370.
62 R. Mancinelli., A. Botti, F. Bruni, M. A. Ricci and A. K. Soper, *J. Phys. Chem. B*, 2007, **111**, 13570–13577.
63 J. D. Smith, R. J. Saykally and P. L. Geissler, *J. Am. Chem. Soc.*, 2007, **129**, 13847–13856.
64 A. W. Omta, M. F. Kropman., S. Woutersen and H. J. Bakker, *Science*, 2003, **301**, 347–349.
65 Y. Levin and J. R. Flores-Mena, *Europhys. Lett.*, 2001, **56**, 187–192.
66 D. E. Otten, P. R. Shaffer, P. L. Geissler and R. J. Saykally, *Proc. Natl. Acad. Sci. U. S. A.*, 2012, **109**, 701–705.
67 J. L. Fulton, S. M. Kathmann, G. K. Schenter and M. Balasubramanian, *J. Phys. Chem. A*, 2009, **113**, 13976–13984.
68 J. L. Fulton, G. K. Schenter, M. D. Baer, C. J. Mundy, L. X. Dang and M. Balasubramanian, *J. Phys. Chem. B*, 2010, **114**, 12926–12937.
69 E. Cauet, S. Bogatko, J. H. Weare, J. L. Fulton, G. K. Schenter and E. J. Bylaska, *J. Chem. Phys.*, 2010, **132**, 194502.
70 W. Kunz, P. Lo Nostro and B. W. Ninham, *Curr. Opin. Colloid Interface Sci.*, 2004, **9**, 1–18.
71 M. D. Baer, C. J. Mundy, M. J. McGrath, I. F. W. Kuo, J. I. Siepmann and D. J. Tobias, *J. Chem. Phys.*, 2011, **135**, 124712.
72 M. D. Baer and C. J. Mundy, *J. Phys. Chem. Lett.*, 2011, **2**, 1088–1093.
73 J. VandeVondele, M. Krack, F. Mohamed, M. Parrinello, T. Chassaing and J. Hutter, *Comput. Phys. Commun.*, 2005, **167**, 103–128.
74 J. VandeVondele and J. Hutter, *J. Chem. Phys.*, 2007, **127**, 114105.

75 S. Goedecker, M. Teter and J. Hutter, *Phys. Rev. B: Condens. Matter*, 1996, **54**, 1703–1710.
76 G. J. Martyna, M. L. Klein and M. E. Tuckerman, *J. Chem. Phys.*, 1992, **97**, 2635–2643.
77 A. D. Becke, *Phys. Rev. A: At., Mol., Opt. Phys.*, 1988, **38**, 3098–3100.
78 C. T. Lee, W. T. Yang and R. G. Parr, *Phys. Rev. B*, 1988, **37**, 785–789.
79 S. Grimme, *J. Comput. Chem.*, 2004, **25**, 1463–1473.
80 M. D. Baer, V.-T. Pham, J. L. Fulton, G. K. Schenter, M. Balasubramanian and C. J. Mundy, *J. Phys. Chem. Lett.*, 2011, **2**, 2650–2654.
81 T. Lewis, B. Winter, A. C. Stern, M. D. Baer, C. J. Mundy, D. J. Tobias and J. C. Hemminger, *J. Phys. Chem. B*, 2011, **115**, 9445–9451.
82 S. Kumar, D. Bouzida, R. H. Swendsen, P. A. Kollman and J. M. Rosenberg, *J. Comput. Chem.*, 1992, **13**, 1011–1021.
83 B. Roux, *Comput. Phys. Commun.*, 1995, **91**, 275–282.
84 D. M. Koch and G. H. Peslherbe, *Chem. Phys. Lett.*, 2002, **359**, 381–389.
85 I.-F. W. Kuo, C. J. Mundy, B. L. Eggimann, M. J. McGrath, J. I. Siepmann, B. Chen, J. Vieceli and D. J. Tobias, *J. Phys. Chem. B*, 2006, **110**, 3738–3746.
86 M.-H. Ho, M. L. Klein and I.-F. W. Kuo, *J. Phys. Chem. A*, 2009, **113**, 2070–2074.
87 T. Lewis, B. Winter, A. C. Stern, M. D. Baer, C. J. Mundy, D. J. Tobias and J. C. Hemminger, *J. Phys. Chem. C*, 2011, **115**, 21183–21190.
88 N. Marzari and D. Vanderbilt, *Phys. Rev. B: Condens. Matter*, 1997, **56**, 12847–12865.
89 P. L. Silvestrelli and M. Parrinello, *Phys. Rev. Lett.*, 1999, **82**, 3308–3311.
90 P. L. Silvestrelli and M. Parrinello, *Phys. Rev. Lett.*, 1999, **82**, 5415–5415.
91 R. F. W. Bader, *Atoms in Molecules: A Quantum Theory*, Oxford University Press, 1990.
92 W. Tang, E. Sanville and G. Henkelman, *J. Phys.: Condens. Matter*, 2009, **21**, 084204.
93 E. R. Batista, S. S. Xantheas and H. Jonsson, *J. Chem. Phys.*, 1999, **111**, 6011–6015.
94 E. Guardia, I. Skarmoutsos and M. Masia, *J. Chem. Theory Comput.*, 2009, **5**, 1449–1453.
95 J. M. Heuft and E. J. Meijer, *J. Chem. Phys.*, 2005, **122**, 094501.
96 J. M. Heuft and E. J. Meijer, *J. Chem. Phys.*, 2005, **123**, 094506.
97 J. M. Heuft and E. J. Meijer, *J. Chem. Phys.*, 2003, **119**, 11788–11791.
98 S. Raugei and M. L. Klein, *J. Chem. Phys.*, 2002, **116**, 196–202.
99 S. Raugei and M. L. Klein, *J. Am. Chem. Soc.*, 2001, **123**, 9484–9485.
100 J. Schmidt, J. VandeVondele, I.-F. W. Kuo, D. Sebastiani, J. I. Siepmann, J. Hutter and C. J. Mundy, *J. Phys. Chem. B*, 2009, **113**, 11959–11964.
101 Y. Levin, *Phys. Rev. Lett.*, 2009, **102**, 147803.
102 Y. Levin, A. P. dos Santos and A. Diehl, *Phys. Rev. Lett.*, 2009, **103**, 257802.
103 W. M. Latimer, K. S. Pitzer and C. M. Slansky, *J. Chem. Phys.*, 1939, **7**, 108–111.
104 M. Baer, C. J. Mundy, T.-M. Chang, F.-M. Tao and L. X. Dang, *J. Phys. Chem. B*, 2010, **114**, 7245–7249.
105 A. Botti, S. E. Pagnotta, F. Bruni and M. A. Ricci, *J. Phys. Chem. B*, 2009, **113**, 10014–10021.
106 P. W. Mason, G. W. Neilson, C. E. Dempsey, A. C. Barnes and J. M Cruickshank, *Proc. Natl. Acad. Sci. U. S. A.*, 2003, **100**, 4557–4561.
107 C. A. Wick and S. S. Xantheas, *J. Phys. Chem. B*, 2009, **113**, 4141–4146.
108 P. B. Petersen, R. J. Saykally, M. Mucha and P. Jungwirth, *J. Phys. Chem. B*, 2005, **109**, 10915–10921.
109 M. Darvas, M. Jorge, M. N. D. S. Cordeiro and P. Jedlovszky, *J. Phys. Chem. C*, 2011, **115**, 11140–11146.
110 R. R. Netz and D. Horinek, *Annu. Rev. Phys. Chem.*, 2012, **63**, 401–418.
111 M. D. Baer, A. C. Stern, Y. Levin, D. J. Tobias and C. J. Mundy, *J. Phys. Chem. Lett.*, 2012, **11**, 1565–1570.

General discussion

Professor Kunz opened the discussion of the Introductory Lecture by Thomas Record: According to the Gibbs–Duhem equation, two independent measurements are required to identify the chemical potential of each component in a ternary system. However, you use only osmometry measurements to infer the chemical potential $\mu_{2,3}$ and the main part of your discussion is based on this potential, even at concentrations above 1 M. Despite the fact that there must therefore be a certain ambiguity in the values of $\mu_{2,3}$, your general picture is remarkable consistent. Could you comment on the possible reasons for that?

Professor Record replied: We are not determining the chemical potential of a solute component, but rather its dependence on the concentration of a second solute component. From the difference ΔOsm between the osmolality of the three component solution Osm(m_2,m_3) and the osmolalities of the two two-component solutions Osm(m_2), Osm(m_3) at the same solute concentrations m_2 and m_3, we determine the chemical potential derivative $\mu_{2,3} = d\mu_2/dm_3$. This chemical potential derivative is the key quantity in the preferential interaction coefficient $\Gamma = -\mu_{2,3}/\mu_{3,3}$ and is the quantity needed to interpret the effect of a solute on a process like protein folding, or dissolving a protein or model compound. Values of Γ and $\mu_{2,3}$ are interpreted in terms of the partitioning of one solute between the surface of the other and bulk water using the solute partitioning model (SPM), and are related to theoretical or experimental determinations of the radial distributions of a solute in the vicinity of another by Kirkwood–Buff integrals.

Professor Goodall said: You mentioned that there was disagreement between your fitting functions for protein and nucleic acids using caffeine as a probe. Caffeine binds to the heterocyclic bases in DNA (see, for example, the references below), but I would not envisage specific interactions with proteins. Might this be a potential cause of the disagreement?

1. Shohreh Nafisi, Firouzeh Manouchehri, Heidar-Ali Tajmir-Riahi and Maryam Varavipour, Structural features of DNA interaction with caffeine and theophylline, *J. Mol. Struct.*, 2008, **875**(1–3), 392–399.
2. V. I. Poltev, T. I. Grokhlina, A. Deriabina and E. González, Caffeine interactions with nucleic acids. Molecular mechanics calculations of model systems for explanation of mechanisms of biological actions, *Theor. Chem. Acc.*, 2003, **110**(6), 466–472.

Professor Record responded: I was referring to unpublished research from our laboratory investigating the effect of urea on the phase partitioning behavior of nucleic acid bases and base analogs like caffeine between hexanol and water. Analysis of the urea–base $\mu_{2,3}$ values obtained from these experiments indicates that the methyl group on the ring of thymine and base analogs like caffeine interacts much more favorably with urea than expected from previously studied model compounds with an aliphatic carbon surface that is not on an aromatic ring (E. Guinn *et al.*, *Proc. Natl. Acad. Sci U. S. A.*, 2011, **108**(41), 16932–16937).

Professor Ben-Amotz opened the discussion of the paper by Phillip Geissler: I am curious about the connection between your results and the exact solvation thermodynamic expressions which may be obtained from the Widom Potential Distribution Theorem. The latter theorem (in its so called inverse form)[1] leads to an expression for the chemical potential (solvation free energy) of any solute in any solvent system purely in terms of the solute–solvent interaction energy probability distribution.[2] This further implies that solvent–solvent contributions to the experimental solvation energy and entropy of any solute must exactly cancel (compensate) when they are combined to obtain the solute's experimental solvation free energy. I am curious about the implications of these results when applied to the interesting capillary wave pinning phenomena that you have identified and described. More specifically, I wonder if you could say more regarding how much of the entropy and energy change associated with capillary wave pinning is compensating (in the above sense)? Note that I am not implying that solute–solvent and solvent–solvent contributions to solvation thermodynamics can be entirely decoupled from each other, as the distribution of solute–solvent energies certainly depend on solvent structure. This coupling is evidenced, for example, by the fact that solute induced changes in the solvent–solvent interaction energy may be obtained from the temperature dependence of the solute–solvent interaction energy distribution function (see for example equations (37) and (38) in ref. 2). Although some of the above expressions can be difficult, or practically impossible, to evaluate using molecular fluid simulations (because of numerical difficulties associated with the high energy tail of the solute–solvent probability distribution function), I expect that no such difficulties would arise in lattice models. Thus, in a lattice simulation, it should be possible to obtain directly all solvation thermodynamic quantities from the solute–solvent energy probability distribution function.

1. B. Widom, *J. Phys. Chem.*, 1982, **86**, 869–872.
2. D. Ben-Amotz, F. Raineri and G. Stell, *J. Phys. Chem. B*, 2005, **109**, 6866–6878.

Professor Geissler replied: The notion of compensating contributions to solvation free energies has been discussed in many contexts.[1] Before addressing the evaluation of such contributions for the lattice model described in our paper, we would like to comment more generally on this idea, which in our view has spurred some unfortunate misconceptions about thermodynamic driving forces for solvation.

A. Solvation thermodynamics and the notion of compensation

The inverse form of Widom's potential distribution theorem[2] can be written

$$\mu_{\text{ex}} = k_{\text{B}}T \ln \langle e^{\beta\psi} \rangle_1 \tag{1}$$

$$= \langle \psi \rangle_1 + k_{\text{B}}T \ln \langle e^{\beta\delta\psi} \rangle_1 \tag{2}$$

where μ_{ex} is the excess chemical potential of an ideally dilute solute, $\psi \equiv \langle \psi \rangle_1 + \delta\psi$ is the fluctuating potential energy of interaction between solute and solvent, $\beta = (k_{\text{B}}T)^{-1}$ is inverse temperature, and $\langle \ldots \rangle_1$ denotes a canonical average in the solute's presence. Eqn (1) and (2) describe exactly the thermodynamic consequences of inserting a solute molecule, even in a dense and complicated liquid like water, without explicit reference to interactions among solvent molecules. While these results are elegant, and may provide an illuminating perspective on certain solvation phenomena, they should not be taken to minimize the importance of solvent–solvent interactions. In anything but a trivially dilute solvent, these interactions profoundly shape the statistics of ψ that determine averages like $\langle e^{\beta\psi} \rangle_1$.

The reversible work μ_{ex} of introducing a stationary solute in a dense liquid generally has both energetic and entropic contributions,

$$\mu_{\text{ex}} = \Delta\varepsilon - T\Delta s, \tag{3}$$

determined by its temperature dependence, $\Delta s = -(\partial\mu_{\text{ex}}/\partial T)_p$ and $\Delta\varepsilon = (\partial\beta\mu_{\text{ex}}/\partial\beta)_p$. We have neglected the distinction between energy and enthalpy, having in mind nearly incompressible solvents at modest pressure. We have also made implicit that composition is held fixed in these partial derivatives. The component $\Delta\varepsilon$ can be written transparently in terms of average energies in the presence and absence of the solute,

$$\Delta\varepsilon = \langle\psi\rangle_1 + \Delta U_{\mathrm{solv}}, \tag{4}$$

where ΔU_{solv} is the change in the solvent's internal energy effected by inserting the solute. Combining this expression with eqn (2) and (3) yields

$$-T\Delta s = k_{\mathrm{B}}T \ln \langle e^{\beta\delta\psi}\rangle_1 - \Delta U_{\mathrm{solv}}. \tag{5}$$

Together, eqn (3)–(5) have inspired a notion of compensation: When $\Delta\varepsilon$ and $-T\Delta s$ are added to obtain μ_{ex}, the explicit contribution ΔU_{solv} from the latter two equations cancels. While this cancellation is mathematically self-evident, its physical implications are less clear. If the remaining contributions were independent of the solvent's internal energetics (which, as Dor notes, they are not), one could conclude that solvent–solvent interactions are thermodynamically irrelevant to solubility. Such a conclusion has been suggested by several authors,[3] and the term $k_{\mathrm{B}}T \ln \langle e^{\beta\delta\psi}\rangle_1$ has even been dubbed an entropy (which it generally is not). Dor's question reflects a more nuanced understanding: this contribution to Δs is typically a very complicated quantity, shaped to some degree by solvent–solvent energetics. In fact, $k_{\mathrm{B}}T \ln \langle e^{\beta\delta\psi}\rangle_1$ can be *dominated* by ΔU_{solv}. Below we will describe a specific example in which this dominance is complete.

B. An illustrative example of linear response

As an important example, consider a linear response model, *i.e.*, the case in which ψ obeys Gaussian statistics. In the presence of solute, the distribution $p_1(\psi)$ of solute–solvent interaction energy is

$$p_1(\psi) \propto p_0(\psi)e^{-\beta\psi}, \tag{6}$$

where $p_0(\psi)$ is the corresponding distribution in the absence of solute. Gaussian fluctuations of ψ in the solute's presence therefore requires that p_0 also be Gaussian:

$$p_0(\psi) \propto \exp[(\psi - \langle\psi\rangle_0)^2/2\langle(\psi - \langle\psi\rangle_0)^2\rangle_0]. \tag{7}$$

The free energy or potential of mean force $F(\psi) = -k_{\mathrm{B}}T \ln p_0(\psi)$ governing pertinent solvent fluctuations must therefore have a form

$$F(\psi) = \psi^2/2\alpha, \tag{8}$$

where $\alpha = \beta\langle(\psi - \langle\psi\rangle_0)^2\rangle_0 = \beta\langle(\psi - \langle\psi\rangle_1)^2\rangle_1$ is a positive parameter and we have set $\langle\psi\rangle_0 = 0$ (without loss of generality).

Dielectric continuum theory is a linear response model of exactly this sort.[4] In that case, the susceptibility α depends on temperature (through the dielectric constant), so that $F(\psi)$ has both energetic and entropic components.[5] For the sake of clarity, let us consider here the simpler case in which α is temperature independent, so that $F(\psi)$ is precisely the solvent's internal energy function,

$$U_{\mathrm{solv}}(\psi) = \psi^2/2\alpha. \tag{9}$$

The various contributions to $\Delta\varepsilon$, Δs, and μ_{ex} can then be simply computed from standard results of Gaussian statistics. In particular, $\langle\psi\rangle_1 = -\alpha$, $\langle e^{\beta\delta\psi}\rangle_1 = \exp(\beta\alpha/2)$, and $\Delta U_{\mathrm{solv}} = \alpha/2$. Most strikingly, for this case of linear response, $\Delta s = 0$ (since the breadth of fluctuations in ψ is insensitive to the solute's presence), *i.e.*, *there is no entropy change to compensate any part of* $\Delta\varepsilon$. The direct (and uncompensated) relevance of the solvent's internal energy change is thus quite clear:

$$\mu_{ex} = \Delta\varepsilon = \langle\psi\rangle_1 + \Delta U_{solv} = -\alpha + \alpha/2 = -\alpha/2. \quad (10)$$

Calculations are slightly more involved when the susceptibility α depends on temperature. The direct and uncompensated role of ΔU_{solv} is nonetheless clear in that case as well.

C. Solvation thermodynamics in lattice gas models

The lattice gas model we have devised to assess basic influences of capillary fluctuations on interfacial solvation is indeed much simpler than detailed molecular models. Decomposing solvation free energies as in eqn (1) and (2) is nonetheless problematic in this context. Specifically, our model solute excludes solvent completely from the lattice site it occupies. As in the case of perfectly hard spheres, the average $\langle e^{\beta\psi}\rangle_1$ has important contributions from configurations with vanishing weight and infinite values of $e^{\beta\psi}$.[6] It therefore cannot be evaluated numerically from simulations including a solute (or calculated analytically within the solid-on-solid approximation).

Calculations of $\Delta\varepsilon$, Δs, and μ_{ex} are presented in our paper (both for simulation and analytical theory). Aside from these quantities, averages that can be evaluated straightforwardly in the lattice model are the same ones accessible to detailed molecular simulations, namely $\langle\psi\rangle_1$ and ΔU_{solv}. These energetic contributions are shown in Fig. 1 (below).

We emphasize again that ΔU_{solv} should not be viewed as a quantity made irrelevant by the cancellation suggested in eqn (3)–(5).

1. D. Ben-Amotz, F. O. Raineri and G. Stell, *J. Phys. Chem. B*, 2005, **109**, 6866.
2. B. Widom, *J. Phys. Chem.*, 1982, **86**, 869.
3. H.-A. Yu and M. Karplus, *J. Chem. Phys.*, 1988, **89**, 2366.
4. J. Jackson, *Classical Electrodynamics*, John Wiley & Sons, 1998, ISBN 9780471309321.
5. R. M. Lynden-Bell and J. C. Rasaiah, *J. Chem. Phys.*, 1997, **107**, 1981.
6. D. A. Kofke, *Mol. Phys.*, 2006, **104**, 3701, and references therein.

Professor Bain said: Professor Geissler has drawn attention to the role of ions in suppressing interfacial fluctuations. Often fluctuations are described within the framework of capillary wave theory, within which the contribution of capillary waves to the surface tension is $-3k_BTq^2_{max}/16\pi$. If we take the upper limit on the wavevector to be π/a, where a is a molecular size (~0.25 nm for a water molecule), we obtain a value of -40 mJ m^{-2} for water (R. A. L. Jones and R. W. Richards, *Polymers at Surfaces and Interfaces*, CUP, Cambridge, 1999, p. 46). The effect on the surface tension of ions at molar concentrations (as shown in Yan Levin's paper) is typically of the order of 1 mJ m^{-2}. Consequently, even a small perturbation to the capillary wave spectrum by adsorbed ions will make a significant contribution to the observed surface tension increment/decrement. Within capillary wave theory, the influence of adsorbed ions could be accommodated either through a change in the high-q cut-off or through the incorporation of an interfacial bending modulus, κ, which has the effect of suppressing short wavelength fluctuations.

Professor Geissler answered: The overall extent and diversity of capillary fluctuations are indeed considerable, as therefore is their potential thermodynamic impact. Our lattice model demonstrates significant influence on solute adsorption through a mechanism that is somewhat distinct from the two Professor Bain mentioned, namely solute-imposed constraints on the local amplitude of interfacial fluctuations. The other mechanisms are likely to be important in more detailed molecular models.

Professor Jungwirth asked: Your model works with ions with reduced charges, which means that you are implicitly accounting for polarization effects. It was shown by your colleagues at UC Davis, Leontyev and Stuchebrukhov (*Phys.*

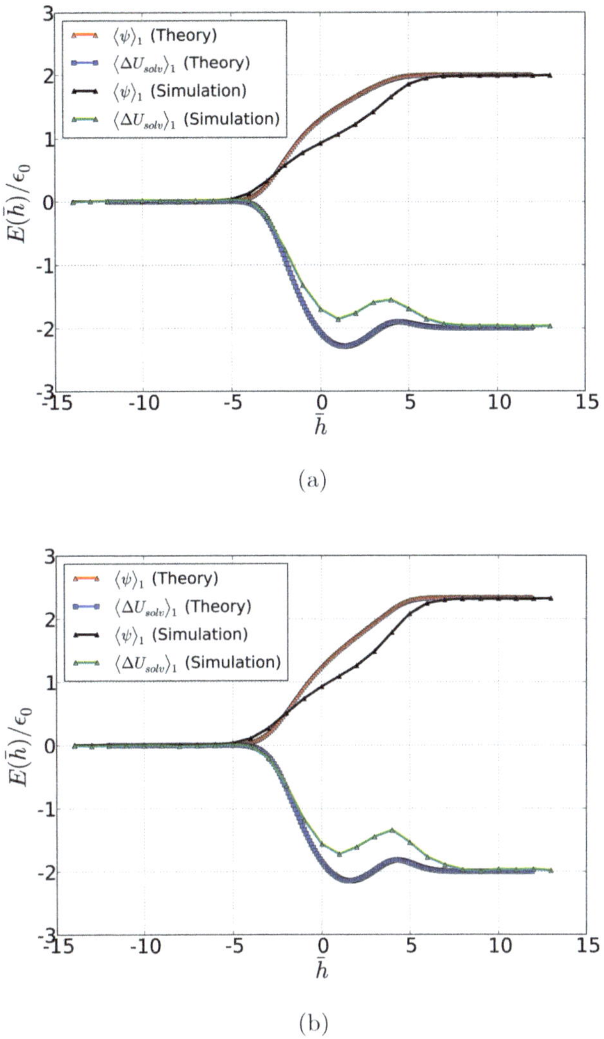

Fig. 1 Estimates of $\langle \psi \rangle_1$ and ΔU_{solv} for $\varepsilon_1 = \varepsilon_0/2$ (a) and for $\varepsilon_1 = 7\varepsilon/12$ (b) obtained from both analytical theory and computer simulations as a function of the height of the solute. The estimates provided are with respect to their values in bulk liquid.

Chem. Chem. Phys., 2011, **13**, 2613 and later papers by the same authors) that reducing the ionic charge by 25% (*i.e.* dividing it by the square root of the electronic part of the dielectric constant of water, which equals to 1.78) is mathematically equivalent to accounting for polarization effects in the electrolyte solution in a mean-field continuum way. We have shown recently, that this works well at the water/oil interface (which is electronically almost continuous), while the interfacial affinities of ions are somewhat overestimated at the (electronically discontinuous) water/vapour interface (see Vazdar *et al.*, *J. Phys. Chem. Lett.*, 2012, **3**, 2087).

Professor Geissler responded: Our motivation for studying fractionally charged ions was to highlight the interplay between different contributions to solvation thermodynamics at the air–water interface. This interplay is relevant to a wide range of ion charges (including the bare charge of a real ion) but is quite pronounced for values near 0.75*e*.

That such a charge reduction could be viewed as accounting for electronic polarizability is a very interesting observation. Leontyev and Stuchebrukhov have indeed provided a compelling argument to this effect in bulk solution, which as Pavel notes should also apply to the oil–water interface. In the electronically heterogeneous environment of the air–water interface, the grounds for this argument are less justified.

Professor Ben-Amotz added: As a follow-up to my previous question and comment to you, Phil, I wonder whether the Widom Potential Distribution Theorem, combined with Linear Response Theory, might help to resolve apparent contradictions between the results in your paper and those in the paper presented by Yan Levin. More specifically, there seems to be a difficulty associated with the fact that the surface capillary waves phenomena which plays a prominent role in your results does not appear at all in Yan's remarkably successful dielectric continuum theory for the influence of salts on the surface tension of water. The latter Linear Response theory is equivalent to assuming that the solute–solvent interaction energy has a strictly Gaussian probability distribution whose width is independent of solute–solvent coupling strength. This also implies various additional exact identities that relate experimental solvation free energies, entropies, and enthalpies, as well as solute–solvent and solvent–solvent contributions to the latter quantities.[1,2] More specifically, although dielectric continuum theory does not directly include energy fluctuations and the associated entropic contribution to solvation free energy, the correspondence to Gaussian solvation theory implies that the latter entropic contribution is related in a simple way to the mean solute–solvent interaction energy (see eqn (8) in ref. 2)* as well as to the entropy arising from solute-induced changes in solvent–solvent interactions (see eqn (12) in ref. 2). More specifically, in the notation of ref. 2, for a dielectric continuum solvation process $TS_{\mathrm{UV}} = -TS_{\mathrm{VV}} = E_{\mathrm{UV}}/2$ (where the subscripts UV and VV pertain to solute–solvent and solvent–solvent contributions to the corresponding entropy S and energy E) and so, in the notation that you have used above, $\langle \psi \rangle_1 = E_{\mathrm{UV}}$ and $-k_{\mathrm{B}}T \ln \langle e^{\beta\delta\psi} \rangle_1 = \mathrm{TS}_{\mathrm{UV}}$ and $\mathrm{DU}_{\mathrm{solv}} = \mathrm{E}_{\mathrm{VV}} = \mathrm{TS}_{\mathrm{VV}}$. The latter result also implies that the solvation free energy ΔG is equivalent to $E_{\mathrm{UV}}/2$, since $\Delta G = E_{\mathrm{UV}} - TS_{\mathrm{UV}}$. Thus, although the experimental solvation entropy ΔS necessarily vanishes in a dielectric solvation process, since $\Delta S = S_{\mathrm{UV}} + S_{\mathrm{VV}} = 0$, the entropic contribution to the solvation free energy, $-TS_{\mathrm{UV}}$, need not be equal to zero in a linear response solvation process. So, getting back to the issue which motivated this comment, I wonder if Yan's dielectric continuum theory may implicitly include the influence of surface capillary wave pinning phenomena, in the sense that the associated solvent–solvent entropy change is implicitly related to the solute–solvent interaction energy as follows: $S_{\mathrm{VV}} = -E_{\mathrm{UV}}/(2T)$.

* Note that in a dielectric continuum solvation process the quantity $E_{\mathrm{UV}}{}^{0}$ in eqn (8) of ref. 2 necessarily vanishes, as this energy pertains to the (hypothetical) interaction energy between the charged solute and the surrounding dielectric continuum which has not yet been polarized by the solute.

1. F. O. Raineri, G. Stell and D. Ben-Amotz, *Mol. Phys.*, 2005, **103**, 3209–3221.
2. D. Ben-Amotz and R. Underwood, *Acc. Chem. Res.*, 2008, **41**, 957–967.

Professor Levin answered: This is an interesting point. It is possible that the capillary contribution is already accounted for in our cavitational energy. Of course the Gibbs adsorption isotherm does already include all the fluctuations in it, since it is an exact thermodynamic relation. The fact that the PMF which we calculate agrees with the *ab initio* simulations and leads to correct surface tensions suggests that all the important factors have been taken into account.

Professor Geissler responded: The physical underpinnings of capillary wave theory are quite distinct from those of Dielectric Continuum Theory (DCT). DCT describes the response of a solvent fluctuating polarization field to electrostatic forces exerted by a solute. In conventional applications, such as those described in ref. 1 & 2, the boundary of the liquid phase is fixed (typically as a sharp, flat plane), which permits exact or nearly exact solutions.

In reality, the interface between liquid and vapor is soft, undergoing substantial fluctuations in shape. Surface tension sets the magnitude of these undulations at large scales; at molecular scales, their amplitude is controlled by more detailed interactions.[3] The adapted version of capillary wave theory discussed in our paper considers the response of such fluctuations to the presence of a solute. The thermodynamics of this response could not be predicted from DCT models that lack the corresponding degrees of freedom.

Put another way, DCT does not by itself provide a means for studying density fluctuations and response. In fact, the internal free energy of a dielectric medium of fixed volume (in the absence of an external field) is insensitive to its shape or topology – the basic physics underlying surface tension is not accounted for. This lack is not a significant issue in the bulk interior of a liquid like water at ambient conditions, where density fluctuations are very restricted. Near the interface, density fluctuations in a microscopic probe volume are unquestionably considerable. How strongly they influence interfacial solvation of real ions is still unknown. Our work demonstrates the possibility for significant contributions to solvation thermodynamics that could not be captured by conventional approaches like DCT.

1. Y. Levin, *Phys. Rev. Lett.*, 2009, **102**, 147803.
2. Y. Levin, A. P. D. Santos, and A. Diehl, *Phys. Rev. Lett.*, 2009, **103**, 257802.
3. A. J. Patel, P. Varilly, S. N. Jamadagni, H. Acharya, S. Garde, and D. Chandler, *Proc. Natl. Acad. Sci. U. S. A.*, 2011, **108**, 17678.

Professor Geissler continued: While capillary wave theory may overestimate quantitative measures of surface roughness, molecular simulations point strongly to the fact that roughness remains considerable down to the molecular scale. This qualitative feature is apparent for both *ab initio* approaches and empirical force fields.

Professor Bain noted: Just to add to Professor Geissler's response, the viscosity does not appear in capillary wave theory. Capillary wave theory has been used to predict the coefficient of ellipticity of water surfaces (an optical property measured by ellipsometry) and the results are in semi-quantitative agreement with experiment – typically overestimating the coefficient of ellipticity by about 20%. (J. Meunier, *J. Phys.*, 1987, **84**, 1819). Interestingly, agreement is particularly poor for water (an error of 40%) implying that capillary wave theory significantly overstates the roughness of a pure water surface.

Dr Vila Verde remarked: The capillary wave spectrum at the air–water interface is different from that at a water–solid hydrophobic interface or at a water–liquid hydrophobic interface. Given this, and the importance of the entropic contribution of capillary waves for the distribution of ions at the air–water interface, it is conceivable that ion distributions differ between these three types of aqueous interface because of differences in their capillary wave spectrum. Would it be interesting/possible to investigate this possibility using your generic model?

Professor Geissler added: This is an interesting idea. In the context of our reduced lattice model, an oil–water interface does not really differ from an air–water interface. Only the meaning of nearest-neighbor interaction energies changes (in the latter case reflecting the total cohesive energy of adjacent liquid cells, and in the former case instead a difference of oil–oil, oil–water, and water–water interactions). Introducing a hard but attractive wall imposes more subtle constraints on capillary fluctuations that are worth exploring. (For a non-attractive wall, the liquid would simply dry and form a free liquid–vapor interface.)

Professor Tiddy opened the discussion of the paper by Colin Bain: In the system with 6% dCTAB present, the neutron scattering data suggest that there is 30–40% CTAB in the surface layer. This seems a bit large – how robust is the model used for the neutron data? I note that signal was only above the background level at low values of q. On a wider subject, what is the relevance of the data presented for understanding Hofmeister effects?

Professor Bain responded: The neutron reflection experiment is at the limit of detection of the deuterated surfactant, and so the precision in the surface composition is not high. What is clear is that the monolayer is enriched in CTAB compared with the bulk, despite the fact that under the conditions of the experiment we do not believe that charged micelles are able to adsorb. The likely explanation is that the exchange rate of surfactant molecules between micelles and solution is more rapid for CTAB than for $C_{16}E_8$. Adsorption of monomers to the surface depletes the subsurface of monomers and causes the micelles to shed more monomers. If the release rate of CTAB is higher than for $C_{16}E_8$, the monolayer will be enriched in CTAB. On the wider question, if our suggestion is correct and only uncharged micelles adsorb in the presence of ionic surfactants, then Hofmeister effects do not play a role in the absence of electrolyte. In the presence of added electrolyte, the adsorption kinetics are ion-specific and so ion-pairing or ion adsorption to the micelle surface play an important role.

Professor Kunz opened the discussion of the paper by Yan Levin: Could you comment on the importance of dispersion forces at the air–water and the oil–water interface and the correlation between surface properties and bulk properties, *i.e.* the viscosity?

Professor Levin responded: Our calculations suggest that dispersion forces do not play significant role at the air–water interface. On the other hand they are very important at the oil–water interface. The reason seems to be that once an ion partially penetrates into oil, there is a significant dispersion interaction with the oil molecules. This explains why the excess surface tensions of salts such as KI are actually negative. For more discussion of this topic please see: A. P. dos Santos, and Y. Levin, Ions at the water–oil interface: Interfacial tension of electrolyte solutions, *Langmuir*, 2012, **28**, 1304.

We find that there is a strong correlation between the kosmotropic/chaotropic nature of the ion at the interface and its viscosity B-coefficient. The kosmotropes ($B > 0$) are repelled from the interface while the chaotropes (B negative) are adsorbed to the interface of the hydrophobic surface.

Professor Kunz asked: As shown in the contribution by Baer and Mundy, in the case of iodate, iodide, and thiocyanate, specificities can be really very specific. I believe that these specificities can even be dependent on the counterion. If this is the case, there is no hope to get a general theory that copes with such subtleties. Therefore, it is important to admit that there is some limit to a general theory of specific ion effects.

Professor Levin answered: For the hydrophobic surfaces and interfaces that we have studied so far, the Hofmeister effect seems to be dominated by the anion. The counterion does not seem to play a significant role. For surface tensions the dependence on the anion is much stronger than it is on the cation. I am not sure that this will always be the case. The Hofmeister seems to be a subtle interplay between ionic hydration, polarizability, electrostatics *etc.* Hydration is very difficult to understand from first principles, and we do not attempt to do this. Instead we observe the correlation between viscosity B-coefficients and hydration of ions near the hydrophobic surfaces. Ions with positive B remain hydrated (and we need to use their hydrated radii in our theory, which are listed in physical chemistry tables) and the ions with negative *B* lose their hydration shell near an interface (and we use their crystallographic size in our theory). This is all we need to calculate very accurately the surface and interfacial tensions of 10 different electrolytes. I am not sure that for proteins it will be possible to come up with a simple theory, however, understanding the physical mechanisms that are behind the Hofmeister effect will help us to come up with better water and ion models which can then be used to simulate complicated protein–electrolyte systems.

Professor Lopes opened the discussion of the paper by Christopher Mundy: We have been developing atomistic force fields for molecular ions for a long time (to be used in the context of classical MD simulations of ionic liquids (*J. Phys. Chem. B*, 2004, **108**, 2038; *J. Phys. Chem. B*, 2004, **108**, 16893; *J. Phys. Chem. B*, 2006, **110**, 19586; *J. Phys. Chem. B*, 2008, **112**, 5039; *J. Phys. Chem. B*, 2010, **114**, 3592). These parameterizations have allowed us and others to describe the static properties and structures of ionic liquids and their mixtures with different solvents in the bulk phase. Recently we are embarking in a refinement process using polarizable models that hopefully will allow us to extend the scope of the model to interfaces and dynamic properties. I was wondering if you could comment on the differences that you have observed between using an *ab initio* based method like the one you have shown and the use of polarizable classical models, specially in the case of molecular ions (like thiocyanate) and also in the case of ions near the liquid/vacuum interface?

Dr Baer replied on behalf of Dr Mundy: In the case of iodide, the only ion we have extensively studied so far, under bulk and interfacial solvation using empirical polarizable force fields and DFT (*J. Phys. Chem. B*, 2010, **114**, 12926–12937; *J. Phys. Chem. Lett.*, 2011, **2**, 1088–1093) we find that the differences in the interaction potentials are brought to full relief in the vicinity of the broken symmetry of an interface. Experimentally (EXAFS) detectable differences are small under bulk solvation. The radial distribution functions, which are often used to judge the performance of pair potentials to reproduce structural properties, for both potential models are nearly indistinguishable under bulk solvation. Surface solvation structures on the other hand are different. Similar findings were already observed for uncharged polyatomic solutes like SO_2 (*J. Phys. Chem. B*, 2010, **114**, 7245–7249).

Professor Geissler asked: Your paper presents calculations of free energy as a function of an ion's distance *r* from the center-of-mass of a small water droplet. These potentials of mean force were estimated from umbrella sampling in a series of ~20 ps molecular dynamics simulations focusing on different ranges of *r*. My question is whether these trajectories are long enough to generate trustworthy free energy profiles. In particular, have you calculated correlation times for quantities like the fluctuating force experienced by the ion?

In analogous simulations of empirical water models (such as SPC/E) in a liquid slab geometry, we have found such correlation times to exceed 1 ns. In that case, potentials of mean force estimated from 20 ps trajectories (amounting to a single statistically independent sample at each ion height) would be very likely to generate misleading results. I recognize that your *ab initio* calculations are extremely taxing by comparison, and I applaud the effort to explore subtle electronic structure effects on ion adsorption. For the purpose of evaluating the success of various theories, however, it might be important to acknowledge the potentially anecdotal nature of these free energy calculations. Alternatively, it would very interesting if correlation times were considerably shorter for small water droplets than for extended interfaces (whose surface shape fluctuations are clearly very different).

Dr Baer responded: The concern with the length of our trajectories is certainly noteworthy. The honest answer is that this is all we can do at this point. The cluster

PMFs for the halides are likely more difficult to converge over the slab geometry for many reasons. The first and foremost reason being the lower temperature (150 K *vs.* 300 K for slab calculations). The force correlation time is a nice idea, especially in comparing the classical models (which are frustrated charge models) to interactions based on DFT. Understanding the different statistics in the forces between these very different models would help in understanding the details of our results. Nevertheless, the results we have presented seem to be consistent with the clear trend in the Hofmeister series. This we believe is robust.

Professor Tiddy said: Your slide shows a region of green for the "hydrophobic" water. Is this really hydrophobic or is it just water with no H-bonds to the solute?

Dr Baer answered: The region marked as hydrophobic illustrates a region in which no hydrogen bonds are formed between water and the sulfur. The solvation structure, orientation of water molecules, around the sulfur resembles that found for classical point charge models around small hydrophobic particles [*Mol. Phys.*, 2007, **104**, 2465–2476].

Professor Tiddy addressed Colin Bain: Is there a "discreteness of charge effect" in the case of $C_{16}EO_6$ micelles where very long rods are formed? The estimated value of 10 for the micelle aggregation number of this surfactant is surprising!

Professor Bain answered: The estimated aggregation number is 1000, not 10 – that was just a typesetting error. Professor Tiddy is right that a mean field model is unlikely to be adequate when the dimension of the micelle is comparable to or greater than the Debye length. We have previously considered the adsorption of vesicles formed from mixtures of cationic and anionic surfactants and bearing a small net charge. If the charges are mobile then, with the aid of the Derjaguin approximation, we can treat the force between the vesicle and the surface in terms of the well-known problem of the force between two charged plates at constant potential. Details may be found in J. P. R. Day, R. A. Campbell, O. P. Russell and C. D. Bain, *J. Phys. Chem. C* 2007, **111**, 8757. A similar approach could be taken here. If the charged surfactant in the micelle is mobile and therefore able to migrate within the micelle to regions of lower potential, the adsorption rate of micelles would increase. Such an effect would explain why our estimate of the aggregation number is less than the value obtained in the SANS measurements of Cummins *et al.* (ref. 31 in our paper).

Professor Tiddy remarked: The surfactant $C_{16}EO_6$ has a cloud point close to room temperature. In the region just below this there are critical concentration fluctuations, with related fluctuations in micelle size and shape. Will these influence your results?

Professor Bain responded: The reported cloud point for $C_{16}E_6$ is 36 °C and the Krafft point is 28 °C (M. Kahlweit, R. Strey and D. Haase, *J. Phys. Chem.*, 1985, **89**, 163 and ref. 31). Our experiment was actually conducted at 25 °C, which is just below the Krafft point, but the solution remained clear throughout the experiments. Ideally, the experiments would be repeated around 30 °C. The solution was prepared by gentle warming overnight and then cooling to room temperature. It is quite common for dilute micellar solutions to remain metastable when cooled just below the Krafft point. We have not studied $C_{16}E_6$ in the immediate vicinity of the cloud point, but have looked at the temperature dependence of $C_{12}E_4$ in the OFC as the cloud point is approached and, surprisingly, found very little effect of temperature (G. Geary and C. D. Bain, unpublished results).

Mr Patko addressed Dr Baer: My question is regarding the results achieved and their dependence upon the methodology chosen, and is in three parts:

1. Were other DFT functionals explored as well as BYLP, particularly some of the newer meta-GGA types?

2. To what extent were the improvements in results attributable to the Grimme dispersion correction? Which Grimme dispersion correction type was used and how was the particular version selected/tested?

3. As the simulation time for this QM/MM type run was shorter than could be achieved with classical MD, is there a reason (*i.e.* prior art) to believe that less simulation time is needed to achieve sufficient convergence for such QM/MM type runs and still get derived thermodynamic data with good confidence?

Dr Baer replied: 1. Even though it is highly desirable to understand the limitations of GGA-functionals in the condensed phase description of water and ion-solvation, we have only investigated GGA-functionals (BLYP and PBE) so far, concentrating mainly on water structure under ambient conditions.

2. We used the second generation dispersion correction. The main reason is the improvement of the water density and the softening of the oxygen–oxygen distribution function which we extensively tested for the used slab configuration (*J. Phys. Chem. B*, 2009, **113**, 11959–11964; *J. Chem. Phys.*, **135**, 124712). Also the estimated self-diffusion coefficient is closer to the experimentally observed one, but it is still not as good as the empirical potentials. There are clearly still a lot of open questions regarding the full phase diagram of DFT water and how the dispersion corrections affect different properties.

3. Of course we are concerned with the length of the runs (see my response to Professor Geissler, above). Your question on whether empirical potentials will converge "faster" or "slower" than DFT based simulations is interesting. At this point we don't have any direct evidence for either (see again my response above); we are currently thinking of ways to test this.

Dr Vila Verde asked: Can you define more precisely what you mean by "flexibility" and "strength of hydration"?

Dr Baer replied: Flexibility describes to what extent the first solvation shell around the ion can structurally change in the vicinity of an interface without changing the coordination number of water molecules, whereas the strength of hydration describes the exchange of water molecules between the first and second solvation shell. We use the solvation structure around the simple spherical halides as a reference. Fluoride is considered a very strong hydrated ion; on the timescales of the DFT simulations there is no exchange between the first and second solvation shell. Even under surface solvation the tetrahedral solvation is kept, therefore the solvation shell is characterized as not flexible. On the other side of the spectrum is iodide, which shows exchange of water molecules and a rather large structural rearrangement under surface solvation without changing the coordination number of the water molecules.

Dr Vila Verde said: Would it be possible to quantify the strength of the hydrogen bonds in your system?

Dr Baer responded: There is no unambiguous way to define the strength of hydrogen bonds. We use the halide anions as reference allowing a characterization of hydrogen bonds based on their radial distribution function. Here we call a fluoride-like hydrogen bond as strong and iodide as weak, whereas chloride hydrogen bonds look very similar to water–water hydrogen bonds. This corresponds quite well with the shift of O–H stretches depending on the donor–acceptor distance that is used in IR-spectra for assigning spectral shifts. In principle it would be possible to estimate the spectral shifts of the O–H stretch (or bend) mode of first shell water molecules to estimate the strength of the ion-water interaction, but this would require a very different simulation protocol and much more sampling, which is not feasible at the moment (*J. Chem. Theory Comput.*, 2011, **7**, 2028–2039; *J. Phys. Chem. A*, 2012, **116**, 2466–2474).

Dr Vila Verde said: Can the residence times of waters in the hydration shells of ions be calculated, and serve as a quantitative indicator of the flexibility of hydration shells?

Dr Baer responded: The residence time of water molecules could be calculated and serve as an indicator for flexibility. Estimating residence times would require very long simulations, as you need to observe several events, not only re-crossing. As a good approximation, the radial distribution function gives an estimator for the barrier height for exchange, which is the reference we are using as discussed in the answer to your initial question.

Dr Henchman addressed Yan Levin: It has been predicted by theory (F. H. Stillinger and A. Ben-Naim, *J. Chem. Phys.*, 1967, **47**, 4431), measured experimentally (Q. Du, E. Freysz and Y. R. Shen, *Science*, 1994, **264**, 826) and determined from molecular dynamics simulations (M. A. Wilson, A. Pohorille and L. R. Pratt, *J. Phys. Chem.*, 1987, **91**, 4873) that hydrogens point preferentially towards the water side at the air–water interface because a smaller number of them point into the vacuum than would be expected in a uniform distribution. This of course occurs to maintain hydrogen bonds. Moreover, this picture is consistent with the negative surface potential at the air–water interface measured electrochemically. How can you maintain therefore that the surface potential could be negligible?

Professor Levin responded: I think there is some confusion about the definition of the electrostatic surface potential.

It is defined as the potential in water minus the potential in vacuum. The fact that SPC/E water produces a negative surface potential of about -0.6 V means that there is a dipole layer that is formed at the surface with the positive charges on top (towards the vacuum) and negative charges on the bottom (towards the bulk water). Since with the point charge models the positive charge comes from the H^+ this means that there has to be an excess of hydrogens pointing towards air. Note that this situation is opposite to what is found using *ab initio* DFT; in which case it is the valence electrons which spill into vacuum producing an inverted dipole layer (negative charge on top, vapor) and positive charge towards the bulk water. This is the reason why the electrostatic surface potential predicted by DFT has an opposite sign to the one predicted by the classical point charge models. There is no electrochemical method for measuring the surface potential. There are only indirect ways to judge its sign. However, the high energy electron holography measurements confirm the predictions of DFT. Maybe Marcel Baer can comment more on this point, since he is one of the people who has done the DFT simulations. Why the large surface potential predicted by the DFT does not seem to translate into a significant contribution to the PMF for ionic adsorption (which it does not, as the DFT simulations show clearly) is a very interesting point which is now under investigation. For a preliminary answer please see my reply to Pavel's question below.

Dr Cwiklik addressed Phillip Geissler: What is the role of model dimensionality and the number of nearest neighbors on the extent of surface undulations? Are you expecting surface undulations to be suppressed when you switch from the two-dimensional to a three-dimensional model?

Professor Geissler answered: The purpose of our paper is to show in general terms how fluctuations at a liquid–vapor interface can impact the thermodynamics of solute adsorption. While we present results only for a two-dimensional lattice system, we have found no qualitative differences in the case of three dimensions. Regardless of dimensionality and lattice geometry, surface undulations can be significantly suppressed when a solute is transferred from the bulk to the interface, leading to a decrease in entropy.

Professor Levin added: One has to take some care when using lattice models. For example how large is the ion compared to the lattice spacing? In lattice models with nearest neighbor exclusions one already finds an entropically driven phase transition, see for example: H. C. Marques Fernandes, J. J. Arenzon and Y. Levin, Monte Carlo simulations of two-dimensional hard core lattice gases, *J. Chem. Phys.*, 2007, **126**, 114508.

Professor Geissler answered: Lattice models have played a prominent role in advancing understanding of the statistical physics of liquids. Their representation of microscopic fluctuations, however, is indeed schematic and in some ways artificial – caution and healthy skepticism are warranted.

The basis for studying fluctuations in interfacial shape using lattice gas models is well established.[1] The standard physical perspective is based on a notional coarse-graining of molecular configurations, in which one resolves in each microscopic region whether the instantaneous density is liquid-like or instead vapor-like. Away from the critical point this distinction is unambiguous, even on scales little larger than a molecular diameter. Adjacent liquid cells contribute a cohesive energy. This simple representation is sufficient to capture the basic physics of surface tension and capillary waves.

In the context of interfacial solvation, the most limiting aspect of such a model is a lack of flexibility for describing a solute's size and shape. Our paper considers solutes that are exactly one lattice cell in size (and that interact only with nearest neighbor cells). We consider this simple case sufficient for an initial exploration of solvation mechanisms. Capturing the detailed behavior of real solutes would certainly require adapting this approach. One route is to take a continuum limit, in which lattice variables represent density fluctuations in ever-smaller physical volumes. A similar limit is implicit in dielectric continuum theory, whose physical basis is most defensible as a coarse-graining of fluctuations in collective molecular polarization. Standard continuum models of capillary waves have in effect already taken this limit. They can be amended to include solutes of arbitrary size and shape, and are better suited to mathematical analysis than numerical simulation.

The entropy associated with fluctuating spatial patterns of liquid density is an important contribution to lattice gas thermodynamics. Beyond the relationship to this physically meaningful contribution, the relevance of entropic phase transitions in systems of lattice-bound volume-excluding particles (such as those described in ref. 2) is not apparent to us. By construction, our lattice gas solvent resides at phase coexistence, so that an interface between liquid and vapor is stable. At low solute concentrations, we do not foresee the emergence of additional equilibrium phases.

1. J. Weeks, *J. Chem. Phys.*, 1977, **67**, 3106.
2. H. C. M. Fernandes, J. J. Arenzon and Y. Levin, *J. Chem. Phys.*, 2007, **126**, 114508.

Professor Dr Fedorov added: The results presented in the paper provide very useful insights into the microscopic mechanisms of solute adsorption at the liquid–vapor interface.

However, I think that (in general) extrapolation of results obtained by models operating with non-smooth Hamiltonian functions (*e.g.* lattice gas or hard sphere models) should be done with some care because real molecules are 'soft' and interact *via* smooth potentials. The non-smooth interaction potentials make the underlying physics behind the lattice gas of hard sphere models *qualitatively* different to the physics of 'soft' molecule interactions. For instance, comparing models with different Hamiltonians we observed that a model system with non-smooth particle–particle interactions can undergo a phase transition at an interface (*e.g.* crystallization near a liquid/solid interface) that is not observed in a similar system with the same geometry and at the same state parameters but with smooth particle–particle interactions (the results were not published). I note that this is just a comment and I overall highly value the presented work.

Professor Dr Fedorov continued: I would like to make a general comment (as a small warning) about using models with non-smooth Hamiltonians for modelling behavior of condensed molecular systems.

Simple models with non-smooth Hamiltonians (like lattice gas models or hard sphere/disk models) are quite popular within the theoretical science community as 'toy models'. In several areas of solvation sciences these models provide very useful insights into microscopic mechanisms of different phenomena in liquid systems and they help experimentalists with qualitative interpretations of their observations.

However, in some cases interpretation of results obtained by these models in a view of *quantitative* description of complex real-world phenomena should be done with some care. The reason is that real molecules are believed to be 'soft' and interact with each other *via* smooth potentials. As I mentioned in my previous comment, the non-smooth interaction potentials make the underlying physics behind the lattice gas of hard sphere models to be in several aspects *qualitatively* different to the physics of models with 'soft' molecule interactions (which are somewhat closer to reality, I believe). I also would like to note that certain averaged properties of interfacial systems and systems in frustrated environments can be quite vulnerable to the non-smoothness of the interaction potentials (particularly in systems with coulomb charges) because the non-smoothness strongly affects interfacial correlations in these systems. Saying all of that, I still think that the simple non-smooth Hamiltonian models have (and I hope will have) a large niche of applications in theoretical/computational physical chemistry and chemical physics of liquid and interfacial systems due to their simplicity and the rigorous mathematical and theoretical background behind many of them.

Dr Ball addressed Yan Levin: It's been suggested recently that hydroxide ions accumulate preferentially at the air–water interface. That's still controversial, but how does hydroxide fit into your picture?

Professor Levin replied: The excess surface tensions of bases are very similar to salts with kosmotropic anions. In this respect, our theory predicts that hydroxide should behave very similarly to a kosmotropic anion and should be repelled from the air–water interface. However, we have not made a detailed study of hydroxide.

Professor Jungwirth asked: How can you ignore the effect of the surface potential of water on the interfacial behavior of ions? For bulk hydration it is well known that anions hydrate better than cations of the same size. This is partly due to the fact that water molecules get closer to anions (to which they point their small H-end) than to cations (to which they point their large O-end), but partly also due to orientation of interfacial water molecules at a cavity boundary. Similarly, at the water surface, there are dangling water H-atoms which will stabilize anions better than cations. This is likely not to be the most important surface effect on ions, but it may still be non-negligible.

Professor Levin responded: This is a very good point. In our polarizable anion dielectric continuum model we neglect the surface potential of the neat air–water interface. This is justified *a posteriori* by the good agreement that we get with the measurements of the surface tensions and the excess surface potential of different electrolytes. Later *ab intitio* DFT simulations of Baer and Mundy have shown that the potential of mean force (PMF) predicted by our theory is in excellent agreement with their simulations for iodide.

For some reason, which is not completely clear at the moment, the electrostatic surface potential of the neat-air–water interface does not seem to play any role in anionic adsorption. This is very different than what happens in classical simulations in which the surface potential does play a very significant role, see: M. D. Baer, A. C. Stern, Y. Levin, D. J. Tobias and C. J. Mundy, Electrochemical surface potential due to classical point charge models drives anion adsorption to the air–water interface, *J. Phys. Chem. Lett.*, 2012, **3**, 1565.

In fact, I think that the surface potential present in point charge water models is one of the reasons why these models produce so much adsorption when ionic polarizability is included.

In these models, the surface potential due to partial orientation of water molecules at the interface produces a strong electric field which favors anionic adsorption. Note that this is a very large energy, ~20 kT. Of course the ion does not feel the full surface potential. The reason is that to transfer an ion from air into water one needs to create a cavity. The surface potential across the cavity–water interface partially cancels out the surface potential across the water–air interface. In SPC/E model this cancellation still results in a 10 kT energy gain when anion moves into air. This is the reason why small variations of the ionic size or a reduction of their charge can make anions adsorb to the interface (even without ionic polarizability). However, the recent simulations of Baer and Mundy show that within the DFT the cancellation (cavity–water and water–air) is almost perfect and the surface potential does not play any role for anionic adsorption.

What drives anions towards the surface is the cavitational free energy gain. For hard non-polarizable ions this is quite small, compared to the electrostatic self energy penalty of exposing the ionic charge to the low dielectric air environment. However, for polarizable ions the cavitational and the electrostatic energies become very similar, producing a small shallow minimum in the potential of mean force, in agreement with the DFT calculations. In any case, both the DFT and our polarizable anion dielectric model seem to indicate that the electrostatic surface potential of the air–water interface does not play any significant role for anionic adsorption.

Professor Geissler noted: The surface potential V, as conventionally defined, is not necessarily relevant to ion adsorption, or more generally to the solvation of any volume-excluding object. In particular, V is determined by an unbiased spatial average, which includes contributions from the electrostatic environment **within** solvent molecules. The solutes under discussion of course do not explore these excluded volumes.

As a more pertinent electrostatic property of the air–water interface, one could consider a potential V' averaged only over spatial regions accessible to solutes of realistic size. Computed values of V' for molecular mechanics models like SPC/E are much smaller in magnitude than those of V. This issue has been discussed recently in depth and detail by Kathmann and coworkers.[1]

1. S. M. Kathmann, I. W. Kuo, C. J. Mundy and G. K. Schenter, Understanding the surface potential of water, *J. Phys. Chem. B*, 2011, **115**, 4369–4377.

Professor Levin remarked: Many classical point charge water models predict existence of a −500 mV surface potential. On the other hand *ab initio* DFT show that the potential is +3 V. However, if coarse graining is performed this potential will actually be close to zero – which does not happen with classical water models, for which the potential remains at a few hundred mV. 500mV translates to about 20kT in free energy. This potential can strongly influence the ionic distribution near hydrophobic surfaces and protein. If the DFT results are correct and this potential is either positive or zero, it will mean that many of the predictions obtained using classical water models such as SPC/E might be wrong. Any comments?

Professor Jungwirth responded: The surface potential contribution is to the free energy of hydration, *i.e.*, to the difference between the free energy of the ion in gas phase and aqueous bulk. For distribution of ions at a water surface it is the difference between the free energy at the surface and in the bulk which is decisive and it is *a priori* not clear how big the contribution of the surface potential is in this case.

Professor Lynden-Bell addressed Professor Levin and Professor Madden: I would like to comment on different electrostatic potentials and their applicability. You have calculated the change in the Poisson potential across the interface by including the ionic concentrations (from the Poisson–Boltzman distribution) and the orientations of the water dipoles. This is the potential that would be felt by an infinitesimal test charge which did not interact with the components of the real liquid. This potential can also be calculated from classical and *ab initio* simulations, but these two methods do not measure the same quantity. Classical force fields are parametrised to give a reasonable description of the electrostatic field outside the molecule, but do not attempt to describe the electrostatic field within the molecule. Thus the change in Poisson potential across an interface corresponds to an attempt to describe the change in potential felt by an infinitesimal non-interacting test charge passing from the gas phase to a point in the liquid between molecules. This is the same, I believe, as the quantity that you have calculated. In an *ab initio* calculation the Poisson potential includes the contributions from within molecules and so will be different. However in most circumstances neither of these potentials is, I believe, relevant. What one needs is the change in potential experienced by a real ion passing into the liquid, interacting with the liquid molecules and changing the local liquid structure. This is what Madden (O. J. Lanning and P. A. Madden, *J. Phys. Chem. B*, 2004, **108**, 11069) termed the Madelung potential, which he showed has a smoother behaviour than the Poisson potential near an interface.

Professor Levin responded: I agree completely with the above comment. The excess electrostatic potential that is obtained from our PB calculations is for a point charge moving across the interface. This, is what was measured in the experiments of Frumkin and Jarvis and Scheiman. On the other hand if one wants to study the influence of the neat water–air surface potential on ionic adsorption one has to worry precisely about the effects mentioned by Professor Lynden-Bell. In fact, we have recently shown that the electrostatic surface potential produced by the partial orientation of water molecules at the interface plays an important role for adsorption of ions in classical water models such as SPC/E, see M. D. Baer, A. C. Stern, Y. Levin, D. J. Tobias and Christopher J. Mundy, Electrochemical surface potential due to classical point charge models drives anion adsorption to the air–water interface, *J. Phys. Chem. Lett.*, 2012, **3**, 1565. The problem is that the *ab initio* DFT simulations show that the real potential of the air–water interface has an opposite sign to the one predicted by the classical water models. This means that this potential, if it really exists, should favor the repulsion of anions from the interface. How to reconcile this with the results of the classical models is not completely clear. What seems to be evident from our calculations is that real ions do not feel the full surface potential. The reason seems to be precisely the kind of interior field produced from the cavity into which the ion is being transferred. This water–cavity potential partially cancels out the water–air potential.

Professor Madden replied: I would like to elaborate on the part of Professor Lynden-Bell's remark which concerns the roles of different electrostatic potentials. As she remarks, the surface potential is the drop in the "Poisson" potential across the interface. This is the potential experienced by a microscopic test charge and is the potential which is needed to connect with macroscopic electrostatic experiments, for example, the electrical capacitance of the interface or the effect of an applied potential on the surface tension. However, it is not the potential needed to calculate the distribution of ions in the liquid. An ion in the liquid responds to the potential caused by all the other ions present, whereas the Poisson potential involves a contribution from all ions, including the one for which we would like to understand the energetic and forces. The significance of the difference can be illustrated by some simulations of molten salts close to a metallic surface (*J. Chem. Phys.*, 2008, **128**, 124701). Because of the packing of the ions at the surface and the need to achieve local electroneutrality the ion densities are significantly oscillatory, which gives the Poisson potential a strongly oscillatory dependence on the distance to the surface with an amplitude of the order of a volt. However, the energy difference between ions of different charge at a given distance from the electrode shows none of this oscillatory character (if the Poisson potential determined the local energies of ionic charges it would be expected to be on the eV scale). Instead of the Poisson potential one can calculate the "Madelung" potential, which is the mean potential experienced by an ion due to all other charges in the system. This shows none of the oscillatory character and is a much better tool for interpreting the energetic of the ions. I would therefore argue that the surface potential at the liquid–vapour interface is a consequence of the ordering of ions there, but should not be thought of as determining that ordering.

Professor Geissler added: The potential described by Professor Lynden-Bell has been evaluated for empirical force fields for liquid water. It is much smaller in magnitude than the conventional surface potential calculated for the same models.

Dr Winter addressed Colin Bain: Interpretation of the diffusion and adsorption mechanism is based on measured surface excess, which however does not reveal the actual composition of the surface monolayer. Have you considered using surface-sensitive and element specific spectroscopy, such as photoemission to identify the nature of adsorbed species?

Professor Bain responded: Clearly, it is desirable to determine the surface composition. The difficulty is that the surface excess of the ionic species is at most a few percent of a monolayer, which poses sensitivity issues for most spectroscopic techniques. The neutron reflection experiments that we report in our paper used deuterated-CTAB and therefore are chemically sensitive, but the signals are close to the limits of detectability and the error on the composition is therefore large. Reflection infrared spectroscopy is possible on the surface of the OFC (see, for example, R. A. Campbell, S. R. W. Parker, J. P. R. Day and C. D. Bain *Langmuir* 2004, **20**, 8740) but the sensitivity is only around 5% of a monolayer. Photoelectron spectroscopy is difficult on an overflowing cylinder owing to the requirement for vacuum conditions, but surfactant adsorption kinetics can also be studied in a liquid jet (T. Battal, C. D. Bain, M. Weiss and R. C. Darton, *J. Colloid Interface Sci.*, 2003, **263**, 250). The timescales of a liquid jet are shorter than an overflowing cylinder, so surfactants have to be selected that show kinetics on the appropriate timescales. With this proviso, it should be possible to study adsorption kinetics by photoelectron spectroscopy on a liquid microjet (R. Seidel, S. Thurmer and B. Winter, *J. Phys. Chem. Lett.*, 2011, **2**, 633) and we would be interested in exploring this avenue. Total reflection XAFS spectroscopy utilising the total ion yield is compatible with ambient conditions and could be applied to an overflowing cylinder. Aratono and coworkers have used this technique to study counterion binding to surfactant monolayers under equilibrium conditions (K. Kashimoto, K. Shibata, T. Matsuda, M. Hoshide, Y. Jimura, I. Watanabe, H. Tanida, H. Matsubara, T. Takiue and M. Aratono, *Langmuir*, 2008, **24**, 6693). One issue when applied to our systems is that the Debye length is greater than the penetration depth of the X-rays, so determining the surface excess of CTAB by counting the bromide counterions is difficult. Consequently, one might have to use special surfactants containing within their chemical structure elements suitable for XAFS measurements.

A related, but even more difficult problem to solve, is the nature of the subsurface – both the chemical composition and the distribution of the surfactant between micelles and monomers. If our model is correct, micelles should be present in the sub-surface layer at higher bulk surfactant concentrations and in the presence of ionic surfactants. We are not aware of any experiments with the sensitivity and spatial resolution to probe the bulk surfactant distribution within the first hundred nanometers of the surface.

How specific are ion specificities? A pilot NMR study

Susanne Dengler,*[a] Angelika Klaus,[a] Gordon J. T. Tiddy†[b] and Werner Kunz[a]

Received 30th April 2012, Accepted 3rd May 2012
DOI: 10.1039/c2fd20090b

NMR quadruple splitting measurements are a useful technique to investigate the adsorption behaviour of ions near to charged surfaces in liquid crystals made of surfactants of different headgroups. In the present paper we show the differences in lithium and sodium adsorption on dodecyl carboxylate and dodecyl sulphate bilayers as a function of surfactant concentration and for two temperatures. To this purpose we mixed the pure surfactants and also mixtures of the lithium and sodium surfactants with octanol as a cosurfactant in appropriate ratios and concentrations to obtain liquid crystals. It turned out that the measured specific ion effects are much more pronounced in the case of carboxylates than for sulphates.

Introduction

Still today many papers start with the statement that specific ion effects and especially the Hofmeister series are not yet understood. However, such a global conclusion is no longer valid. Significant progress has been made over the last twenty years. We quote only few of the new insights:

- Collins' concept of matching water affinities[1] contributed much to a qualitative description of specific ion–ion and ion–headgroup[2] interactions. Even rough predictions of ion behaviours in complex systems are possible.
- Since the landmark work by Jungwirth and Tobias[3] it is clear that ions can adsorb into the first layer near the air–water interface, whereas the overall profile of the ion concentration shows a depletion, in agreement with the Gibbs adsorption isotherm.
- In a recent work[4] it was shown why different Hofmeister series may exist and why they can be partially or fully reversed, depending on the adjacent interface.

It seems to be clear now that the specific behaviour of an ion strongly depends on the ion's environment and especially on the counterions. This is not a trivial statement, since it explains why engineers have difficulties to define single ion parameters.

Having now a basic understanding of specific ion effects and the related subtle balance between different interaction forces, it is tempting to go deeper and deeper into the details of such subtleties. But the danger is that we get lost in these details and that it is more and more difficult to infer generalisable trends. It is evident that spherical (*e.g.* chloride), linear (thiocyanate) and flat (nitrate) ions must behave

[a]*Universität Regensburg - Institut für Physikalische und Theoretische Chemie, Regensburg, Germany. E-mail: susanne.dengler@chemie.uni-regensburg.de*
[b]*UMIST - Department of Chem. Engineering, PO Box 88, Manchester, M60 1QD, United Kingdom*

† Current Address: School of Chemical Engineering and Analytical Sciences, The Mill, The University of Manchester, Oxford Road, Manchester, M13 9PL, UK

differently at very short distances and that the structure of their hydration shells must be different, despite the same charge they bear. This is as true for ions as it is true for any other chemical species. Therefore, there is nothing to generalise when questions are asked that involve these details. As a consequence, related studies are only of limited value, because the results cannot be easily extended to other systems and situations.

Our conclusion is that it is important to answer the following two questions:

- How specific are ion specificities? In which cases are predictions possible from general rules, in which cases does the ion behave so individually that nothing can be predicted? Then only a particular experiment or simulation can help. We can formulate the problems in other terms: how close energetically are the states of different ions in similar systems? If they are very close, it is difficult or impossible to predict *e.g.* if there is a reversal of the ions in the Hofmeister series for that particular case or not.

- When do specific ion effects appear? From numerous studies it is well-known that at very low concentrations ion interactions are dominated by electrostatic interactions and the Debye–Hückel theory is the appropriate description. No or only minor specific ion effects are visible. At very high concentrations electrostatic interactions (now screened and therefore "only" short-ranged) also dominate. For example even chaotropic ions begin salting out at sufficiently high concentrations. It is the intermediate concentration range, in which specific ion effects appear and this is also the concentration range of biological relevance.

Whereas this fact is well-known for the concentration range of the ions, it is much less investigated for the concentration range of the charged counterparts. For example, it would be interesting to study the relevance of specific ion effects close to charged membranes or layers as a function of their charge density—with possible impact on the understanding of biological systems.

In the present paper we try to contribute to answer the first question. It is commonly accepted that according to Collins' ideas more cosmotropic ions interact more strongly with more cosmotropic (oppositely charged) headgroups and more chaotropic ions with more chaotropic headgroups. For example lithium ions should have a stronger propensity to the headgroups of alkylcarboxylate (cosmotropic) than sodium, and the contrary can be expected for alkylsulphate (a chaotropic headgroup).[1] The question is if the differences are comparably strong or if in one case the specificity is more pronounced than in the other case. A problem in the experimental determination of ion binding is to identify a well defined system where the measured parameters are directly related to ion binding, without other changes to the system.

To answer this question we performed NMR experiments in which we determined the quadrupole splitting of sodium (Δ_{Na}) and lithium (Δ_{Li}) in lamellar liquid crystals. Amphiphile lamellar (L_α) phases consist of alternating surfactant and water layers, see Fig. 1.

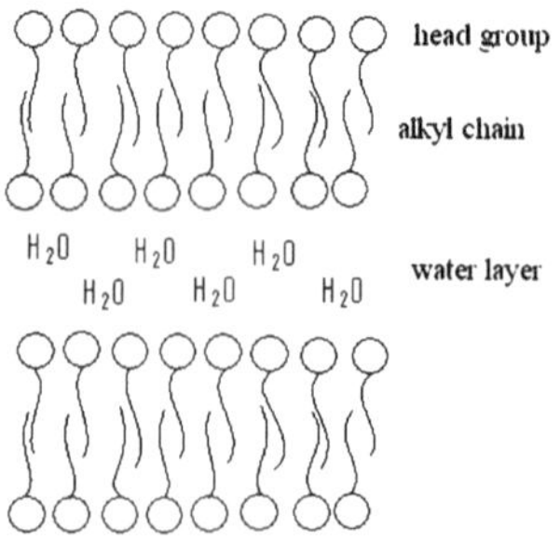

Fig. 1 A schematic picture of a lamellar phase.

The surfactant can be ionic, zwitterionic or non-ionic, while the water layers can vary in thickness from *ca.* 1–2 nm up to >10 nm. Within the surfactant layers it is easy to mix different amphiphiles such as ionic surfactants and uncharged cosurfactants. Thus it is possible to vary both the charge density of the amphiphile layers and the distance between opposing charged layers. NMR spectroscopy is a possible method to monitor the behaviour of ions in liquid crystalline phases. In anisotropic media (such as a lamellar phase) the spectrum of ions like Na^+, Li^+ and Cs^+ (and many more) which possess a nuclear quadrupole moment is split into a number of peaks rather than the single peak observed in isotropic solutions. In the simplest case, for "well behaved" systems, the magnitude of the separation between the peaks (the quadrupole splitting, Δ) is proportional to the fraction of ions in contact with the head groups (p_b).[5,6] Thus by measuring Δ it is possible to monitor how this fraction changes with composition, for example when competitive binding occurs (*e.g.* the displacement of sodium ions by lithium ions at an anionic surface). Both sodium and lithium ions possess a quadrupole moment, hence give multiple lines in the NMR spectra of anisotropic phases. Thus by selecting an appropriate lamellar phase system it should be possible to quantify the displacement of sodium ions by lithium ions from the changes in the Δ values of both. The objective of this initial study is to establish if a suitable system can be identified, where the changes in Δ reflect changes in ion-binding. If such a system can be found, then it should be possible to extend the studies to a very wide range of ions.

It is well known that lamellar phases are formed from mixtures of ionic surfactants with a long chain cosurfactant such as an alcohol.[7,8] We have selected two ionic surfactants for study, dodecyl sulphate and dodecyl carboxylate, with sodium and lithium as counterions and octanol as the cosurfactant. Our first step was to determine composition regions where lamellar phases occur in dodecyl sulphate/octanol and dodecyl carboxylate/octanol mixtures for a reasonable range of water concentrations (45–75%). Then the ^{23}Na and ^{7}Li Δ values were measured as a function of water concentration at a fixed ratio of octanol : surfactant. Finally we selected several fixed values of the water concentration and measured the ^{23}Na and ^{7}Li Δ values as a function of the Na/Li ratio to investigate the binding competition.

Theoretical background to NMR of counterions

The environment of ions in lyotropic liquid crystals is anisotropic, and in non-cubic phases this gives rise to the observation of resonance splittings for quadrupolar nuclei. The anisotropy causes an electric field gradient at the nucleus and some orientations of this field gradient (with respect to the liquid crystal axis) are favoured over others. Theoretical descriptions of NMR resonance splittings due to this orientation have been given in a number of articles[5,6,9] so only a brief outline is given here. The orientation gives rise to the presence of non-averaged electric field gradients, and nuclei having electric quadrupole moments (spin quantum numbers $I > 1/2$) interact with the field gradient at the nucleus, resulting in the splitting of the NMR resonance into $2I$ peaks. The frequency difference between adjacent peaks is termed the quadrupole splitting (Δ), and its magnitude is given by eqn (1)[6] for a powder sample (one with no alignment of the liquid crystal phase).

$$\Delta = \left| \sum_{\mathrm{i}} p_{\mathrm{i}} \nu_{\mathrm{Qi}} S_{\mathrm{i}} \right| \tag{1}$$

S_i, the order parameter describing the orientation of the fraction of molecules at site i (p_i), is given by $S_i = \langle 3\cos^2\theta_{DMi} - 1\rangle/2$. θ_{DMi} is the angle between the liquid crystal axis (the director) and the electric field gradient. $4\nu_{Qi}$ is equal to the quadrupole coupling constant. Since the organisation of the head groups on the surface of the bilayer is locally disorganised, at least to some extent, there will be a range of values for S_i. But the values of Δ are a weighted average of the values at the different i sites because rapid exchange on the NMR time scale (10^{-5} s) is found to occur. For

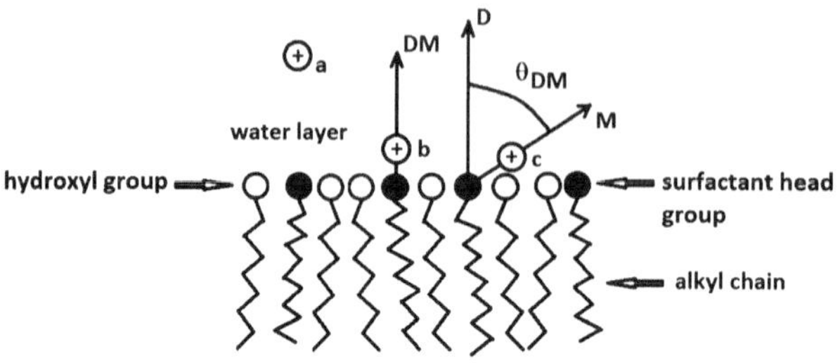

Fig. 2 Schematic drawing of the counter-ion binding at the lamellar surface according to the discussion in the text. The following three possibilities are shown: (a) the counter-ion is moving freely in the water layers. This location is characterized by $S = 0$. (b) The counter-ion is located symmetrically with respect to the amphiphile polar end-group. In this case the average angle between the director (D) and the electric field gradient (M) is equal to 0°. (c) The counter-ion is located between amphiphile polar head-groups.[11]

species which in the unperturbed state have a cubic or higher symmetry, such as hydrated ions, the electric field gradient at the nucleus can arise from the electric fields due to the charged amphiphile head groups and the distortions of the hydration sheath. Note that Δ can be positive or negative, according to the value of θ_{DMi} (see Fig. 2). The quadrupole coupling constant is a key parameter for analysing the absolute values of the Δ. This is not known. However, changes in the relative values can be taken to indicate changes in ion binding.[3,4]

In previous papers it has been shown that a simple model suffices to fit the observed Δ values for $^{23}Na^+$ in various liquid crystal phases. Sodium ions in contact with head groups ("bound", fraction p_b) have a finite Δ value (Δ_b) while those more than *ca.* 3–4 Å from the surface have Δ values of zero, ("free" fraction p_f) ($p_f + p_b = 1$). Previous studies of ionic surfactants have shown that the Δ values are broadly consistent with the "ion condensation" model, where ion binding (and Δ) is invariant over a fairly large water concentration. Because of the difficulties in obtaining the absolute values of the quadrupole coupling constant the method has fallen into disuse. But it should be ideal for monitoring competitive ion binding. In a mixture of two ions, A & B, if B displaces A then the Δ values of A will decrease on addition of B. Those of B will also decrease because the highest fraction of bound B ions occurs with small additions of B. Fortunately there are several cations that possess nuclear quadrupole moments. We have selected ^{23}Na and ^{7}Li for study because they have high sensitivity and can easily be measured with conventional multi-frequency high resolution spectrometers. Also, they are reported to have different specific binding capabilities with different anions.

We have chosen two surfactant systems for study, dodecyl sulphate/octanol and dodecyl carboxylate/octanol. These are thought to have very different specificities for Li or Na.[1] Octanol is employed as a cosurfactant because it is necessary for the formation of a lamellar phase. Also, changing the surfactant/octanol ratio allows the charge density on the surfactant bilayer to be varied. From the previous studies it is known that the octyl sulphate/decanol lamellar phase gives Δ values in the range 9–16 kHz, while the octyl sulphonate/decanol system gives 10–14 kHz. We selected dodecyl surfactants rather than the C_8 homologues to obtain a better defined bilayer. However, we had to employ octanol rather than decanol to prevent the occurrence of gel (L_β) phases.[10]

In contrast to the roughly constant Δ values for the sulphate and sulphonate systems, previous data for sodium Δ values in the sodium octanoate/decanol system demonstrated that the simple bound/free model did not fit the data, particularly at higher concentrations. Some compositions gave ^{23}Na Δ values that were much smaller, even approaching zero. These were highly temperature dependent, both increasing and decreasing with temperature (although not at the same time).

Changes in the fraction of "bound" ions could not account for the behaviour. It was necessary to invoke a second binding site with a very different value of Δ_b, one that was of opposite sign to that of the first binding site. The data were fitted with Δ_b values of +12 kHz for the normal ion binding site and −7 kHz for the second, which was thought to be from Na ions located between the head groups (Fig. 2). Low temperatures and higher surfactant concentrations favoured the second site. The data indicated that as the water layer thickness became smaller, the location of the bound ions altered, with the more ions moving into positions between the head groups. In the present study it is of interest to find out if any trace of this behaviour occurs for the sulphate system and if the Li ions show the same behaviour.

Lamellar phase—structure, charge density

The lamellar phase has a regular structure of repeating surfactant and aqueous layers, with the surfactant being organized in bilayers. Except for the restrictions on molecular configurations imposed by the layer structure, the system has liquid-like disorder. The repeat dimension of the structure can be measured by X-ray diffraction, hence the dimensions of the two layers can be calculated, provided that their densities are known.[12] In addition, the area per amphiphile can be calculated. Thus the charge density on the surface can be calculated precisely. Whilst we have not carried out X-ray measurements on the systems studied here, there is a substantial body of data available on similar systems.[7,8,13,14] These show that the area per amphiphile (a) for both sodium and potassium soaps in mixtures with decanol is of order 25–30 Å^2, and changes but little with water content. We were unable to find similar data for alkyl sulphates with octanol or decanol, The phase behaviours of alkylsulphates and carboxylates in water alone are known to be very similar[6] as they are in the presence of cosurfactants.[7,8] Thus we take the area per amphiphile for the sulphate system to be similar to that of the carboxylate—probably slightly larger by 3–5 Å^2.[13] We have studied two different ratios of surfactant : cosurfactant, these being 1 : 1 for the dodecyl sulphate/octanol system and 1 : 3 for the dodecyl carboxylate/octanol system. The lower charge density in the latter system was necessitated by the insolubility of the carboxylates. Taking the upper value of a = 30 Å^2 gives an area per charge of 60 Å^2 for the sulphate system, while for the carboxylate system we take the lower value of a = 27 Å^2 (area per charge = 108 Å^2).

The ionic strength of the samples is very high and an estimate is given in Table 1 for the carboxylate system. Only the cations have been considered, whereas the carboxylate head group was ignored, hence the concentration of the ions in the water was halved. Since about 80% of the cations are bound to the head groups the effective ionic strength of the aqueous region is only about 20% of the

Table 1 Ionic strength of the carboxylate samples for all compositions and all concentrations

	Ionic strength of Na (mol L^{-1})				Ionic strength of Li (mol L^{-1})				Ionic strength (mol L^{-1})			
n(SDC) : n(LiCOO)	25 wt%	35 wt%	45 wt%	55 wt%	25 wt%	35 wt%	45 wt%	55 wt%	25 wt%	35 wt%	45 wt%	55 wt%
0 : 1					0.23	0.33	0.22	0.55	0.23	0.33	0.43	0.54
1 : 4	0.05	0.05	0.09	0.11	0.19	0.27	0.35	0.42	0.24	0.33	0.43	0.52
2 : 3	0.09	0.13	0.17	0.22	0.14	0.19	0.26	0.34	0.23	0.33	0.44	0.55
1 : 1	0.12	0.17	0.22	0.27	0.11	0.17	0.22	0.27	0.23	0.33	0.44	0.54
3 : 2	0.14	0.20	0.26	0.32	0.09	0.13	0.18	0.21	0.24	0.33	0.44	0.54
4 : 1	0.19	0.27	0.35	0.43	0.05	0.07	0.09	0.11	0.24	0.33	0.43	0.54
1 : 0	0.23	0.33	0.43	0.54					0.23	0.33	0.43	0.55

calculated values. Similar, but somewhat higher values are estimated for the sulphate samples.

Materials and methods

Materials

Dodecanoic acid (COOH, 99.6%) was received from Alfa Aesar, lithium dodecyl sulphate (LiDS, ≥99%) was obtained from AppliChem (Germany), sodium dodecyl sulphate (SDS, ≥99%) and lithium hydroxide (LiOH, 98%) were purchased from Merck (Germany) and SDC sodium dodecanoic acid (SDC, ≥99%) and octanol were received from Sigma Aldrich (Germany). Deuterated water was purchased from Deutero (Kastellaun, Germany) with a purity of 99.95%. All components were used as received.

Methods

^{23}Na- and ^{7}Li-NMR measurements were performed using a Bruker Avance 400 spectrometer operating at 105.84 MHz in the case of sodium and operating at 155.51 MHz in the case of lithium. For ^{23}Na resonance a 30° pulse length of 8.70 μs and for ^{7}Li resonance a 30° pulse length of 7.10 μs was used. The number of scans varied from 128 to 1024, but most of the spectra were collected with 512 scans. The measurements with the sulphates SDS and LiDS were performed at 300 K, the measurements of the carboxylates SDC and LiDC were performed at 300 K as well as at 310 K. This small temperature range was selected in order to ensure the presence of liquid crystals.

Each sample (3 g) was prepared in a tube, heated to 60 °C and mixed with a magnetic stirrer for at least 1 week. Afterwards about 800 mg was transferred in a NMR tube (5 mm o.d.). LiDC was prepared *in situ* by mixing dodecanoic acid and lithium hydroxide.

Each sample contains D_2O, octanol and surfactant. The surfactants are either carboxylate surfactants (SDC and LiDC) or sulphate surfactants (SDS and LiDS). The molar ratio of the sodium surfactant to lithium surfactant varies from 0 to 100% sodium surfactant in 20% steps. In each sample the molar ratio of SDS + LiDS to octanol was 1 : 1, and due to the lower solubility of carboxylates in water, the molar ratio of SDC + LiDC to octanol was reduced to 1 : 3. The concentrations of SDS + LiDS + octanol and SDC + LiDC + octanol in D_2O were 35 wt% to 75 wt% in 10 wt% steps.

In order to identify the lamellar phase, the samples were observed visually *via* crossed polar filters. All the samples appeared as single phases with typical lamellar textures.[15]

Results

Phase structure and NMR spectra

All of the samples examined appeared as single lamellar phases by polarising microscopy. However, there were problems with some compositions. Samples with 15% SDC/LiDC/octanol gave broad ^{7}Li NMR spectra without clear quadrupole splitting. This is discussed further below. For 65 and 75 wt% SDC/LiDC/octanol samples the spectra indicated that the samples were incompletely mixed. This is the composition region where solubility is most limited, and the samples were not examined further because of time constraints. In the ^{23}Na- and ^{7}Li-NMR spectra of the sulphate samples at 300 K and of the carboxylate samples at 300 K and 310 K a single quadrupole splitting was visible, except for the sample with 25 wt% with a composition of SDC : LiDC = 1 : 4, which shows no sodium splitting at 300 K.

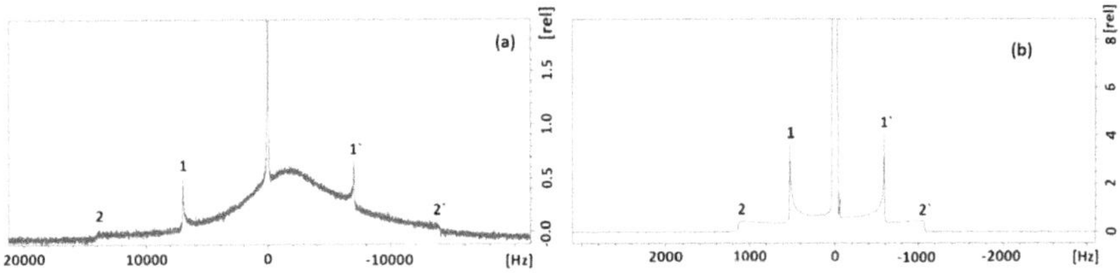

Fig. 3 (a) ^{23}Na-NMR spectrum of a 35 wt% sample with a composition of SDC/LiDC = 2/3 at 300 K. The relative intensity of the signals is plotted against the frequency in [Hz]. (b) ^{7}Li-NMR spectrum of a 35 wt% sample with a composition of SDS/LiDS = 2/3 at 300 K. The relative intensity of the signals is plotted against the frequency in [Hz].

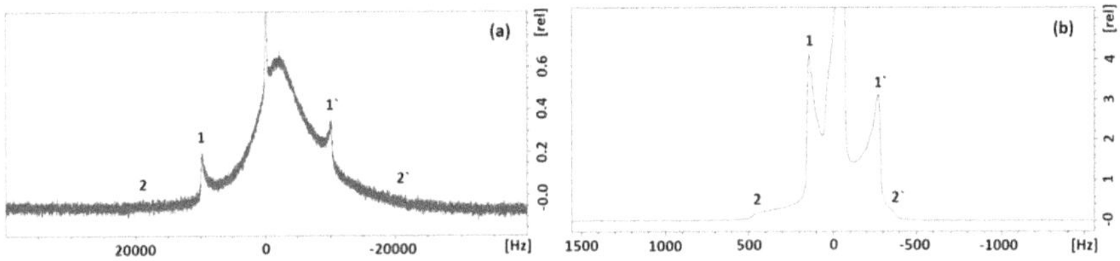

Fig. 4 (a) ^{23}Na-NMR spectrum of a 65 wt% sample with a composition of SDS/LiDS = 4/1 at 300 K. The relative intensity of the signals is plotted against the frequency in [Hz]. (b) ^{7}Na-NMR spectrum of a 65 wt% sample with a composition of SDS/LiDS = 4/1 at 300 K. The relative intensity of the signals is plotted against the frequency in [Hz].

In the Fig. 3a and 3b typical ^{23}Na and ^{7}Li spectra of the carboxylate system can be seen. In the Fig. 4a and 4b typical ^{23}Na and ^{7}Li spectra of the sulphate system are shown. (Note that the broad signal underlying the Na spectra arises from sodium in the glass NMR tubes.) All of these are typical powder patterns with a single quadrupole splitting.

The quadrupole splitting (Δ) is equal to one half the distance between the signals 1 and 1′ and one quarter the distance between 2 and 2′. These distinct features can be seen for nearly all mixtures. In the sulphate as well as in the carboxylate systems the ^{23}Na splitting is much larger than the ^{7}Li splitting, the values of both being in good agreement with those in the literature.[16–18] In all of the spectra having well-defined powder patterns there is an asymmetry in their appearance (clearly visible in Fig. 4b). The outer features (2,2′) are off-set compared to the inner peaks (1,1′). In addition, the central peak is asymmetric in shape. The cause is the well-known chemical shift anisotropy ($\Delta\sigma$) that occurs for liquid crystals.[19] This effect is very small, only *ca.* 100Hz, but because the ^{7}Li splitting are small then it is easily observed. We are currently examining the dependence of $\Delta\sigma$ for both the ^{23}Na and ^{7}Li spectra. It appears that $\Delta\sigma$ varies with the bulk compositions of the samples, rather than giving information on the molecular behaviour.

Influence of surfactant concentration on the quadrupole splitting

In Fig. 5 the sodium and the lithium Δ values with increasing concentration are shown for the single ion systems. The lithium splittings are roughly constant for the carboxylate samples, as expected. However, the sodium splitting decreases with concentration in both the carboxylate and the sulphate samples, up to 65% after which it increases. Moreover, in the sulphate samples the lithium splitting also decreases with increasing concentration up to 65%. It seems highly unlikely that the ion binding would decrease with increasing surfactant concentration. Hence, as described above, we attribute this decrease to the existence of two distinct

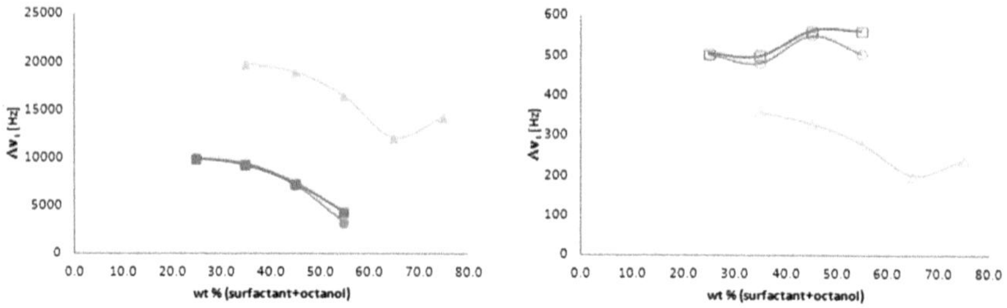

Fig. 5 Sodium splitting with increasing concentration of surfactant and octanol (left) and lithium splitting with increasing concentration of surfactant and octanol (right). ● pure SDC at 300 K, ■ pure SDC at 310 K, ▲ pure SDS at 300 K, ❍ pure LiDC at 300 K, □ pure LiDC at 310 K, △ pure LiDS at 300 K.

locations for the bound ions, one on the surface of the head groups (positive Δ values, Δ_{bs}) and the second, in between the head groups (negative Δ values, Δ_{bb}). Of course, these are average values since both sites will include ions having a distribution of positions. The increase in Δ values for the highest sulphate concentration simply represents the movement of ions from site bb to site bs because the head groups move closer together.

Ion specifity of carboxylate

The Δ values of ^{23}Na and ^{7}Li for a mixture of 25 wt% SDC/LiDC/octanol in D_2O in Fig. 6 show only a marginal increase at the higher temperature. Thus the fraction of ions in site bb is small. However, the sodium Δ values increase with increasing amount of sodium, while the lithium Δ values decrease with an increasing amount of lithium. Moreover, the relative magnitude of the changes is the same for both, but of opposite sign. That is strong evidence that the changes in Δ values reflect changes in ion binding and that Li binding to carboxylate is stronger than Na. The addition of Li ions displaces Na ions from the surface, decreasing the Na Δ values. For the Li ions, the largest fraction of bound ions occurs at the lowest Li concentration, and this decreases with added Li ions because a larger fraction must replace the free Na ions. The data suggest that the fraction of bound Li ions is *ca.* 28% larger for the 0.2 fraction Li mixture than for the pure Li system.

Similarly to the 25 wt% sample, at 35 wt% the sodium splittings increase with increasing amount of sodium and the lithium splitting decreases with increasing fraction of lithium (Fig. 7). Hence, these results confirm the stronger binding of lithium

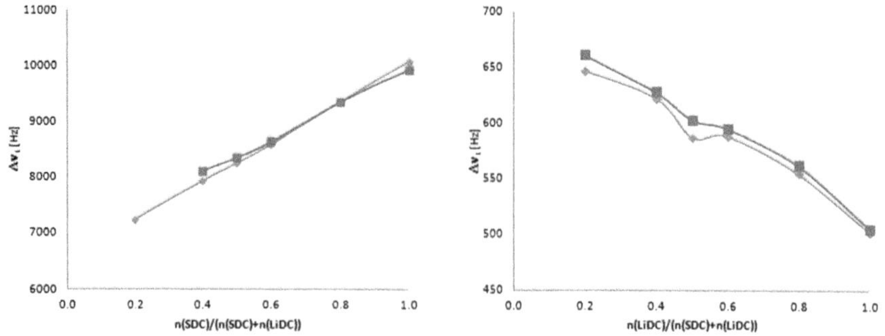

Fig. 6 Sodium splitting with increasing amount of sodium (left) and lithium splitting with increasing amount of lithium (right) at 300 K and 310 K for 25 wt% SDC/LiDC/octanol in D_2O. ◆ 300 K, ■ 310 K.

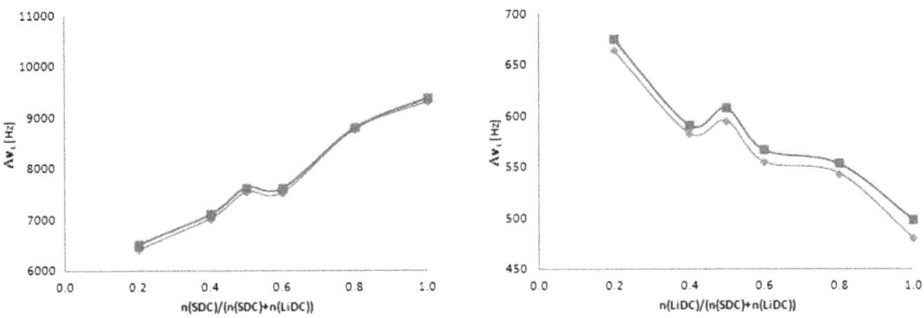

Fig. 7 Sodium splitting with increasing amount of sodium (left) and lithium splitting with increasing amount of lithium (right) at 300 K and 310 K for the 35 wt% samples. ◆ 300 K, ■ 310 K.

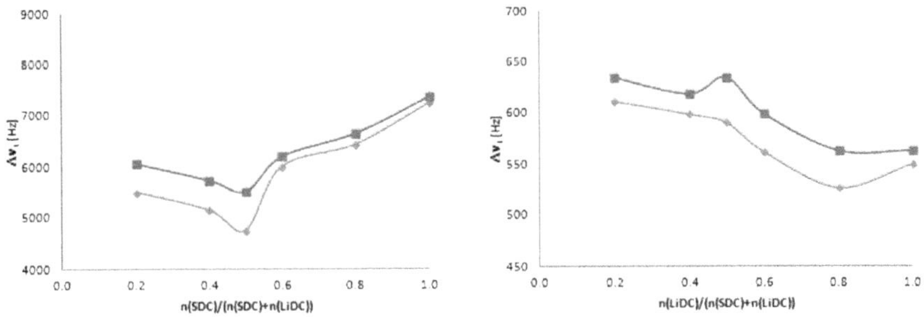

Fig. 8 Sodium splitting with increasing amount of sodium (left) and lithium splitting with increasing amount of lithium (right) at 300 K and 310 K for the 45 wt% samples. ◆ 300 K, ■ 310 K.

to the carboxylate head group. The fraction of free sodium appears to increase by about 31% from pure SDC to 20% SDC, in agreement with the increased fraction of bound lithium ions.

The quadrupole splittings of the 45 wt% samples (Fig. 8) show the same trend as in the 25 and 35 wt% samples. The sodium splitting is lower than for the 35 wt% samples, while the lithium splitting is still the same. The amount of released sodium ions for this concentration is 24%, whereas the fraction of bound lithium ions apparently increases only by about 11%. However, note that as the concentration of surfactant increases, the temperature dependence of the Δ values increases, along with a decrease in the overall Δ values. The Δ values no longer show a monotonic change. Clearly, as well as the change in Δ values due to the displacement of Na by Li we also have a growth in the fraction of both ions in the bb site. With three sites and a single parameter we are in the fortunate position that almost any interpretation cannot be disproved. Note that irregular behaviour is shown for *ca.* 1 : 1 Li/Na mixtures. It appears that the Na ions favour the bb site over the Li ions, but that the Li ions still bind preferentially.

This fortunate position continues with the trends of the 55 wt% samples (Fig. 9). The sodium splitting is lower than in the more dilute samples and is nearly the same for all compositions, except the 40% SDC sample at 300 K. The lithium splitting (Fig. 6) is the same as the more dilute samples. The splitting decreases with increasing fraction of lithium in the samples. The increase in bound lithium is about 26%.

In summary, all measurements have explicitly shown that lithium binds more strongly to the carboxylate head group than sodium. In addition, the Li ions favour the bs-site over the bb-site.

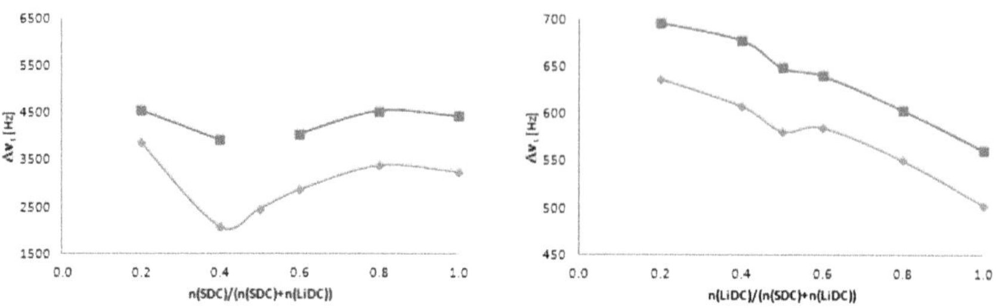

Fig. 9 Sodium splitting with increasing amount of sodium (left) and lithium splitting with increasing amount of lithium (right) at 300 K and 310 K for the 55 wt% samples. ◆ 300 K, ■ 310 K.

Ion specificity of sulphate

For this system the lamellar phase concentration range with a ratio of surfactant : octanol of 1 : 1 was 35–75 wt% SDS/LiDS/octanol in D_2O.

In Fig. 10 the sodium and the lithium splitting of the 35 wt% samples with an increasing fraction of sodium and lithium respectively is shown. Both splittings remain nearly constant for all compositions. Hence sulphate does not show any ion specificity towards sodium or lithium for this composition.

Similarly, for the 45, 55 and 65 wt% samples (Fig. 11), the Δ values show only small changes with varying amounts of sodium and lithium. The marginal change of the splittings is a hint only of a slight preference of the sulphate head group to sodium. If the three site model is used, the non-monotonic behaviour points to a slight preferential location of Na ions in the bb site. Rather than at 1 : 1, the most irregular behaviour occurs at *ca.* 1 : 4 Li/Na mixtures.

For the 75 wt% SDS/LiDS/octanol in D_2O samples (Fig. 12) the sodium splitting decreases with increasing amount of sodium except for pure SDS in octanol and water. In the 80% sodium sample the fraction of bound sodium in the bb site is 24% higher than in the 20% sodium sample. Admittedly the trend of the lithium splitting is not coincident with the sodium splitting ones.

The ion specificity of the sulphate head group towards sodium and lithium is much less pronounced than the ion specificity of the carboxylate head group.

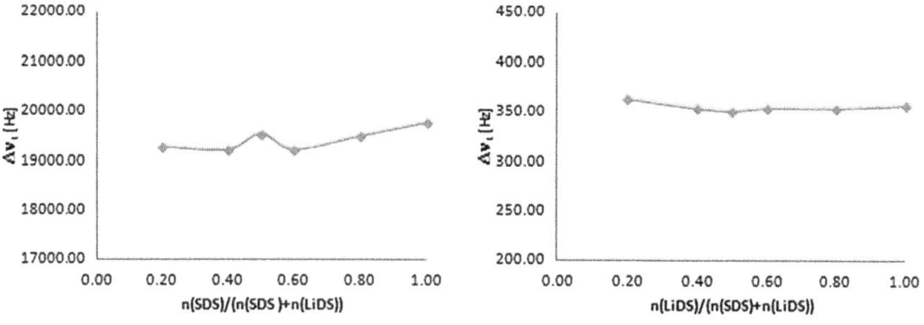

Fig. 10 Sodium splitting with increasing amount of sodium (left) and lithium splitting with increasing amount of lithium (right) at 300 K for 35 wt% samples.

Discussion

From the results presented in the last section, several conclusions can be drawn:

The quadrupole splitting clearly reflects differences in the local environment of lithium and sodium in liquid crystalline systems made up with carboxylate

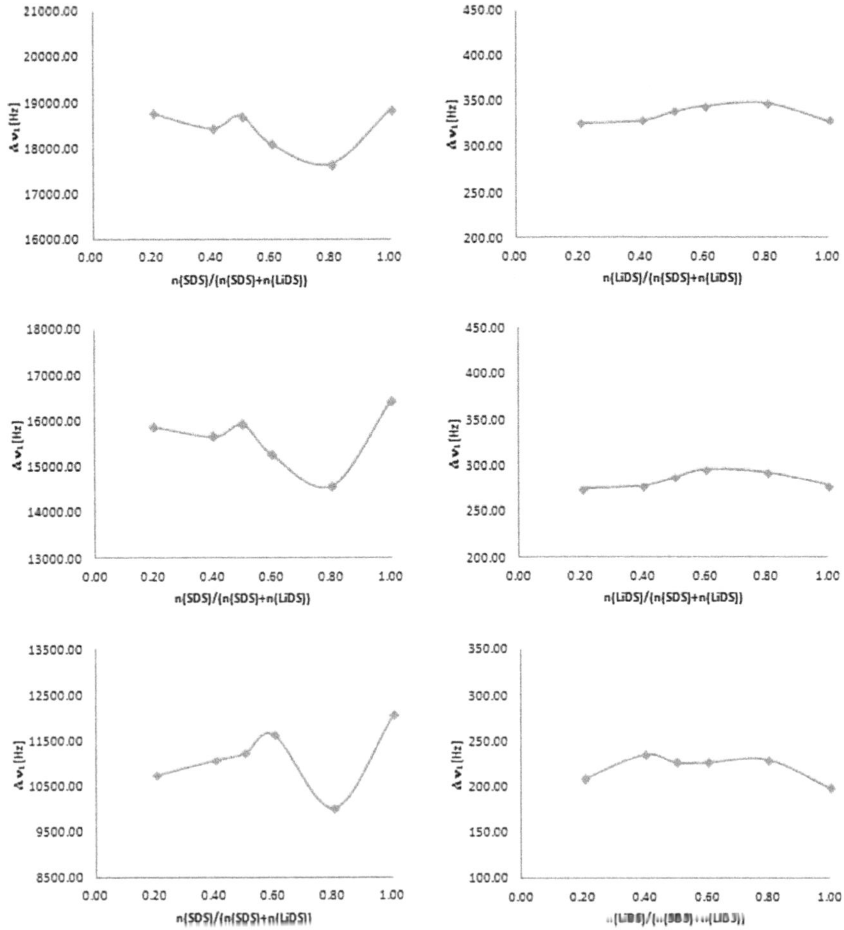

Fig. 11 Sodium splitting with increasing amount of sodium (left) and lithium splitting with increasing amount of lithium (right) at 300 K for 45, 55 and 65 wt% samples.

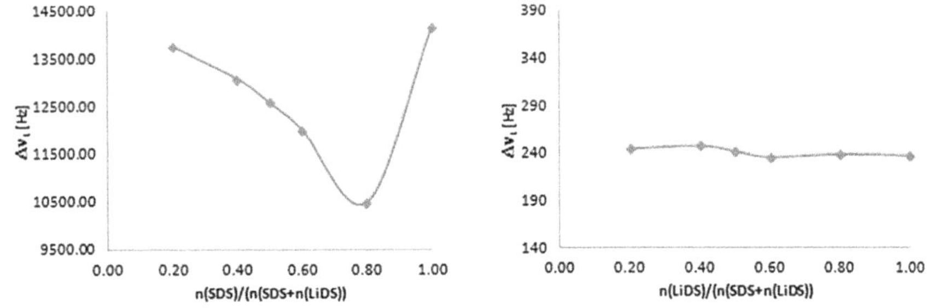

Fig. 12 Sodium splitting with increasing amount of sodium (left) and lithium splitting with increasing amount of lithium (right) at 300 K for 75 wt% samples.

surfactants and octanol as cosurfactant. The specific ion effects can be interpreted as a preferred propensity of lithium towards carboxylate as compared to sodium. The specificity is more pronounced at smaller absolute surfactant/cosurfactant concentration (35 wt%), whereas the temperature effect is more pronounced at higher surfactant/cosurfactant concentration. In the case of the corresponding sulphate system it turns out that specific ion effects are less pronounced.

For a more detailed interpretation several other factors have to be taken into account. The first is the total ionic strength. Depending on the models and estimations for bound ions, it varies between 0.1 and 0.3 M, which is typically the concentration range for which specific ion effects can be expected.

The second parameter is the area per charge, which is about 108 Å^2 for the carboxylate systems and about 60 Å^2 for the sulphate systems. Note that the surfactant-to-cosurfactant ratio is 1 : 1 for the sulphate and 1 : 3 for the carboxylate system. It may be that the higher charge density in the case of the sulphate system cancels out all ion specificity. Such a consequence of high charge density was already observed several times.[20,21] But it may be also that the behaviour of lithium and sodium in the neighbourhood of sulphate headgroups is not different enough to cause a noticeable difference in the quadrupole splittings. Only further experiments can remove this ambiguity.

The third parameter that has to be taken into account is the available space for the ions and for "bulk" water. Even in the most concentrated systems the thickness of the water layer is at least 1 nm. This is enough to have ions sufficiently far away from the headgroup so that they can be considered to be unbound. Consequently, there is always an equilibrium possible between bound and unbound ions. In any case, the NMR quadrupole is very sensitive to even small distortions of the local environment around the ions. That means that even a slight displacement of sodium by lithium ions (or *vice versa*) will be visible in the spectra.

Concerning the amount of counterion condensation onto the negatively charged double layers, the NMR spectra cannot deliver absolute values. However, from classical theories and numerous experiments it can be inferred that about 70 to 80% of the counterions are attached to headgroups. According to the present NMR results, there is roughly a replacement of around 25% out of these 70–80% of sodium by lithium close to carboxylic headgroups whereas no noticeable change is detectable near sulphate headgroups.

Two further points are worth considering. First, the temperature dependence of the quadrupole splittings (measured for carboxylate systems only): obviously a slight raise of temperature of 10 K increases the quadrupole splitting of both ions, whereas the specific ion effects essentially remain unchanged. A plausible explanation could be the assumption that at higher temperatures the ions have a higher tendency to bind at the bs site rather than at the bb site. Of course, these effects are very subtle, but on the other hand, as stated previously, this NMR technique is sensitive enough to subtle effects.

The second point is the negative quadrupole splitting in the case of sodium–carboxylate. As mentioned in the introduction this behaviour can be ascribed to a second binding site of the sodium ions and this site could be deeper within the carboxylate headgroup. Probably, the average distance of sodium to the carboxylate and sulphate headgroups and lithium to the sulphate headgroups depends on the presence of other ions either of the same type or of different type and on the concentration of both, *cf.* Fig. 5. Note that the influence of the ion concentration on the average ion headgroup distance can be of the same order of magnitude as the influence of the competition of lithium and sodium ions.

In any case this preliminary study has shown that NMR quadrupole splitting measurements are a valuable technique to investigate even tiny differences in ion–headgroup interactions and that this method is therefore suitable for the study of specific ion effects in colloidal systems. It is known that the chloride ion quadrupole splitting for cationic surfactant liquid crystals shows similar behaviour to that described above, hence the method may also be applicable to anions.[22] We hope that more of these experiments will be performed in the future to get a more general and concise picture of these subtle effects, especially concerning the relative importance of ion effects close to different headgroups.

References

1 K. D. Collins, Ions from the Hofmeister series and osmolytes: effects on proteins in solution and in the crystallization process, *Methods*, 2004, **34**, 300.

2 N. Vlachy, B. Jagoda-Cwiklik, R. Vacha, D. Touraud, P. Jungwirth and W. Kunz, Hofmeister series and specific interactions of charged headgroups with aqueous ions, *Adv. Colloid Interface Sci.*, 2009, **146**(1–2), 42.

3 P. Jungwirth and D. J. Tobias, Specific Ion Effects at the Air/Water Interface, *Chem. Rev.*, 2006, **106**, 1259.

4 N. Schwierz, D. Horinek and R. R. Netz, Reversed Anionic Hofmeister Series: The Interplay of Surface Charge an Surface Polarity, *Langmuir*, 2010, **26**, 7370.

5 H. Wennerstrom, G. Lindblom and B. Lindman, Theoretical aspects on the NMR of quadrupolar ionic nuclei in micellar solutions and amphiphilic liquid crystals, *Chemica Scripta*, 1974, **6**(3), 97.

6 G. Lindblom, B. Lindman and G. J. T. Tiddy, Ion binding studied using quadrupole splittings of sodium-23(1+) ions in lyotropic liquid crystals. The dependence on surfactant type, *J. Am. Chem. Soc.*, 1978, **100**(8), 2299.

7 P. Ekwall, L. Mandell and K. Fontell, Ternary Systems of Potassium Soap, Alcohol and Water; I Phase Equilibria and Phase Structures, *J. Colloid Interface Sci.*, 1969, **31**(4), 508.

8 P. Ekwall, L. Mandell and K. Fontell, Ternary Systems of Potassium Soaps, Alcohol and Water; II Structure and Composition of the Mesophases, *J. Colloid Interface Sci.*, 1969, **31**(4), 530.

9 I. D. Leigh, M. P. McDonald, R. M. Wood, G. J. T. Tiddy and M. H. Trevethan, Structure of Liquid-crystalline Phases formed by Sodium Dodecyl Sulphate and Water as Determined by Optical Microscopy, X-ray Diffraction and Nuclear Magnetic Resonance Spectroscopy, *Journal of the Chemical Society, Faraday Transaction 1: Physical Chemistry in Condensed Phases*, 1981, **77**(12), 2867.

10 S. Hassan, W. Rowe, and G. J. T. Tiddy, Surfactant Liquid Crystals, in *Handbook of Applied Surface and Colloid Chemistry*, K. Holmberg, Editor. 2002, John Wiley & Sons, LtD: West Sussex. p. 465.

11 G. Lindblom, B. Lindman and G. J. T. Tiddy, Counter-ion Quadrupole Splittings in Lyotropic Liquid Crystals. Determination of the Sign of the Order Parameter, *Acta Chem. Scand., Ser. A*, 1975, **A 29**(9), 876.

12 V. Luzzati, X-ray diffraction studies of lipid–water systems, *Biological Membranes*, 1968, 71.

13 P. Ekwall, Compositions, properties and structures of liquid crystalline phases in systems of amphiphilic compounds, *Advances in Liquid Crystals*, 1975, **1**, 1.

14 P. Ekwall, L. Mandell and K. Fontell, Solubilization in micelles an mesophases and the transition from normal to reversed structures, *Mol. Cryst.*, 1969, **8**, 157.

15 F. B. Rosevear, The microscopy of liquid crystalline neat and middle phases of soaps and synthetic detergents, *J. Am. Oil Chem. Soc.*, 1954, **31**, 628.

16 G. Lindblom, Ion binding in liquid crystals studied by NMR: III. Sodium-23 quadrupolar effects in a model membrane system, *Acta Chem. Scand.*, 1971, **25**(7), 2767.

17 O. Soderman, G. Arvidson, G. Lindblom and K. Fontell, The Interactions between Monovalent Ions and Phosphatidyl Cholines in Aqueous Bilayers, *European Journal of Chemistry*, 1983, **134**, 309.

18 E. Everiss, G. J. T. Tiddy and B. A. Wheeler, Phase Diagram and NMR Study of the Lyotropic Liquid Crystalline Phases Formed by Lithium Perfluoro-Octanoate and Water, *J. Chem. Soc., Faraday Trans. 1*, 1976, **72**, 1747.

19 H. Wennerstrom, N. O. Persson, G. Lindblom and B. Lindman, Chemical Shift Anisotropies of $^{133}Cs^{+}$ Counterions in Lyotropic Liquid Crystals, *Journal of Magnetic Resonance*, 1978, **30**, 133.

20 R. Komsa-Penkova, R. Koynova, G. Kostov and B. G. Tenchov, Thermal stability of calf skin collagen type I in salt solutions, *Biochim. Biophys. Acta, Protein Struct. Mol. Enzymol.*, 1996, **1297**, 171.

21 A. Kubicova, T. Krizek, P. Coufal, M. Vazdar, E. Wernersson, J. Heyda and P. Jungwirth, Overcharging in Biological Systems: Reversal of Electrophoretic Mobility of Aqueous Polyaspartateby Multivalent Cations, *Phys. Rev. Lett.*, 2012, **108**(18), 186101.

22 E. S. Blackmore and G. J. T. Tiddy, Optical microscopy, multinuclear NMR (^{2}H, ^{14}N and ^{35}Cl) and X-Ray studies of dodecyl- and hexadecyl-trimethylammonium chloride/water mesophases, *Liq. Cryst.*, 1990, **8**(1), 131.

Simulation study of ion pairing in concentrated aqueous salt solutions with a polarizable force field†

Yun Luo,[ab] Wei Jiang,[b] Haibo Yu,‡[a] Alexander D. MacKerell, Jr.[c] and Benoît Roux*[ab]

Received 10th April 2012, Accepted 14th May 2012
DOI: 10.1039/c2fd20068f

The accuracy of empirical force fields is critical for meaningful molecular dynamics simulations of concentrated ionic solutions. Current models are typically developed on the basis of single ion properties such as the monohydrate energy in the gas phase, or the absolute hydration free energy at infinite dilution. However, the failure of these models to represent accurately the properties of concentrated solutions cannot be excluded. Here, these issues are illustrated for a polarizable potential based on classical Drude oscillators. To model accurately concentrated ionic solutions, the parameters of the potential functions are optimized to reproduce osmotic pressure data. The sodium-chloride potential of mean force in solution calculated from the empirically-adjusted model is consistent with the results from that calculated from *ab initio* CPMD simulations.

1. Introduction

Computer simulations of atomic models are increasingly used to study a broad range of physical and biological processes. Ions play a fundamental role in many of these complex molecular systems, and for meaningful computer simulations studies, accurate force fields are absolutely critical. Modeling ions and strong electrolytes accurately is particularly challenging because of the long-range nature and strength of electrostatic interactions that are involved. The ion models widely used in molecular dynamics (MD) simulations are generally developed on the basis of single ion properties, such as the energy and geometry of the monohydrate ion in the gas phase and the absolute hydration free energy of the ion at infinite dilution.[1–5] It has been reported that ion models developed with this strategy can dramatically fail to represent accurately the properties of concentrated solutions, such as the solubility[3,4] and osmotic pressure.[6] Even if a model is optimized to match the gas-phase monohydrate energy and geometry, the hydration free energies in the bulk phase at infinite dilution, as well as dynamic transport properties, may be in good agreement with experimental data and high-level *ab initio* calculations,[5] but it is not guaranteed that the thermodynamic properties of concentrated solutions will be accurately represented.

[a]*Department of Biochemistry and Molecular Biology, The University of Chicago, Gordon Center for Integrative Science, 929 East 57th Street, Chicago, Illinois 60637, USA*
[b]*Argonne National Laboratory, Argonne, Illinois, 60439, USA*
[c]*Department of Pharmaceutical Sciences, School of Pharmacy University of Maryland, Baltimore, Maryland, 21201, USA*

† Electronic supplementary information (ESI) available. See DOI: 10.1039/c2fd20068f
‡ Present address: School of Chemistry, University of Wollongong, Wollongong NSW 2522 Australia, Email address: roux@uchicago.edu

The thermodynamic properties of concentrated solutions depend sensitively on the propensity of ion pairs. However, exploiting this information to test, validate and improve the force fields is difficult because the available data provide only indirect information about ion pairing. Neutron and X-ray scattering yield the average structure factor of a solution and multiple experiments involving isotopic substitutions as well as extensive analysis are needed to extract individual solvent–solvent, ion–solvent, and ion–ion pair correlation functions.[7–9] The amount of ion pairing can indirectly be related to conductivity measurements,[10] although this involves additional assumptions. Ion-pairing in aqueous solution has been studied using a wide range of computational and theoretical methods, including classical force field molecular dynamics (MD)[6,11–14] and Monte Carlo (MC) simulations,[15,16] simple models based on a polarizable continuum,[17,18] and quantum mechanical Car–Parinello MD (CPMD) simulations.[19] Nevertheless, in spite of these efforts, our ability to assess computationally the accuracy of force field models of electrolyte solutions at very high concentrations remains fairly limited.

One experimental observable that is potentially of great interest to test, validate, and refine force field models used in computer simulations of concentrated ionic solutions is the osmotic pressure. It is one of the key thermodynamic properties that can be measured for a wide range of conditions and it offers a measure of the strength of the effective solvent-mediated interaction between ions in the solution. For non-ideal solutions the osmotic pressure is equal to $\Pi = \phi\Pi^*$, where ϕ is the osmotic coefficient defined with respect to the osmotic pressure Π^* for an ideal solution by the Van't Hoff equation $\Pi^* = cRT$ (c is the total concentration of mobile ions and R is the gas constant). Theoretical studies of thermodynamics of ion-pairing, such as activity coefficient and osmotic coefficients of electrolyte solutions have often been based on statistical mechanical frameworks such as the McMillan–Mayer (MM)[20] and Kirkwood–Buff (KB)[21] theories. Some studies have aimed at calculating Kirkwood–Buff Integrals (KBIs) directly using computer simulations.[22–29] Statistical mechanical integral equation theories, such as the mean spherical approximation (MSA)[30,31] or the hypernetted chain (HNC) equation[32] were used to derive pair correlation functions needed to calculate ϕ. The multicomponent reference interaction site model HNC (RISM-HNC) theory was used to incorporate effects arising from the molecular nature of water.[33] For a review see ref. 6.

While these approaches are useful, they do not provide a simple route for establishing a quantitative relationship between the force field used in a computer simulation and the osmotic pressure. It is largely to address this issue that we previously developed a simple and practical approach allowing the direct calculation of the osmotic pressure from all-atom MD simulations of concentrated aqueous solutions.[6] The basic concept is to introduce "virtual" walls to represent the effect of ideal semipermeable membranes, separating the high concentration region from the pure water region. The walls keep the solutes confined but the water molecules are allowed to pass through freely, permitting an equalization of their chemical potential throughout the entire system. The mean force per unit area exerted on the solutes by the virtual walls during the simulations directly reports the osmotic pressure.

In the present study, the Drude polarizable force field for K^+, Na^+ and Cl^- is tested by calculating the osmotic pressure over a wide range of concentrations. A polarizable force field is crucial for the realistic simulations of charged or highly polar species, and also for describing accurately the electrostatic properties of the hydrophobic region in protein and membranes. In the Drude force field, the induced electronic polarization is introduced by an auxiliary particle attached to each polarizable atom *via* a harmonic spring. The Drude ion models were previously developed to be consistent with the SWM4-NDP water model.[5,34] Here, the osmotic pressure of NaCl and KCl aqueous solution are calculated over a wide range of concentrations (up to 5 m) and compared with the experimental values.[35] The results show that the osmotic pressure depends sensitively on ion pair interactions, in agreement with previous studies.[36,37] The present study demonstrates that the osmotic pressure is a powerful route for validating and parameterizing accurate atomistic force fields.

2. Methods

2.1. Classical Drude polarizable model

The present simulations of ionic solutions are based on the SWM4-NDP polarizable water model[34] and the set of polarizable models developed recently for the alkali-halide series.[5] The SWM4-NDP model correctly reproduces the static dielectric constant of liquid water, which makes it appropriate for simulating systems where water-mediated electrostatic interactions have dominant effects,[34] such as concentrated electrolyte aqueous solutions. The gas-phase experimental geometry of water is maintained rigidly, and the static charge distribution of the isolated molecule gives rise to a dipole of 1.85 D, in accord with experiment. In the classical Drude oscillator model, electronic induction is represented by introducing a mobile auxiliary charged particle attached to a polarizable atom *via* a harmonic spring of force constant k_D. The relative displacement of the atom–Drude pair gives rise to an electric dipolar moment. Assuming that the Drude particle carries a q_D and the atom carries an opposite charge $-q_D$, the molecular polarizability α is equal to q_D^2/k_D.[38] In the present models, the charge carried by the Drude particles is always negative, to mirror the physical effects of the electronic density around positively charged nuclei. An anharmonic restoring force was introduced to damp atom–Drude separations larger than 0.2 Å to prevent numerical instabilities with the highly polarizable anions.[5] Atomic dispersion and electronic overlap effects are represented in a pair-wise additive way using the Lennard-Jones (LJ) 6–12 potential. The electrostatic interaction energy of a single monovalent ion with a water molecule is comprised of a sum of charge-charge Coulomb interactions running over all nuclei and Drude particles.

All dynamical simulations were performed by considering the dynamics of an extended Lagrangian in which a small mass and kinetic energy is attributed to the Drude particles. The equations of motion of the atom–Drude pair are separated into the global motion of the center-of-mass, and the relative internal motion of the oscillator (reduced mass). To ensure that the time course of the induced dipoles stays close to the self-consistent field (SCF) solution, the amplitude of the atom–Drude oscillations away from the local energy minimum is controlled with a low-temperature thermostat acting in the local center-of-mass reference frame of each atom–Drude pair.[38] The trajectories generated according to this extended Lagrangian are very close to those generated by the SCF regime of induced polarization.[38]

For the present study, MD trajectories were generated with the program CHARMM[39] and NAMD2.8.[40] In the simulations generated with the program CHARMM,[39] the two-temperature isothermal-isobaric ensemble was simulated using a Nosé–Hoover thermostat scheme[41,42] and the modified Anderson–Hoover barostat.[43] The simulations were propagated using the velocity-Verlet algorithm with a 0.5 fs time step. The internal geometry of the SWM4-NDP water molecule was kept rigidly fixed using the SHAKE/Roll and RATTLE/Roll algorithm.[44,45] Long-range electrostatic interactions were computed using particle mesh Ewald summation (PME).[46] A smooth real space cutoff is applied between 10 and 12 Å with an Ewald splitting parameter of 0.34 Å^{-1}, a fast-Fourier grid density of ≈ 1 Å^{-1}, and a sixth-order interpolation of the charge to the grid. The LJ potential is smoothed with a switching function with the same cutoff scheme, and a long range correction of the energy and pressure from the LJ potential is included.[47]

In the simulations generated with the program NAMD2.8,[40] a dual-Langevin scheme is employed to propagate the trajectory of the polarizable Drude model.[48] While the concept is similar to the dual Nosé–Hoover thermostat in the program

CHARMM,[38] the dual-Langevin scheme in NAMD2.8 enables large-scale simulations of the Drude model on massively parallel supercomputers, such as Blue-Gene/P.[48] To ensure near-SCF conditions, the thermostat for the relative motion of the Drude particles with respect to their parent atoms was set at a temperature of 1 K with a damping coefficient of 0.1 ps^{-1}. The thermostat for the center-of-mass was set at a temperature of 298 K with a damping coefficient of 0.05 ps^{-1}. The geometry of the SWM4-NDP water molecule was maintained using SETTLE.[49] A multiple time-step integration scheme was used, with 0.5 fs for the calculation of the internal energy terms, and 1.0 fs for the calculation of the nonbonded LJ and electostatics terms. The shorter time-step 0.5 fs time step is necessary for stable integration due to the anharmonic restoring force introduced to prevent the hyper-polarization of anions at the finite concentrations.[5] The electrostatic interactions were treated using PME.[46,50] A smooth real space cutoff is applied between 10 and 12 Å with grid density of ≈ 1 $Å^{-1}$, and a sixth-order interpolation of the charge to the grid. The same cutoff is used for the LJ potential with a switching function.

2.2. Single ion pair potential of mean force (PMF) simulation details

The 1D-PMF of single ion pair in bulk was calculated as a function of the cation–anion separation, $r = \|\mathbf{r}_c - \mathbf{r}_a\|$, using umbrella sampling.[51] A total of 26 independent biasing windows in which the two ions were harmonically restrained by the potential $u_i(r) = k(r - R_i)^2/2$, were set up at different separation R_i and force constant k to ensure good sampling and overlap of neighboring windows (0.2 Å increments between 2.5 Å to 3.5 Å with force constant of 30 kcal $mol^{-1}Å^{-2}$, 0.25 Å increments between 3.5 Å to 7.0 Å with force constant of 30 kcal $mol^{-1}Å^{-2}$, and 0.5 Å for the rest with the force constant of 20 kcal $mol^{-1}Å^{2}$). The simulations were carried out using the program CHARMM,[39] with one ion pair placed in a periodic cubic box containing 395 explicit SWM4-NDP water molecules. The chloride ion was harmonically restrained near the center of water box to prevent any significant drift of the ion pair. Each window simulation of 500 ps was initially started from a configuration with the ion near the restraint position and equilibrated for 50 ps. The results from the 26 windows were unbiased using the weighted histogram analysis method (WHAM),[52] with a bin size of 0.05 Å and a stringent tolerance of 0.0001 kcal mol^{-1} on every point in the PMF. The radial PMF resulting directly from umbrella sampling simulations biased by a simple distance restraint implicitly contains an additional hidden term, $+2k_BT\ \ln(r)$, arising from the increasing volume element in the configurational space at large ion–ion separation.[53] This can be seen by converting the two Cartesian coordinates of the cation and anion, $\mathbf{r}_c$ and $\mathbf{r}_a$, into the relative spherical coordinates in the biased Boltzmann configurational integral, yielding the Jacobian $4\pi r^2$. To obtain the PMF that is a simple function of ion–ion separation, we write $W(r) = W_{radial}(r) + 2k_BT\ln(r) + C$, where the constant C is chosen such that $W(r) \rightarrow 0$ as $r \rightarrow \infty$. In practice, the offset constant was chosen such that $W(r)$ matches the value expected in a dielectric continuum with $\varepsilon = 79$[34] at an ion–ion distance of 7 Å.

2.3. Osmotic pressure simulation details

The osmotic pressure for KCl and NaCl aqueous solutions was calculated by introducing an idealized virtual semi-permeable membrane separating a salt solution from pure water, according to a method described previously.[6] Here, the idealized virtual semi-permeable membrane is made of two half-harmonic walls acting only on the ions to keep them within the salt solution region. The water molecules can freely cross the virtual semi-permeable membrane and move throughout the system. Averaging the force acting on the ions from the half-harmonic walls under these conditions directly yields the osmotic pressure in the system. The osmotic pressure MD simulations were carried out with periodic boundary conditions using the program NAMD2.8.[40] The shape of the entire simulation cell is orthorhombic with dimension 96 × 96 × 96 Å^3 and is comprised of approximately 150 000 atoms. The central region corresponds to a salt solution with dimensions 96 × 96 × 48 Å^3. A flat-bottom half-harmonic planar restraint with a force constant of 10 kcal mol^{-1}Å^{-2} was applied to the ions at $z = \pm 24$ Å to keep them inside the region corresponding to the salt solution. The virtual walls acting on the ions were setup *via* a Tcl script in NAMD. No restraint was applied to the water molecules. The molality scale (m ≡ mol kg^{-1}) is used throughout the manuscript to compare directly to the experimental osmotic pressure data. The atomic systems were first setup at different molarity concentrations (1, 2, 3, 4, and 5, mol l^{-1}), as in our previous simulations of nonpolarizable models,[6] and the effective molality concentrations of the solution were estimated after equilibration. After a 400 ps equilibrium run under NPT conditions at 298.15 K and constant area, 6 independent 800 ps simulations (different initial configuration and initial velocity) were carried out under NVT condition for each concentration and the coordinates of the ions were saved every 0.1 ps.

3. Results and discussion

Fig. 1(a) and 1(b) show the calculated osmotic pressure for the NaCl and KCl aqueous solutions using the Drude polarizable force field with standard LJ parameters. Without any additional adjustments, the force field for KCl is able to reproduce the experimental osmotic pressure quite well, with a slight deviation from the experimental value at higher concentrations (above 2 m). For the NaCl solution, however, the osmotic pressure is systematically underestimated, even at concentrations as low as 1 m. Although the discrepancy is larger than previously observed for the nonpolarizable NaCl ion models, the analysis showing that such a systematic

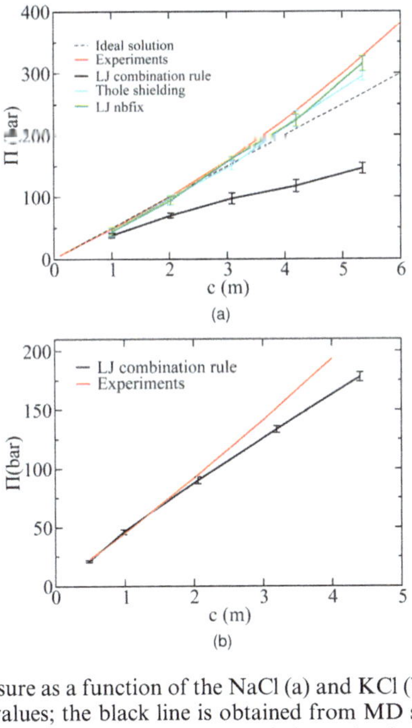

Fig. 1 The osmotic pressure as a function of the NaCl (a) and KCl (b) salt concentration. The red line is experimental values; the black line is obtained from MD simulations with the Lorentz–Berthelot combination rule. In (a), the cyan line is from simulations with adjusted Na–Cl R_{min} 4.09 Å LJ parameters using NBFIX; the green line is from simulations with a Thole shielding screening parameter t_{ij} 2.10. The error bars were estimated from the variance amongst 6 independent 800 ps simulations.

underestimation of the osmotic pressure in electrolyte solution reflects an excess of ion pairing[6] remains valid. A simple explanation is that the osmotic pressure is directly affected by the actual concentration of mobile species in solution, cations and anions. For example, if all cations and anions were bound together to form stable pairs, the resulting osmotic pressure would be roughly reduced by a factor of two compared to a fully dissociated electrolyte solution. Thus, excessive ion pairing caused by inaccuracies in the force field model results in an underestimated osmotic pressure. To improve the accuracy of the model, it is necessary to alter and optimize the potential function. This is possible by using the experimental osmotic pressure as target data.

The origin of the problem with the model with standard LJ parameters starts with the over-binding of the Na–Cl pair in vacuum. Within the Drude polarizable force field, the potential energy between two ions is written as a sum of Coulombic interactions summed over all charges (from the atom and the Drude particles), and a Lennard-Jones (LJ) 6–12 potential to account for the core–core repulsion and the van der Waals dispersive attraction between the two atoms. The default parameters of the LJ potential for the interactions between two ions i and j, that is, $R^{ij}_{\rm min}$ and $E^{ij}_{\rm min}$, are generated with the standard Lorentz–Berthelot combining rule. The latter uses the arithmetic mean for

$$R^{ij}_{\rm min} = (R^{ii}_{\rm min} + R^{jj}_{\rm min})/2$$

and the geometric mean for the well depth $E^{ij}_{\rm min} = \sqrt{E^{ii}_{\rm min} E^{jj}_{\rm min}}$. Fig. 2(a) shows the potential energy of an NaCl ion pair in vacuum as a function of distance. The energy minimum obtained from the Drude model with default LJ parameters is 25 kcal mol^{-1} lower than the *ab initio* value. The *ab initio* quantum chemical computation

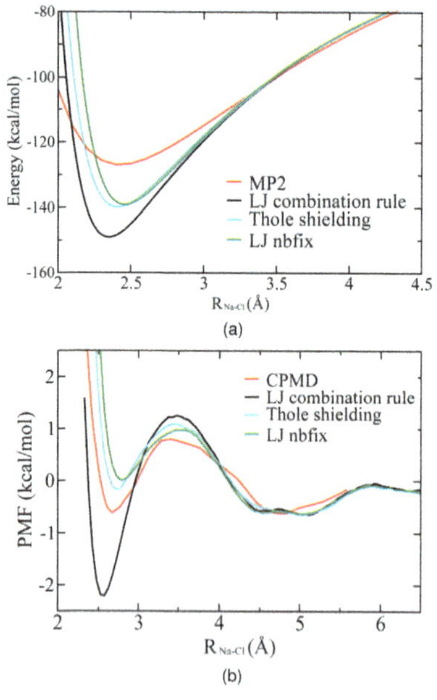

Fig. 2 (a) Potential energy of NaCl ion pair in vacuum as a function of distance. The *ab initio* quantum chemical computation is carried out at MP2/aug-cc-pVDZ level (BSSE corrected) using Gaussian03 (54). (b) Potential of mean force (PMF) of the single ion pair in bulk calculated from MD simulation with different Na–Cl pair parameters. The PMF from CPMD simulation is taken from ref. 19.

shown here was carried out at the MP2/aug-cc-pVDZ level (BSSE corrected) using Gaussian03.[54] Therefore, there is considerable over-binding of the cation–anion pair in vacuum according to the polarizable Drude model with default LJ parameters.

The over-binding of Na^+ and Cl^- in vacuum represents a failure of the complete model, with default LJ parameters. It is important to realize, however, that the LJ 6–12 potential is only an approximation to the quantum mechanical cation–anion interactions, and the usage of the Lorentz–Berthelot combination rule is merely a convenient route for generating pairwise parameters between unlike atoms.[55] The LJ parameters of the model were previously optimized to reproduce accurately single-ion properties such as the geometry and energy of the monohydrate, and the solvation free energy in bulk water at infinite dilution.[5] Thus, the current model is not constrained by any experimental information at finite concentrations where ion–ion interactions become important. For example, it is perfectly reasonable to imagine that by increasing the radius, $R^{ij}_{\rm min}$, the cation and anion will be pushed further apart, decreasing the favorable association that leads to an excess of ion pairing and an under-estimated osmotic pressure. Ultimately, several different methods may be considered to alter and improve the polarizable Drude model. In the following, two different methods were considered.

First, we developed pair-specific LJ parameters between Na^+ and Cl^-. This is perhaps the simplest correction to the model, as it keeps all other features of the potential function unchanged. To override the values generated from the standard combination rule, the pair-specific LJ parameters are assigned using the NBFIX option of the CHARMM parameter file. Importantly, the LJ parameters controlling the direct interaction between the ions and water remain untouched by the NBFIX option. A second approach was to introduce a shielding function to alter the bare Coulombic interactions between the cation and the anion,

$$\left(\frac{q_i q_j}{r_{ij}}\right) \rightarrow \left(\frac{q_i q_j}{r_{ij}}\right) S(r_{ij}) \tag{1}$$

where $S(r_{ij})$ is a screening function given by,

$$S(r_{ij}) = 1 - \left(1 + \frac{r_{ij} t_{ij}}{2(\alpha_i \alpha_j)^{1/6}}\right) e^{-r_{ij} t_{ij} / (\alpha_i \alpha_j)^{1/6}} \tag{2}$$

where α_i and α_j are the polarizability of atom i and atom j, respectively, and t_{ij} is the Thole parameter modulating the strength of the electrostatic screening for the ij pair. This functional form is consistent with the general idea that the Coulomb interaction between delocalized clouds of electronic charges does not diverge as $1/r$ at short distances. It was first introduced by Thole to account for intramolecular polarization.[56] In the Drude polarizable force field, the $S(r)$ shielding function has been utilized to treat the intramolecular 1–2 and 1–3 bonded interactions.[57] More recently, it was used to "tame" the very strong electrostatic interactions between divalent ions (Zn^{2+}, Mg^{2+}, Ca^{2+}, Sr^{2+}, Ba^{2+}) and the induced dipole component of the oxygen of the water molecules.[5] Such shielding is unnecessary for the interactions of monovalent ions with the water molecules, but it could be needed to treat accurately the cation–anion pairs. Here, a pair-specific Thole parameter was introduced between Na^+ and Cl^- using the option NBTHOLE of the CHARMM potential function parameter file.

Using either the pair-specific LJ parameter (NBFIX) modification and the Thole shielding (NBTHOLE), it ought to be possible to correct for the over-binding of the Na–Cl pair. However, in trying to optimize and improve the force field parameters, it is difficult to rely directly on the *ab initio* data in vacuum in a meaningful way. The interactions of the ions in vacuum is simply too different from the effective interactions in bulk solution, and the impact of any modifications to the vacuum interaction on the thermodynamic properties in the bulk is difficult to predict. A more productive route to assess the default model is to start from considerations of the effective solvent-mediated potential of mean force (PMF) of the single ion pair in bulk water. Here, a key piece of information is provided by the PMF between an Na^+ and a Cl^- in bulk water previously calculated by Kuyucak and co-workers using CPMD simulations.[19] In Fig. 2(b), the PMF from the Drude model with default LJ parameters is compared with the PMF calculated from CPMD simulations. While the two PMFs are in near-quantitative agreement at large and moderate distances they differ strongly at short distances (*i.e.*, less than 3 Å) in the region where the core repulsion begins to be important. The discrepancy at short distances affects the depth of the first free energy well in the PMF, which corresponds to the situation where the cation and anion form a contact pair. As shown in Fig. 2(b), the Drude model with default LJ parameters gives the first minimum of NaCl PMF 1.6 kcal mol^{-1} deeper than the first minimum obtained from CPMD.[19] Obviously, the depth of the first free energy well directly affects the propensity for cation–anion association.

To clarify the features of the model having the most direct impact on ion pairing, it is useful to separate the total PMF from the default model into a long-range and short-range contributions,

$$W_{\rm lr}^{\rm MD}(r) = \Delta W_{\rm lr}^{\rm MD}(r) + U_{\rm LJ}(r) \quad (3)$$

where $\Delta W_{\rm lr}^{\rm MD}(r)$ represents all the long-range electrostatic contributions along with the effects of solvent, other than the cation–anion short-range repulsion and van der Waals dispersion, which are modeled by the LJ potential, $U_{\rm LJ}(r)$, with default parameters generated from the standard combining rule. The function $W_{\rm lr}^{\rm MD}(r)$ extracted from the PMF that was calculated from the simulations with the default LJ parameters for NaCl is shown in Fig. 3(a). This contribution to the PMF bears some similarities with the indirect cavity potential that is sometimes considered in integral equation studies.[33] At moderate distances, the oscillations in $\Delta W_{\rm lr}^{\rm MD}(r)$ reflect the complexity of the hydration shell around the cation and anion. At short distances, in the region where the core repulsion begins to be important, the function $\Delta W_{\rm lr}^{\rm MD}(r)$ becomes strongly attractive. This strong attraction is compensated by the short-range repulsion, which essentially controls the depth of the free energy well in the PMF at ion–ion contact. It is worth pointing out that the steepness of the cavity potential largely undermines our ability to be accurately predictive about the opposing short-range repulsive core potential, a problem noted long ago by Dang *et al.*[58] For this reason, relying on the osmotic pressure data to adjust empirically the force field is absolutely necessary and unavoidable to obtain accurate models.

Although the separation given in eqn (3) is trivially satisfied by construction, it seems plausible on physical grounds that the indirect cavity potential contribution to the PMF should be fairly insensitive to the details of the short-range contribution of the model. This suggests that the function $\Delta W_{\rm lr}^{\rm MD}(r)$ evaluated from the default model could be used to extract the short-range cation–anion repulsive potential that is needed to best-match the PMF from CPMD. Following this argument, one may write,

$$U_{\rm sr}^{\rm CPMD}(r) \equiv W^{\rm CPMD}(r) - \Delta W_{\rm lr}^{\rm MD}(r) \quad (4)$$

where $\Delta W_{\rm lr}^{\rm MD}(r)$ is the long-range contribution shown in Fig. 3(a). The short-range potential $U_{\rm sr}^{\rm CPMD}(r)$ extracted from the CPMD simulations *via* eqn (4) is shown in Fig. 3(b). The function $U_{\rm sr}^{\rm CPMD}(r)$ can then be used as a "target" to help design an optimal model for Na^+ and Cl^- that is able to reproduce the PMF from *ab initio* $W^{\rm CPMD}(r)$. The default value of the LJ parameter $R_{\rm min}^{ij}$ for the Na–Cl pair is 3.94 Å. By visual inspection of Fig. 3(b), the pair-specific radius $R_{\rm min}^{ij}$ must be adjusted to 4.03 Å to match the short-range repulsion extracted from the PMF calculated from CPMD. Further simulations showed that the experimental osmotic pressure were obtained with a Na–Cl $R_{\rm min}^{ij}$ of 4.09 Å. As shown in Fig. 1(a), $R_{\rm min}^{ij}$ 4.09 Å

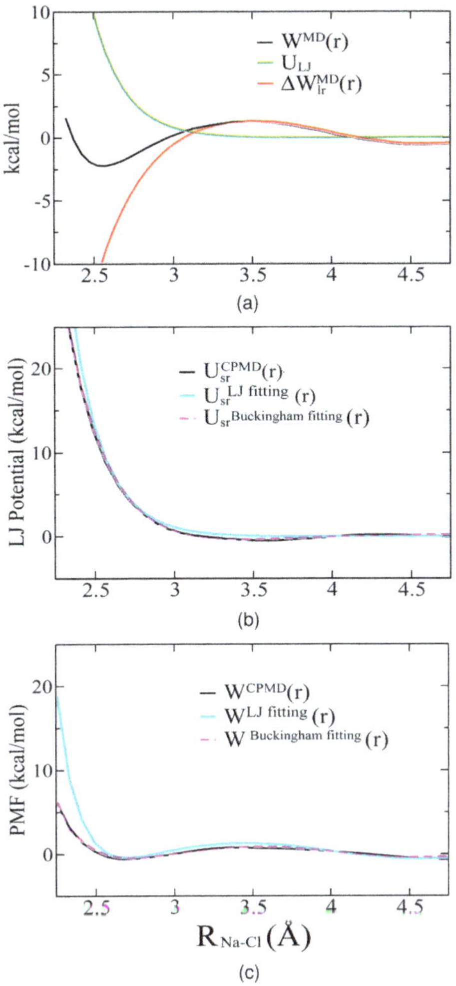

Fig. 3 (a) The $\Delta W(r)$ (red line) is extracted from the default PMF profile (black line) subtracted by the default LJ pair potential (green line); (b) the black line is the target LJ potential needed to fit the CPMD PMF; the cyan line is the fitted LJ potential using the standard 6–12 potential giving an estimation of $R_{\min}$ as 4.03 Å (the default value is 3.94 Å); the magenta line corresponds to the Buckingham potential fitting; (c) the black line is the PMF calculated from CPMD; the cyan line is the PMF from the LJ potential fitting; the magenta line is the PMF from the Buckingham potential fitting.

can reproduce the experimental osmotic pressure very well for a broad range of concentrations, up to the solubility limit of 6 m. Fig. 2(a) shows the Na–Cl potential with this empirically adjusted pair-specific LJ parameters. The energy minimum for the ion pair in vacuum is increased by 10 kcal mol^{-1}, but is still about 15 kcal mol^{-1} less than the *ab initio* value. Alternatively, it is also possible to derive an accurate model with Na–Cl Thole shielding. This was achieved by trying to match the vacuum energy and PMF calculated from the optimized NBFIX model (with $R^{ij}_{\min}$ = 4.09 Å). Following this procedure, an optimal value of 2.10 was obtained for the Thole shielding parameter t_{ij}. All pair-specific parameter results are summarized in Table 1.

Based on the above results, it is clear that the two approaches, the pair-specific LJ parameter modification and the Thole shielding, can both be utilized to empirically

Table 1 LJ parameter of ions and pair-specific parameters

Ion	$\frac{1}{2}R_{min}^{ii}$ (Å)	$\frac{1}{2}E_{min}^{ii}$ (kcal/mol)
Na^+	1.4616800	−0.0315100
Cl^-	2.4811139	−0.0719737
Na^+–Cl^-	R_{min}^{ij} (Å)	t_{ij}
Combination rule	3.94	—
LJ NBFIX	4.09	—
Thole shielding	3.94	2.10

improve the model in order to reproduce the experimental osmotic pressure data in a broad concentration range. It is worth pointing out that the optimized models obtained here are not concentration dependent and that the models are, thus, transferable to any finite concentrations. From this point of view, the current strategy departs from approaches based on a concentration-dependent dielectric constant used to account for the many-body interactions in implicit solvent models.[14,36,59] However, comparing the PMF and also the interaction energy in vacuum with the corresponding *ab initio* data (Fig. 2(a) and 2(b)), the electrostatic Thole shielding strategy produces the energy minimum at a similar position as the *ab initio* data, while increasing the radius (R_{min}) has pushed the energy minimum about 0.6 Å further apart in both vacuum energy and bulk phase PMF profiles. Also, it is known that the classical Drude model does not allow charge transfer between atoms at the short distance. This approximation may cause the electrostatic interaction at short distances to be too favorable, which causes the excess ion pairing. In both cases, the Thole shielding strategy seems more reasonable and precise than LJ parameter modification for the Drude ion pair parameters. As observed in Fig. 3(b), it is of interest to note that the Buckingham potential is able to reproduce accurately the target short-range cation–anion repulsion better than a standard LJ 6–12 potential, particularly for distances shorter than 2.75 Å. This is expected, as it is well understood that the r^{-12} repulsion from the LJ potential is much too steep at short distances and that the exponential functional form of the Buckingham potential is a better physical model for the core repulsion. Nevertheless, the LJ potential, which remains a simple and popular approximation that is widely used in the field of biomolecular simulations, can also yield reasonably accurate results for the concentrated ionic solutions.

The osmotic pressure of concentrated electrolyte solution appears to result from a delicate balance of strongly attractive and repulsive interactions. A useful framework to deepen our understanding of these systems is provided by the McMillan–Mayer (MM) theory.[60] According to MM theory, the osmotic virial series for a non-ideal solution can be expressed as,

$$\Pi = (1 + B_2^{tot}c + \ldots)cRT \tag{5}$$

where c is the concentration of dissociated ionic species and B_2^{tot} is the second osmotic coefficient. The latter is related to the effective inter-particle potential *via*,

$$B_2^{tot} = -\frac{1}{2}\sum_{ij}\int d\mathbf{r}\left[e^{-W_{ij}(r)/k_BT} - 1\right] \tag{6}$$

where $W_{ij}(r)$ is the PMF between ion i and j (the indices run over the cation and anion). The osmotic virial coefficient B_2^{tot} is, thus, directly related to the Na–Cl, Na–Na and Cl–Cl inter-ionic PMFs. Long-range ionic screening is ignored in this discussion focused on the direct ion–ion PMF. The running integrals for the second osmotic coefficient B_2^{tot} are shown in Fig. 4 for a few different models. It is observed that the coefficient B_2^{tot} determined from the default model based on the LJ parameters determined from the combination rule is extremely large, with an estimated value at a distance of 6.5 Å of −420 Å^3. In contrast, the second osmotic coefficient B_2^{tot} for the two optimized models remain much smaller. Their estimated value at a distance of 6.5 Å the running integral for the model with an optimized pair-specific LJ parameters (NBFIX) yields −74 Å^3, and the model with the optimized cation–anion Thole shielding yields −54 Å^3 (Fig. 4(b)). It is interesting to note that the Na–Cl contribution from the two optimized models are both very close to the CPMD estimates, further strengthening the present optimized models for NaCl aqueous solutions at finite concentration.

While eqn (6) is truly valid only at low concentration c, it can nevertheless serve as a useful guide to highlight the large opposing contributions that are at play in the electrolyte. An important observation is that the osmotic pressure of the NaCl solution does not markedly deviate from the ideal relation, $\Pi = cRT$ (see Fig. 1(a)). For this reason, it is imperative that the second osmotic coefficient B_2^{tot} in eqn (6) be relatively small. The indirect implication is that the contribution from the cation–anion pair and that from the like-charged ion pairs (Na–Na, Cl–Cl) must almost cancel each other out. The near-ideal behavior of the NaCl aqueous solution implies that

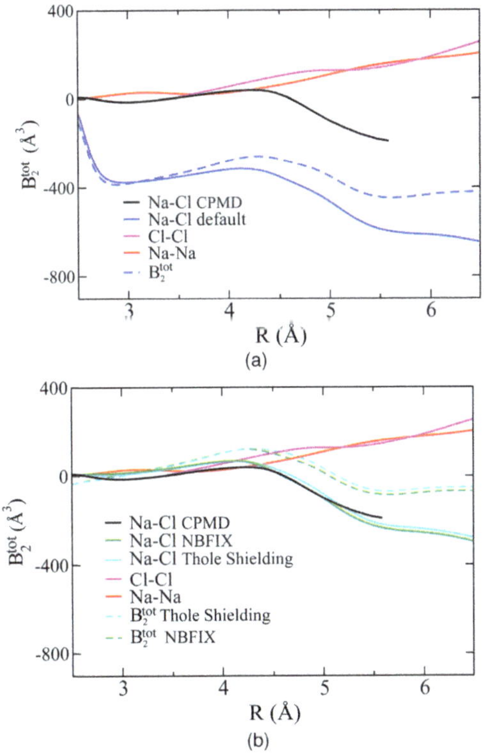

Fig. 4 The second virial coefficient B_2 as a function of ion pair distance. (a) The black line is B_2 from CPMD PMF; the blue solid line is B_2 of Na–Cl ion pairs from LJ combination rule; the blue dashed lines are the sum of B_2 from all ion pairs; the red line is B_2 of Na–Na pairs and the magenta line is B_2 of Cl–Cl pairs, calculated from the radial distribution of like-charged ion pairs in bulk solution (see supplementary material†). (b) The black, red and magenta lines are the same as in (a); the green and cyan solid lines are B_2 coefficients of Na–Cl pairs from LJ nbfix and Thole shielding respectively; the green and cyan dashed lines are the sum of B_2 from all ion pairs from LJ nbfix and Thole shielding respectively.

there is an underlying balance in the Na–Na, Cl–Cl and Na–Cl pairs in this system. For the present model, the contributions for like-charged ion pairs are on the order of +200 $Å^3$ and +230 $Å^3$ for Na–Na and Cl–Cl, respectively (these estimates were obtained from the radial distribution function directly extracted from the unbiased simulations of the NaCl salt solution; see Supplementary Information†). As the latter are dominated by the long-range Coulombic repulsion, these values are expected to be fairly insensitive to the details of the models. Thus, the total contribution from the like-charge pairs can be used to set an empirical upper-bound on the Na–Cl contribution. This means in practice that the Na–Cl contribution should not considerably exceed −430 $Å^3$ to be in agreement with the near-ideal solution experimental osmotic pressure. This shows that the estimated contribution for the Na–Cl pair of the default model based on the LJ parameters determined from the combination rule, which amounts roughly to −840 $Å^3$, is clearly much too large due to overbinding of the cation–anion pair. This cannot be compensated by the opposite Na-Na and Cl–Cl repulsive contributions, hence resulting in an osmotic pressure that deviates strongly from an ideal solution. Despite the limitations of the BLYP density functional theory used in the CPMD simulations.[19] it is comforting to note the relatively good agreement with the PMF calculated from the optimized Drude force field that was empirically adjusted to match the osmotic pressure data. The good accord between the two optimized models, as well as with the CPMD calculations, is a strong indication that the association of Na^+ and Cl^- in aqueous solution should be consistent with these 3 different approaches.

4. Conclusion

The osmotic pressure simulation is used to validate the polarizable classical Drude oscillator ionic parameters at finite concentrations. The ion models were previously optimized in conjunction with the polarizable SWM4-NDP water model, to be accurate in the infinite dilution limit.[5] However, there's no guarantee that the model is accurate at finite concentrations. The osmotic pressure MD simulation method offers a direct way to measure the thermodynamic properties of electrolyte solution at a wide range of concentrations, up to the solubility limit. In this study, the osmotic pressure of NaCl and KCl aqueous solutions are calculated from all-atom MD simulations for a variety of concentrations. It was found that the NaCl parameters underestimate the osmotic pressure significantly even at 1 m. Two sets of pair-specific parameters, electrostatic Thole shielding and the pairwise LJ parameter, are introduced to modify the ion pair potential in order to reproduce the experimental osmotic pressure for a wide range of concentrations. The fitting procedure is done carefully and hierarchically, from vacuum pair potentials, single ion pair PMFs in bulk to the osmotic pressure at finite concentrations. The fitting results show that the experimental osmotic pressure can be reproduced by either increasing the LJ radius by 0.15 Å from the default combination rule or by assigning a Thole shielding factor of 2.10. The PMF profiles of a single ion pair in the bulk and the ion pair interaction energy in vacuum show that the electrostatic Thole shielding strategy is able to shift the magnitude of the interaction energy and preserve the position of the energy-well at the same time, while increasing the pair-specific LJ radius ($R_{\min}$) pushes the most favorable interaction position further apart in both vacuum energy and bulk phase PMF profiles. Also, introducing the electrostatic Thole shielding at short distances seems more reasonable to take into account the lack of the charge transfer effect between atoms in the Drude polarizable models.

The second virial coefficients B_2^{ij}, which are directly related to the pair interactions, were also calculated from the ion pair PMF profiles. It is clear that the B_2^{ij} coefficients from the fitted pair-specific parameters are very close to the CPMD data, and much smaller than the one from the default Drude model. The sum of the B_2^{ij} coefficients of Na–Cl, Na–Na and Cl–Cl ion pairs is close to zero with the pair-specific parameters, but around −400 $Å^3$ with the default Drude parameters. The fact that B_2^{tot} is small suggests that the non-ideal contributions from the like-charged ion pairs and unlike-charged ion pairs cancelled each other out. It is again consistent with the fact that the osmotic pressure of NaCl solution is very close to the ideal solution.

This work illustrated the important role of the osmotic pressure simulation in validating and calibrating the empirical force field under realistic conditions. Small adjustments of the pair-specific interactions can improve the thermodynamic properties of electrolyte solution at finite concentrations significantly. Relying on the experimental osmotic pressure data to adjust empirically the force field gets around the problems arising from our limited ability to accurately predict these interactions from first principles. It is worth pointing out that the osmotic pressure MD simulation method can be used for a large variety of molecular fragments to validate and adjust the empirical force field for lipids or other biomolecular systems. It offers a novel strategy for advanced parameterization.

Acknowledgements

This work was supported by the National Institute of Health (NIH) through grants GM-072558 and GM-051501. Computational resources were provided by the Extreme Science and Engineering Discovery Environment (XSEDE), which is generously supported by grant CAO 93577 from NIH/NSI and by grant MCB-0920261 from NSF.

References

1 J. Åqvist, Ion–water interaction potentials derived from free energy perturbation simulations, *J. Phys. Chem.*, 1990, **94**(21), 8021–8024.
2 D. Beglov and B. Roux, Finite representation of an infinite bulk system: Solvent boundary potential for computer simulations, *J. Chem. Phys.*, 1994, **100**(12), 9050–9063.
3 A. A. Chen and R. V. Pappu, Quantitative characterization of ion pairing and cluster formation in strong 1 : 1 electrolytes, *J. Phys. Chem. B*, 2007, **111**(41), 6469–6478.
4 I. S. Joung and T. E. Cheatham, III, Determination of alkali and halide monovalent ion parameters for use in explicitly solvated biomolecular simulations, *J. Phys. Chem. B*, 2008, **112**, 9020–9041.
5 H. Yu, T. W. Whitfield, E. Harder, G. Lamoureux, I. Vorobyov, V. M. Anisimov, A. D. MacKerell, Jr and B. Roux, Simulating monovalent and divalent ions in aqueous solution using a Drude polarizable force field, *J. Chem. Theory Comput.*, 2010, **6**(3), 774–786.
6 Y. Luo and B. Roux, Simulation of osmotic pressure in concentrated aqueous salt solutions, *J. Phys. Chem. Lett.*, 2010, **1**(1), 183–189.
7 J. E. Enderby, Neutron scattering from ionic solutions, *Annu. Rev. Phys. Chem.*, 1983, **34**, 155.
8 R. Leberman and A. K. Soper, Effect of high salt concentrations on water structure, *Nature*, 1995, **378**, 364–366.
9 G. W. Neilson, P. E. Mason, R. Ramos and D. Sullivan, Neutron and X-ray scattering studies of hydration in aqueous solution, *Philos. Trans. R. Soc. London, Ser. A*, 2001, **359**, 1595–1591.
10 A. R. Altenberger and H. L. Friedman, Theory of conductance and related isothermal transport coefficients in electrolytes, *J. Chem. Phys.*, 1983, **78**, 4162–4173.
11 S. Weerasinghe and P. E. Smith, A Kirkwood–Buff derived force field for sodium chloride in water, *J. Chem. Phys.*, 2003, **119**(21), 11342.
12 B. Hess and N. F. A. van der Vegt, Cation specific binding with protein surface charges, *Proc. Natl. Acad. Sci. U. S. A.*, 2009, **106**(32), 13296–13300.
13 P. Ganguly, P. Schravendijk, B. Hess and N. F. A. van der Vegt, Ion pairing in aqueous electrolyte solutions with biologically relevant anions, *J. Phys. Chem. B*, 2011, **115**(13), 3734–3739.
14 J. Shen, C. Li, N. F. van der Vegt and C. Peter, Transferability of coarse grained potentials: Implicit solvent models for hydrated ions, *J. Chem. Theory Comput.*, 2011, **7**(6), 1916–1927.
15 A. Banerjee, A. Quigley, R. F. Frey, D. Johnson and J. Simons, Monte Carlo simulations of small ion-pair hydrate clusters: hydrated lithium nitrite (NO_2–:Li+ $(H_2O)_n$), *J. Am. Chem. Soc.*, 1987, **109**(4), 1038–1043.

16 S. Boudon, G. Wipff and B. Maigret, Monte Carlo simulations on the like-charged guanidinium–guanidinium ion pair in water, *J. Phys. Chem.*, 1990, **94**(15), 6056–6061.
17 K. T. No, K. Nam and H. A. Scheraga, Stability of like and oppositely charged organic ion pairs in aqueous solution, *J. Am. Chem. Soc.*, 1997, **119**(52), 12917–12922.
18 K. Cho, K. T. No and H. A. Scheraga, Ion pair interactions in aqueous solution: Self-consistent reaction field (SCRF) calculations with some explicit water molecules, *J. Phys. Chem. A*, 2000, **104**(27), 6505–6509.
19 J. Timko, D. Bucher and S. Kuyucak, Dissociation of NaCl in water from *ab initio* molecular dynamics simulations, *J. Chem. Phys.*, 2010, **132**(11), 114510.
20 W. G. McMillan and J. E. Mayer, The statistical thermodynamics of multicomponent systems, *J. Chem. Phys.*, 1945, **13**, 276–305.
21 J. G. Kirkwood and F. Buff, The statistical mechanical theory of solutions I, *J. Chem. Phys.*, 1951, **19**, 774–777.
22 R. Chitra and P. E. Smith, Molecular association in solution: A Kirkwood–Buff analysis of sodium chloride, ammonium sulfate, guanidinium chloride, urea, and 2,2,2-trifluoroethanol in water, *J. Phys. Chem. B*, 2002, **106**, 1491–1500.
23 S. Weerasinghe and P. E. Smith, A Kirkwood–Buff derived force field for the simulation of aqueous guanidinium chloride solutions, *J. Chem. Phys.*, 2004, **121**, 2180–2186.
24 S. Weerasinghe and P. E. Smith, A Kirkwood–Buff derived force field for methanol and aqueous methanol solutions, *J. Phys. Chem. B*, 2005, **109**, 15080–15086.
25 M. Kang and P. E. Smith, A Kirkwood–Buff derived force field for amides, *J. Comput. Chem.*, 2006, **27**, 1477–1485.
26 V. Pierce, M. Kang, M. Aburi, S. Weerasinghe and P. E. Smith, Recent applications of Kirkwood–Buff theory to biological systems, *Cell Biochem. Biophys.*, 2007, **50**, 1–22.
27 J. Rösgen, B. M. Pettitt and D. W. Bolen, Uncovering the basis for nonideal behavior of biological molecules, *Biochemistry*, 2004, **43**, 14472–14484.
28 H. Kokubo, J. Rösgen, D. W. Bolen and B. M. Pettitt, Molecular basis of the apparent near ideality of urea solutions, *Biophys. J.*, 2007, **93**, 3392–3407.
29 J. Rösgen, B. M. Pettitt and D. W. Bolen, An analysis of the molecular origin of osmolyte-dependent protein stability, *Protein Sci.*, 2007, **16**, 733–743.
30 J. L. Lebowitz and J. K. Percus, Mean spherical model for lattice gases with extended hard cores and continuum fluids, *Phys. Rev.*, 1966, **144**(1), 251.
31 R. Triolo, J. R. Grigera and L. Blum, Simple electrolytes in the mean spherical approximation, *J. Phys. Chem.*, 1976, **80**(17), 1858–1861.
32 A. R. Allnatt, Integral equations in ionic solution theory, *Mol. Phys.*, 1964, **8**, 533.
33 B. M. Pettitt and P. J. Rossky, Alkali halides in water: Ion–solvent correlations and ion–ion potentials of mean force at infinite dilution, *J. Chem. Phys.*, 1986, **84**(10), 5836–5844.
34 G. Lamoureux, E. Harder, I. V. Vorobyov, B. Roux and A. D. MacKerell, Jr, A polarizable model of water for molecular dynamics simulations of biomolecules, *Chem. Phys. Lett.*, 2006, **418**(1–3), 245–249.
35 R. A. Robinson, R. H. Stokes, *Electrolyte Solutions*, Dover Publications, Inc, 2002.
36 B. Hess, C. Holm and N. F. A. van der Vegt, Osmotic coefficients of atomistic NaCl (aq) force fields, *J. Chem. Phys.*, 2006, **124**, 164509.
37 A. P. Lyubartsev and A. Laaksonen, Osmotic and activity coefficients from effective potentials for hydrated ions, *Phys. Rev. E*, 1997, **55**(5), 5689–5696.
38 G. Lamoureux and B. Roux, Modeling induced polarization with classical Drude oscillators: Theory and molecular dynamics simulation algorithm, *J. Chem. Phys.*, 2003, **119**(6), 3025.
39 B. R. Brooks, C. L. Brooks III, A. D. MacKerell, Jr, L. Nilsson, R. J. Petrella, B. Roux, Y. Won, G. Archontis, C. Bartels, S. Boresch, A. Caflisch, L. Caves, Q. Cui, A. R. Dinner, M. Feig, S. Fischer, J. Gao, M. Hodoscek, W. Im, K. Kuczera, T. Lazaridis, J. Ma, V. Ovchinnikov, E. Paci, R. W. Pastor, C. B. Post, J. Z. Pu, M. Schaefer, B. Tidor, R. M. Venable, H. L. Woodcock, X. Wu, W. Yang, D. M. York and M. Karplus, CHARMM: the biomolecular simulation program, *J. Comput. Chem.*, 2009, **30**(10), 1545–1614.
40 J. C. Phillips, R. Braun, W. Wang, J. Gumbart, E. Tajkhorshid, E. Villa, C. Chipot, R. D. Skeel, L. Kale and K. Schulten, Scalable molecular dynamics with NAMD, *J. Comput. Chem.*, 2005, **26**, 1781–1802.
41 S. Nosé, A unified formulation of the constant temperature molecular dynamics methods, *J. Chem. Phys.*, 1984, **81**(1), 511.
42 W. G. Hoover, Canonical dynamics: Equilibrium phase–space distributions, *Phys. Rev. A: At., Mol., Opt. Phys.*, 1985, **31**(3), 1695.
43 G. J. Martyna, D. J. Tobias and M. L. Klein, Constant pressure molecular dynamics algorithms, *J. Chem. Phys.*, 1994, **101**(5), 4177–4189.

44 J. Ryckaert, G. Ciccotti and H. J. C. Berendsen, Numerical integration of the Cartesian equations of motion of a system with constraints: molecular dynamics of n-alkanes, *J. Comput. Phys.*, 1977, **23**(3), 327–341.
45 G. Martyna, M. Tuckerman, D. Tobias and M. Klein, Explicit reversible integrators for extended systems dynamics, *Mol. Phys.*, 1996, **87**(5), 1117–1157.
46 U. Essmann, L. Perera, M. L. Berkowitz, T. Darden, H. Lee and L. G. Pedersen, A smooth particle mesh Ewald method, *J. Chem. Phys.*, 1995, **103**(19), 8577.
47 P. Lagüe, R. W. Pastor and B. R. Brooks, Pressure-based long-range correction for Lennard-Jones interactions in molecular dynamics simulations: application to alkanes and interfaces, *J. Phys. Chem. B*, 2004, **108**(1), 363–368.
48 W. Jiang, D. J. Hardy, J. C. Phillips, A. D. MacKerell, Jr, K. Schulten and B. Roux, High-performance scalable molecular dynamics simulations of a polarizable force field based on classical Drude oscillators in NAMD, *J. Phys. Chem. Lett.*, 2011, **2**(2), 87–92.
49 S. Miyamoto and P. A. Kollman, Settle: An analytical version of the SHAKE and RATTLE algorithm for rigid water models, *J. Comput. Chem.*, 1992, **13**(8), 952–962.
50 T. Darden, D. York and L. Pedersen, Particle-mesh ewald—an *n*log(*n*) method for Ewald sums in large systems, *J. Chem. Phys.*, 1993, **98**, 10089–10092.
51 G. M. Torrie and J. P. Valleau, Nonphysical sampling distributions in Monte Carlo free-energy estimation: Umbrella sampling, *J. Comput. Chem.*, 1977, **23**(2), 187–199.
52 S. Kumar, J. M. Rosenberg, D. Bouzida, R. H. Swendsen and P. A. Kollman, The weighted histogram analysis method for free-energy calculations on biomolecules. I. The method, *J. Comput. Chem.*, 1992, **13**(8), 1011–1021.
53 D. Trzesniak, A. E. Kunz and W. F. van Gunsteren, A comparison of methods to compute the potential of mean force, *ChemPhysChem*, 2007, **8**(1), 162–169.
54 M. J. Frisch, G. W. Trucks, H. B. Schlegel, G. E. Scuseria, M. A. Robb, J. R. Cheeseman, J. A. Montgomery, Jr., T. Vreven, K. N. Kudin, J. C. Burant, J. M. Millam, S. S. Iyengar, J. Tomasi, V. Barone, B. Mennucci, M. Cossi, G. Scalmani, N. Rega, G. A. Petersson, H. Nakatsuji, M. Hada, M. Ehara, K. Toyota, R. Fukuda, J. Hasegawa, M. Ishida, T. Nakajima, Y. Honda, O. Kitao, H. Nakai, M. Klene, X. Li, J. E. Knox, H. P. Hratchian, J. B. Cross, V. Bakken, C. Adamo, J. Jaramillo, R. Gomperts, R. E. Stratmann, O. Yazyev, A. J. Austin, R. Cammi, C. Pomelli, J. W. Ochterski, P. Y. Ayala, K. Morokuma, G. A. Voth, P. Salvador, J. J. Dannenberg, V. G. Zakrzewski, S. Dapprich, A. D. Daniels, M. C. Strain, O. Farkas, D. K. Malick, A. D. Rabuck, K. Raghavachari, J. B. Foresman, J. V. Ortiz, Q. Cui, A. G. Baboul, S. Clifford, J. Cioslowski, B. B. Stefanov, G. Liu, A. Liashenko, P. Piskorz, I. Komaromi, R. L. Martin, D. J. Fox, T. Keith, M. A. Al-Laham, C. Y. Peng, A. Nanayakkara, M. Challacombe, P. M. W. Gill, B. Johnson, W. Chen, M. W. Wong, C. Gonzalez, J. A. Pople, *Gaussian 03*, Revision C.02, Gaussian, Inc., Wallingford, CT, 2004.
55 A. Stone, The Theory of Intermolecular Forces, *International Series of Monographs in Chemistry*, vol. 32, Clarendon Press, Oxford, 1996.
56 B. Thole, Molecular polarizabilities calculated with a modified dipole interaction, *Chem. Phys.*, 1981, **59**(3), 341–350.
57 E. Harder, V. M. Anisimov, T. Whitfield, A. D. MacKerell, Jr and B. Roux, Understanding the dielectric properties of liquid amides from a polarizable force field, *J. Phys. Chem. B*, 2008, **112**(11), 3509–3521.
58 L. Dang, B. M. Pettitt and P. J. Rossky, On the correlation between like ion-pairs in water, *J. Chem. Phys.*, 1992, **96**(5), 4046–4047.
59 B. Hess, C. Holm and N. van der Vegt, Modeling multibody effects in ionic solutions with a concentration dependent dielectric permittivity, *Phys. Rev. Lett.*, 2006, **96**(14), 147801.
60 J. McMillan and J. E. Mayer, The statistical thermodynamics of multicomponent systems, *J. Chem. Phys.*, 1945, **13**(7), 276–305.

Understanding ion–ion interactions in bulk and aqueous interfaces using molecular simulations

Liem X. Dang,* Xiuquan Sun, Bojana Ginovska-Pangovska, Harsha V. R. Annapureddy and Tai Ba Truong

Received 1st May 2012, Accepted 4th May 2012
DOI: 10.1039/c2fd20093g

In addition to its scientific significance, the distribution of ions in the bulk and at aqueous interfaces is also very important for practical reasons. Providing a quantitative description of the ionic distribution, and describing interactions between ions in different environments, remains a challenge, and is the subject of current debate. In this study, we found that interionic potentials of mean force (PMFs) and interfacial properties are very sensitive to the ion–ion interaction potential models. Our study predicted a Sr^{2+}–Cl^- PMF with no contact ion-pair state and a shallow solvent-separated ion-pair state. In addition, we were able to quantitatively capture the experimental X-ray reflectivity results of the aqueous salt interface of the Sr^{2+}–Cl^- ion-pair, and provided a detailed physical description of the interfacial structure for this system. We also predicted the X-ray reflectivity results for $SrBr_2$ and SrI_2 systems.

1. Introduction

In addition to their scientific significance, the interactions/distributions of ions in the bulk and at aqueous/air interfaces are also very important for practical applications.[1,2] Significant progress has been made in the research focused on quantifying the distribution of ions at the aqueous/air interface, and this research field remains very active. Molecular dynamics (MD) simulations have contributed significantly to the understanding of the behavior of ions at the aqueous/vapor interface.[3–23] However, because MD simulation results are often very sensitive to interaction models, the major challenge of this approach is to accurately describe the molecular interactions.[24,25] The interfacial structures predicted from MD simulations are still uncertain, even though the calculated interfacial properties, such as surface tension and surface potential, could reproduce corresponding values obtained from experimental efforts. The reason for this uncertainty is the indirect coupling of the experimental measurements and the interfacial structures.

X-Ray reflectivity measures the electron density distribution across the interface, which is directly related to the interfacial structure. Furthermore, the electron density distribution at the interface is relatively easy to obtain from an MD simulation. Therefore, combining experimental and theoretical results may significantly improve our understanding of the molecular nature of the ion aqueous/vapor interface. Because of its ability to directly measure the electron density distribution across the interface, X-ray reflectivity is a significant experimental methodology used in this field of research.[15] However, interpretation of the experimental results at the molecular scale is still an open question.

Chemical and Materials Sciences Division, Pacific Northwest National Laboratory, Richland, WA 93352. E-mail: liem.dang@pnnl.gov

In this report, we study the sensitivity of ion–ion interactions, and use a PMF approach to demonstrate the consequence of this sensitivity.[24,25] We also suggest ways to improve our understanding of the relationship between interactions of ions and interfacial structures. To this end, we describe our MD simulation studies of interionic PMFs for the Sr^{2+}–Cl^- ion pair in bulk water and at the aqueous interface. Because of the availability of experimental data on the X-ray reflectivity of aqueous interfaces, we believe this is an ideal system for benchmarking the ion–ion interactions. We extended the same approach to the $SrBr_2$ and SrI_2 systems, and due to the unavailability of experimental data, we present only our simulation results for comparison.

2. Computational method

In our MD simulations, the DC97 polarizable water model was used.[26] Atomic polarizabilities were present on Sr^{2+} and Cl^- for the polarizable simulations, and all of the ion parameters are given in Table 1. The ion–water and ion–ion models were constructed to reproduce experimental ion-hydration enthalpies, as well as the quantum chemical (QM) calculations of the minimum energy for the Sr^{2+}–Cl^- interactions. For the QM calculations, the Møller–Plesset 2 (MP2) level of theory[27] was used, with the DGPZVP basis set[28,29] and the Gaussian 98 code.[30] The ion–ion cross terms were calculated either using the Lorentz–Berthelot (LB) combining rules[31] or adjusted to match the QM calculations. The results for this exercise are summarized in Fig. 1, which shows that the LB approach gives slightly higher energy values for the ion-pair minima compared to the minima calculated with the parameters matching the QM calculations.

The system under investigation consisted of 700 H_2O molecules and one Sr^{2+}–Cl^- ion pair. Simulations were carried out under ambient conditions in a cubic box with dimensions of $27.6 \times 27.6 \times 27.6$ Å^3. The particle mesh Ewald summation technique was used to evaluate long-range electrostatic interactions.[32] The following expression was used to calculate the solute–solute mean force as an average over the different solvent configurations:

$$F(r) = \frac{1}{2}\left\langle \vec{r}_u \cdot \left(\vec{F}_A - \vec{F}_B\right)\right\rangle. \quad (1)$$

In this expression, F_A and F_B are the forces acting on the solutes. The term r_u, which is a unit vector along the AB (*i.e.* the distance between the two solutes) direction, is defined as follows:

$$\vec{r}_u = \vec{r}_{AB}/|r_A - r_B| \quad (2)$$

Table 1 The potential parameters for water and ions used in the study[a]

Molecule	Atom type	σ (Å)	ε (kcal mol^{-1})	q (e)	α (Å^3)
H_2O	O	3.221	0.1825	0.000	1.444
	H	0.000	0.0000	0.519	0.000
	M	0.000	0.0000	−1.038	0.000
Sr^{2+}	Sr	3.167	0.1000	2.000	0.860
Cl^-	Cl	4.410	0.1000	−1.000	3.690

[a] The modified interaction parameters between Sr^{2+} and Cl^- are 3.9912 Å and 0.1 kcal mol^{-1} for σ and ε, respectively.

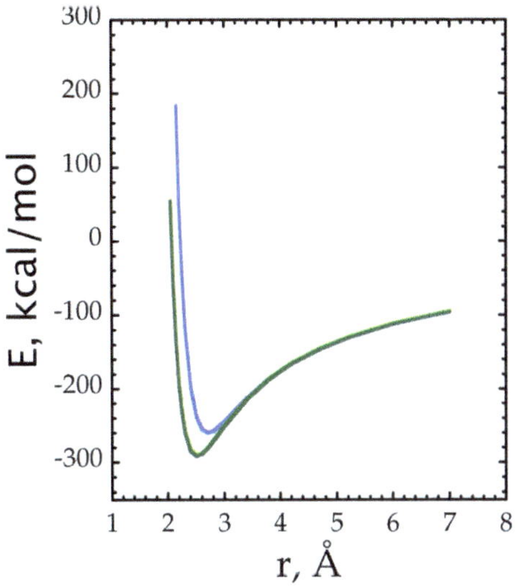

Fig. 1 Calculated potential energy surfaces for Sr^{2+}–Cl^- ion pairs using the LB combining rules (green) and a modified potential model (blue).

and PMF, $W(r)$, is calculated as

$$W(r) = \int_{r_o}^{r_s} \langle F_r \rangle dr. \tag{3}$$

The distance between the ions was increased in 0.25 Å increments. At each separation, the average $F(r)$ was determined from 1.5 ns of simulation time, preceded by 500 ps of equilibration. The uncertainties of the PMFs were ±0.2 kcal mol^{-1}, which were estimated by calculating the averaged force (the corresponding PMFs) over four equally spaced time periods during the production.

The physical property that we use for comparison between experiments and simulations is the structure factor. Below, we describe the methods used to calculate the structure factors from both experimental and simulation results. For simulation results, the structure factor is defined as follows:

$$\Phi(q_z) = \frac{1}{\rho_b} \int_{-\infty}^{\infty} \left[\frac{d\rho(z)}{dz}\right] \exp(iq_z z) dz \tag{4}$$

Here, $\rho(z)$ is the electron density profile along the surface normal direction z, and ρ_b is the electron density of the bulk solution. The electron density profile can be obtained from the number density profile, $n(z)$, according to the following expression:

$$\rho(z) = \sum_i n^i(z) Z_i \tag{5}$$

where Z_i is the atomic number of species i. The structure factor calculated directly from simulated electron density profile includes the thermally induced amplitudes. Using the following equation, a correction factor needs to be applied to obtain the intrinsic structure factor:

$$|\Phi_{int}(q_z)|^2 = |\Phi(q_z)|^2 \exp(\sigma_{sim}^2 q_z^2) \tag{6}$$

where σ_{sim} is the thermal roughness of the surface, which can be estimated from the Capillary Wave Theorem (see below):[33]

$$\sigma_{sim}^2 \approx \frac{k_B T}{2\pi\gamma} \ln \frac{k_{max}}{k_{sim}} \tag{7}$$

In eqn (7), k_B is the Boltzmann constant, T is the temperature of the system, γ is the surface tension, $k_{max} = \pi/a$, and $k_{sim} - \pi/S$ are the cutoffs of the capillary wave's spectrum, due to the finite size of molecule a and the finite size of lateral box size S.

3. Results and discussion

In Fig. 2, our computed PMF that used the LB combining rules (green line) shows stable contact-ion pair (CIP) and solvent-separated ion-pair (SSIP) states centered at 2.75 and 5.00 Å, respectively. The CIP and SSIP calculated from this simulation have free energies of −3.2 and −0.6 kcal mol^{-1}, respectively, with a 5.5 kcal mol^{-1} barrier to CIP dissociation. The CIP well depth is much deeper than the corresponding SSIP well depth, which is indicative of a fairly stable contact ion-pair. The computed association constant, k_a = 2.6 L mol^{-1}, and the corresponding binding free energy, −0.6 kcal mol^{-1}, are significantly smaller than the corresponding association constant, k_a of 43 L mol^{-1} and the binding free energy of −2.3 kcal mol^{-1} for the Ca^{2+}–Cl^- system.[24] This can be attributed to the significantly larger interaction energy for the Ca^{2+}–Cl^- system compared to the Sr^{2+}–Cl^- system.

There is no experimental result available to directly compare our PMF results, but we can use this set of potential models to predict the X-ray reflectivity of the aqueous interface of an electrolyte solution, and evaluate the potential model by comparing this property against measured data. Our simulations consisted of 4000 H_2O molecules, 180 Sr^{2+}, and 360 Cl^- ions, which corresponded to a 2.7 M $SrCl_2$ aqueous solution under ambient conditions. The simulations were carried out in boxes with dimensions of 52 × 52 × 150 Å^3, with liquid occupying approximately 50 Å of the simulation box along the z-axis and vapor occupying the remainder of the box. This theoretical configuration resulted in two liquid/vapor interfaces. We employed Langevin dynamics with a 5 ps^{-1} collision frequency to keep the temperature constant near 298 K.[34] After a 10 ns equilibration period with a time step of 2 fs, data for analysis were averaged over 10 ns in the NVT ensemble. The structure factor calculated from simulation results and corresponding experimental data are illustrated in Fig. 3.

It shows a broad peak centered at ~0.45 Å^{-1}. The peak is a result of the single maximum of the electron density profile across the interface in the MD simulation,

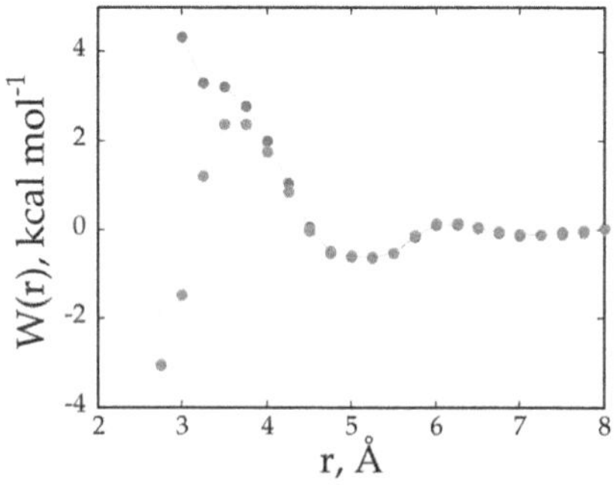

Fig. 2 Computed PMFs for the Sr^{2+}–Cl^- ion pair interaction in water at 300 K. The green circles are results obtained using LB combining rules. The blue circles are results obtained with a modified Sr^{2+}–Cl^- interaction.

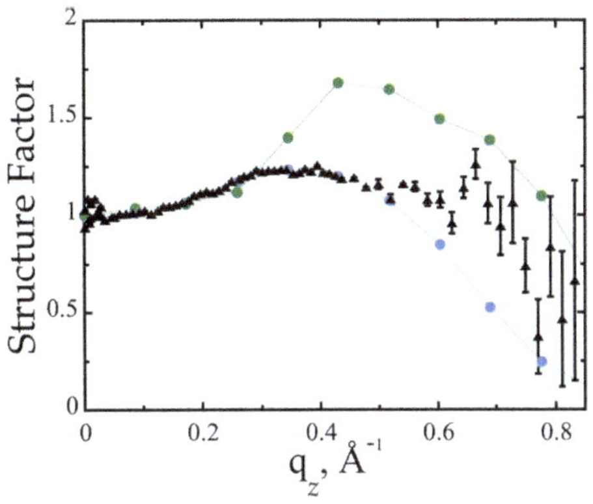

Fig. 3 Structure factors from experiments (black triangles), MD simulations using LB combining rules (green circles), and modified Sr^{2+}–Cl^- interaction potential (blue circles), respectively.

and the resonance frequency is determined by the width of the peak. The amplitude of the peak of the structure factor is prescribed to the degree of enhancement of the electron density at the interface corresponding to the bulk value. Compared to the structure factor calculated from X-ray reflectivity experiments, the resonance frequency is higher and the amplitude is larger for the one calculated from the MD simulation with the LB parameters for the ion–ion potential (green circles in Fig. 3). The comparison indicates that the distribution of electron density is too compact and the electron density is too high at the interface in the MD simulation. Consequently, the number densities of ions at the interface are overestimated in this MD simulation.

As stated above, the cross-term Lennard–Jones parameters for the ion–ion potential model were initially constructed using the LB combining rules.[31] However, we found that this approach usually overestimates the minimum interaction energy when compared to accurate, high-level electronic structure calculations. We are interested in the accuracy of the intermolecular interaction and what effect it has on the computed ion-pair PMFs in aqueous solutions. To this end, we performed electronic structure calculations on the Sr^{2+}–Cl^- complex to determine its minimum energy and distance. *Ab initio* calculations using the MP2 method[27] in conjugation with DGDZVP basis sets[28,29] gave a minimum potential energy of 258.4 kcal mol^{-1}† for the $SrCl^+$ ion pair, and a distance $r_{(Sr-Cl)} = 2.6$ Å. We adjusted the Sr^{2+} and Cl^- Lennard–Jones parameters to match the QM calculations (reported in Table 1). Additional PMF calculations were carried out using this new set of parameters for the Sr^{2+}–Cl^- potential model to assess its effect on the computed PMF. The resultant PMF also is displayed in Fig. 2 (blue line). This PMF is significantly different from the original PMF, and with these parameters, we find that the CIP is greatly diminished. There is a residual of the CIP near 3 Å, but it is very small and is repulsive. It also is interesting to note that beyond the barrier to the CIP (at $r = 3.75$ Å), the computed PMFs for both models are essentially identical.

Fig. 3 compares the structure factors calculated from simulations that used LB combining rules, the new set of parameters, and experimental data. As mentioned in the theoretical background section and in previous studies, the structure factor

† This value was compared and is in excellent agreement with calculations carried out at a CCSD(T) level of theory with a larger basis set (*i.e.* ECP28MDF [I. S. Lim, H. Stoll and P. Schwerdtfeger, *J. Chem. Phys.*, 2006, **124**, 034107) for Sr and aug-cc-pVQZ (*i.e.*, D. E. Woon and T. H. Dunning, Jr., *J. Chem. Phys.*, 1993, **98**, 1358) for Cl, which gives a minimum potential energy of 254.8 kcal mol^{-1}.

can be calculated accurately from both experimental and simulation results. Furthermore, the structure factor links the experimental measured spectrum directly to the molecular level interfacial structure determined from the MD simulations. As such, it is the appropriate property for comparing results obtained from experiments and simulations, which can improve our understanding of the relationship between the X-ray reflectivity and the interfacial structure. However, in the computational procedure, there are some uncertainties associated with calculating the intrinsic structure factors from both experiments and simulations. The uncertainties are caused mainly by estimations of the induced wave amplitudes, but they have little effect on the characteristics of the structure factors. Accordingly, to fit the experimental value, a Debye–Waller-like factor is used to optimize the calculated structure factors from MD simulations. Therefore, the comparison between experimental and simulation results focuses on the features of the structure factor, such as resonance frequency and the curvature trend. The calculated structure factors from both experiments and MD simulations are less accurate in the higher q_z regions. In simulations, the electron density could be accurately computed from the atomic number of species only when q_z is small. In experiments, the signal is reduced significantly with increased q_z, and the fluctuations increase significantly, introducing large error bars in the measured values. Consequently, the comparison of structure factors calculated from experiments and MD simulations at the low q_z value regions is more meaningful and accurate.

Because of the large fluctuation when q_z is larger than 0.6 Å^{-1}, the most distinguishable feature of the structure factor calculated from experimental X-ray reflectivity measurement is the broad peak centered at ~0.35 Å^{-1}, ranging from 0.1 to 0.6 Å^{-1}. The calculated structure factors from MD simulations using both old and new potentials are optimized (with only one Debye–Waller-like factor) to fit to the experimental results, and the fits are weighted more at smaller values of q_z. The structure factor calculated from simulation using an old potential shows a broad peak centered at ~0.45 Å^{-1}.

The structure factor calculated from a simulation using a modified potential shows features that are very similar to those obtained from experiment. Especially when q_z is smaller than 0.5 Å^{-1}, the simulation quantitatively reproduces the experimental results. This is strong evidence that the electron density profile calculated from the MD simulation demonstrates the same features as the experimental one, and the corresponding ion distribution and interfacial structure from the MD simulation using the new potential may reasonably represent the experimental results. The resonance frequency at 0.35 Å^{-1} of the structure factor corresponds to a wavelength of ~18 Å^{-1} for the electron density profile. This long wavelength is attributed to the fluctuation of the number density profile across the interface in the simulation, including both the strong ion double layer at the interface and the weak ion double layer towards the bulk phase. The simulation results suggest that enhancement of the electron density at the interface, compared to bulk phase, is prerequisite to reproducing the experimental results. However, the precise fine structure of the liquid/vapor interface for the aqueous ion solution in MD simulation is essential for the structure factor to be consistent with the experimental results. In principle, one electron density profile could be derived from different sets of number densities as the contribution to the total electron density profile is from three species, and the structure factor can be calculated with different combinations of electron density profiles, interface thickness, and surface roughness. However, to the best of our knowledge, this MD simulation study, using a modified ion–ion potential model, provides the most reasonable prediction of the ions distribution across the liquid/air interface in an $SrCl_2$ aqueous solution.

We performed additional interfacial calculations using the new force field. The normalized number density profiles of all species calculated from MD simulations are shown in Fig. 4a.

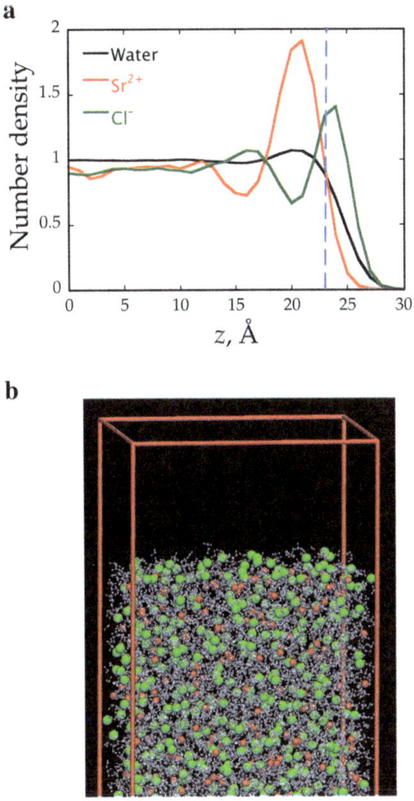

Fig. 4 (a) Normalized number density profiles of water (black), Sr^{2+} (red), and Cl^- (green) across the liquid/vapor interface in MD simulations carried out using a modified Sr^{2+}–Cl^- interaction potential at 298 K. The blue dashed line denotes the Gibbs dividing surface. (b) A snapshot was taken from a MD simulation that used a modified Sr^{2+}–Cl^- potential model. Water molecules are represented as cyan lines, Sr^{2+} ions are represented as red spheres, and Cl^- ions are represented as green spheres.

The results clearly show an enhanced density for Cl^- at the interface. The density of Cl^- at the interface is about 1.5 times higher than the bulk value. No enhancement for Sr^{2+} density was observed at the interface. In contrast, a strong increase of the density for Sr^{2+} compared to the bulk value (by a factor of approximately two) was observed about 3 Å from the maximum enhanced density of Cl^- inside the bulk solution. The position of maximum enhancement for Sr^{2+} density, corresponds to a decrease of the Cl^- density. A strong double-layer ion structure is formed at the interface, and this double-layer structure vanishes very quickly away from the interface and towards the bulk solution. Only a very weak double-layer structure was observed next to the first strong double layer formed at the interface, and then the densities for both cations and anions converge to the bulk values. The new computed ion density profile for $SrCl_2$ is very similar to the one for the monovalent cation NaCl aqueous solution, including the double-layer structure, and the degree of enhancements for both cations and anions. This indicates a similar surface structure for divalent and monovalent cation aqueous solutions, and the anions are the major species present at the liquid/air interfaces. A snapshot from the simulation using the new potential is shown in Fig. 4b. The anions are evident at the interface, and cations are repelled from the interface. Both ions are distributed uniformly in the bulk phase, and few contact ion pairs were formed in the solution. A snapshot that illustrates our discussion is shown in Fig. 4b.

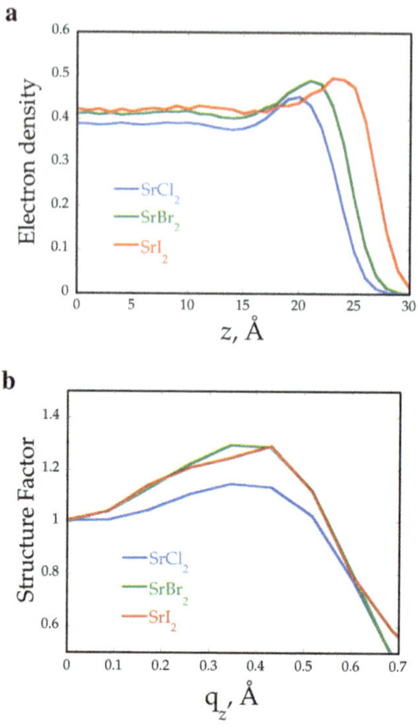

Fig. 5 (a) Calculated electron density profiles for $SrCl_2$ (blue line), $SrBr_2$ (green line), and SrI_2 (red line). (b) Calculated structure factors for $SrCl_2$ (blue line), $SrBr_2$ (green line), and SrI_2 (red line).

We have applied our approach to the $SrBr_2$ and SrI_2 systems, and the predicted electron density profiles and the corresponding X-ray reflectivity for these systems are presented in Fig. 5a and 5b.

We did not make comparisons with experimental results for the $SrBr_2$ and SrI_2 system, because to the best of our knowledge, no experimental structure factors are available. However, we used the same approach and accurately predicted the structure factor for the $SrCl_2$ as shown in Fig. 3. Thus, our predicted values for the $SrBr_2$ and SrI_2 systems could be used for comparisons with future experimental work.

4. Conclusions

Molecular dynamics simulations using different potential parameters were carried out to study the liquid/vapor interface of a divalent cation salt, $SrCl_2$, aqueous solution. The potential parameters for the ions were optimized initially to reproduce the experimental structural and thermodynamic properties between the ions and water. The number density profiles of the ions at the normal direction of the surface indicate that both Sr^{2+} and Cl^- are highly enhanced at the liquid/vapor interface, and many contact ion pairs are observed in the system. These simulation results suggest an overestimation of the interaction between Sr^{2+} and Cl^-. The calculated number density profiles of ions in the simulation using a modified potential for the ion–ion interaction show formation of an ion double-layer structure at the liquid/vapor interface, where the density of anion Cl^- is enhanced at the interface, compared to the bulk value, because of the large polarizability. In contrast, the cation Sr^{2+} is repelled from the interface, and the density reaches a maximum value several angstroms away from the interface, where the density of Cl^- decreases. This structure is similar to the one observed in the monovalent cation salt aqueous solutions. Consequently, the total electron density profiles at the surface normal direction calculated from the number densities in the MD simulations using two different potentials show dissimilar features. The peak of the electron density profile at the interface in the simulation using an old potential is more abrupt, and the ratio of the maximum electron density to the bulk value is much higher than the one calculated from the simulation using the new potential. Compared to experimental results, the resonance frequency of the structure factor computed from the electron density profile in the simulation using the LB combining rule approach is higher, and the amplitude of the peak is much larger. In contrast, the structure factor obtained from the simulation using the new potential quantitatively agrees with experimental results in the low-frequency region. This work suggests that the ion distributions and liquid/vapor interfacial structure of the $SrCl_2$ aqueous solution obtained from these MD simulations can be used to characterize those obtained experimentally.

Acknowledgements

This work was supported by the U.S. Department of Energy (DOE), Office of Basic Energy Sciences (BES), Division of Chemical Sciences, Geosciences and Biosciences. Pacific Northwest National Laboratory is a multiprogram national laboratory operated for DOE by Battelle. The calculations were carried out using computer resources provided by BES.

References

1 I. Benjamin, *Chem. Rev.*, 1996, **96**, 1449.
2 P. Davidovits, C. E. Kolb, L. R. Williams, J. T. Jayne and D. R. Worsnop, *Chem. Rev.*, 2006, **106**, 1323.
3 I. Langmuir, *J. Am. Chem. Soc.*, 1917, **39**, 1848.
4 J. W. Gibbs. *The Collected Works of J. Willard Gibbs*; Longmans: New York, 1928.
5 G. Jones and W. A. Ray, *J. Am. Chem. Soc.*, 1937, **59**, 187.
6 M. Dole and J. A. Swartout, *J. Am. Chem. Soc.*, 1940, **62**, 3039.
7 B. Case and R. Parsons, *Trans. Faraday Soc.*, 1967, **63**, 1224.
8 J. E. B. Randles, *Phys. Chem. Liq.*, 1977, 7, 107.
9 C. G. Barraclough, P. T. Mctigue and Y. L. Ng, *J. Electroanal. Chem.*, 1992, **329**, 9.
10 D. F. Liu, G. Ma, L. M. Levering and H. C. Allen, *J. Phys. Chem. B*, 2004, **108**, 2252.
11 P. B. Petersen, J. C. Johnson, K. P. Knutsen and R. J. Saykally, *Chem. Phys. Lett.*, 2004, **397**, 46.
12 P. B. Petersen and R. J. Saykally, *Chem. Phys. Lett.*, 2004, **397**, 51.
13 G. Andersson, T. Krebs and H. Morgner, *Phys. Chem. Chem. Phys.*, 2005, **7**, 2948.
14 O. Hofft, A. Borodin, U. Kahnert, V. Kempter, L. X. Dang and P. Jungwirth, *J. Phys. Chem. B*, 2006, **110**, 11971.
15 E. Sloutskin, J. Baumert, B. M. Ocko, I. Kuzmenko, A. Checco, L. Tamam, E. Ofer, T. Gog, O. Gang and M. Deutsch, *J. Chem. Phys.*, 2007, **126**.
16 S. Ghosal, J. C. Hemminger, H. Bluhm, B. S. Mun, E. L. D. Hebenstreit, G. Ketteler, D. F. Ogletree, F. G. Requejo and M. Salmeron, *Science*, 2005, **307**, 563.
17 L. X. Dang and T. M. Chang, *J. Phys. Chem. B*, 2002, **106**, 235.
18 P. Jungwirth and D. J. Tobias, *J. Phys. Chem. B*, 2002, **106**, 6361.
19 B. C. Garrett, *Science*, 2004, **303**, 1146.
20 P. Jungwirth and D. J. Tobias, *Chem. Rev.*, 2006, **106**, 1259.
21 C. D. Wick and L. X. Dang, *Chem. Phys. Lett.*, 2008, **458**, 1.
22 X. Q. Sun and L. X. Dang, *J. Chem. Phys.*, 2009, **130**, 124709.
23 X. Q. Sun, C. D. Wick and L. X. Dang, *J. Phys. Chem. B*, 2009, **113**, 13993.
24 L. X. Dang, T. B. Truong and B. Ginovska-Pangovska, *J. Chem. Phys.*, 2012, **136**, 126101.
25 J. Timko, A. De Castro and S. J. Kuyucak, *Chem. Phys.*, 2011, **134**, 204510.
26 L. X. Dang and T. M. Chang, *J. Chem. Phys.*, 1997, **106**, 8149.
27 C. Moller and M. S. Plesset, *Phys. Rev.*, 1934, **46**, 0618.
28 N. Godbout, D. R. Salahub, J. Andzelm and E. Wimmer, *Can. J. Chem.*, 1992, **70**, 560.
29 C. Sosa, J. Andzelm, B. C. Elkin, E. Wimmer, K. D. Dobbs and D. A. Dixon, *J. Phys. Chem.*, 1992, **96**, 6630.

30 M. J. Frisch; *et al. Gaussian 98, Revision A.11.4*; Gaussian, Inc., Pittsburgh PA, 2002. Gaussian 98, Revision A.11.4.
31 J. P. Hansen, I. P. McDonald, *Theory of Simple Liquid* Academic, London, 1986.
32 U. Essmann, L. Perera, M. L. Berkowitz, T. Darden, H. Lee and L. G. J. Pedersen, *Chem. Phys.*, 1995, **103**, 8577.
33 F. P. Buff, R. A. Lovett and F. H. Stilling, *Phys. Rev. Lett.*, 1965, **15**, 621.
34 M. P. Allen; D. J. Tildesley. *Computer Simulation of Liquids*; Oxford University Press Inc.: New York, 1987.

A new structural technique for examining ion-neutral association in aqueous solution†

Philip E. Mason,*[a] George W. Neilson,[c] David L. Price,[e] Marie-Louise Saboungi[d] and John W. Brady[b]

Received 23rd April 2012, Accepted 29th May 2012
DOI: 10.1039/c2fd20081c

The method of intramolecular coordination number concentration invariance (ICNCI) is used on neutron diffraction with isotopic substitution (NDIS) measurements of aqueous solutions to separate the intra- and intermolecular contributions to the total intensities. Molecular dynamics simulations of corresponding systems are then used to interpret the ICNCI function. It is found that the ICNCI function (characterized by two concentration measurements) is sensitive specifically to intermolecular association and that the molecular dynamics can successfully replicate this function in the cases of the neutral species xylose and pyridine in aqueous solution. ICNCI functions can also be obtained by the addition of a cosolute (such as adding GdmCl or Gdm_2SO_4 to pyridine solutions). In that case it is found that molecular dynamics can replicate the ICNCI function for the addition of GdmCl to pyridine, but fails to successfully replicate the same function for the addition of Gdm_2SO_4. This result implies that the interaction of pyridine with guanidinium sulfate is over-estimated in these MD simulations, and is of significant importance to the use of molecular dynamics simulations to elucidate an atomic level understanding of the Hofmeister series.

Introduction

Key to obtaining an atomic level understanding of the Hofmeister series is acquiring the structural changes that such ions effect on the solutions. Obtaining insight into the structure of aqueous solutions is hampered by the fact that what is measured is rarely what is desired. The most frequently required structural insight into biochemical systems relates to their intermolecular associations. Here we describe the application of a new structural technique, internal coordination number concentration invariance (ICNCI), to a variety of solutions of both xylose and pyridine and show how this technique can yield useful structural information on the intermolecular associations in these solutions.

Direct structural techniques (*e.g.* neutron diffraction) are clearly better for examining the structure of condensed matter than the less direct inferences obtained from spectroscopic methods. Neutrons offer several advantages over X-rays in diffraction

[a]*Institute of Organic Chemistry and Biochemistry, Academy of Sciences of the Czech Republic, Flemingovo nám. 2, 16610 Prague 6, Czech Republic. E-mail: phil.mason@marge.uochb.cas.cz*
[b]*Cornell University, Dept Food Science, Ithaca, NY 14853, USA*
[c]*H.H. Wills Physics Laboratory, University of Bristol, BS8 1TL, UK*
[d]*Centre de Recherche sur la Matière Divisée, 1 bis rue de la Férollerie, 45071 Orléans, France*
[e]*Conditions Extrêmes et Matériaux: Haute Température et Irradiation, 45071 Orléans Cedex 2, France*

† Electronic supplementary information (ESI) available. See DOI: 10.1039/c2fd20081c

measurements, with the primary one being that they are scattered by the nuclei themselves, which are essentially point sources, while X-rays are scattered by the diffuse electron clouds. Another advantage is that hydrogen atoms are very visible in neutron diffraction while they are almost invisible to X-rays. Furthermore, hydrogen and deuterium are both readily available and clearly distinguishable in neutron diffraction experiments. Nonetheless, neutron diffraction from an aqueous solution, which involves correlations from every atom to every other atom, is typically so complicated as to defy easy interpretation. For example, even with a relatively simple solution like pyridine in water, there are five types of nuclei: the oxygen and the exchangeable hydrogens (H_{ex}) of the water, and the non-exchangeable hydrogens (H_{non}) and carbon and nitrogen atoms of the solute, producing 15 different pairwise correlation functions, so that the neutron diffraction pattern from the solution is a weighted sum of all 15 pairwise distribution functions. As a result, the interpretation of the diffraction pattern in terms of the individual atom–atom correlations presents a formidable challenge. Moreover, as the number of types of atoms in the sample increases, these correlations become progressively more difficult to disentangle.

Neutron diffraction with isotopic substitution (NDIS) was developed to simplify the interpretation of the structural data obtained from diffraction experiments.[1] This method involves performing paired experiments in which one atom type is replaced by an isotope of the same element with a different neutron scattering length, under the important assumption that the overall structure is not altered by the substitution. Subtracting the structure factors for the two experiments then eliminates all correlations except those involving the substituted atom.[2] In the pyridine example cited above, such a first-order difference experiment can be performed using H/D substitution on the H_{non} atoms of the pyridine, reducing the 15 correlations in the total scattering measurement to just five: H_{non}–C, H_{non}–N, H_{non}–H_{ex}, H_{non}–O and H_{non}–H_{non}. Such experiments have been effective in the study of the solution structure of many simple solutes such as rare gases,[3] mono-atomic ions,[4,5] and approximately spherical molecular solutes.[6] In recent years, NDIS methods has been applied to much more complex molecular species including *tert*-butyl alcohol,[7,8] ethanol, phenol, D-glucose,[9] amino acids,[10–12] and polymer electrolytes.[13,14] However, the interpretation of these experiments is impeded by the intramolecular correlations, which tend to produce very sharp and large correlations compared with the intermolecular (solute–water and solute–solute) correlations that are of primary interest. To circumvent this difficulty we developed the technique of internal coordination number concentration invariance (ICNCI).[15] ICNCI uses the fact that the coordination numbers of the intramolecular correlations, particularly for rigid solutes such as pyridine, are relatively independent of concentration, while intermolecular correlations are strongly dependent on it. It is therefore possible to perform two NDIS measurements at different concentrations and numerically separate the components that vary with concentration from those that do not. Interpretation of this ICNCI function using information from molecular dynamics simulations of the same solution can lead to insight into the solvation structure of the species in solution. In this work we use the ICNCI technique on aqueous solutions of D-xylopyranose (doubly-substituted with deuterium at the non-exchangeable C5 hydrogen positions, Fig. 1), and pyridine (H/D substituted on all the non-exchangeable hydrogens, Fig. 1), and also on solutions of pyridine in GdmCl and Gdm_2SO_4.

Procedures

While the method of ICNCI is described in detail elsewhere,[15] it is necessary to declare the functions measured in this paper. The total measured intensity for the sample ($F(Q)$) is given by

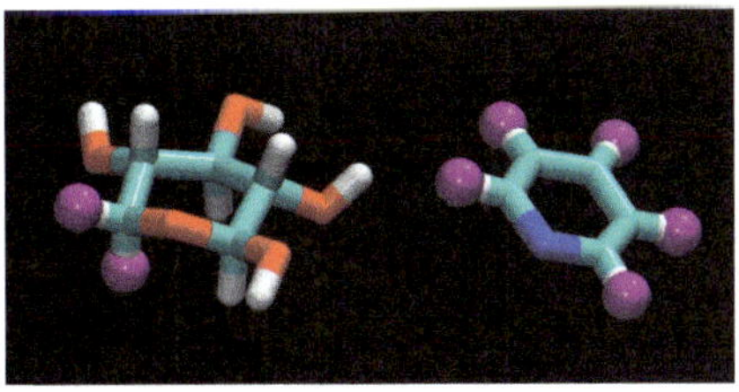

Fig. 1 Left: β-D-xylopyranose in the $^4C_1$ conformation; right: pyridine. In both cases the substituted hydrogen atoms are shown as purple.

$$F(Q) = \sum_{\alpha} \sum_{\beta} c_\alpha c_\beta b_\alpha b_\beta (S_{\alpha\beta}(Q) - 1) \tag{1}$$

where c_α is the atomic concentration of species α, b_α is the coherent neutron scattering length, and the sums are over all the atomic species in solution. In this equation, $S_{\alpha\beta}(Q)$ is the partial structure factor for atoms α and β. The partial radial distribution function $g_{\alpha\beta}(r)$ is derived from $S_{\alpha\beta}(Q)$ by Fourier transformation.

While, in principle, it does not make any difference what the substituted nucleus is, we will first discuss the example of xylose. The aqueous xylose solution in this study contains 6 types of nuclei: H_{ex} (exchangeable hydrogen atoms on water and xylose); O_w (water oxygen); C (carbon in xylose); O_x (oxygen in xylose); H_{non} (the non-exchangeable hydrogen atoms of xylose, excluding the substituted nuclei); and H5 (the substituted nuclei, see below). Note that while each of the unlabeled non-exchangeable protons H_{non} is in a unique stereochemical environment, they are considered to be effectively equivalent here. Thus, the $F(Q)$ of this solution is the sum of 21 weighted partial structure factors. Two $F(Q)$ measurements are then made on chemically identical solutions, the first containing xylose with the H5 positions containing natural abundance hydrogen, the second with these hydrogens substituted with deuterium. When the first order difference is applied to the $F(Q)$s of these two solutions, all of the correlations exactly cancel except those that involve structure factors containing the substituted nuclei H5. This difference reduces the number of summed correlations from 21 in $F(Q)$ to 6 in $\Delta S^X_{H5}(Q)$,

$$\begin{aligned}\Delta S^X_{H5}(Q) &= F_{d_5,d_5\text{-xylose}}(Q) - F_{\text{neutral-xylose}}(Q)\\ &= A[S_{H5Hex}(Q) - 1] + B[S_{H5Ow}(Q) - 1] + C[S_{H5Ox}(Q) - 1] + D[S_{H5C}(Q) - 1] +\\ &\quad E[S_{H5Hnon}(Q) - 1] + F[S_{H5H5}(Q) - 1]\end{aligned} \tag{2}$$

The superscript X indicates that the function contains only correlations from the substituted nuclei H5 to all other nuclei in the system. The transform of this function gives the single atom radial distribution function $G^X_{H5}(r)$,

$$\begin{aligned}G^X_{H5}(r) = &A\, g_{H5Hex}(r) + B\, g_{H5Ow}(r) + C\, g_{H5Ox}(r) + D\, g_{H5C}(r)\\ &+E\, g_{H5Hnon}(r) + F g_{H5H5}(r) - G\end{aligned} \tag{3}$$

where G = A + B + C + D + E + F and $g_{H5X}(r)$ is the radial distribution function for atoms X around the substitution-labeled position H5, and A–F are the neutron scattering prefactors for each atom type (which in each case is equal to $c_{H5}c_X\Delta b_H b_X$, where c_X is the atomic concentration of each atom, b_X is the coherent neutron scattering length of each atom type X, and Δb_H is the contrast in the coherent scattering lengths of the substituted nuclei. In this case $\Delta b_H = b_D - b_H$ (deuterium and protium, respectively). In this study, two first order differences on xylose were obtained, the first at 2 molal and the second at 5 molal. The neutron scattering prefactors *A*–*F* for first order differences at these concentrations are given in Table 1.

The method of ICNCI was then applied to these solutions. If the conformation of the sugar does not change significantly with concentration, then it can be shown[15] that an appropriately scaled subtraction of these two data sets will yield a difference function that contains no correlations that are not sensitive to the concentration of the sugar. For reasons that are explained elsewhere,[15] this scaled subtraction is slightly different for the real space data than for the reciprocal space data. For the real-space data this difference can be expressed as:

$$\Delta G^{X}_{H5}(r) = \frac{\rho_2}{c_{H5_2}} G^{X}_{H5}(r)_2 - \frac{\rho_1}{c_{H5_1}} G^{X}_{H5}(r)_1 \cdot \tag{4}$$

and for the Q space data it is expressed as:

$$\Delta\Delta S^{X}_{H5}(Q) = \frac{1}{c_{H5_2}} \Delta S^{X}_{H5}(Q)_2 - \frac{1}{c_{H5_1}} \Delta S^{X}_{H5}(Q)_1, \tag{5}$$

This technique may be applied equally well with the addition of a co-solute to the solution. For example, if a first order difference is measured with 3 m pyridine, and a second measured for 3 m pyridine in solution with 3 m GdmCl, application of both eqn (4) and (5) will yield an ICNCI function that contains no intramolecular correlations. This ICNCI function then gives information on the difference in the solvation of pyridine in water and in 3 m GdmCl solution.

Experimental

Seven first-order difference experiments are conducted in the present study (summarized in Table 2). These consisted of aqueous solutions of 2 and 5 m xylose (with H/D substitutions on both of the C5 hydrogen atoms); 1, 3 and 5 m pyridine (with H/D substitutions on all of the non-exchangeable hydrogen atoms); 3 m pyridine with 3 m GdmCl; and 3 m pyridine with 1.5 m Gdm_2SO_4.

Samples of D-xylose doubled-labeled with deuterium (>95% enrichment) at the H5 positions were purchased from Omicron, Inc., and natural abundance D-xylose was purchased from Sigma. Xylose solutions were prepared by placing about 1 g (carefully weighed) in a flask of known mass. About 3 ml of D_2O was then added and was mostly removed under vacuum. This process was repeated four times in total before the correct amount of D_2O was gravimetrically added to make the solution up to 5 molal (about 1.3 g). A 1 ml aliquot of this sample was then placed in the neutron cell (a null scattering Ti/Zr alloy) and diffraction measurements were carried out for

Table 1 The neutron scattering prefactors for the first order differences of D′xylose solutions substituted with deuterium at the H5 nuclei for the 2 and 5 molal solutions in D_2O

		Value in mb	
Prefactor	Correlation	2 molal	5 molal
A	H_5H_{ex}	15.51	25.63
B	H_5O_w	6.29	9.45
C	H_5O_x	1.13	4.25
D	H_5C	1.30	4.87
E	H_5H_{non}	−0.47	−1.76
F	H_5H_5	−0.58	−2.19
Total (*G*)		23.76	42.44

Table 2 The details of the solutions examined in this study and the ICNCI functions calculated from the experimental and molecular dynamics simulation data

Solution/ concentration	Xylose 2 m	Xylose 5 m	Pyridine 1 m	Pyridine 5 m	Pyridine 3 m	Pyridine 3 m	Pyridine 3 m
Substitution	H/D both H_{non} on C5		H/D all H_{non}		H/D all H_{non}		
Composition of solution	D_2O	D_2O	Null water (average scattering length of water hydrogen = 0)		D_2O	3 m GdmCl D_2O	1.5 m Gdm_2SO_4 D_2O
ICNCI function calculated	ICNCI 1		ICNCI 2		ICNCI 3		
					ICNCI 4		ICNCI 4

about 4 h. The sample was then transferred back to the flask of known mass, and the sample cell washed out with D_2O. Further D_2O was then gravimetrically added to the flask of known mass to dilute the sample to the desired concentration of 2 m. After thorough mixing, 1 ml of this solution was transferred to the neutron scattering cell and data were collected for about 4 h. Solutions of pyridine/water and pyridine/solution were prepared by direct mixing of pyridine with the solution in question (total solution volume was typically ~2 ml). Solutions of guanidinium salts were prepared by the addition of D_2O to the salt (a known amount in a flask of known weight) to achieve approximately the correct concentration. The D_2O was removed under vacuum and more D_2O added. This procedure was repeated 4 more times and finally the correct amount of D_2O added to effect the correct concentration. These solutions were then used to prepare the Gdm salt–pyridine solution. All data were corrected for multiple scattering and absorption, and normalized *versus* a vanadium rod using standard procedures.[14,16] All measurements were performed on the D4C diffractometer at the Institut Laue-Langevin (ILL, Grenoble, France).[16]

Molecular dynamics (MD) simulation procedures

Neutral and periodic cubic systems were created at 2 and 5 molal concentrations containing a number of independent xylose and water molecules interacting according to the CSFF sugar force field[17,18] and the modified TIP3P model[19,20] for water. All simulations were performed with the CHARMM program,[21,22] with chemical bonds to hydrogen atoms kept fixed using SHAKE[23] and with a time step of 1 fs. The starting coordinates for the 2 m system were created by placing 33 randomly orientated xylose molecules (consisting of 22 β and 11 α anomers) in a 34 Å cube. The ratio of 2 β : 1 α is close to that of a D-xylose solution at anomeric equilibrium. These coordinates were superimposed on a box of 1296 water molecules, and solvent molecules that overlapped with the solutes were deleted, using an arbitrary separation cutoff, to produce the correct concentration (33 xylose molecules and 917 TIP3P water molecules, 1.999 m). Finally, the box length was rescaled to 32.0613 Å, which yielded the correct physical number density (0.1035 atoms $Å^{-3}$). The 5 m solution, created in the same manner, contained 60 xylose molecules and 667 TIP3P water molecules, giving a concentration of 4.998 m. The box length was then rescaled to 30.9471 Å, which yielded the correct physical number density (0.1080 atoms $Å^{-3}$). All van der Waals and electrostatic interactions were smoothly truncated on a neutral group basis using switching functions from 13 to 15 Å. Initial velocities were assigned from a Boltzmann distribution (300 K) followed by 10 ps of equilibration dynamics with velocities being reassigned every 0.2 ps. The simulation was then run as an NVE ensemble for 1.0 ns with no further velocity modification. The first 0.33 ns of the trajectory was used as equilibration, and the remaining 0.67 ns for analysis.

Similar procedures were used for the simulation of 1 and 5 m pyridine solutions, and for 3 m pyridine, (10 ns, 56 pyridines, 1036 waters, box size 33.6771 Å, number density 0.0975 atoms Å^{-3}), 3 m pyridine/3 m GdmCl (10 ns, 36 pyridines and 36 GdmCl, 667 waters, box size, 30.6312 Å, number density 0.0972 atoms Å^{-3}), and 3 m pyridine/1.5 m Gdm_2SO_4 solutions (50 ns, 114 pyridine, 57 Gdm2SO4, 2114 waters, box size 44.8489, number density 0.100).

Results and discussion

To demonstrate the validity of this approach (even for a more conformationally flexible solute such as xylose), the ICNCI function was calculated from the molecular dynamics data. The functions $G^{X}_{H5}(r)$ at both concentrations (2 and 5 m) and the ICNCI function $\Delta G^{X}_{H5}(r)$ as calculated from the MD simulations are shown in Fig. 2. The difference function $\Delta G^{X}_{H5}(r)$ shows no evidence of intermolecular correlations, suggesting the validity of the application of the technique of ICNCI to potentially more flexible solutes such as xylose. This result is not surprising given that molecular dynamics simulations of the similarly conformationally-flexible glucose molecule showed no detectable ring conformational changes over the concentration range 1–5 m.[24] The contrast (the sum of the scattering prefactors A through F (see Table 1)) in the 2 m NDIS experiment was 24 mb, while it was found that the contrast of the function $\Delta G^{X}_{H5}(r)$ was about 2 mb (estimated from the high r limit from Fig. 2). This is near the limit of what can be experimentally measured, both due to the accuracy with which solutions of the correct atomic concentrations can be made, and due to the statistical counting error using current diffractometers. However the limitation of working so close to the statistical counting limit of the diffractometer is offset by the fact that the measurement is of direct relevance to the intermolecular association of the species in solution. It is topologically inevitable that as the concentration of the solute is increased, the solute–solute coordination number will increase, while the solute–solvent coordination number will decrease. Therefore, the function $\Delta G^{X}_{H5}(r)$ contains information about how the solute–solute and solute–solvent interactions vary with concentration. That is the function $\Delta G^{X}_{H5}(r)$ now contains exclusively structural data that is of relevance to the competition of the water–xylose/xylose–xylose interactions.

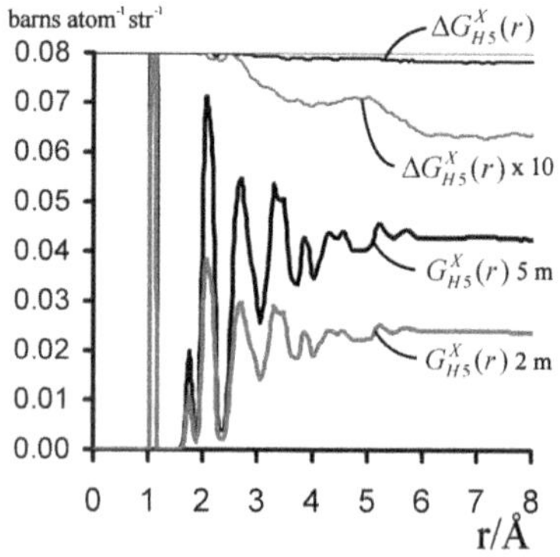

Fig. 2 The MD predictions for the function $G^{X}_{H5}(r)$ at xylose concentrations of 2 and 5 m (lower grey and black lines, respectively). The upper black line shows the function $\Delta G^{X}_{H5}(r)$ (translated for clarity by 0.08 barns atom^{-1} str^{-1}), and the upper grey line the same function scaled by a factor of 10 (prior to translation).

The MD simulations are invaluable for examining the constitution and general features of such functions. Since the pair correlation components of $\Delta G^{X}_{H5}(r)$ are additive, we can examine each component from the simulations separately in an 'augmentation/exclusion', or AE, plot (see Fig. 3). Not surprisingly, it was found that water is excluded as the concentration of sugar increases. Therefore, the H5-O_w and H5-H_{ex} components are both negatively weighted. Conversely, the coordination number of the sugar increases with concentration. However, the prefactor for H5-H_{non} is negative, and consequentially the curves corresponding to this increase in sugar coordination number are positively weighted for the H5-C and H5-O_x correlations, but negatively weighted for the H5-H_{non} correlation.

While the signal in $\Delta G^{X}_{H5}(r)$ is broad and featureless at higher r, the directly measurable function $\Delta\Delta S^{X}_{H5}(Q)$ (as predicted from MD, and directly related to $\Delta G^{X}_{H5}(r)$) shows a very characteristic signal at $Q < 2$ Å^{-1}. This is the key diagnostic feature of the experimental data. Principally for this reason, the comparison of the ICNCI functions is better performed in Q-space than r-space (Fig. 3 and 4).

The raw scattering data from the experiments, corrected for multiple scattering and absorption, and normalized to a standard vanadium rod, produces the total scattering patterns ($F(Q)$s) shown in Fig. 4. The most obvious observation from these data is that the solutions that contain more hydrogen have a higher scattering level, and appear to be on a higher slope. The higher level is a consequence of the very high incoherent scattering cross section of hydrogen, while the slope is due to the inelastic recoil of the protium nucleus (the Placzek effect[25,26]). For both the 2 and the 5 m solution the first order difference was obtained by the subtraction of the $F(Q)$ for the natural abundance xylose solution from the $F(Q)$ for the $d5,d5$-substituted xylose solution. Both of these first order differences, $\Delta S^{X}_{H5}(Q)$, are shown in Fig. 4. There are two conspicuous differences between these two $\Delta S^{X}_{H5}(Q)$. First,

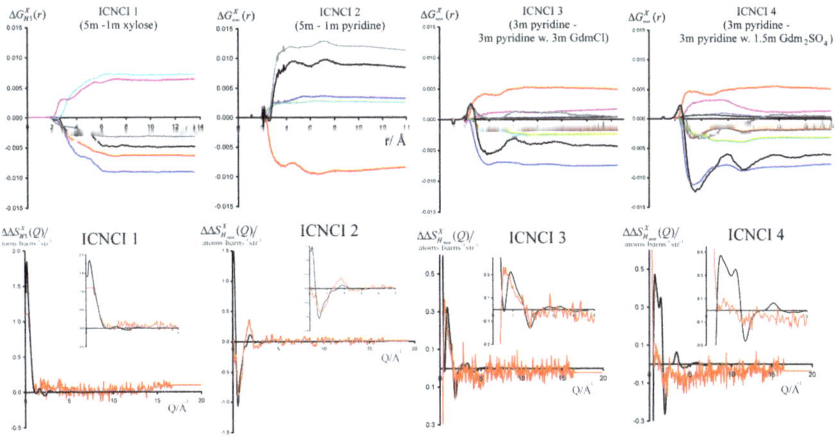

Fig. 3 Upper: the real-space augmentation/exclusion plots; lower: the reciprocal-space ICNCI functions (as derived from eqn (5). red = experiment, black = molecular dynamics). In both the upper and lower plots the black line is the MD prediction of the ICNCI function. The color schemes for the AE plots, ICNCI 1, blue = exchangeable hydrogen, red = oxygen on water, grey = non-exchangeable hydrogens on xylose (negative weighting prefactor), purple = oxygen on xylose, cyan = carbon on xylose. ICNCI 2 (performed in null water so there is no exchangeable hydrogen correlation), red = oxygen, grey = carbon, blue = nitrogen, green = non-exchangeable hydrogens. ICNCI 3 (defined by pyridine–pyridine/GdmCl), red = oxygen, grey = exchangeable hydrogens, purple = carbon on pyridine, brown = carbon on Gdm^+, blue = nitrogen on Gdm^+, green = chloride. ICNCI 4, as with ICNCI 3 except green = oxygen on sulfate. For both ICNCI 3 and 4 components that make very minor contributions to these functions have been omitted for clarity. Minor corrections were made to the experimental ICNCI 1 and 2 (see ESI†). ICNCI 3 and 4 are the raw experimental functions.

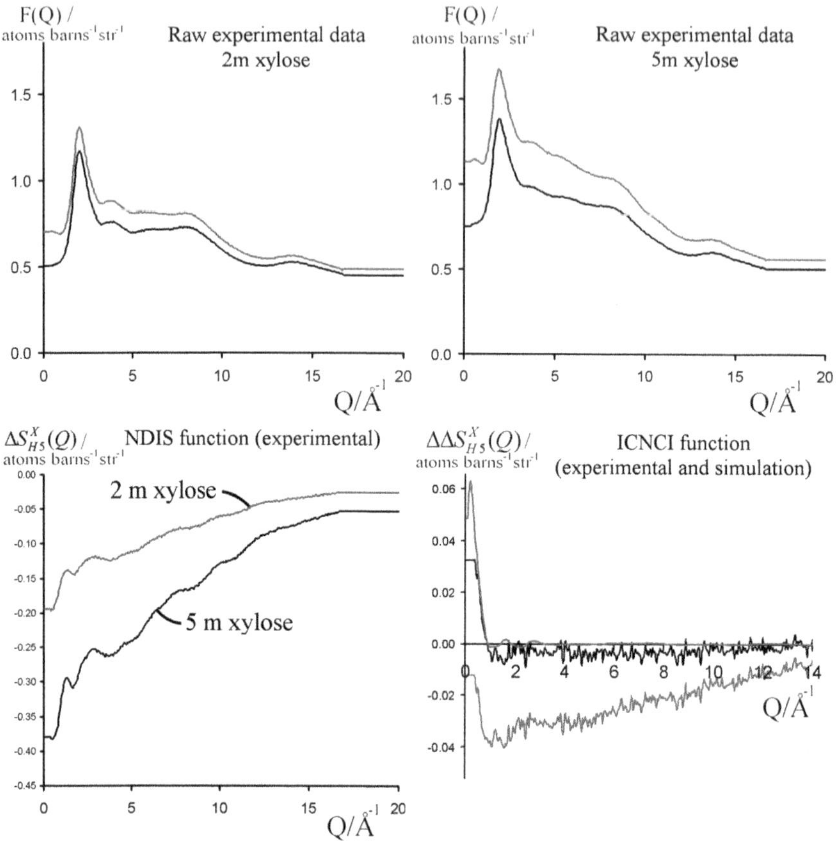

Fig. 4 Top left: the $F(Q)$s for 2 m xylose in D_2O (grey, natural xylose; black, *d5,d5* xylose); top right: the $F(Q)$s for 5 m xylose in D_2O (grey, natural xylose; black, *d5,d5* xylose). Lower left: the first order difference functions ($\Delta S^{X}_{H5}(Q)$) for the previous two plots (grey, 2 m xylose; black, 5 m xylose). Lower right: the function $\Delta\Delta S^{X}_{H5}(Q)$ as calculated from MD (upper grey), and directly determined by experiment (grey). The black line shows the fit obtained by altering the weighted subtraction necessary to obtain the function $\Delta\Delta S^{X}_{H5}(Q)$ by 9%. This figure graphically shows the size of the function $\Delta\Delta S^{X}_{H5}(Q)$ *versus* the original $F(Q)$s. For details of a specific correction applied to ICNCI 1 see the ESI†.

the oscillations in the higher concentration function are larger due to the higher contrast of this difference, and second, the higher concentration function has a greater slope. This second point is a combination of the fact that the substituted nucleus is hydrogen, which contains significant contributions from the incoherent scattering, and of the Placzek effect. While these artifacts may seem large, it should be born in mind that this effect yields an extremely broad feature (wavelengths ~80 $Å^{-1}$) and does not contribute to the higher frequency components that are of interest in this study (wavelengths 0–10 $Å^{-1}$).

For the other ICNCI differences in this study, similar AE plots can be derived from the MD data (Fig. 4). As expected, the AE plots reflect the geometry of the molecule on which the NDIS substitution has been performed (xylose or pyridine). Further the AE plots show clearly different structural responses to either the changing concentrations of solute or co-solute. In order to obtain an accurate ICNCI function, the force field in question must be able to reproduce the balance of water–solute and solute–solute interactions. For ICNCI 1, the force field used satisfactorily reproduces the balance of the water–xylose and xylose–xylose interactions. For ICNCI 2 (the balance of pyridine–water *versus* pyridine–pyridine) this balance is also good, although not as satisfactory as with xylose. For ICNCI 2 the AE plot reveals that the majority of this function is due to pyridine replacing water in the pyridine solvation shell at higher concentration. The form of this interaction has been shown to be mostly comprised of T-type interactions.[15,27] For ICNCI 3, the molecular dynamics force field seems to reproduce the relative balance of the associations in the pyridine–pyridine and pyridine–(pyridine + GdmCl) solutions acceptably. This solution has been studied in significant detail, where it can be shown that the majority of this ICNCI function is due to the stacking of the guanidinium ion with the pyridine.[27] However for ICNCI 4, the correlation between the experiment and the molecular dynamics is poor. In this solution it was found that there is strong association between the guanidinium and sulfate ions, leading to nanometer sized clusters with lifetimes of tens of nanoseconds, similar to those observed previously.[25,28] The pyridine shows a strong association to these clusters. While multiple separate lines of evidence support the suggestion that the association between guanidinium and high charge density anions such as sulfate is stronger than lower charge density ions such as chloride,[28–30] it seems reasonable to suggest that the molecular dynamics in this case is significantly over-estimating the level of association of the Gdm^+–SO_4^{2-} ion pair, and the interaction of the pyridine with these clusters. A possible explanation for this may be that in MD simulations with non-polarizable force fields such as those used here, the effects of the electronic polarizibility of the water on the effective charge of the ion are not taken into account.[31,32]

Conclusions

We have shown that the functions derived from ICNCI are useful for studying both flexible solutes such as xylose and ones like pyridine, and can be used either by varying the concentration of the substituted solute, or by adding a co-solute. We find that the ICNCI function is surprisingly sensitive to the solute in question, showing different responses to the variation of concentration in xylose and pyridine, and a difference in solvation of pyridine between the addition of GdmCl or Gdm_2SO_4. These results suggest that the interaction of pyridine with the Gdm^+–SO_4^{2-} clusters in these simulations is significantly overestimated. Previously it has been suggested that the level of this ion pairing is important for the position of the ions in the Hofmeister series, and is also important in the formation of the salt bridges that contribute to the stability of proteins. This study suggests that there are some significant limitations in this type of molecular dynamics in obtaining the correct level of ion pairing in these simulations.

Acknowledgements

This project was supported by Grant GM63018 from the National Institutes of Health and by a grant of beam time from the Institute Laue-Langevin. We are also grateful to Dr Henry E. Fischer (ILL) for his help with the neutron diffraction experiments.

References

1 G. W. Neilson and J. E. Enderby, *J. Phys. Chem.*, 1996, **100**, 1317.
2 I. Bako, G. Palinkas, J. C. Dore, H. Fischer and P. Jovari, *Chem. Phys. Lett.*, 2004, **388**, 468.
3 D. M. Sullivan, G. W. Neilson and H. E. Fischer, *J. Chem. Phys.*, 2001, **115**, 339.
4 G. W. Neilson and J. E. Enderby, *Proceedings of the Royal Society of London*, 1983, **A390**, 353.
5 J. E. Enderby and G. W. Neilson, *Rep. Prog. Phys.*, 1981, **44**, 593.
6 N. G. Polydorou, J. D. Wicks and J. Z. Turner, *J. Chem. Phys.*, 1997, **107**, 197.

7 D. T. Bowron, J. L. Finney and A. K. Soper, *J. Phys. Chem. B*, 1998, **102**, 3551.
8 D. T. Bowron, A. K. Soper and J. L. Finney, *J. Chem. Phys.*, 2001, **114**, 6203.
9 K. S. Sidhu, J. M. Goodfellow and J. Z. Turner, *J. Chem. Phys.*, 1999, **110**, 7943.
10 Y. Kameda, H. Ebata, T. Usuki, O. Uemura and M. Misawa, *Bull. Chem. Soc. Jpn.*, 1994, **67**, 3159.
11 M. Sasaki, Y. Kameda, M. Yaegashi and T. Usuki, *Bull. Chem. Soc. Jpn.*, 2003, **76**, 2293.
12 Y. Kameda, K. Sugawara, T. Usuki and O. Uemura, *Bull. Chem. Soc. Jpn.*, 2003, **76**, 935.
13 G. Mao, M.-L. Saboungi, D. L. Price, Y. S. Badyal and H. E. Fischer, *Europhys. Lett.*, 2001, **54**, 347.
14 A. C. Barnes, J. E. Enderby, J. Breen and J. C. Leyte, *Chem. Phys. Lett.*, 1987, **142**, 405.
15 P. E. Mason, G. W. Neilson, D. L. Price, M.-L. Saboungi and J. W. Brady, *J. Phys. Chem. B*, 2010, **114**, 5412.
16 H. E. Fischer, G. J. Cuello, P. Palleau, D. Feltin, A. C. Barnes, Y. S. Badyal and J. M. Simonson, *Appl. Phys. A: Mater. Sci. Process.*, 2002, **74**, S160.
17 R. Palma, M. E. Himmel and J. W. Brady, *J. Phys. Chem. B*, 2000, **104**, 7228.
18 M. Kuttel, J. W. Brady and K. J. Naidoo, *J. Comput. Chem.*, 2002, **23**, 1236.
19 W. L. Jorgensen, J. Chandrasekhar, J. D. Madura, R. W. Impey and M. L. Klein, *J. Chem. Phys.*, 1983, **79**, 926.
20 E. Neria, S. Fischer and M. Karplus, *J. Chem. Phys.*, 1996, **105**, 1902.
21 B. R. Brooks, R. E. Bruccoleri, B. D. Olafson, S. Swaminathan and M. Karplus, *J. Comput. Chem.*, 1983, **4**, 187.
22 B. R. Brooks, C. L. Brooks, A. D. MacKerell, J.r, L. Nilsson, R. J. Petrella, B. Roux, Y. Won, G. Archontis, C. Bartels, S. Boresch, A. Caflisch, L. Caves, Q. Cui, A. R. Dinner, M. Feig, S. Fischer, J. Gao, M. Hodoscek, W. Im, K. Kuczera, T. Lazaridis, J. Ma, V. Ovchinnikov, E. Paci, R. W. Pastor, C. B. Post, J. Z. Pu, M. Schaefer, B. Tidor, R. M. Venable, H. L. Woodcock, X. Wu, W. Yang, D. M. York and M. J. Karplus, *J. Comput. Chem.*, 2009, **30**, 1545.
23 J.-P. Ryckaert, G. Ciccotti and H. J. C. Berendsen, *J. Comput. Phys.*, 1977, **23**, 327.
24 P. E. Mason, G. W. Neilson, A. C. Barnes, J. E. Enderby, J. W. Brady and M.-L. Saboungi, *J. Chem. Phys.*, 2003, **119**, 3347.
25 P. E. Mason, G. W. Neilson, C. E. Dempsey and J. W. Brady, *J. Am. Chem. Soc.*, 2006, **128**, 15136.
26 P. E. Mason, S. Ansell and G. W. Neilson, *J. Phys.: Condens. Matter*, 2006, **18**, 8437.
27 P. E. Mason, C. E. Dempsey, G. W. Neilson, S. R. Kline and J. W. Brady, *J. Am. Chem. Soc.*, 2009, **131**, 16689.
28 P. E. Mason, G. W. Neilson, S. R. Kline, C. E. Dempsey and J. W. Brady, *J. Phys. Chem. B*, 2006, **110**, 13477.
29 P. E. Mason, C. E. Dempsey, G. W. Neilson and J. W. Brady, *J. Phys. Chem. B*, 2005, **109**, 24185.
30 C. E. Dempsey, P. E. Mason, J. W. Brady and G. W. Neilson, *J. Am. Chem. Soc.*, 2007, **129**, 15895.
31 I. V. Leontyev and A. A. Stuchebrukhov, *J. Chem. Theory Comput.*, 2010, **6**, 1498.
32 I. V. Leontyev and A. A. Stuchebrukhov, *J. Chem. Theory Comput.*, 2010, **6**, 3153.

Femtosecond study of the effects of ions and hydrophobes on the dynamics of water

Sietse T. van der Post, Klaas-Jan Tielrooij, Johannes Hunger, Ellen H. G. Backus and Huib J. Bakker*

Received 3rd May 2012, Accepted 6th June 2012
DOI: 10.1039/c2fd20097j

We study the effects of ions and hydrophobic molecular groups on the orientational dynamics of water using THz dielectric relaxation (THz-DR) and polarization-resolved femtosecond infrared (fs-IR) pump–probe spectroscopy. We measure the dynamics of water in solutions of NaI, NaCl, CsCl, guanidinium chloride (GndCl) and tetramethyl guanidinium chloride (TMGndCl) of different concentrations. With THz-DR we observe that strongly hydrated cations align the static dipoles of their surrounding water molecules. With fs-IR we find that the OD groups that form hydrogen bonds to halide anions reorient with two distinct time constants of 2 ± 0.3 ps and 9 ± 1 ps. The fast process is assigned to a wobbling motion of the OD group that keeps the hydrogen bond with the anion intact. The amplitude of this wobbling motion depends on the nature of both the anion and the counter cation. The replacement of four of the six hydrogen atoms of the weakly hydrated cation guanidinium by hydrophobic methyl groups leads to an exceptionally strong slowing down of the water dynamics. Hydrophobic groups thus appear to have a much stronger effect on the dynamics of water than ions. These findings give new insights in the mechanism of protein denaturation by GndCl and TMGndCl.

I. Introduction

The conformation of macromolecules in aqueous media is crucial for their functioning in biological systems. Many studies have been devoted to the understanding of how these structures are formed and how this formation can be influenced with co-solutes like salts and sugars. These studies go back to the pioneering work of Franz Hofmeister.[1] In his 1888 paper Hofmeister discussed the influence of different salts on the precipitation of proteins, a process called salting out. Currently, the well known Hofmeister series comprises two series of cations and anions, ordered by their ability to salt out proteins. This property has been interpreted in terms of structure-making and structure-breaking concepts,[2] *i.e.* a presumed ion-induced strengthening or weakening of the hydrogen-bond network of water over long distances. Although these concepts offer an explanation of the observed (de)naturation effects according to the Hofmeister series, experimentally no evidence was found of the supposed long range structuring influence of ions on water.[3–6] Most ions influence the structure and dynamics of only their first solvation shell[7] and only in the (rare) case of very strongly hydrated ions, water molecules were found to be affected over a somewhat larger distance.[8] The exact nature of ion hydration and its relation to the (de)stabilization of the conformation of macromolecules thus remain not well understood.

FOM Institute AMOLF, Science Park 104, Amsterdam, The Netherlands. E-mail: bakker@amolf.nl

Guanidinium (see inset Fig. 1) is on the far end of the Hofmeister series and its chloride salt (GndCl) is among the strongest denaturants known, causing almost all proteins to unfold completely at a concentration of 6 mol l^{-1}.[9] There is an ongoing debate about the underlying microscopic principles of its working mechanism. Some argue that denaturants work indirectly by affecting the properties of water.[10–12] Dielectric relaxation spectroscopy and neutron diffraction studies on the water structure around the guanidinium ion suggest that the ion lacks a significant hydration shell.[13,14] From these results it was concluded that the denaturation process likely involves a direct interaction with proteins, as has been proposed for urea.[14,15] However, molecular dynamics (MD) simulations show surprising differences between urea and guanidinium chloride:[10,11,16] it was found that urea binds to peptide groups but guanidinium does not.[11] In line with this finding, it was shown in other studies that the interaction between guanidinium and water might be crucial in understanding its denaturing properties.[12] Finally, it should be noted that both microscopic mechanisms for denaturation are not mutually exclusive and might in fact both be active.[17]

The substitution of the hydrogen atoms of urea by methyl groups yields the osmolyte tetra-methyl urea (TMU). Compared to urea, TMU is an even more effective denaturant[18] and shows a profound (hydrophobic) effect on the water dynamics of its solvation shell.[19] It is therefore not surprising that previous work indicated that denaturation induced by TMU may have a different microscopic origin to urea.[18] Given the fundamental differences in the nature of guanidinium chloride and urea it is interesting to study the effects of the substitution of the hydrogen atoms of the guanidinium ion by methyl groups on the molecular properties of water and the denaturation of proteins.

Over the past ten years THz dielectric relaxation (THz-DR) and polarization-resolved femtosecond infrared pump–probe (fs-IR) spectroscopy have been established as powerful tools to study water dynamics.[20,21] Recently it was shown that these techniques can provide complementary information, as they probe different axes of the water molecule.[8] In this paper we use THz-DR and fs-IR to study the effects of electrostatic and hydrophobic interactions on the dynamics of water. We compare solutions of the strong denaturing salt guanidinium chloride and tetramethyl guanidinium chloride. To bring this into the perspective of denaturation, we compare the effectiveness of guanidinium chloride, tetramethylguanidinium chloride, urea and TMU in the denaturation of the model protein Photoactive Yellow Protein.

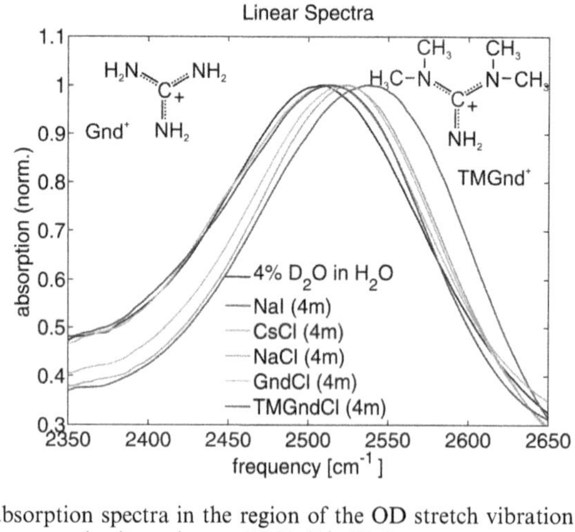

Fig. 1 Linear absorption spectra in the region of the OD stretch vibration for concentrated solutions of the measured salts and pure 4% D_2O in H_2O.

II. Experiment

A. THz dielectric relaxation

In THz dielectric relaxation spectroscopy the dynamics of the static dipoles of water are probed by characterizing a THz pulse after transmission through the sample. Generally, an applied electric field tends to align the permanent dipoles of the water molecules against the thermal fluctuations. The proportionality of the resulting induced macroscopic polarization and the applied field is given by the frequency-dependent dielectric function. At frequencies lower than the characteristic timescale at which reorientation processes of water molecules take place, the dipoles can follow the electric field oscillations and the built-up polarization is only limited by the thermal fluctuations of the dipole orientations. At high frequencies, the dipoles fail to follow the oscillations of the applied electric field. The transition from the low-frequency to the high-frequency domain is marked by a strong response in the imaginary dielectric function due to a phase lag in the reorienting dipoles relative to the externally applied oscillating electric field.

For pure water the maximum of the out-of-phase response is found at 20 GHz, corresponding to a timescale τ_{bulk} of $\approx$8 ps.[22] This time constant is often referred to as the Debye relaxation time and has been assigned to the timescale of the spontaneous restructuring of the hydrogen-bond network. A contribution with a much lower amplitude has been found at higher frequencies (~0.9 THz, $\tau_{\text{fast}} \approx 350$ fs), and has been assigned either to quick jumps of undercoordinated water,[23,24] interaction-induced components in the water relaxation mechanism[25,26] or a small angular rotation preceding a large angle jump.[27] The low-frequency dielectric function $\hat{\varepsilon}(\nu)$ of water is thus modeled as a sum of two Debye modes with different time constants τ_n,

$$\hat{\varepsilon}_{H_2O}(\nu) = \frac{S_{\text{bulk}}}{1 + 2\pi i\nu\tau_{\text{bulk}}} + \frac{S_{\text{fast}}}{1 + 2\pi i\nu\tau_{\text{fast}}} + \varepsilon_{\infty}, \qquad (1)$$

where ε_{∞} is the permittivity at infinite frequency. The fraction of water molecules that reorient with time constant τ_n is proportional to the amplitude S_n. For pure water $S_{\text{bulk}} \approx 70$ and $S_{\text{fast}} \approx 2$.

In the case of salt solutions the conductivity of the sample will contribute to the dielectric response, which is incorporated by extending eqn (1) according to,

$$\hat{\varepsilon}(\nu) = \hat{\varepsilon}_{H_2O}(\nu) + \frac{\sigma}{2\pi i\varepsilon_0\nu}, \qquad (2)$$

where the conductivities σ are taken from the literature.[13,28–31] The addition of ions leads to a decrease of $\hat{\varepsilon}(\nu)$ (depolarization), as a result of three contributions. First, in a constant volume there is a decreased number of water molecules that contribute to the signal because of dilution when a solute is added. Second, the water molecules that bind to the ions cannot reorient in the applied electric field and no longer contribute to $\hat{\varepsilon}(\nu)$. Finally there is the effect of kinetic depolarization that results from the movement of charges in an electric field. The procedures to take the dilution effects and kinetic depolarization into account are described in ref. 32.

The far-infrared THz pulse is generated by optical rectification in a ZnTe crystal pumped by a part of the output of a Ti:Sapphire pulsed laser (800 nm, ~150 fs, ~70 μJ). The resulting pulse typically comprises one cycle of the electric field and has a duration of ~1 ps and a frequency spectrum between 0.4 and 1.2 THz. We obtain the dielectric function $\hat{\varepsilon}(\nu)$ from the change in amplitude (absorption) and delay (refraction) of the THz pulse propagated through the sample.[19] The frequency window of the generated THz pulses does not reach down to the maximum of the main Debye response of water at 20 GHz, but covers the high frequency wing of this response that extends to THz frequencies. The decrease of the amplitude (depolarization) of the main Debye mode can therefore be measured by THz-DR. The samples were placed in an Infrasil quartz cuvette with a pathlength of 103 ± 0.5 μm.

B. fs-IR pump–probe

In the fs-IR pump probe experiment we excite and probe the OD stretch vibration of HDO molecules in isotopically diluted water with an ultrafast laser pulse that is resonant with the $\nu = 0 \rightarrow 1$ transition (4 μm). The pulses are generated *via* frequency-conversion processes that are pumped with the output of a Ti:Sapphire 110 fs 800 nm pulsed laser (Hurricane, Spectra-Physics).[32] The generated pulses have a central wavelength of 4 μm, a duration of 150 fs, and a pulse energy of 5 μJ. The pulses are used in a polarization-resolved pump–probe experiment. The excitation by an intense pump pulse (4 μJ) labels a subset of water molecules that have their OD stretch transition dipole aligned preferentially in the direction of the pump polarization, thus creating an anisotropically excited ensemble. After a tunable delay, a probe pulse (200 nJ), with its polarization either parallel or perpendicular to the pump polarization, interrogates the sample. At short delay times after the excitation, the probe with its polarization parallel to the pump will give a larger signal than the perpendicular probe, due to the pump-induced anisotropy. With increasing delay times the excited ensemble will decay to an isotropic state in which both probe signals are identical. The time scale at which this happens is a measure for the characteristic time at which the molecules reorient in the liquid. The transient absorption spectra measured with the parallel probe ($\Delta\alpha_{\parallel}$) and the perpendicular probe ($\Delta\alpha_{\perp}$) for pump–probe delay times t are used to construct the anisotropy parameter $R(t)$,

$$R(t) = \frac{\Delta\alpha_{\parallel}(t) - \Delta\alpha_{\perp}(t)}{\Delta\alpha_{\parallel}(t) + 2\Delta\alpha_{\perp}(t)}, \tag{3}$$

The denominator represents the isotropic signal. In the expression for $R(t)$ all isotropic effects like the vibrational relaxation are divided out so that the dynamics of $R(t)$ represents the reorientation of the excited vibration only. For a sample consisting of a randomly oriented collection of dipolar oscillators the maximum anisotropy $R = 0.4$. For a pure sample of 4% D_2O in H_2O the anisotropy decays with a time constant of 2.5 ps.[33]

fs-IR and THz-DR measure intrinsically different observables.[34,35] THz-DR is sensitive to the first-order time dipole–dipole correlation function, while fs-IR measures the second-order time-correlation function of the OD transition dipole orientation. In addition to that, while fs-IR measures the reorientation of – a labeled subset of – single molecules, THz-DR measures the macroscopic polarization to which also local field and dipole–dipole correlations contribute. For the main reorientation (Debye) component of bulk liquid water the ratio between the two time-scales was found to be 3.4.[36]

C. Samples

The alkali-halide salts, guanidinium chloride (GndCl, see Fig. 1) and 1,1,3,3-tetramethylguanidine (TMG) were purchased from Sigma Aldrich. TMG is a strong base ($pK_a = 13.6$) and was neutralized by adding a stoichiometric amount of hydrochloric acid to create the tetramethylguanidinium chloride (TMGndCl) salt. Fig. 1 shows a graphical representation of the (tetramethyl)guanidinium cations. Solutions at different concentrations were prepared with a mixture of 4% D_2O in H_2O (Sigma Aldrich) as the solvent. The measured concentrations range from 0.25 to 4 m (mol kg^{-1} solvent) for the THz-DR experiment, and from 1 to 6 m for the fs-IR experiments. The deuterium atoms of D_2O exchange both with water to form HDO molecules and with the NH_2 groups of the guanidinium salts to form NHD. The resonance frequency of the thus formed ND is very close to that of the OD, but its cross section – which enters quadratically in the pump–probe signal – is twice as small. In addition, for a 4 M GndCl solution the concentration of ND is 3 times lower than the OD concentration. Therefore the ND groups will contribute negligibly to the pump probe signal. In the fs-IR measurements we used two square CaF_2 windows (thickness 4 mm) with a 50 or 100 μm spacer in between. The concentrations of H_2O in the solutions are determined using a density meter (Mettler Toledo DM40).

III. Effects of ions on water dynamics

A. THz dielectric relaxation

Fig. 2A shows the imaginary part of the dielectric function $\hat{\varepsilon}(\omega)$ for different salt solutions. The dielectric functions are calibrated using a reference salt solution with known dielectric properties that was measured in parallel. The solid lines in Fig. 2A are fits of eqn (2) to the data. For stronger hydrated cations we observe a larger depolarization (decrease of S_1 in eqn (1)). Fig. 2B shows the depolarization caused by hydration shell water molecules that do not participate in dielectric relaxation for a series of concentrations. The depolarizations are corrected for the kinetic depolarization term,[32] using conductivity values from the literature.[13,28–31] The slope of the fitted lines represents the hydration number N_p, the number of (dipoles of) water molecules that show much slower orientational dynamics than bulk water due to the electrostatic interactions with the ions. The weakly hydrated cation cesium has a very small hydration number ($N_p \approx 0$), meaning that this ion hardly affects the orientational dynamics of water. For smaller cations (Na^+ or Li^+) or increasing cation charge (Mg^+) the slope becomes much larger, indicating a larger hydration number. In line with previous results,[37] the anions show very little effect, as is illustrated by the results for Cs_2SO_4 which combines the weakly hydrated cation Cs^+ with the strongly hydrated anion SO_4^{2-}.

B. fs-IR pump–probe

Fig. 3 shows transient absorption spectra at different pump–probe delay times measured for a 4 mol kg^{-1} NaCl solution. The excitation of a few percent of the OD vibrations to the first excited state leads to transient spectra that consists of two main contributions. At frequencies >2450 cm^{-1} a bleaching signal is observed due to the bleaching of the ground state and stimulated emission from the excited state. At frequencies <2450 cm^{-1} induced absorption from the first to the second excited state is observed. The latter absorption is red-shifted with respect to the fundamental transition because of the anharmonicity of the OD vibration.

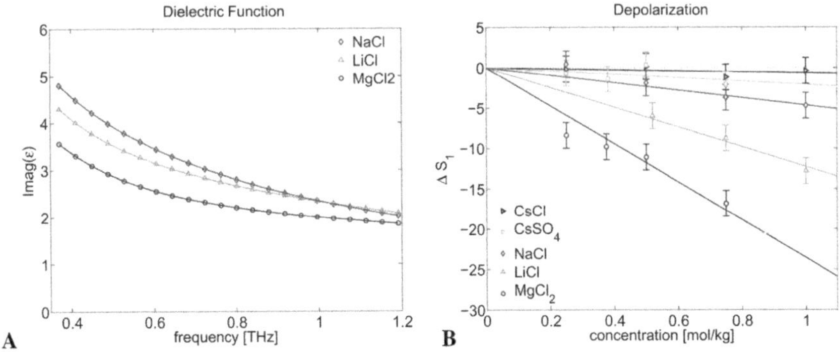

Fig. 2 (A) The imaginary part of the dielectric function measured for different salts. (B) The fraction of slow water deduced from the depolarization as a function of concentration for a number of salts. The weakly hydrated cesium hardly affects the dynamics of water, in contrast to the strongly hydrated magnesium.

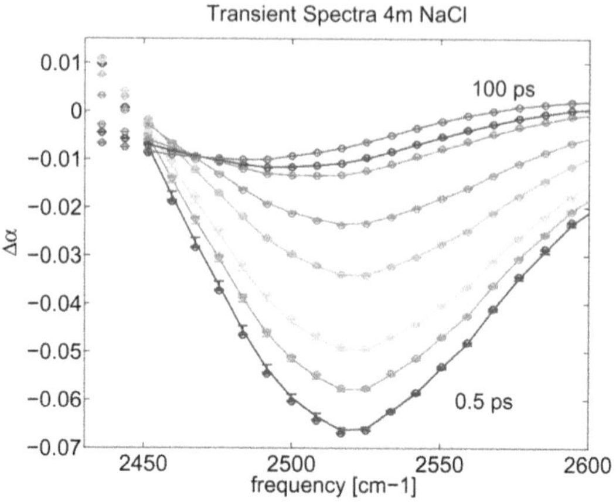

Fig. 3 Transient absorption spectra at different delay times for a 4 mol kg^{-1} NaCl solution. The solid lines are the results from the fit to the model described in the text.

Vibrational relaxation from the first excited state to the ground state will lead to a decay of the transient spectral changes. At long delay times all OD oscillators have completely relaxed and the energy has become thermally equilibrated over the sample. From those times on, the transient spectrum represents the effect of a small increase in temperature (by a few kelvin) on the absorption spectrum of the sample.

As the dissolution of NaCl introduces a blue-shift of the OD absorption band in the linear spectrum (Fig. 1), we model the isotropic data with two independent excited states representing the absorption of the bulk-like HDO:H_2O molecules and a blue-shifted absorption band of HDO molecules hydrating ions.[38] The results of the fit to the data obtained for a 4 mol kg^{-1} NaCl solution are shown in Fig. 3 and 4. The experimental spectra are well described by a sum of two spectral components centered around 2500 and 2530 cm^{-1}. The decay time of the blue-shifted component is 4.5 ps, which is considerably slower than the bulk-like component (1.8 ps). We assign the blue component to water solvating the chloride (OD···Cl$^-$). It has been shown before that around chloride and most other halide ions the decay of the OD/OH-stretch excitation is much slower than in bulk liquid water.[38,39] For a

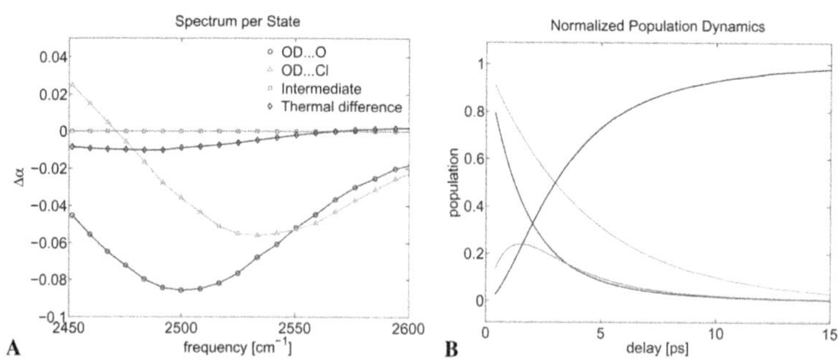

Fig. 4 (A) Spectral components obtained from the transient absorption spectra presented in Fig. 3. The response of the OD groups bound to the chloride is blue-shifted with respect to the OD···O band. The intermediate state (squares) accounts for a slightly delayed ingrowth of the heat and has no associated transient spectrum. (B) Time dependence of the normalized fitted populations in the different states. The OD···Cl$^-$ band shows a much slower vibrational relaxation.

number of chloride solutions it was found that independent of the cation, a 30 cm^{-1} blue shifted component is present.[38,40,41] The blue-shift and the slower relaxation can be explained from the fact that the OD···Cl^- hydrogen bond is weaker than the average water–water hydrogen bond (OD···O).[40] In case of larger halide anions the surface charge density is smaller, the hydrogen-bond even weaker and hence the blue-shift larger (for example ~70 cm^{-1} in case of OD···I^-, Fig. 1). Since all salt solutions under study with fs-IR contain either chloride or iodide, the same kinetic model is used for all fs-IR data shown in this paper. We use the fitted spectra and their dynamics to correct the data for the ingrowth of the thermal difference spectrum in order to calculate the anisotropy decay of purely the excited OD oscillators. Since the vibrational lifetime of the OD···Cl^- component is different from that of the OD···O component, we have to decompose the data such that we can determine the anisotropy decays of both components separately. The total difference transient spectra $D(t,\omega)$ is given by:

$$D(t,\omega) = \Delta\alpha_{\parallel}(t,\omega) - \Delta\alpha_{\perp}(t,\omega) = R_O(t,\omega)N_O(t)\sigma_O(\omega) + R_A(t,\omega)N_A(t)\sigma_A(\omega), \quad (4)$$

where $R_i(t,\omega)$ denotes the delay time-dependent anisotropy of the OD oscillators bound to other water molecules (i = O) or the anion (i = A) probed at frequency ω. $N_i(t)$ denotes the excited population at pump–probe delay time t and $\sigma_i(\omega)$ is the excited state spectrum of species OD···O or OD···A, obtained from the isotropic fit (Fig. 4). For most of the systems under study spectral diffusion is fast enough such that the anisotropy within one state can be treated as behaving uniformly, so that the frequency dependence of R_i can be neglected.[39,42] This enables us to obtain the time dependent amplitude $R_i(t)N_i(t)$ for every time t by fitting the spectra $\sigma(\omega)$ of the different components to the difference transient spectrum $D(t,\omega)$. For every component i the time traces resulting from the decomposition contain the anisotropy decay $R_i(t)$ multiplied by the isotropic population decay $N_i(t)$. The latter is known from the isotropic fit and can thus be divided out. The anisotropy curves for the OD···O, OD···Cl^- and OD···I^- components of 4 mol kg^{-1} solutions of NaCl and NaI are shown in Fig. 5A.

The anisotropy decay of the OD···O species ($R_O(t)$) can be fitted well with a mono-exponential function with a time constant of ≈2.5 ps and thus resembles the dynamics observed for pure HDO:H_2O.[33] It should be noted that $R_O(t)$ also contains the dynamics of the water molecules in the hydration shell of the cations.

The anisotropy decay of anion bound OD groups ($R_A(t)$) shows a very different behavior compared to $R_O(t)$. Within the first few hundred femtoseconds an extremely rapid decay makes the anisotropy drop to about 75% of its theoretical

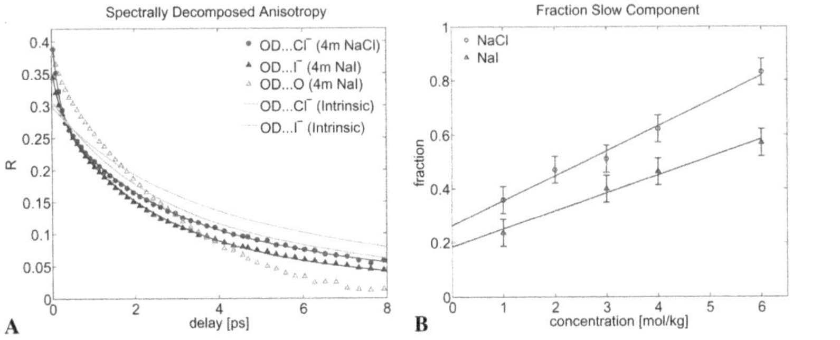

Fig. 5 (A) The anisotropy dynamics of the OD···A^- and OD···O components. The intrinsic anisotropy decays are obtained by fitting eqn (5) to the data. In the fits the exchange effects are included. The solid lines through the data points represent the fits including the initial drop due to librational motions. (B) The amplitude of the slow component of the intrinsic OD···A^- anisotropy curves at different concentrations.

maximum of 0.4, likely due to fast librational motions.[43] This contribution can be fitted separately for early delays with a mono-exponential function with a time constant of 170 ± 40 fs.

To interpret the behavior of $R_A(t)$ for different salts, we should take into consideration that the OD oscillators in the bulk exchange with those in the ion hydration shell. These exchange events occur on a time scale of ~9 ps.[44] As a result, there will be oscillators that are initially excited as OD···A⁻, and that are probed after having rotated *out* of the anion hydration shell. Since their spectral response also changed, these oscillators are no longer contributing to $R_A(t)$. However, there are also oscillators that are excited as OD···O and that are probed after having rotated into the anion hydration shell. This class of oscillators contributes to the spectral response of the OD···A⁻ component but will have lost almost all of their directionality in the exchange event, thereby lowering $R_A(t)$.

We model the anisotropy of the anion-bound water by assuming a bi-exponential form for the intrinsic anisotropy $R_A^{intr}(t)$, *i.e.* the anisotropy of the ion hydration shell in case there would be *no* exchange:[45]

$$R_A^{intr}(t) = R_0[(1 - B_0)e^{-k_1^A t} + B_0 e^{-k_2^A t}] \tag{5}$$

Here R_0 is the initial amplitude, B_0 a proportionality factor between the two exponentials and k_1^A and k_2^A the rates of the reorientation processes. The initial drop due to librational motions is here neglected, as we only consider the data after 0.5 ps in the fit. We include exchange effects by modeling the population dynamics of OD groups that remain hydrogen bonded and the rotated-in OD groups separately using the exchange rate and the vibrational decay rate.[45] It is not possible to determine the time scale of the exchange for chloride and iodide solutions from our experiments because the OD···O, OD···Cl⁻ and OD···I⁻ bands strongly overlap. Therefore we fixed the exchange time constant to 9 ps, as found by Gaffney *et al.* for perchlorate solutions.[44] Rotating into the anion hydration shell involves a jump over an angle of ~55 degrees,[44] causing the OD-oscillators to lose their orientation completely in the event. The solid lines in Fig. 5A represent the intrinsic $R_A^{intr}(t)$ for the NaCl and NaI solutions that follows from a fit to the data including the exchange effects. For both NaCl and NaI we find $(k_1^A)^{-1} = 2 \pm 0.3$ ps and $(k_2^A)^{-1} = 9 \pm 1$ ps. The 2 ± 0.3 ps component has a quite large amplitude. Although $R_A(t)$ only represents the reorientation of anion-bound OD groups, it not only depends on the nature of the anion but also on the nature of the cation. In Fig. 6 we compare the dynamics of $R_{A=Cl}(t)$ for chloride salts with a strongly hydrated cation (NaCl) and a weakly hydrated cation

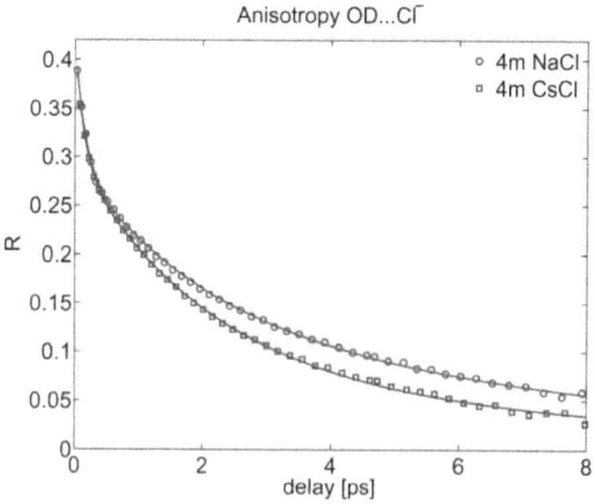

Fig. 6 The anisotropy decay of OD groups bound to chloride depends on the nature of the cation. The stronger hydrated Na⁺ induces a smaller amplitude for the wobbling component than the weakly hydrated Cs⁺.

(CsCl). We observe that the 2 ps decay component of $R_{Cl}(t)$ has a smaller amplitude for a solution of NaCl than for a solution of CsCl. In Fig. 5B the relative amplitude of the slow component of $R_A^{intr}(t)$ (one minus the fraction of the fast component) is shown as a function of concentration. This amplitude increases with concentration and is larger for NaCl than for NaI.

C. Discussion

1. Effect of cations. Fig. 2 shows that the effect of salts on the orientation dynamics of the water dipoles, as measured with THz-DR, is dominated by the nature of the cation. The fraction of slow water is larger going from CsCl to NaCl to $MgCl_2$, which can be explained from the increase of the surface charge density of the cation. When the surface charge density becomes larger, the electric field exerted on the surrounding water molecules also increases. Mg^{2+} is doubly charged and has an even smaller radius than Na^+, making its alignment effect on the water dipoles very strong. This results in the observation of a large fraction of slow water for solutions containing Mg^{2+}.

Previous fs-IR experiments showed that cations have a relatively small effect on the reorientation of the water hydroxyl groups. The anisotropy decay of solutions of $Mg(ClO_4)_2$ is observed to be hardly different from that of pure water,[7] meaning that the strong electric field exerted by Mg^{2+} hardly influences the orientational mobility of the hydroxyl groups. Cations thus primarily limit the reorientation of the water dipoles but leave the reorientation of the hydroxyl groups largely unaffected.

These findings point at the following picture of the cation hydration shell. The positive charge of a cation attracts the negatively charged oxygen atom of the water molecule and repels the positively charged hydrogen atoms, thereby fixing the dipole moment of the water molecule in a radial direction.[8] In this configuration the OD/OH groups of the water molecules remain relatively free to reorient. The dynamics of the OD/OH groups will thus not be strongly different from the dynamics of bulk liquid water. Of course, one could argue that the restricted rotation in a cone (with the dipole moment as the primary axis) constitutes a limitation compared to the case of bulk water. However, a water molecule in the bulk experiences a strong restriction in its rotational mobility resulting from the hydrogen bonds that it accepts from other water molecules. These accepted hydrogen bonds fix the positions of the lone pairs of the oxygen atom of the reorienting water molecule, thus limiting the orientational space of the hydroxyl groups at the other side of the molecule. For a water molecule with its oxygen atom adjacent to a cation this limitation is less severe. The two restrictions (cone-like reorientation and fixing of the lone pairs) may have a similar effect on the orientational mobility of the water molecule, thus making the reorientation rates of hydroxyl groups in the cation hydration shell and in bulk quite similar.

2. Effect of anions. The effect of anions on the orientational mobility of the water dipoles probed with THz-DR is quite small, as is illustrated by the weak effect of Cs_2SO_4 in Fig. 2. The fact that anions do not fix the water dipoles shows that the interactions of anions with water is not purely electrostatic in nature. Otherwise, the effects of cations and anions should have been quite similar, and a strongly hydrated anion would have fixed its surrounding water dipoles in a similar manner (but opposite direction) as a strongly hydrated cation.

fs-IR probes the transition dipole of the OD-stretch vibration, and thus measures the reorientation dynamics of the hydroxyl groups of the water molecules. In this case the effects of salts on the water dynamics are dominated by the nature of the anion. If the anion strongly binds the water molecules, a corresponding large fraction of water molecules is observed for which the orientational dynamics of the OH groups is slow.[8]

There exists an interesting asymmetry in the effects of anions and cations on the dynamics of water. This asymmetry is due to the fact that most anions accept hydrogen bonds from the OH groups of the solvating water molecules. Hydrogen bonds have as a special property that they are much stronger in cases where the OH bond is aligned with the $O\cdots A^-$ coordinate.[46] The formation of such a directional hydrogen bond prevents a configuration in which the electrostatic interaction would have been optimized. If the interaction between the anion and water had been purely electrostatic, the anion would have been closest to the middle of the two hydrogen atoms of the water molecules, and the OH groups would thus be positioned in a straddling configuration. However, in such a configuration the OH bonds make a quite large angle with the $O\cdots A^-$ coordinate and the hydrogen-bond interaction would be very weak. It is thus energetically favorable to form a directional $OH\cdots A^-$ with only one of the OH groups of the water molecule. The other OH group points away from the anion and will form a hydrogen bond to another water molecule. The directional character of this hydrogen bond limits the mobility of the bonded OH group, but leaves the other OH group and the dipole moment of the water molecule relatively free to reorient.

The fast reorientation process with a time constant of ~2 ps that we find for $R_A(t)$ can be assigned to a wobbling motion of the OD groups that keep their hydrogen bond with the anion intact. The HOD molecules that donate these OD groups can form up to three hydrogen bonds to water molecules outside the anion hydration shell. The reorganization of the hydrogen-bond network outside the first hydration shell is therefore reflected in the orientation of the HOD molecule and thus the wobbling motion of the anion bound OD group. The fitted fast time constant of 2 ± 0.3 ps is in good agreement with this interpretation. For larger anions, for example I^-, the angular spread over which OD groups have the freedom to wobble is larger (Fig. 7), and thus the amplitude of the fast component is larger (Fig. 5, note that in this figure the fraction of the slow component is shown). Since the maximal anisotropy decay due to the wobbling motion is determined by the cone angle, any residual anisotropy will have to decay through other processes like the reorganization of the hydration shell or the reorientation of the complete anion hydration structure. The time constant that we find for this process is 9 ± 1 ps, in agreement with previous studies on salt solutions, in which only the dynamics at longer delays was considered.[38] This time constant should not be confused with the time constant of exchange, which is coincidentally also 9 ps.

The presence of a quite large wobbling component in the orientational dynamics of the OD groups bonded to the halide anions will enlarge the orientational space for the dipole moment of the water molecules to which the OD groups belong. Hence, the wobbling component forms part of the reason why halide anions have a negligible effect on the orientational mobility of water dipoles. In fact, the average orientational mobility of the water dipoles has been observed to be even somewhat faster for solutions of halide anions than for bulk water,[37] which may be due to the fact that the wobbling motion in the hydration shell of the halide anions shows a somewhat shorter time constant than the reorientation time of bulk liquid water.

3. Cooperative effects of cations and anions. Although cations primarily affect the reorientation of the water dipoles and anions the reorientation of the water hydroxyl

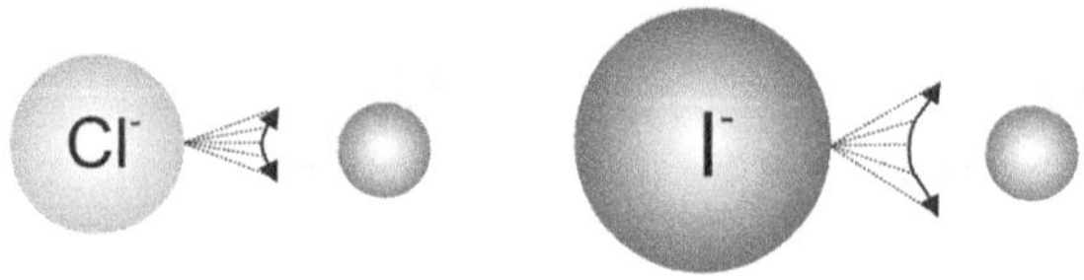

Fig. 7 The hydrogen bond between a water molecule and an anion has the freedom to wobble. For larger anions, the angular spread of the wobbling is larger.

groups, there are indications of cooperative effects. For instance, the nature of the cation influences the reorientation of the water molecules in the anion hydration shell (Fig. 6). Probably, the electric field exerted by the cation restricts the cone over which the anion-bound OD groups can wobble. This effect increases with increasing charge density of the cation, *i.e.* in the series $Cs^+ < Na^+ < Li^+ < Mg^{2+}$.

The cooperative effect of cations and anions on the dynamics of water increases with increasing hydration strength of both ions. For $MgSO_4$ solutions the hydration number is ~18 both in dielectric relaxation and in femtosecond IR measurements.[8] This large hydration number implies that the hydration structures must extend well beyond the first hydration shells of the ions. The formation of the extended rigid hydration structures probably arises as a result of the combination of a strong fixing of the dipole vectors by the cations and a strong fixing of the direction of the hydroxyl groups by the anions. The Mg^{2+} and SO_4^{2-} ions thus cooperate in impeding the water mobility over relatively long ranges and probably induce the formation of a locked hydrogen-bonded structure of water molecules in between the ions. This latter picture is in line with the observation that solutions of strongly hydrated ions, like $MgSO_4$, give rise to the formation of solvent-separated ion pairs.[47]

IV. Hydrophobic effects on water dynamics

We investigate the combined effects of electrostatic and hydrophobic interactions by comparing the effects of guanidinium chloride (GndCl) and tetramethylguanidinium chloride (TMGndCl) on the dynamics of water.

A. THz dielectric relaxation

In Fig. 8 we show the imaginary part of the dielectric function measured for different concentrations of GndCl and TMGndCl. TMGndCl shows a much stronger depolarization effect than GndCl. This cannot be due to electrostatic alignment of the water molecules in the solvation shell of the cation, since the electric field exerted by the $TMGnd^+$ cation is not stronger than the field exerted by the Gnd^+ cation. More likely, a large fraction of water in TMGndCl solutions reorients slowly because of the presence of the hydrophobic methyl groups, as has been observed before for other amphiphilic solutes.[19,48] We model these water molecules with an additional Debye mode, resulting in the model function:

$$\hat{\varepsilon}(\nu) = \frac{S_{\text{slow}}}{1 + 2\pi i\nu\tau_{\text{slow}}} + \frac{S_{\text{bulk}}}{1 + 2\pi i\nu\tau_{\text{bulk}}} + \frac{S_{\text{fast}}}{1 + 2\pi i\nu\tau_{\text{fast}}} + \frac{\sigma}{2\pi i\epsilon_0\nu} + \varepsilon_\infty. \tag{6}$$

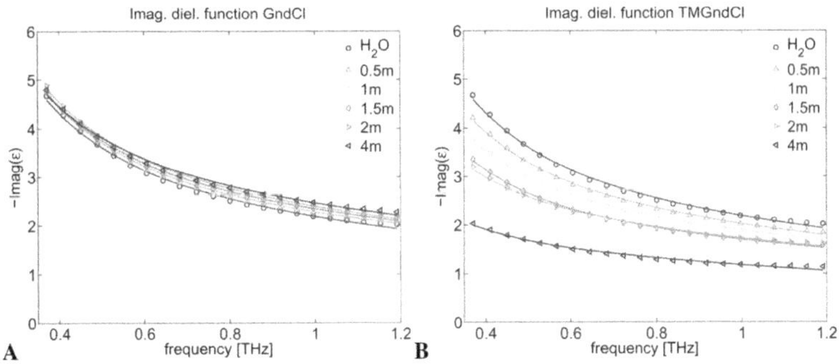

Fig. 8 The imaginary part of the dielectric function of (A) GndCl and (B) TMGndCl solutions of different concentrations.

As pointed out in the Experimental section, the THz-DR experiment is not very sensitive to the frequency dependence of the dielectric relaxation modes that peak at much lower frequencies. The THz-DR experiment primarily probes the decrease in amplitude of the bulk mode, as a decrease in this amplitude leads to depolarization. Hence, in fitting the observed dielectric response of the TMGndCl, we fix the values for τ_{slow} of the slow water mode and τ_{bulk} of the bulk water to the parameters found for TMU in the GHz regime.[19] The fits to the data are presented as the solid lines in Fig. 8.

After correcting for dilution effects and the kinetic depolarization, we obtain the values of the slow water fractions for the different solutions, shown in Fig. 12A. For comparison we also show the slow fractions observed for TMU at the same concentrations.[19]

B. fs-IR pump–probe

The isotropic transient absorption data obtained for guanidinium chloride (GndCl) and tetramethylguanidinium chloride (TMGndCl) were modeled according to the description in section 3.2. The transient spectra of GndCl showed the presence of a third spectral component (in addition to the OD···O and OD···Cl$^-$ components) on the red side of the OD···O spectrum (<2470 cm^{-1}) with a very fast vibrational decay, possibly due to coupling to vibrational modes in the guanidinium cation.[12,49] This component will be subject of a future study. In the present study we minimize the contribution of this fast component by using a pump–pulse frequency centered around 2525 cm^{-1}.

Fig. 9 shows the total anisotropy measured for solutions of GndCl and TMGndCl of different concentrations. The curves shown in Fig. 9 represent the averaged reorientation behavior of all OD···O and OD···Cl$^-$ groups in solution, weighted by their respective time-dependent populations. For the GndCl solutions the anisotropy decays are observed to be only slightly different from the anisotropy decays of pure HDO:H_2O. In contrast, the anisotropy curves of the TMGndCl solutions show a much slower decay, indicating that a large amount of the water molecules show very slow orientational dynamics. The slowing down effect increases with concentration.

The dynamics of the OD···O and OD···Cl$^-$ component of the TMGndCl solutions cannot be determined using the same analysis method that was used for the alkali-halide salts, as this analysis requires the anisotropy dynamics to be frequency independent within the two components. This condition is fulfilled for the alkali-halide solutions, because these solutions show quite fast spectral diffusion. However, for TMGndCl we observe a frequency dependence of the anisotropy that persists at

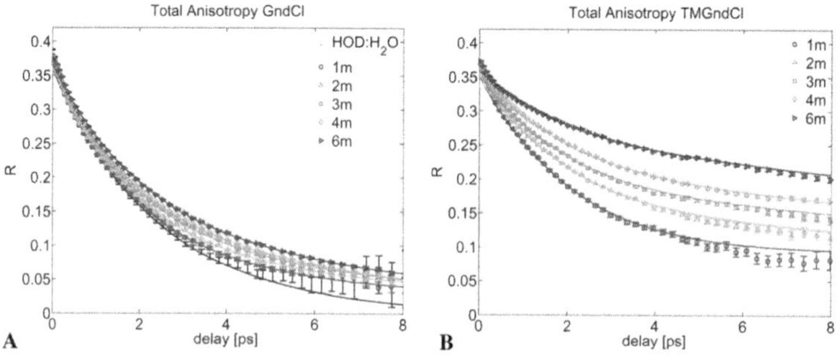

Fig. 9 The total anisotropy decay for solutions containing different concentrations of GndCl and TMGndCl. The substitution of hydrogen atoms by methyl group has a dramatic slowing down effect on the anisotropy decay.

longer delays, as shown in Fig. 10, indicating that the solution is heterogeneous and that the spectral diffusion is quite slow. Hence $R_A(t,\omega)$ should be taken dependent on frequency in eqn (4). To fit the data we use the following functional forms for the anisotropy values of the two components. For $R_O(t)$ we take a bi-exponential function representing bulk water and water with slower dynamics due to the presence of hydrophobic groups,

$$R_O(t) = R_0[(1 - A)e^{-k_1^O t} + Ae^{-k_2^O t}] \quad (7)$$

where A is the fraction of water with slower dynamics, k_1^O the bulk water reorientation rate and k_2^O the reorientation rate of the water with slower dynamics. We fit the data starting at 0.5 ps, thus neglecting any contribution due to fast libration motions.

$R_{Cl}(t)$ was already found to possess a fast and a slow component (eqn (5)). The dynamics of the fast (wobbling) component is governed by the hydrogen-bond dynamics of the surroundings. Hence, this component will be slowed down as a result of the presence of the hydrophobic molecular groups in solution. In addition, there is a persistent frequency dependence of the relative amplitudes $1 - B_\omega$ of the wobbling component and B_ω of the slow component because the spectral diffusion is slow. The functional form of the anisotropy parameter for chloride-bound water is thus:

$$R_{Cl}(\omega,t) = R_0[(1 - B_\omega)((1 - A)e^{-k_1^{Cl} t} + Ae^{-k_2^{Cl} t}) + B_\omega e^{-k_3^{Cl} t}] \quad (8)$$

where the rates k_1^{Cl} and k_2^{Cl} are analogous to the rates from eqn (7). The relative amplitude between the fast and slow component B_0 is, due to the strong inhomogeneous character of the solution, replaced by B_ω, for which we take a linear frequency dependence. The amplitudes of the wobbling motion (~2 ps) and the slower rotational diffusion (~9 ps) thus show an opposite linear dependence on frequency that does not depend on the delay time. We do not include the effects of the exchange between the bulk and the hydration shell in the anisotropy dynamics of the $OD\cdots Cl^-$ component. This exchange is expected to be quite slow in view of the high viscosity of the samples and the slow spectral dynamics of the TMGndCl solutions.

To limit the fit parameters in the fit, we fix $(k_1^O)^{-1}$ to a value of 2.5 ps, as observed for a pure 4% D_2O in H_2O sample.[33] Furthermore we fix $(k_1^{Cl})^{-1}$ to 2 ps and the amplitude of B_ω at 2530 cm^{-1} to the value found for a CsCl solution of the same concentration.[45] The rates k_2^{Cl} and k_3^{Cl} in eqn (8) both represent a slow timescale and we assume these to be the same. This is of course not necessarily the case, but

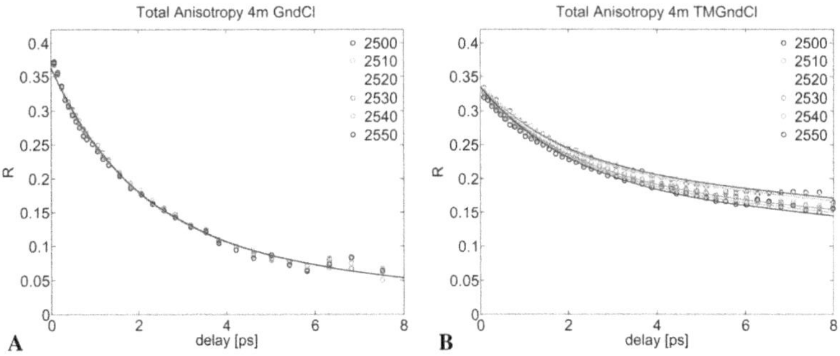

Fig. 10 Anisotropy decay curves for solutions of GndCl (A) and TMGndCl (B). The solutions of TMGndCl show a frequency dependence in their total anisotropy which is absent for solutions of GndCl. This frequency dependence shows that spectral diffusion in the TMGndCl solutions is very slow.

as we can only measure the anisotropy dynamics up to 10 ps, we cannot distinguish between relaxation time constants that are similar to or larger than 10 ps. The fitted values of $k_2^{Cl} = k_3^{Cl}$ lie in the range of 30 to 40 ps. The amplitude of the slow component of $R_{Cl}(t)$ is thus the sum of the slow component resulting from the binding to the anion and the fraction of the fast (wobbling) component that is slowed down as a result of the hydration of the hydrophobic molecular groups. Consistent with the THz-DR analysis, we fix k_2^O in $R_O(t)$ to the values found for TMU with GHz-DR,[19] taking into account the conversion factor ~3.4 discussed at the end of section 2.1. The results of the fits are presented as the solid lines in Fig. 11 and the slow water fractions calculated from the fitted amplitudes A are plotted in Fig. 12B.

C. Discussion

The addition of GndCl to water is observed to have a very small effect on the reorientation dynamics of the water molecules. Guanidinium is a rather large cation which means that the charge density of the positive charge will be low. Hence, the electrostatic interaction between water and this ion will be rather weak. The interaction of water with guanidinium will be dominated by hydrogen bonding to the NH_2 groups. These hydrogen-bond interactions will be quite similar to the hydrogen bonds between water molecules and the hydrogen bonds between water and urea (CH_4N_2O).[13,14] Urea was also observed to show very little effect on the reorientation dynamics of water. The small slowing down of the total anisotropy decay observed for solutions of GndCl, as shown in Fig. 9A, can be fully assigned to the presence of the chloride anion. Since Cl^- anions bind the solvating water molecules by making a hydrogen bond to their OH/OD groups, the dipole moments of these water molecules remain quite free in their reorientation, as was discussed in the previous section. The water dipole moments thus show a similar mobility as in bulk water, as is indeed observed in the THz-DR measurements (Fig. 12A).

TMGndCl is observed to have a large effect on the dynamics of water. The fs-IR measurements show a strong slowing down of the water dynamics in both the OD···O and the OD···Cl^- component. The increase of the slow component in the chloride-bound water compared to GndCl can be completely ascribed to the effects of the hydrophobic molecular groups of $TMGnd^+$ on the dynamics of the wobbling component of the chloride-bound OD groups. The THz-DR results show a similar slowing down effect of the water dynamics as is observed in the fs-IR experiments, which indicates that the $TMGnd^+$ cation affects the dynamics of the OD groups and the water dipole moments in a similar manner. For solutions of TMU it was also observed that the slowing down effect is quite similar in dielectric relaxation and in femtosecond mid-infrared measurements.[19] It is illustrative to compare the results

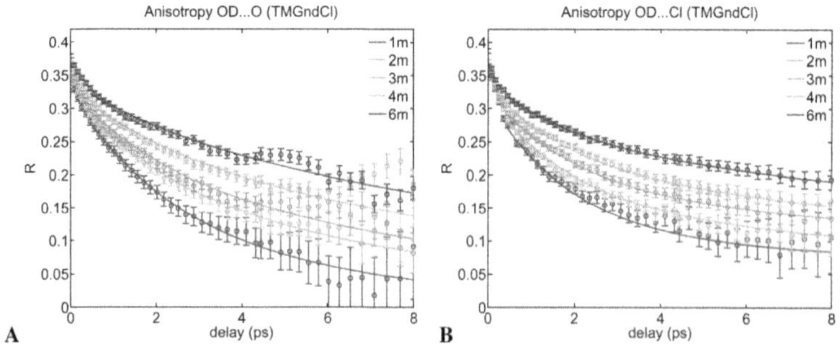

Fig. 11 Anisotropy decays of OD···O (A) and OD···Cl^- (B) for different concentrations of TMGndCl. The hydrophobic groups of the cation slow down the dynamics of the water hydrogen-bond network outside the anion hydration shell, and thereby also decelerate the wobbling motion of the chloride bound OD groups.

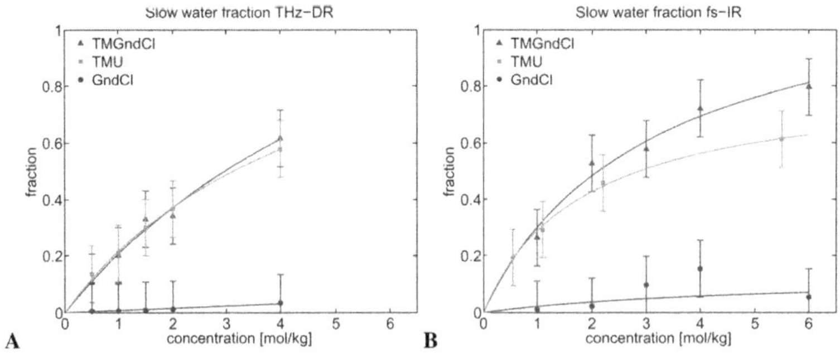

Fig. 12 (A) Slow water fractions obtained from the THz-DR measurements. For both experiments hardly any effect is seen for GndCl, while the effects for TMGndCl and TMU are almost identical. The slow water fractions for GndCl and TMU were calculated with data from ref. 13 and 19 respectively. (B) Slow water fractions of the OD···O anisotropy decays obtained from the fs-IR measurements. The data for TMU are taken from ref. 63.

obtained for TMGndCl with the results obtained for solutions of TMU at the same concentration, as TMU also possesses four methyl groups in a similar configuration as the $TMGnd^+$ cation. The THz-DR data of TMGndCl agree very well with the results of TMU (Fig. 12), confirming that the effects of $TMGnd^+$ and TMU on the dynamics of water are completely dominated by the methyl groups.

In Fig. 12B the slow water fraction of the TMU and TMGndCl solutions shows a saturation effect that has its origin in overlapping hydration shells and aggregation. TMU is known to aggregate at high concentrations.[19,50,51] This Figure shows that the aggregation effect for the $TMGnd^+$ cation is quite similar as for TMU. This is somewhat surprising because the positive charge of the $TMGnd^+$ cations is expected to disfavor aggregation. Apparently, the electrostatic effects of this charge are very weak, as is also demonstrated by the very weak effects of the Gnd^+ cation on the dynamics of water.

In recent two-dimensional infrared (2D-IR) spectroscopy studies it was found that water molecules show very slow spectral dynamics around the hydrophobic groups of tertiary butylalcohol (TBA), trimethylamine-*N*-oxide (TMAO) and (TMU).[52,53] The fraction of water showing slow spectral diffusion was found to be highly correlated to the fraction of water showing slow reorientation dynamics.[53] This correlation points at a common origin. The molecular reorientation of water involves a jump mechanism in which the hydrogen atom of the reorienting hydroxyl group forms a bifurcated hydrogen bond to the oxygen atoms of two nearby water molecules.[23,54,55] The evolution from a linear hydrogen bond to the bifurcated transition state is accompanied by a large change in vibrational frequency and thus leads to strong spectral diffusion, irrespective of whether this evolution leads to a new hydrogen bond and thus reorientation (successful switch) or to the restoring of the original hydrogen bond (unsuccessful switch). The correlated slowing down of the spectral diffusion and the reorientation indicates that water near hydrophobic groups is strongly hindered in its evolution to a bifurcated hydrogen bond structure. A possible explanation for this observation is that the hydrophobic solute methyl groups fill up the cavities of the hydrogen bond network of water, thereby preventing the local collapses of the network that are required for the formation of bifurcated hydrogen bonds.

In NMR studies it is also observed that hydrophobic molecular groups have a strong effect on the reorientation dynamics of liquid water that increases with increasing size of the hydrophobic group.[56–59] For instance, in the NMR study by Shimizu *et al.*,[58] the slowing down of the water reorientation dynamics was found to scale with the number of substituted methyl groups of the urea molecule, urea showing only a negligible effect and TMU the strongest effect. In the same study

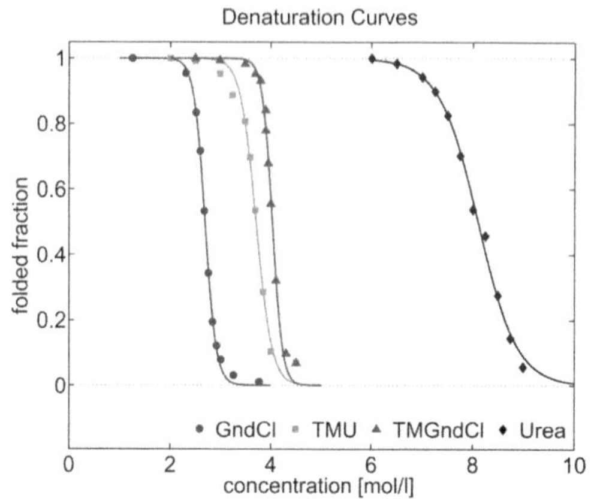

Fig. 13 Fraction of folded protein PYP as a function of denaturant concentration for different denaturants. The measurements were performed for a buffered solution at pH 7.

GndCl was also found to have a negligible effect on the reorientation dynamics of water,[58] in line with the present observations. The reorientation and spectral diffusion dynamics of water in solutions of amphiphiles have also been studied with classical molecular dynamics simulations.[51,60,61] In these studies a slowing down of the reorientation and spectral diffusion of the water molecules around the hydrophobic groups is found, but not to the extent as is observed in the fs-IR and the 2D-IR experiments. In order to relate the effects of GndCl and TMGndCl on the dynamics of water to their role in the denaturation of proteins, we measured the fraction of unfolded Photoactive Yellow Protein (PYP) as a function of concentration GndCl, TMGndCl, urea and TMU. The measurements, of which the results are shown in Fig. 13, were performed according to the procedure described in ref. 62 at pH 7 (using a potassium phosphate buffer). GndCl is clearly most effective, as it completely unfolds all PYP at a concentration of 3.5 mol l^{-1}. Urea is not only the least effective, its shallow slope indicates that more urea molecules per protein are needed for the denaturation process. These results indicate that the mechanisms for denaturation by GndCl and urea differ.

When the four hydrogen atoms of urea are replaced by methyl groups to form TMU, the effectiveness in denaturing PYP becomes strongly enhanced. One may expect the same enhancement for TMGndCl relative to GndCl, but TMGndCl is less effective than GndCl and shows a similar denaturation behavior as TMU. The denaturation mechanisms of urea and GndCl compounds appear to be completely replaced by denaturation effects associated with the methyl groups of TMU and TMGndCl. As TMGndCl and TMU also show a quite similar effect on the dynamics of water, the similar denaturation effect of TMGndCl and TMU may well be related to the hydrophobic hydration of their methyl groups.

V. Conclusions

We studied the effects of ions and hydrophobic molecular groups on the reorientation dynamics of water using THz dielectric relaxation (THz-DR) and polarization-resolved femtosecond mid-infrared spectroscopy (fs-IR). THz-DR probes the reorientation of the water dipole moments while fs-IR probes the reorientation of the water OH/OD groups.

We find that cations primarily slow down the reorientation dynamics of the water dipole moments. This effect increases with decreasing size and increasing charge of the cation. Anions primarily slow down the reorientation dynamics of the water

OH/OD groups. Most anions form a directional hydrogen bond with one of the OH/OD groups of the solvating water molecules.

The reorientation of OD groups that are hydrogen bonded to the halide anions Cl^- and I^- is found to consist of two components: a fast component with a time constant of 2 ± 0.3 ps and a slow component with a time constant of 9 ± 1 ps. The fast component represents a wobbling motion of the OD groups that keeps the hydrogen bond intact. The dynamics of this motion are governed by the dynamics of the water hydrogen-bond network outside the anion hydration shell. The slow component is associated with the diffusion of OD groups over the surface of the anion and/or with the reorientation of the complete anion hydration structure.

The amplitude of the wobbling motion increases with increasing size of the anion and also depends on the nature of the cation. For a strongly hydrated cation like sodium, the wobbling cone is restricted to smaller angles. When very strongly hydrated anions and cations are combined, as in a solution of $MgSO_4$, the restrictions on the wobbling cone angle and the water dipole moments are large enough to lock water molecules in both degrees of freedom outside the first solvation shells.

A comparison of the dynamics of water molecules in solutions of guanidinium chloride (GndCl) and tetramethyl guanidinium chloride (TMGndCl) salts shows that the effect on the dynamics of water are dominated by the hydrophobic methyl groups of the TMGnd cation. For GndCl solutions there is a small slow component in the reorientation dynamics associated with water molecules that are hydrogen bonded to the Cl^- anion. The guanidinium ion has no effect on the reorientation dynamics of water. The four methyl groups of $TMGnd^+$ lead to a strong slowing down of the reorientation dynamics of water, that is quite similar to what is observed for solutions of tetramethylurea (TMU). The water molecules that form hydrogen bonds to Cl^- show, in addition to the anionic effect, a strong slowing down of their wobbling motion due to the nearby presence of the hydrophobic $TMGnd^+$ cations.

Finally, we put these results in the perspective of the capabilities of GndCl, TMGndCl, urea and TMU to denature proteins. While urea and GndCl have very different mechanisms of denaturation, TMGndCl and TMU both show very similar behavior. As TMGndCl and TMU also show a similar effect on the dynamics of water, the denaturation mechanism of TMGndCl and TMU may well be associated with their hydrophobic effect on the dynamics of water.

Acknowledgements

This work is part of the research program of the "Stichting voor Fundamenteel Onderzoek der Materie (FOM)", which is financially supported by the "Nederlandse organisatie voor Wetenschappelijk Onderzoek (NWO)". The authors thank K. J. Hellingwerf and the University of Amsterdam (UvA) for the denaturation experiments. J. H. thanks the *Deutsche Forschungsgemeinschaft* (DFG) for funding through a research fellowship.

References

1 F. Hofmeister, *Arch. Exp. Pathol. Pharmakol*, 1888, **24**, 247–260.
2 Y. Marcus, *Chem. Rev.*, 2009, **109**, 1346–1370.
3 J. D. Smith, R. J. Saykally and P. L. Geissler, *J. Am. Chem. Soc.*, 2007, **129**, 13847–13856.
4 R. Mancinelli, A. Botti, F. Bruni, M. A. Ricci and A. K. Soper, *J. Phys. Chem. B*, 2007, **111**, 13570–13577.
5 D. A. Schmidt, O. Birer, S. Funkner, B. P. Born, R. Gnanasekaran, G. W. Schwaab, D. M. Leitner and M. Havenith, *J. Am. Chem. Soc.*, 2009, **131**, 18512–18517.
6 Y. A. Lin, B. M. Auer and J. L. Skinner, *J. Chem. Phys.*, 2009, **131**, 144511.
7 A. W. Omta, M. F. Kropman, S. Woutersen and H. J. Bakker, *Science*, 2003, **301**, 347–349.
8 K. J. Tielrooij, N. Garcia-Araez, M. Bonn and H. J. Bakker, *Science*, 2010, **328**, 1006–1009.
9 F. Rashid, S. Sharma and B. Bano, *Protein J.*, 2005, **24**, 283–292.

10 C. Camilloni, A. G. Rocco, I. Eberini, E. Gianazza, R. A. Broglia and G. Tiana, *Biophys. J.*, 2008, **94**, 4654–4661.
11 W. K. Lim, J. Rosgen and S. W. Englander, *Proc. Natl. Acad. Sci. U. S. A.*, 2009, **106**, 2595–2600.
12 J. Scott, N. V. Nucci and J. M. Vanderkooi, *J. Phys. Chem. A*, 2008, **112**, 10939–10948.
13 J. Hunger, S. Niedermayer, R. Buchner and G. Hefter, *J. Phys. Chem. B*, 2010, **114**, 13617–13627.
14 P. E. Mason, G. W. Neilson, C. E. Dempsey, A. C. Barnes and J. M. Cruickshank, *Proc. Natl. Acad. Sci. U. S. A.*, 2003, **100**, 4557–4561.
15 A. Das and C. Mukhopadhyay, *J. Phys. Chem. B*, 2009, **113**, 12816–12824.
16 P. A. Mason, J. W. Brady, G. W. Neilson and C. E. Dempsey, *Biophys. J.*, 2007, **93**, L4–L6.
17 M. C. Stumpe and H. Grubmuller, *J. Am. Chem. Soc.*, 2007, **129**, 16126–16131.
18 H. Wei, Y. Fan and Y. Gao, *J. Phys. Chem. B*, 2010, **114**, 557–568.
19 K. J. Tielrooij, J. Hunger, R. Buchner, M. Bonn and H. J. Bakker, *J. Am. Chem. Soc.*, 2010, **132**, 15671–15678.
20 H. J. Bakker and J. L. Skinner, *Chem. Rev.*, 2010, **110**, 1498–1514.
21 U. Heugen, G. Schwaab, E. Bründermann, M. Heyden, D. M. Leitner and M. Havenith, *Proc. Natl. Acad. Sci. U. S. A.*, 2006, **103**, 12301–12306.
22 T. Fukasawa, T. Sato, J. Watanabe, Y. Hama, W. Kunz and R. Buchner, *Phys. Rev. Lett.*, 2005, **95**, 197802.
23 D. Laage and J. T. Hynes, *Science*, 2006, **311**, 832–835.
24 H. Yada, M. Nagai and K. Tanaka, *Chem. Phys. Lett.*, 2008, **464**, 166–170.
25 C. J. Fecko, J. D. Eaves and A. J. Tokmakoff, *J. Chem. Phys.*, 2002, **117**, 1139–1154.
26 D. A. Turton, J. Hunger, G. Hefter, R. Buchner and K. Wynne, *J. Chem. Phys.*, 2008, **128**, 161102.
27 A. Y. Zasetsky, *Phys. Rev. Lett.*, 2011, **107**, 117601.
28 H. Bianchi, H. R. Corti and R. Fernandez-Prini, *J. Solution Chem.*, 1988, **17**, 1059–1065.
29 R. Buchner, S. G. Capewell, G. Hefter and P. M. May, *J. Phys. Chem. B*, 1999, **103**, 1185–1192.
30 T. Chen, G. Hefter and R. Buchner, *J. Phys. Chem. A*, 2003, **107**, 4025–4031.
31 W. Wachter, S. Fernandez, R. Buchner and G. Hefter, *J. Phys. Chem. B*, 2007, **111**, 9010–9017.
32 K. J. Tielrooij, S. T. van der Post, J. Hunger, M. Bonn and H. J. Bakker, *J. Phys. Chem. B*, 2011, **115**, 12638–12647.
33 Y. L. A. Rezus and H. J. Bakker, *J. Chem. Phys.*, 2005, **123**, 114502.
34 C. Böttcher, *Theory of Electrical Polarization Vol. I*, Elsevier, Amsterdam, 1973.
35 C. Böttcher, *Theory of Electrical Polarization Vol. II*, Elsevier, Amsterdam, 1978.
36 K. J. Tielrooij, C. Petersen, Y. L. A. Rezus and H. J. Bakker, *Chem. Phys. Lett.*, 2009, **471**, 71–74.
37 W. Wachter, W. Kunz and R. Buchner, *J. Phys. Chem. A*, 2005, **109**, 8675–8683.
38 M. F. Kropman and H. J. Bakker, *Science*, 2001, **291**, 2118–2120.
39 S. Park and M. D. Fayer, *Proc. Natl. Acad. Sci. U. S. A.*, 2007, **104**, 16731–16738.
40 P. A. Bergstrom, J. Lindgren and O. Kristiansson, *J. Phys. Chem.*, 1991, **95**, 8575–8580.
41 M. F. Kropman and H. J. Bakker, *J. Am. Chem. Soc.*, 2004, **126**(29), 9135–9141.
42 S. T. Roberts, J. J. Loparo and A. Tokmakoff, *J. Chem. Phys*, 2006, **125**(084502).
43 B. Nigro, S. Re, D. Laage, R. Rey and J. T. Hynes, *J. Phys. Chem. A*, 2006, **110**, 11237–11243.
44 K. J. Gaffney, M. Ji, M. Odelius, S. Park and Z. Sun, *Chem. Phys. Lett.*, 2011, **504**, 1–6.
45 S. T. van der Post and H. J. Bakker, *Phys. Chem. Chem. Phys.*, 2012, **14**, 6280–6288.
46 K. B. Moller, R. Rey and J. T. Hynes, *J. Phys. Chem. A*, 2004, **108**, 1275–1289.
47 R. Buchner, T. Chen and G. Hefter, *J. Phys. Chem. B*, 2004, **108**, 2365–2375.
48 R. Buchner, C. Hölzl, J. Stauber and J. Barthel, *Phys. Chem. Chem. Phys.*, 2002, **4**, 2169–2179.
49 D. Y. Vorobyev, C. H. Kuo, D. G. Kuroda, J. N. Scott, J. M. Vanderkooi and R. M. Hochstrasser, *J. Phys. Chem. B*, 2010, **114**, 2944–2953.
50 L. Almasy, A. Lenb, N. K. Szekelyb and J. Pletilc, *Fluid Phase Equilib.*, 2007, **257**, 114–119.
51 G. Stirnemann, F. Sterpone and D. Laage, *J. Phys. Chem. B*, 2011, **115**, 3254–3262.
52 A. A. Bakulin, C. Liang, T. la Cour Jansen, D. A. Wiersma, H. J. Bakker and M. S. Pshenichnikov, *Acc. Chem. Res.*, 2009, **42**, 1229–1238.
53 A. A. Bakulin, M. S. Pshenichnikov, H. J. Bakker and C. Petersen, *J. Phys. Chem. A*, 2011, **115**, 1821–1829.
54 J. J. Loparo, S. T. Roberts and A. Tokmakoff, *J. Chem. Phys.*, 2006, **125**, 194522.
55 H. J. Bakker, Y. L. A. Rezus and R. L. A. Timmer, *J. Phys. Chem. A*, 2008, **112**, 11523–11534.

56 Y. Ishihara, S. Okouchi and H. Uedaira, *J. Chem. Soc., Faraday Trans.*, 1997, **93**, 3337–3342.
57 K. Fumino, K. Yukiyasu, A. Shimizu and Y. Taniguchi, *J. Mol. Liq.*, 1998, **75**, 1–12.
58 A. Shimizu, K. Fumino, K. Yukiyasu and Y. Taniguchi, *J. Mol. Liq.*, 2000, **85**, 269–278.
59 J. Qvist and B. Halle, *J. Am. Chem. Soc.*, 2008, **130**, 10345.
60 D. Laage, G. Stirnemann and J. T. Hynes, *J. Phys. Chem. B*, 2009, **113**, 2428–2435.
61 G. Stirnemann, J. T. Hynes and D. Laage, *J. Phys. Chem. B*, 2010, **114**, 3052–3059.
62 B. C. Lee, M. Kumauchi and W. D. Hoff, *J. Biol. Chem.*, 2010, **285**, 12579–12586.
63 Y. L. A. Rezus and H. J. Bakker, *J. Phys. Chem. A*, 2008, **112**, 2355–2361.

Ion cooperativity and the effect of salts on polypeptide structure – a molecular dynamics study of BBA5 in salt solutions†

Wen Jun Xie and Yi Qin Gao*

Received 7th April 2012, Accepted 15th May 2012
DOI: 10.1039/c2fd20065a

Using molecular dynamics simulations, we investigated effects of inorganic salts on the structure and dynamics of a short α,β-polypeptide, BBA5. The simulations showed that three model salts, NaI, NaF, and Na_2SO_3, have very different effects on the structure of the polypeptide. The addition of NaI to the aqueous solution caused denaturation and significantly weakened hydrogen bonds of the polypeptide. Na_2SO_3 strengthened the hydrophobic interactions and increased hydrogen bonding of the polypeptide. Preferred binding of Na^+ to the backbone carbonyl groups of BBA5 occurred in the NaI solution, consistent with the weakened protein backbone hydrogen bonds, whereas Na^+ is excluded more from the vicinity of the protein backbone in the Na_2SO_3 solution. This difference in Na^+ binding correlates well with the different propensities of the counter ions approaching the protein surface: SO_3^{2-} is much more strongly expelled from the protein apolar surface than I^-, and demonstrates the importance of cation–anion cooperativity in affecting protein structures. The binding of the two salts to and their effects on the hydration of the protein surface depends strongly on the polarity of the latter. However, both salts reduce the flexibility of the polypeptide and the fluctuation of its hydration layer. These simulations showed that the chaotropic NaI affects protein structure mainly through a direct binding of Na^+ to the backbone and I^- to the protein surface. The main effect of Na_2SO_3 manifests in strengthening the hydrophobic interaction and consequently the hydrogen bonding of the protein, more likely through an "indirect" mechanism. In addition, the simulations showed that NaF has a similar effect as Na_2SO_3 (but weaker than the latter, consistent with their positions in the Hofmeister series).

Introduction

It has been found that the salting out effects of anions on proteins generally follow the Hofmeister series: $CO_3^{2-} > SO_4^{2-} > F^- > Cl^- > Br^- > I^- \sim NO_3^-$.[1–8] The underlying physical mechanism behind such a series has been the subject of a very active research area in recent years and has been interrogated by both experimental and theoretical/computational approaches.[9–16] The mechanisms through which electrolytes (in general, all components of the aqueous solutions) affect protein structure and solubility were categorized into direct and indirect effects. Through extensive measurements of low critical solution temperature (LCST), Cremer and coworkers[13,15,16] used both synthetic polymers and polypeptides/proteins to study

Institute of Theoretical and Computational Chemistry, College of Chemistry and Molecular Engineering, Beijing National Laboratory for Molecular Sciences, Peking University, Beijing, 100871, China. E-mail: gaoyq@pku.edu.cn

† Electronic supplementary information (ESI) available: See DOI: 10.1039/c2fd20065a

how various cations and, especially, anions affect their solubility. It was found that electrolytes' salting out effects correlate well with the solvation entropy of the ions and such a correlation breaks down for "salting in" ions. Instead, the salting in effects were well explained by the direct binding of ions to the polymer[13] or protein.[15,16] Based on these observations, it was proposed that the salting in effect is more likely a result of the direct binding of the ions to the polar domain of the protein or polymer, *e.g.*, the CO or NH moiety), the salting out effect, on the other hand, is more likely due to the change of the water/polymer (or water/protein) interactions by the addition of ions. An elegant analysis performed by Pegram and Record[14] on inorganic salts affecting water/air surface tension revealed the distribution of ions and showed that chaotropic anions are enriched and kosmotropic ones are depleted near such surfaces. They also compared the difference between water/air and water/protein interfaces and suggested that the salting in/salting out effects of the cations and anions correlate well with their preferred binding/exclusion at the interfaces. These results are consistent with earlier thermodynamics characterization of the preferred binding behavior of electrolytes (*e.g.* ref. 3).

As for the indirect effect, earlier studies suggested[17] that salts have little effect on the water dynamics and the effects tend to be localized to the first solvation shell of the ions. A more recent study[18] of different salts, however, shows that water re-orientation dynamics could be significantly affected by salt ions. This result is in accordance with NMR,[19] neutron diffraction,[20] the optical Kerr effect,[21] and viscosity measurements.[22] In particular, the addition of strong kosmotropes (structure makers) generated "slow rotating" water molecules, the population of which increases almost linearly with ion concentrations. These results again suggested that one can not simply rule out the indirect mechanism for ion effects on protein solubility. The effects of tested inorganic salts not only depend on the type of individual dissociated ions, but also show a strong correlation effect between the positively and negatively charged ions. This correlation appears to follow the empirical "matching water affinity"[23] rule. For example, salts consisting of both "hard" and "soft" ions (such as $LiClO_4$) tend to have a smaller effect than those consisting of matching pairs.[18] It is important to point out that a comparison between the optical Kerr effect and dielectric relaxation spectroscopic measurements showed that different techniques could be detecting different dynamics of water. It is possible that[21,24] ions can slow down water rotational relaxation without a strong perturbation to the translation motions, the latter is more likely being measured by the optical Kerr effect.[24] The comparison of different experiments shows the complexity of the relation between structural, dynamic, and thermodynamic properties of water and, therefore, one should be more specific in relating the dynamic effects of ions to their effects on water structure and the solvation of proteins.

Recently,[25] using K-edge X-ray diffraction measurements combined with computational studies and triglycine as a model system, Saykally and coworkers studied the ion/water/polypeptide interaction. This study clearly showed that there is significant direct binding of SO_3^{2-} to the polypeptide and this binding resulted in an increase of the rigidity of the tri-polypeptide. The latter is consistent with a lowered solubility of the polypeptide in the sulfite solution. These results were used to argue against the indirect effects of kosmotropes in affecting the water/protein interaction.[25] However, as pointed out by these authors,[25] the polypeptide used in the study is very short. It is also highly charged, with at least two protonated sites playing the most important roles in anion binding. Therefore, it might not be the best example as a protein and could differ significantly from the polymer model used in Cremer's study.[13] The motivation of the present study is thus to investigate the interactions of kosmotropes (using Na_2SO_3 as an example following ref. 25) and chaotropes (the supposedly more chaotropic NaI instead of NaBr was used as an example) with a longer polypeptide that has a well defined structure, and to understand how they affect the structure of the latter.

Molecular dynamics simulations have played a very important role in understanding the salt effects on various properties of water and its biologically related solutes (*e.g.*, see some recent ref. 26–29). Jungwirth, Tobias, Mundy, Dang and others[29–31] have studied the distribution of cations and anions on water interfaces, and showed that chaotropic anions tend to bind to the water/air surface as well as hydrophobic interfaces while the cations and negatively charged kosmotropes are expelled from the interfaces. Although the exact reasons for such behaviors are still under debate,[32] and could at least partially originate from the intrinsic properties of water itself,[33,34] these findings also correlate well with the effects of anions on protein solubility. Elcock and coworkers[35] thoroughly studied the effects of salts on various water structural properties and they found a reasonable correlation between the Hofmeister series and the hydrogen bonding of water in the presence of salts. Simulations were also performed to investigate directly how ions interact with biomolecules and affect the water interaction with the latter. For example, an MD study on NaCl and KCl interacting with polyalanine[36] showed that the cations prefer binding to the protein backbone carbonyl group and this binding is responsible for the electrolytes' ability of salting in the protein. In addition, the stronger binding capability of Na^+ makes NaCl a stronger denaturant than KCl. Consistently, a more recent MD simulation by Jungwirth and coworkers on salt/phospholipid interactions revealed the preferred binding of the cations to the negative charges of the lipids.[37] These studies have pointed to the importance of cation binding, as well as the ion cooperativity[37,38] in affecting the structure of biomolecules.

To study the salt effects on protein structures, especially on protein hydrogen bonds and on hydrophobic hydration, in our molecular dynamics simulations, we chose a longer polypeptide BBA5[39] which has well defined secondary structures and a hydrophobic core. BBA5 is a relatively small polypeptide and has both alpha and beta structural elements. It was shown to have a stable aqueous solution structure and has been shown[40] to be sensitive to the solution conditions. These properties make it a very good model system for our study of salt effects on protein structures. Two model sodium salts, one "kosmotrope" Na_2SO_3, and one "chaotrope" NaI, were used in this study to test how they affect the different protein structure elements. Consistent with expectations, it was found that NaI denatures BBA5 and the addition of Na_2SO_3 with a similar ion concentration increases hydrophobic interactions. NaI weakens the protein hydrogen bonds and Na_2SO_3, on the contrary, stabilizes the hydrogen bonds of BBA5. It was observed that Na^+ shows a preferred binding with the backbone carbonyl groups of BBA5 in the NaI solution but is excluded from the vicinity of the protein backbone in the Na_2SO_3 solution. It was also found, consistent with earlier studies, that the direct binding of the cation to the protein is more prominent than that of the two anions. In addition, SO_3^{2-} but not I^- is strongly expelled from the protein apolar surface. Although the binding of the two salts to the protein surface and their effects on the hydration of the latter depend strongly on the polarity of the local protein surface, both salts decrease the fluctuation of the protein hydration layer and the solvent accessible surface area (SASA). These results suggest that the chaotropic NaI affects protein structure mainly through a direct binding of Na^+ to the backbone and I^- to the protein surface, whereas the main effect of Na_2SO_3 manifests in strengthening the hydrophobic interaction and consequently the hydrogen bonding of the protein. Since both ions of Na_2SO_3 are expelled from the protein hydrophobic surface, such an effect is more likely an "indirect" one and is thus expected to correlate with its ability in affecting water structure and dynamics, as suggested by Cremer and coworkers.[13,15,16] As discussed earlier, Na_2SO_3 was used here to model a system with strongly hydrated anions. The hydration and interaction of SO_3^{2-} in water can be complicated. To further validate the results, we also used NaF as another (but weaker) model system for kosmotropes. Some of the results on the NaF systems are given in the ESI†.

Simulation details

All classical MD simulations were performed in explicit solvent using the AMBER 9.0 suite of programs with the parm99 additive force field.[41] Water is described with the SPC/E model.[42] For sodium and iodide we used parameters from Joung and Cheatham.[43] The charge distributions on all atoms in the sulfite ion were obtained using the RESP-fit method[44,45] based on B3LYP/cc-pVTZ calculations[46–49] with the solvent effect ($\varepsilon = 4$) taken into account using the polarizable continuum model (PCM).[50–52] Before charge fitting, the geometry was fully optimized in the gas phase at the same theoretical level. For each polypeptide in a Na_2SO_3/NaI solution, three independent trajectories were run, respectively. Each trajectory was ~200 ns long. The concentration of NaI was ~3.9 mol L^{-1} and that of Na_2SO_3 wa ~2.3 mol L^{-1}. The initial structure of BBA5 was taken from its NMR structure.[39] The polypeptide was immersed into a cubic box containing both Na_2SO_3 (or NaI) and water molecules. In the meanwhile, each polypeptide in the pure water solution was also studied and the simulation result was compared with those in the Na_2SO_3 or NaI solutions. NPT ensemble calculations were performed for each system. Periodic boundary conditions were used in the energy minimization and MD simulations. The SHAKE algorithm was used to constrain all bonds involving hydrogen atoms. A cutoff of 10.0 Å was applied for nonbonding interactions. The Particle Mesh Ewald method was applied to treat long-range electrostatic interactions.

The system was first heated to 360 K and equilibrated for 1 ns at 360 K, followed by 200 ps of cooling from 360 K to 300 K performed with harmonic restraints (force constant = 10.0 kcal mol^{-1} $Å^{-2}$) applied to the backbone atoms. Finally, the long production runs were performed and used for the data analysis.

Hydrogen bond definition: a hydrogen bond is considered as formed only if the-distance between the hydrogen donor and acceptor is less than 3.5 Å and the N–H–O angle is greater than 135°.

Hydrophobic contact definition: if the center of the two residues is separated in the sequence by more than 3 residues and the pair is less than 5.0 Å away, we define it has a hydrophobic contact. We found 8 such possible hydrophobic contacts in BBA5. We listed them as follows: Y1-L14, Y1-L18, V3-F8, V3-L17, V3-L18, F8-L17, L14-L18, L18-A22.

Results

1. BBA5 takes different structures in different aqueous solutions

To visualize the effects of the two salts on the protein structure of BBA5, we show in Fig. 1 its representative structure after relatively long equilibration (200 ns), in pure water, in 2.3 M Na_2SO_3 solution and in 3.9 M NaI solution, respectively. It is clearly seen from these figures that the salts affect the structure of BBA5 in different ways. Differences exist in both secondary structures and the assembly between them. Fig. 2

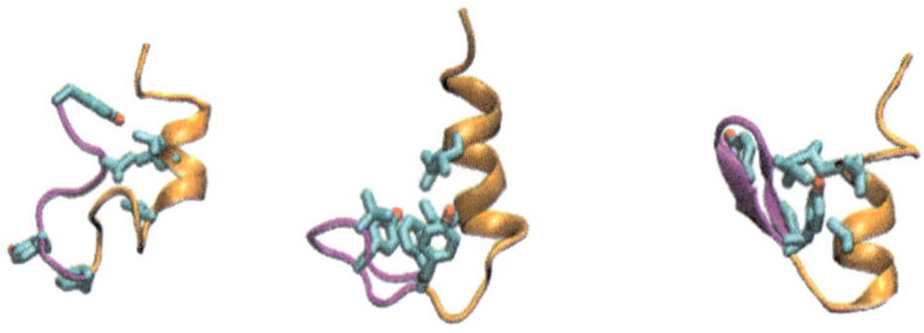

Fig. 1 Representative structures of BBA5 after equilibration in pure water (middle), 3.9 M NaI (left) and 2.3 M Na_2SO_3 (right) solutions, respectively. In the representative structures, the hairpin (residues 1–8) of the native structure is drawn in purple and the helix (residues 9–23) in gold. The side chains of residues that form the hydrophobic core are in cyan.

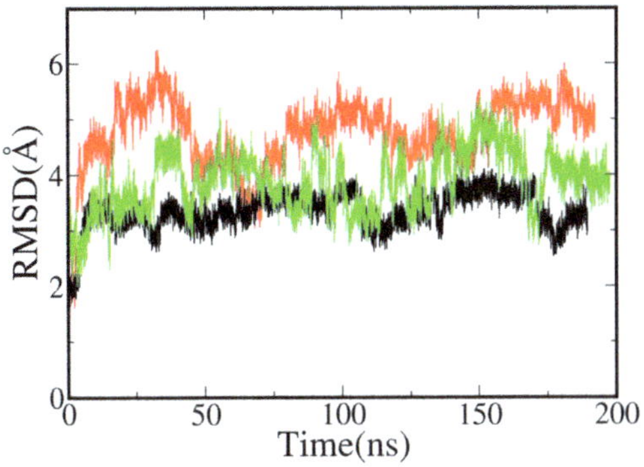

(a)

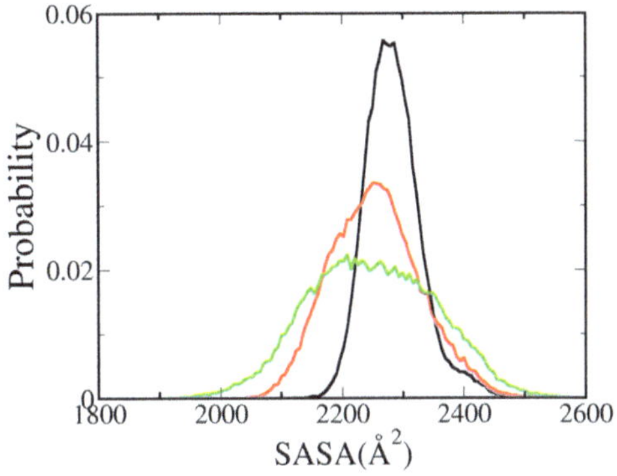

(b)

Fig. 2 Time evolution of the (a) RMSD and (b) SASA of BBA5 in the three different solutions (green, pure water; red, NaI solution; and black, Na_2SO_3 solution, this color code is used throughout this paper).

shows the different effects of the two salts on the root mean square displacement (RMSD) and the solvent accessible surface area of BBA5. On average, Na_2SO_3 increases the SASA of BBA5, although as seen in Fig. 1, the polypeptide has the smallest RMSD value in the Na_2SO_3 solution consistent with a small change compared to the NMR structure. On the other hand, the effect of NaI on the average SASA is small, but the overall RMSD is increased in this salt solution, indicating a change of the BBA5 structure without a significantly increased exposure to the solvent. As seen later, NaI has a strong effect on the BBA5 secondary structure.

To further locate and characterize the structural changes induced by the ions, we calculated SASA for atoms that are defined as polar and apolar, (SASA was also calculated for polar and apolar residues separately, the results are very similar and are not shown here). These results are shown in Fig. 3. It is seen from this figure that Na_2SO_3 has opposite effects on the polar and apolar components of BBA5: it causes an increase of the polar SASA but a decrease of the apolar SASA. The observed overall increase of the SASA of BBA5 is thus a result of the further exposure of its polar surface, suggesting the strong interactions of the ions with the polar or charged protein groups, consistent with ref. 25. NaI, on the other hand, induces a slight increase of the apolar SASA, and has a small effect on the polar SASA (a slight increase).

Next, to characterize the change of the intra-molecular interactions of the protein that resulted from the addition of salts, the average number of protein hydrogen

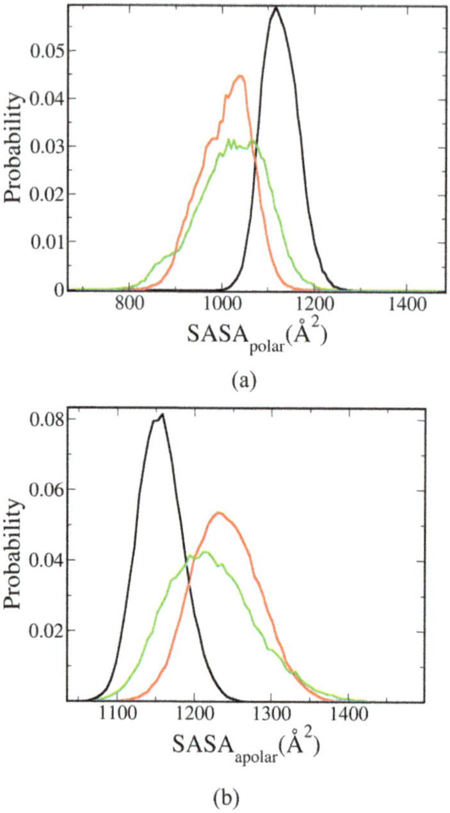

Fig. 3 SASA of (a) polar and (b) apolar atoms of BBA5 in the three different solutions. (green, pure water; red, NaI solution; and black, Na_2SO_3 solution).

bonds in the three different systems were calculated and listed in Table 1. It is seen that NaI quite noticeably weakens the native protein hydrogen bonding interactions, whereas the protein hydrogen bonds are strengthened overall in the Na_2SO_3 solution. As a result, the former but not the latter destabilizes the polypeptide secondary structure, consistent with the RMSD values in different solutions (as can also be seen from the structure comparison given in Fig. 1). Finally, we also calculated the average number of hydrophobic contacts. They are 3.1, 7.6, 5.2 and 2.6, for pure water, Na_2SO_3, NaF, and NaI solutions, respectively. These numbers show that Na_2SO_3 (as well as NaF) but not NaI induces more stable association of the apolar residues of the polypeptide, as seen in the increase of the number of hydrophobic contacts in the Na_2SO_3 solution but not when NaI is present. This result is also consistent with the changes in polar and apolar SASA discussed above.

In summary, these quantitative analyses demonstrate that both Na_2SO_3 and NaI affect the structure of BBA5 in water, but in very different ways. Na_2SO_3 and NaI have the overall effects of a protectant and denaturant on the secondary structure of BBA5, respectively. In the next section, we try to identify the responsible molecular (ionic) interactions for the different response of BBA5 to the addition of the two salts. In particular, we try to distinguish between direct ion binding and indirect effects.

2. Interactions of water and ions with proteins

To understand the effects of an inorganic salt on the protein structure, it is crucial to understand the interactions of the various solution components with the protein. We

Table 1 The average number of backbone hydrogen bonds of BBA5 formed in NaI solution, Na_2SO_3 solution, and pure water. The average is over the last 50 ns of the simulations. Overall, hydrogen bonds are strengthened (**bold**) in the Na_2SO_3 solution and weakened (*italic*) in the NaI solution, compared to those in pure water. The strengthening of hydrogen bonds in the Na_2SO_3 solution is seen where strong hydrophobic contacts are formed between the alpha and the beta domains (residues 3, 6 and 10–15)

Backbone HB	In NaI solution	In Na_2SO_3 solution	In Water
F8O-Y1H	0.00	0.00	0.00
Y1O-F8H	*0.00*	*0.00*	0.68
Y6O-V3H	*0.00*	0.96	0.96
V3O-Y6H	*0.00*	**0.75**	0.27
R10O-L14H	**0.27**	**0.89**	0.07
S11O-A15H	*0.00*	**0.62**	0.19
D12O-16H	0.00	0.10	0.09
E13O-L17H	0.00	0.00	0.07
L14O-L18H	0.41	**0.96**	0.35
A15O-19H	**0.40**	**0.38**	0.27
K16O-20H	*0.00*	*0.00*	0.12
L17O-H21H	**0.45**	*0.00*	0.30
L18O-A22H	*0.19*	0.28	0.36
R19O-23H	0.00	0.00	0.00

analyzed the binding of water, cation, and anion with the different parts of the protein by calculating various radial distribution functions (RDFs), which are shown in Fig. 4 to 6. It is clearly seen from these figures that the interactions between the ions, the protein backbone and side chains depend heavily on the ion type and the polarity of the protein. The most noticeable difference between the two salt solution systems exists in the Na^+/carbonyl oxygen radial distribution function, given in Fig. 4a. This RDF shows a high peak for the NaI system but not for the other. The high peak in the NaI solution indicates the important role of the Na^+–backbone interaction in breaking the protein hydrogen bonds and/or in stabilizing the denatured protein structure. Furthermore, consistent with earlier simulation studies,[36] the binding of both anions to the protein backbone NH group is weak (Fig. 4b).

The difference between Na^+ binding to the carbonyl oxygen in the presence of I^- and SO_3^{2-} suggests that the counter ion has an important influence on the cation interaction with the protein backbone. In particular, the stronger interaction between the ions in Na_2SO_3 solution (as seen from the number of SO_3^{2-} around each Na^+, Fig. 5) and the exclusion of the anion from the protein hydrophobic surface (Fig. 6a) keeps Na^+ from approaching the hydrophobic part of the protein surface and reduces Na^+/protein binding (although the ions preferentially bind the charged and polar protein groups). Namely, although Na^+ by itself preferentially interacts with the carbonyl oxygen, due to the exclusion of SO_3^{2-}, both cations and anions are expelled to a large extent from the protein surface in this electrolyte aqueous system. In contrast, the interactions between the anions and cations are weaker in the case of NaI (Fig. 5), presumably due to the lower charge density of I^-. As seen from the example given in Fig. 6a (for the non-polar residue L14), SO_3^{2-} but not I^- is strongly expelled from the protein hydrophobic surface. Such a more favored binding of I^- to the protein surface compared to SO_3^{2-} (also clearly seen from the binding data given in Table 2) results in a more favorable environment for the cations (as seen from Fig. 6b). Na^+ is expelled from the hydrophobic side chains to a much less extent in NaI than in Na_2SO_3 solution. We also want to point out here that both anion and cation binding to the side chains are sensitive to the charge of the residue (for example, Fig. 6c and d for various RDF diagrams for the charged and polar residues show a large variation of RDF).

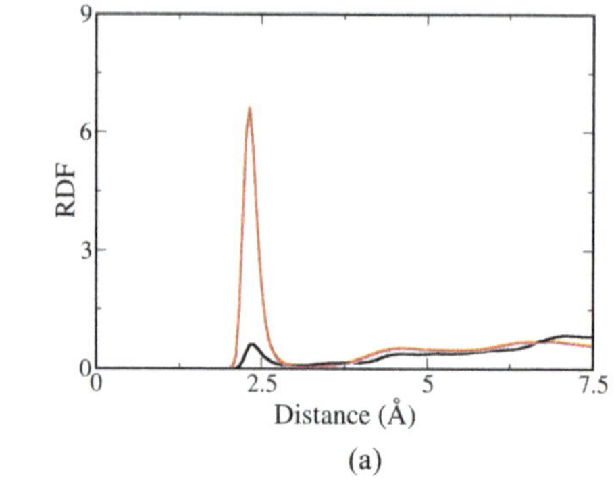

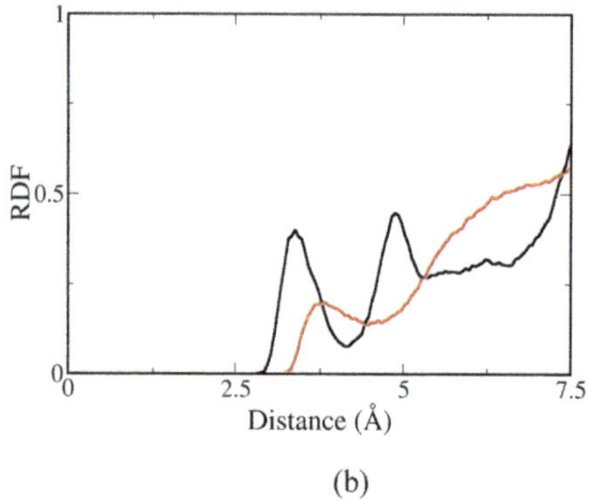

Fig. 4 (a) The RDF of Na^+ around the carbonyl oxygen averaged for all residues in Na_2SO_3 (black) and NaI (red) solutions. (b) The RDF of SO_3^{2-} (black) and I^- (red) around the polypeptide backbone NH groups.

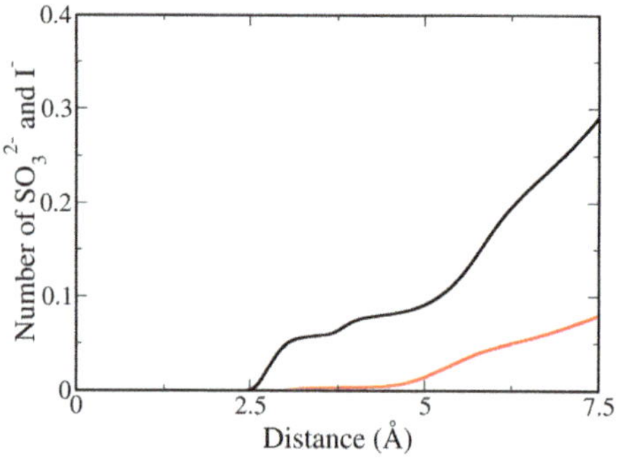

Fig. 5 The averaged number of SO_3^{2-} (black) and I^- (red) around Na^+ as a function of center-to-center distances between the ions.

The importance of Na^+ interaction with the protein backbone carbonyl oxygen can be further quantified by the calculated number of cations and water at given distances to the carbonyl oxygen. From the comparisons given in Fig. 7, it is clearly seen that in the case of the NaI solution, the number of Na^+ adjacent to the carbonyl oxygen is significantly larger than that of water, while in the Na_2SO_3 solution, water dominates over Na^+ near the carbonyl groups. This result strongly suggests that the preferred binding of Na^+ to the protein backbone plays a very important role in weakening the protein hydrogen bonds in denaturing the protein structure. Its

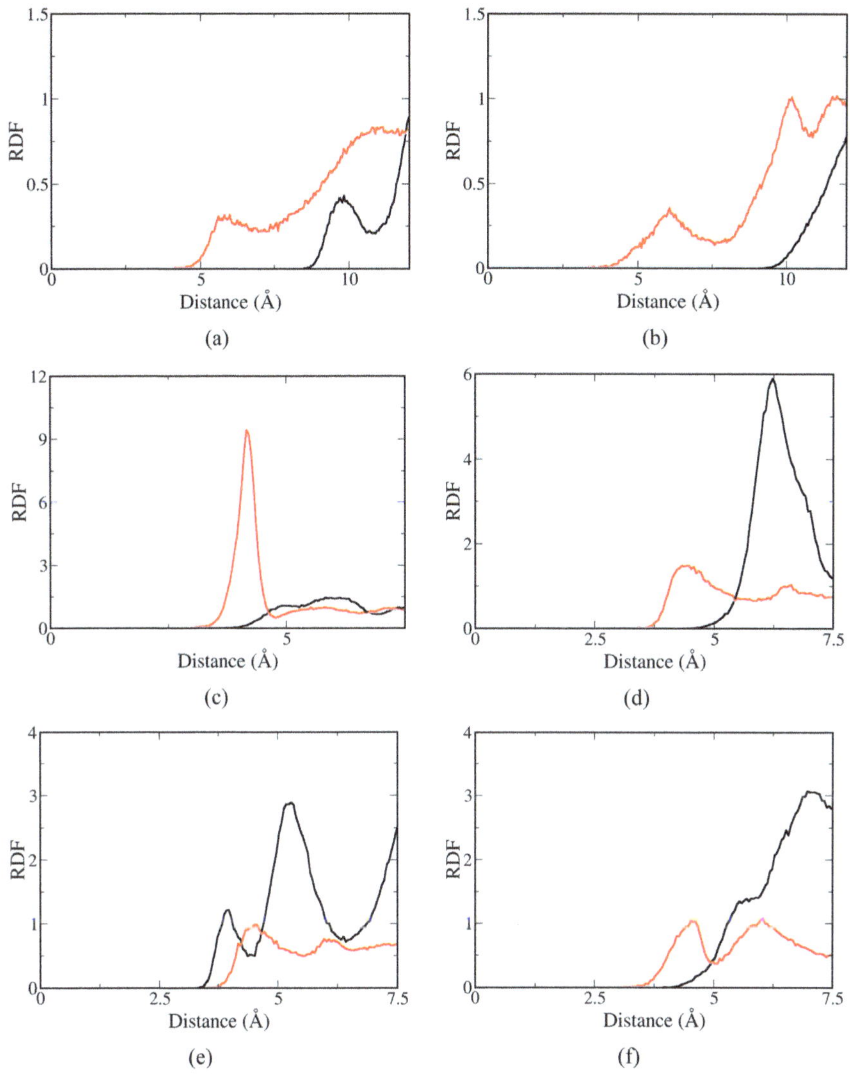

Fig. 6 The RDF of ions around the side chain of residues (ions in Na_2SO_3/NaI solutions are colored black and red, respectively): (a) anions (SO_3^{2-} in black and I^- in red) around the apolar Leu14. (b) Na^+ in different solutions around Leu14, (c) Na^+ around negatively charged Glu13, (d) anions around the positively charged residue Arg19, (e) anions around polar Gln20, (f) Na^+ around Gln20.

denaturing ability, however, is strongly affected by the counter ions present in the solution. In summary, the observed denaturation of BBA5 in NaI solution is mainly a direct effect of Na^+ binding to the protein backbone, and the I^- anion plays a role that is more appropriately described as indirect.

To further quantify the direct and indirect effects of ions on protein solvent/solute interactions, in Table 2, we list the average number of various ions residing in the first hydration shell of the protein surface. Consistent with the RDFs given earlier, the ions of Na_2SO_3 are more likely expelled from the protein hydrophobic surface, and its effects on the association of apolar residues are thus more likely resulted from an indirect mechanism. The indirect "dewetting" effect of this salt can be seen from the decrease of the number of water molecules, N_{hyd}, in the first hydration layer of

Table 2 The number of ions in the Na_2SO_3 solution and the NaI solution near the BBA5 surface

	BBA5 surface	# of cations (within 2.5 Å)	# of anions (within 2.5 Å)	# of cations (within 5 Å)	# of anions (within 5 Å)
NaI	apolar	0.57	0.66	4.9	6.9
	polar	3.55	3.19	11.7	13.8
Na_2SO_3	apolar	0.35	0.23	1.72	1.50
	polar	3.84	6.05	21.8	19.1

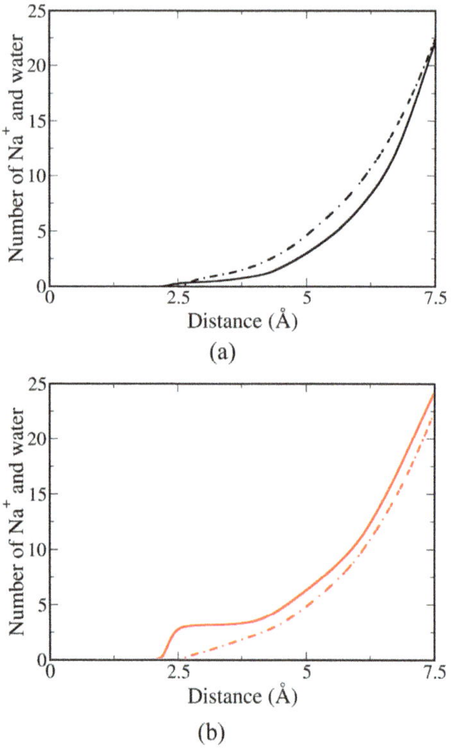

Fig. 7 The number of Na^+ (line) and water molecules (dash dot) around the BBA5 backbone carbonyl oxygen as a function of center-to-center distances in (a) Na_2SO_3 and (b) NaI solutions.

the hydrophobic residues of BBA5 in its solution. In addition, we also calculated the number of water molecules per unit SASA apolar of the protein, $N_{s,ap}$.

$$N_{s,ap} = \frac{N_{hyd,apolar}}{SASA_{apolar}}$$

The values are 3.20, 3.29, and 3.09 for every 100 $Å^2$ of apolar SASA for pure water, NaI and Na_2SO_3 aqueous solutions, respectively. These values suggest that Na_2SO_3 not only reduces the SASA for the apolar part of the protein but also reduces its hydration (NaI, on the other hand, increases this hydration). The observation that ions of Na_2SO_3 are expelled from the protein apolar surface and also cause a weaker hydration of the latter suggests that this salt has an indirect effect on protein hydrophobic interactions.

Next, we analyze the effects of the two inorganic salts on the polar residues of BBA5. Fig. 3b shows that NaI has a small effect, whereas Na_2SO_3 induces an increase of the polar SASA, consistent with the preferred binding of the SO_3^{2-} to the polar residues, as seen from Table 2. To estimate the effects of ion binding on the hydration of polar protein residues, we calculated the number of water molecules per unit SASA for the polar residues. The calculated values are 6.62, 5.95, and 5.78 for every 100 Å^2 of SASA for pure water, Na_2SO_3 and NaI aqueous solutions, respectively. First, we notice that the hydration of the polar residues per unit area is about two times of that of the apolar residues. Secondly, it is seen that the number of water molecules per unit protein polar surface decreases in the order of pure water, Na_2SO_3 and NaI solutions. This decrease in the case of salt solutions, however, does not reflect the dewetting of the protein apolar surface by the inorganic salts, but rather is a result of ion distribution near the protein surface, as seen from Table 2: due to the relative strong binding between ions and the protein polar surface water molecules are substituted to a large extent by ions at such positions.

In Fig. 8 we give the detailed distribution functions of the hydration numbers. It is seen from both Fig. 3 and Fig. 8, consistent with the values provided in Table 2, that NaI and Na_2SO_3 affect the hydration of the protein and its SASA in a manner strongly dependent on the ion type (as well as on the property of the protein surface). The distribution functions for both NaI and Na_2SO_3, however, are all narrower compared to the corresponding properties in pure water. These results indicate that both salts, although one induces protein denaturation and the other increases protein structure stability, have similar effects of reducing the fluctuation of the system. For example, if we define a hydration entropy

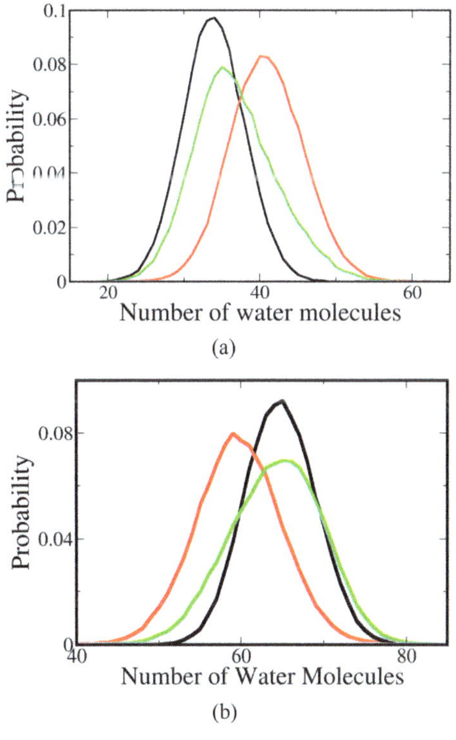

Fig. 8 The distribution function of the number of water molecules in the first hydration shell of the (a) polar and (b) apolar protein surface in the three aqueous solutions (green, pure water; red, NaI solution; and black, Na_2SO_3 solution).

$$S_h = -k_B \sum_i f(i) \ln f(i)$$

where i is the number of hydration water molecules, defined as the number of water molecules residing in the first hydration shell of protein, $f(i)$ the probability for the hydration number to be i, and k_B the Boltzman constant. The calculated S_h is $3.09k_B$, $2.82k_B$ and $2.98k_B$ for pure water, Na_2SO_3 and NaI aqueous solutions, respectively, again showing that the addition of salts decreases the fluctuation of the corresponding properties the system and that Na_2SO_3 has a stronger effect than NaI. Next, to show the different dynamic properties of BBA5 in different solutions, we calculated the time correlation function of RMSD and SASA,

$$c(t) = \frac{\langle (x(t) - \langle x \rangle)(x(0) - \langle x \rangle) \rangle}{\langle (x(0) - \langle x \rangle)^2 \rangle}$$

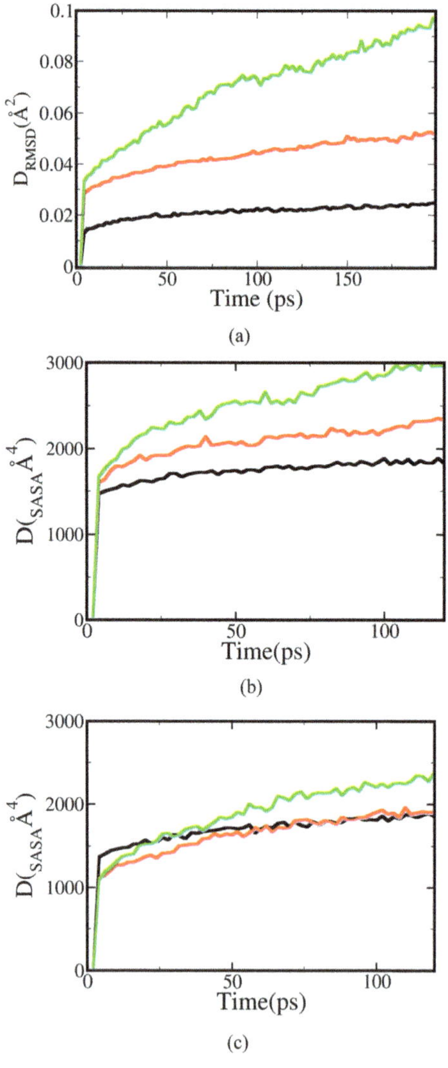

Fig. 9 Time evolution of the correlation function $c(t)$ for (a) RMSD, (b) SASA (polar), and (c) SASA (apolar). (green, pure water; red, NaI solution; and black, Na_2SO_3 solution).

where x is RMSD or SASA (polar and apolar) values. The calculated $c(t)$ is shown in Fig. 9. It is seen from these figures that the time evolution of $c(t)$ for both protein RMSD and SASA become slower in the presence of salts, and especially so in the Na_2SO_3 solution. These results show that both NaI and Na_2SO_3 slow down the dynamics of the protein, consistent with the narrower distribution functions of RMSD and SASA of these two systems, as discussed in the previous section. It is thus concluded from these studies that the addition of salts to the aqueous solution of BBA5 increases the rigidity of the polypeptide, consistent with the observations of Saykally and coworkers.[25]

Discussion

In this molecular dynamics simulation study, we used BBA5, a helix-hairpin two-domain simple protein, as an example to study how electrolytes affect protein structures and dynamics. Two model inorganic salts, NaI, with a low charge density anion to represent a chaostrope, and, Na_2SO_3, to represent a salt with a kosmotropic anion, were employed in the simulations. Due to uncertainties in force field parameters, these systems should be considered only as qualitative in representing chaotropes and kosmotropes, respectively. The current study showed that these two salts differ significantly in their interactions with BBA5 as well as in their effects on the protein structure. As expected, NaI rather strongly denatures BBA5, as manifested from the disruption of protein hydrogen bonds in its aqueous solution, whereas Na_2SO_3 enhances the protein structure, especially evident from the increased hydrophobic interactions in its solution. NaI induces a slight expansion of the protein structure, as seen from the increased SASA for both polar and apolar groups; Na_2SO_3 induces more evident exposure of the protein surface. However, it increases only the SASA of the polar residues but reduces that of the apolar ones. NaI also increases the hydration of both polar and apolar residues as well as the number of water molecules per surface area in the first-hydration layer, whereas Na_2SO_3 significantly reduces the hydration of apolar residues and the number of first solvation shell water molecules per surface area.

However, somewhat surprisingly, although they affect BBA5 structures differently, the addition of NaI or Na_2SO_3 to the aqueous solution of protein both slows down the protein dynamics and reduces and slows down the fluctuation of SASA and the hydration number. This finding that both salts reduce the fluctuation of the protein structure, is consistent with their effects of increasing water surface tension (and viscosity). The indirect "cavity formation" effect[2,45] of both salts through affecting water surface tension is to salt out the protein, although the indirect effect of NaI is significantly weaker than that of Na_2SO_3. What makes the two salts so different in affecting BBA5's structure lies in the preferential binding of the cations to the protein in the NaI system, which is not seen in the other salt solution. I^- but not SO_3^{2-} binds both polar and apolar residues of the protein surface, which facilitates Na^+ binding to the protein backbone. The latter in turn induces protein hydrogen bond breaking and protein denaturation. As a result of the direct binding of ions, one observes an expansion of the SASA for both polar and apolar residues. In particular, in an apparent contrast to the case of Na_2SO_3, NaI increases the SASA and hydration of apolar residues, as a result of I^- binding. The dominance of the direct binding of I^- to the protein surface and thus the binding of Na^+ to the protein backbone results in an overall "salting in" effect. This analysis leads us to conclude that the chaotropic behavior of NaI is mainly a manifest of its direct interactions with the protein BBA5 and the kosmotropic effect of Na_2SO_3 is, on the other hand, mainly an indirect effect.

The marked difference of Na^+ in the NaI and Na_2SO_3 solutions clearly demonstrates the importance of ion cooperativity in affecting aqueous solution properties. In two recent papers,[33,34] we discussed the cooperative effects between cations and anions in affecting the water surface tension and electrolyte osmotic coefficients. This argument provides a connection of the complex trends of salt affecting surface tension to the "matching water affinity" rule proposed by Collins. Recent experimental measurements of water dynamics in various aqueous solutions also revealed the importance of ion cooperativity. Furthermore, due to the association and cooperativity between cations and anions, it is instructive to discuss separately the effects of anions and cations in terms of "direct" and "indirect" mechanisms. In particular, anions not only affect the protein structure through direct and indirect mechanisms as mentioned above, they also affect the cation binding to the protein. For example, the direct binding of I^- but not SO_3^{2-} to the hydrophobic protein side chains reduces protein hydrophobic interactions and increases its solubility, and also allows cation binding, in particular, to the backbone carbonyl. It is conceivable that such an effect is protein specific and depends on the charge distributions of the protein. Therefore, the separation of the salt effects into different inter-dependent terms can become complicated and deserves more careful and quantitative studies.

Certainly, as observed by Saykally,[25] if the protein or polypeptide is rich of positive charges, direct interaction between the anion with a high charge density would become important. The current molecular dynamics simulation study shows that, consistent with the findings of Saykally and coworkers, the ions of the komostropic salt Na_2SO_3 bind strongly the polar (or charged) protein surface. The study also showed that the preferred binding of strongly solvated ions does decrease the fluctuation and slows down the dynamics of the BBA5 structure, corresponding to the increased rigidity (as shown in Fig. 7 and 8) of the protein, consistent with the conclusions drawn in ref. 25.

However, different from the simulations of ref. 25, the polypeptide studied in the present paper is significantly longer and has more hydrophobic residues as well as a better defined structure. Through the analysis of the hydration of these hydrophobic protein components in the absence and presence of ions, it was found that the strongly solvated ions tend to be excluded from the apolar protein surface. These ions not only significantly decrease the hydration of the hydrophobic surfaces but also reduce the fluctuation of the number of hydrating water molecules. (The main difference between the two salts, as stated above, is that I^- but not SO_3^{2-} binds the hydrophobic surface and increases the hydration of the latter.) Especially in the case of Na_2SO_3, both ions are strongly expelled from the apolar protein surface and attracted to the polar one. The exclusion of the ions from the protein apolar residues (as well as from the backbone) strengthens the hydrophobic interactions and helps protect the protein hydrogen bonds. Therefore, the strengthening of the protein structure by this salt is more likely an indirect effect.[13]

In summary, this simulation study showed that NaI denatures BBA5 mainly through a direct binding of ions to the protein, especially to its backbone, and the "kosmotropes", modeled in this study by both Na_2SO_3 and NaF strengthen protein structure mainly through an indirect mechanism: although they both appear to make the protein structure more "rigid," Na^+ in NaI but not in Na_2SO_3 is capable of strongly interacting with the protein backbone and causes denaturation of the protein hydrogen bonds. Interestingly, this observed difference in the roles of "direct" and "indirect" mechanisms in denaturation and "salting effects" of NaI and Na_2SO_3 suggests that the effects of the latter but not the former correlates with their effects on water structure or dynamics. This observation is consistent with the conclusions of Cremer and coworkers that effects of kosmotropes but not chaotropes on PNIPAM LCST correlated with their solvation entropy.[13] For a better understanding of the direct and indirect effects, in the future study it is interesting and important to compare the effects of salts with those of organic cosolvents, such as the denaturant urea and the protectant trimethylamine *N*-oxide (TMAO).[53]

Acknowledgements

The research was partially supported by the National Science Foundation of China (21125311 and 91027044). Y.Q.G. thanks Peking University and the Chinese Ministry of Education for financial support. He is a 2008 Changjiang Scholar.

References

1 F. Hofmeister, *Naunyn-Schmiedebergs Arch. Pharmacol.*, 1888, **24**, 247.
2 R. L. Baldwin, *Biophys. J.*, 1996, **71**, 2056.
3 S. N. Timasheff, *Adv. Protein Chem.*, 1998, **51**, 355.
4 H. S. Frank and F. J. Franks, *Chem. Phys.*, 1968, **48**, 4746.
5 K. A. Dill, T. M. Truskett, V. Vlachy and B. Hribar-Lee, *Annu. Rev. Biophys. Biomol. Struct.*, 2005, **34**, 173.
6 W. Kunz, J. Henle and B. W. Ninham, *Curr. Opin. Colloid Interface Sci.*, 2004, **9**, 19.
7 F. Franks, *Biophys. Chem.*, 2002, **96**, 117.
8 T. Arakawa and S. N. Timasheff, *Biochemistry*, 1982, **21**, 6545.
9 T. Arakawa, R. Bhat and S. N. Timasheff, *Biochemistry*, 1990, **29**, 1914.
10 M. V. Athawale, S. Sarupria and S. J. Garde, *J. Phys. Chem. B*, 2008, **112**, 5661.
11 M. T. Record, W. Zhang and C. F. Anderson, *Adv. Protein Chem.*, 1998, **51**, 281.
12 Y. Zhang and P. J. Cremer, *Annu. Rev. Phys. Chem.*, 2010, **61**, 63.
13 Y. Zhang and P. J. Cremer, *J. Am. Chem. Soc.*, 2005, **127**, 14505.
14 L. M. Pegram and M. T. J. Record Jr., *J. Phys. Chem. B*, 2007, **111**, 5411.
15 Y. Cho, Y. Zhang, T. Christensen, L. B. Sagle, A. Chilkoti and P. J. Cremer, *J. Phys. Chem. B*, 2008, **112**, 13765.
16 Y. Zhang and P. S. Cremer, *Proc. Natl. Acad. Sci. U. S. A.*, 2009, **106**, 15249.
17 A. W. Omta, M. F. Kropman, S. Woutersen and H. J. Bakker, *Science*, 2003, **301**, 347.
18 K. J. Tielrooij, N. Garcia-Araez, M. Bonn and H. Bakker, *Science*, 2010, **328**, 1006.
19 R. Struis, J. Debleijser and J. C. Leyte, *J. Phys. Chem.*, 1989, **93**, 7943.
20 J. E. Enderby, *Chem. Soc. Rev.*, 1995, **24**, 159.
21 W. Wachter, S. Fernandez, R. Buchner and G. J. Hefter, *J. Phys. Chem. B*, 2007, **111**, 9010.
22 K. D. Collins, *Methods*, 2004, **34**, 300.
23 K. D. Collins and M. W. Q. Washabaugh, *Q. Rev. Biophys.*, 1985, **18**, 323.
24 D. A. Turton, J. Hunger, G. Hefter, R. Buchner and K. J. Wynne, *Chem. Phys.*, 2008, **128**, 161102.
25 C. P. Schwartz, J. S. Uejio, A. M. Duffin, A. H. England, D. N. Kelly, D. Prendergast and R. J. Saykally, *Proc. Natl. Acad. Sci. U. S. A.*, 2010, **107**, 14008.
26 L. I. N. Tome, M. Jorge, J. R. B. Gomes and J. A. P. J. Coutinho, *J. Phys. Chem. B*, 2010, **114**, 16450.
27 C. P. Schneider, D. Shukla and B. L. J. Trout, *J. Phys. Chem. B*, 2011, **115**, 7447.
28 C. H. Wang, Y. Ge, J. Mortensen and P. Westh, *J. Phys. Chem. B*, 2011, **115**, 9955; B. Hess and N. F. A. van der Vegt, *Proc. Natl. Acad. Sci. U. S. A.*, 2009, **106**, 13296.
29 P. Jungwirth and D. J. Tobias, *Chem. Rev.*, 2006, **106**, 1259–1281.
30 C. D. Wick, I. F. W. Kuo, C. J. Mundy and L. X. Dang, *J. Chem. Theory Comput.*, 2007, **3**, 2002.
31 J. Heyda, J. C. Vincent, D. J. Tobias, J. Dzubiella and P. J. Jungwirth, *J. Phys. Chem. B*, 2010, **114**, 1213.
32 C. Caleman, J. S. Hub, P. J. van Maaren and D. van der Spoel, *Proc. Natl. Acad. Sci. U. S. A.*, 2011, **108**, 6838.
33 L. J. Yang, Y. Fan and Y. Q. J. Gao, *J. Phys. Chem. B*, 2011, **115**, 12456.
34 Y. Q. Gao, *J. Phys. Chem. B*, 2011, **115**, 12466–12472.
35 A. S. Thomas and A. H. Elcock, *J. Am. Chem. Soc.*, 2007, **129**, 14887.
36 J. J. Dzubiella, *J. Am. Chem. Soc.*, 2008, **130**, 14000.
37 R. Vacha, P. Jurkiewicz, M. Petrov, M. L. Berkowitz, R. A. Boeckmann, J. Barucha-Kraszewska, M. Hof and P. J. Jungwirth, *J. Phys. Chem. B*, 2010, **114**, 9504.
38 P. Ganguly, P. Schravendijk, B. Hess and N. F. A. J. van der Vegt, *J. Phys. Chem. B*, 2011, **115**, 3734.
39 M. Struthers, J. J. Ottesen and B. Imperiali, *Folding Des.*, 1998, **3**, 95.
40 S. Hwang, Q. Shao, W. Howard, C. Hilty and Y. Q. J. Gao, *J. Phys. Chem. B*, 2011, **115**, 6653.
41 J. M. Wang, P. Cieplak and P. A. Kollman, *J. Comput. Chem.*, 2000, **21**, 1049.
42 H. J. C. Berendsen, J. R. Grigera and T. P. Straatsma, *J. Phys. Chem.*, 1987, **91**, 6269.
43 I. S. Joung and T. E. Cheatham III, *J. Phys. Chem. B*, 2008, **112**, 9020.
44 U. C. Singh and P. A. J. Kollman, *J. Comput. Chem.*, 1984, **5**, 129–145.

45 B. H. Besler, K. M. Merz and P. A. J. Kollman, *J. Comput. Chem.*, 1990, **11**, 431–439.
46 A. D. Becke, *Phys. Rev. A: At., Mol., Opt. Phys.*, 1988, **38**, 3098–3100.
47 C. T. Lee, W. T. Yang and R. G. Parr, *Phys. Rev. B*, 1988, **37**, 785–789.
48 A. D. J. Becke, *Chem. Phys.*, 1993, **98**, 1372–1377.
49 R. A. Kendall, T. H. Dunning and R. J. J. Harrison, *Chem. Phys.*, 1992, **96**, 6796–6806.
50 E. Cances, B. Mennucci and J. J. Tomasi, *Chem. Phys.*, 1997, **107**, 3032–3041.
51 M. Cossi, V. Barone, B. Mennucci and J. Tomasi, *Chem. Phys. Lett.*, 1998, **286**, 253–260.
52 B. Mennucci and J. J. Tomasi, *Chem. Phys.*, 1997, **106**, 5151–5158.
53 H. Y. Wei, Y. B. Fan and Y. Q. J. Gao, *Phys. Chem. B*, 2010, **112**, 9020.

General discussion

Professor Bakker opened the discussion of the paper by Werner Kunz: You argued that the lack of ion selectivity of the 1 : 1 surface of dodecyl sulphate and octanoldodecyl sulphate would be due to the charge density being too high. Could it be that the lack of ion selectivity is not due to the fact that the average charge density is high, but rather due to the fact that the charge distribution will be quite homogeneous for this surface? For comparison, for the other system you studied, the dodecyl carboxylate mixed with octanol, the ratio was 1 : 3 and the negative charge distribution will be much more inhomogeneous, showing highly concentrated sites of negative charge. This inhomogeneity of the charge distribution may be the reason that this surface does show a selectivity for Li^+ over Na^+.

Professor Kunz responded: You are absolutely right. This may well be also the reason that for the moment we cannot distinguish between both effects.

Dr Wheatcroft said: You mention two binding sites for metal ions with the carboxylate groups. Would the difference in orbital overlap between the sodium and lithium ions with the carboxylate groups contribute to the competition in sodium and lithium ion binding in these locations?

Also, how would you expect the difference in hydration numbers[1] for the lithium ions (6.6) and the sodium ions (3.9) to have a significant effect on their propensity to bind to the carboxylic acids from the aqueous environment as the lithium ions would be more strongly solvated than the sodium ions?

1. A. Zavitsas, *J. Phys. Chem. B*, 2001, **105**, 7805–7817.

Professor Kunz replied: On the orbital overlap – both Na and Li have the first hydration shell mainly intact otherwise the NMR splittings would be enormous (at least 10 times or more bigger than we see). The water will prevent orbital overlap between the metal ions and the carboxylate groups, so this cannot be a factor.

On the hydration numbers (n), a value of 6.6 for Li suggests that there is a distribution of numbers governed by equilibrium constants. *i.e.* species exist with $n = 6$ and $n = 7$, but also with other numbers. But, at high surfactant concentrations the n value will decrease because of lower water activity. It seems possible that the smaller the value of n, the more likely it is for the "negative" binding site to be occupied because the hydrated ion is smaller. However, from the current NMR data alone it is difficult to say to which extent the ion headgroup interactions are mediated by hydration water. This information could be obtained by simulating the whole system and to calculate from these results the quadrupole splittings.

Professor Jungwirth asked: You see stronger interaction of Li^+ than Na^+ with carboxylic groups in your liquid crystal, while X-ray photoemission data as well as calculations indicate the opposite ordering of these two cations for acetate in bulk water (E. F. Aziz *et al.*, *J. Phys. Chem. B*, 2008, **112**, 12567). Do you think it is mainly due to the differences between the two systems (liquid crystal *vs.* bulk solution) or rather experimental techniques with different sensitivities to contact *vs.* solvent separated ion pairs?

Professor Kunz responded: This is a delicate question. First, let me stress that we are sure of what we saw, the NMR quadrupole splitting is a very sensitive technique. I do not know very well how sensitive X-ray photoemission is. Second, there must be a pronounced difference between liquid crystals and bulk solutions, because it is well known that a high percentage of counterions is usually bound to highly ordered carboxylate groups showing a high charge density (at least 70% is bound). This is certainly not the case in the bulk. So, both systems are really different and this is most exciting for specific ion effects. Finally, in our experiments, we investigated directly the competition of Li^+ and Na^+ by mixing them at different ratios, whereas in the X-ray photoemission study, the ions were considered separately. For me, the second point is the most important one. This means that interactions between ions and headgroups also depend on the precise system, the ordering of the headgroups and the local charge density.

Professor Hugel asked: Do you expect any temperature dependence if the temperature is varied in a larger range – for example up to 350 K?

Professor Kunz responded: I have to say that I have no answer for the moment. The reason is that we may assume different bridging mechanisms between Li^+ and Na^+ with the carboxylate headgroups and we know nothing about their respective temperature dependencies. However, from the very small temperature dependence that we observed in a narrow temperature range we may assume that larger temperature variations are required to see significant changes.

Professor Kunz asked: Would it be possible to do such NMR quadrupole splitting experiments in liquid crystals at higher temperatures, say between 80 and 90 °C?

Professor Tiddy answered: The main problem with making measurements at high temperature is that the lamellar phase might melt to form a micellar solution. This could be avoided by the replacement of octanol by a longer chain cosurfactant – say oleoyl alcohol. High temperatures will give a larger surfactant solubility, making measurements possible at very high concentrations. This is one of many further experiments under consideration.

Professor Halling asked: I notice that on many of your plots there is a deviation from the trend at exactly equimolar Li and Na. Are these significant, and do you have any suggestions about what special occurrence may be happening at this composition?

Professor Kunz responded: I think that indeed these deviations are significant, we detected them several times. However, for the moment we have no explanation for this phenomenon.

Professor Bain remarked: I have a related question to the one raised by Prof. Halling. Can you treat your data with a competitive Langmuir isotherm, or do you need to consider interactions between the Li^+ and Na^+ ions, either directly or indirectly through the effect of ion binding on the structure of the surfactant bilayers? If a competitive Langmuir model is appropriate, could you study the binding of ions that are not NMR-active by observing their competition with Li^+ or Na^+?

Professor Kunz answered: I am not sure that a Langmuir model with a single binding constant would fit the data, but qualitatively it should give the general features that we see. We could definitely try to see the competition effect of a non NMR-active ion with Na^+ as the NMR-active ion. If the Na^+ quadrupole splitting is lowered, this could be interpreted as a higher propensity of the competing ion to the headgroups.

Professor van der Vegt said: In your system, Li^+ binding to carboxylate is stronger than Na^+. In bulk solution (formate and acetate salts), experiments and simulations have shown a reversed order. Would it be plausible to assume that you observe

Li-mediated bridging of carboxylate groups at the interface? Molecular simulations of methacrylic acid based polyelectrolyte brushes in the paper by myself and Dr. Rodriguez-Ropero show that with both Li^+ and Na^+ salt bridging occurs. In these systems, the life time of Li^+ bridges is larger than Na^+.

Professor Kunz replied: Yes, indeed, this is definitely a possibility. As I mentioned in the answers to preceding questions, bridging could explain the reversed order, although with our present experiments we are not able to prove that. In any case it seems clear to me that ion specificities towards headgroups may strongly depend on the aggregation state of the molecules bearing these headgroups.

Professor Lopes remarked: In order to get the liquid crystal structures necessary for the experimental determinations that you have presented, you have used quite different quantities of co-surfactant in the case of the carboxylate and sulfate systems. Could you comment on the possible influence of the co-surfactant quantity on the specificity of the interactions with the two types of cation studied? Wouldn't it be possible to control this variable by producing liquid crystals with similar quantities of co-surfactant (even if this would imply changing the co-surfactant or the temperature of the measurements)?

Professor Kunz responded: You are right, the different quantities of co-surfactants may well influence the different selectivities for several reasons. As I mentioned this could be the surface charge density or, as Huib Bakker proposed, the reason could be the different inhomogeneities in the charge distribution. The reason why we chose these surfactant-to-co-surfactant ratios was that we wanted to have a maximum range of comparable liquid crystalline structures at the same temperatures. Please note that this was only a preliminary study as a proof of concept. Much work remains to be done to elucidate the factors influencing the different ion specificities.

Professor Lopes asked: As a follow up to the previous question: would it be possible to "build" liquid crystal structures where the surface concentration of the head groups (either carboxylate or sulfate or others) was controlled so as to be always the same (by the addition of more or less important fraction of co-surfactant) and so the specificity of the counter-ions (lithium, sodium or others) could be compared in a more robust way?

Professor Kunz responded: Yes, it should be fairly simple to do this because the range of surfactant/co-surfactant compositions over which the lamellar phase is stable is quite large. If the octanol systems do not allow this (unlikely) it will be possible using alkanols with other chain lengths.

Dr Ottosson remarked: Pavel Jungwirth made a comment concerning our previous X-ray absorption spectroscopic investigations of ion pairing in aqueous acetate salts which I would like to clarify somewhat: in our study (E. F. Aziz *et al.*, *J. Phys. Chem. B*, 2008, **112**, 12567) we looked at the low-energy O1s absorption feature specifically related to the carboxylate group of the acetate anion and found that the intensity of this feature varied with the counterion, something which could be attributed to the degree of ion pairing in the solution. Somewhat peculiarly, we found that the efficiency for pairing decreased in the order $Na^+ > Li^+ > K^+ > NH_4^+$, *i.e.* something looking like a Hofmeister cation series but with sodium and lithium swapped. In relation to the here presented work it is important to point out that X-ray absorption spectroscopy in total fluorescence yield mode, as we performed, is a bulk-sensitive technique so comparison with surface specific data can be of great interest.

Professor Kunz answered: Nevertheless, this discrepancy is surprising, because thermodynamic data, *e.g.* activity coefficients, point also at a stronger ion pairing of Li–carboxylate compared to Na–carboxylate in the bulk. Maybe, as Pavel Jungwirth suggests, the type of ion-pairs are different in the bulk and at the interface, with solvent-separated ion pairs in the bulk and contact ion pairs at the layered liquid crystal interface.

Professor Bakker asked: There is an interesting difference between the ion selectivity of carboxylate in bulk water and carboxylate at the surface you studied. In bulk water, carboxylate preferentially binds Na^+ over Li^+, whereas at the surface it is the other way round.

Could this difference be due to the much stronger hydration of carboxylate in bulk liquid water, thus leading to a different, probably more diffuse, distribution of its negative charge?

Professor Kunz responded: The difference in hydration between the bulk and the surface may well play an important role. A further and very important difference is that in the bulk carboxylate groups do not aggregate, whereas at the interface they are closely packed together so that cations could make a bridge between them. Maybe this could even introduce packing of surfactants and a non-homogeneous distribution of co-surfactants and surfactants at the interface, much as lipid rafts in membranes.

Dr Eisenberg opened the discussion of the paper by Benoît Roux by communicating: I look forward to the accurate calculation of the activity of ions in the mixtures that make up intracellular and extracellular solutions. That is an admirable goal and an important one that has eluded physical chemists for more than one hundred years. I hope Benoît and his colleagues and community can reach it. That is, however, only an intermediate goal, in my opinion. The most important goal is the accurate calculation of ions in the highly concentrated solutions that are so important for the function of industrial electrochemical systems (like batteries) and so important for the function of biological systems. The activity of ionic solutions is most important biologically in and near active sites (of enzymes and binding proteins), ion channels, transporters, and nucleic acids. Number densities of ions there usually exceed 10 M and the types of ions present are strongly influenced by the special properties of the concentrated environment and nearby protein, or nucleic acid, or electrode (in industrial batteries for example). Ions in these places have large excess free energies that are different for each type of ion. In other words, the system is specific as well as concentrated and important. I look forward to the accurate calculation of the free energy per mole of ions in the highly concentrated mixtures that are vital for function.

Dr Roux communicated in reply: This comment emphasizes an important point. While the concentration of ionic solutions appropriate for biological systems is typically within a range of 1–200 mM, this can be deceptive. In fact, the local density of ions may far exceed this in the neighbourhood of charged macromolecular assemblies. One example is the concentrated double layers of ions that accompany bilayer membranes formed with negatively charged lipids. Similarly, the local concentration of cations near DNA, a negatively charged polymer, can reach values that are significantly larger than the bulk concentration. As pointed out by Dr Eisenberg, ions in these systems have large excess free energies that are different for each type of ion. Most of the molecular mechanical force fields used currently in MD simulations have not been calibrated to guarantee that they will be accurate in all such situations. As shown in the present study, issues of calibration and optimization of a force field, whether the latter incorporates induced polarization or not, are not automatically resolved. Simulations of complex systems are not necessarily correct just because a polarizable force field is utilized. Tests with highly concentrated model systems must be conducted to verify the accuracy of the force field, and if needed, optimize the parameters or the functional forms. In principle, we could have accurate models of all these systems once we are able to validate all the ion pair parameters against experimental and thermodynamic data. This is certainly feasible but it requires a lot of work.

Professor Kunz communicated: How would these results compare to the force fields inferred from thermodynamic quantities as done by D. Horinek *et al.*, *J. Chem. Phys.*, 2009, **130**, 124507 and I. Kalcher and J. Dzubiella, *J. Chem. Phys.*, 2009, **130**, 134507, which are the two most relevant papers.

Dr Roux communicated in reply: In the study of Horinek *et al.* (2009), the solvation free energy and the solvation entropy were used in conjunction to optimize the Lennard-Jones parameters for the alkali and halide ions. The nonpolarizable SPC and SPC/E water models were used. While this is a very sound approach, the models were optimized using only single ion properties (the free energy and entropy). For this reason, the models are designed only for low and intermediate concentrations. As shown here, the results for very concentrated ionic solutions might, or might not be accurate when the ion–ion force field relies on the standard Lorentz–Berthelot combining rule for the Lennard-Jones parameters. In fact, it has been previously shown that there can be problems with such models (I. N. Joung and T. E. Cheatham, *J. Phys. Chem. B*, 2008, **112**, 9020–9041; Y. Luo and B. Roux, *J. Phys. Chem. Lett.*, 2010, **1**, 183–189). Ultimately, the models must be tested by simulating concentrated ionic solutions.

The study of Kalcher and Dzubiella (2009) goes beyond optimization based on single ion properties by examining the structure of aqueous LiCl, NaCl, KCl, CsCl, KF, and NaI solutions at finite concentrations. They also estimated the osmotic coefficients of the electrolytes from an integration of the average distribution functions *via* the compressibility route, which is also referred to as the Kirkwood–Buff theory. The simulations were based on the nonpolarizable SPC/E water model and the Dang Lennard-Jones parameters for the ions. In the present study based on polarizable models, we have chosen to avoid using such an indirect route to the osmotic coefficient, and rather calculate the osmotic pressure directly *via* a novel simulation setup using "virtual semipermeable walls" (Y. Luo and B. Roux, *J. Phys. Chem. Lett.*, 2010, **1**, 183–189). As pointed out by the authors, the thermodynamic properties such as the osmotic pressure are strongly sensitive to small changes in the potential of mean force between the ions. For this reason, we believe that our direct "virtual walls" approach reduces the accumulation of errors due to statistical uncertainty in the average distribution functions when trying to estimate the osmotic coefficients of the electrolyte solutions.

Dr Reif communicated: Can you comment on the magnitude of possible finite-size and artificial-periodicity artifacts in your estimation of osmotic pressure and ion–ion PMFs?

Dr Roux communicated in reply: In the osmotic pressure simulations, a central salt solution is sandwiched between the two bulk regions in the 3D system with full periodic boundary conditions. The shape of the entire simulation cell was orthorhombic with dimension 96 × 96 × 96 Å^3, and the embedded salt solution was 96 × 96 × 48 Å^3. Thus the closest distance of ion and image ion is 144 Å in the *xy* direction, and 96 Å in the *z* direction. The box size is significantly bigger than the real-space cutoff (12 Å) and the correlation length of the radial distribution function, $g(\mathrm{r})$ of the screened ion–ion interactions. In McMillan–Mayer theory, the osmotic pressure is expressed as a sum over integrals involving multi-ion distribution functions, and the latter would be expected to decay to zero well within the size of the finite system simulated here. Thus size and periodicity artifacts are not expected to be an issue given the size of the system being used. This could be verified and validated in practice, by repeating the same calculations with a larger system, although this was not attempted here due to the computational cost involved. Regarding the ion–ion PMF, it should be noted that they were used only to estimate the initial specific ion-pair parameters in this study. The final parameters were determined from direct osmotic pressure calculations. For the ion–ion PMF calculations, a cubic box size of 24 Å was used. The maximum distance between ion pair is 8 Å, which is much smaller than half box length. In this condition the influence of finite-size and periodic boundary condition are expected to be relatively small due to the high permittivity of the solvent.

Professor Madden opened the discussion of the paper by Phil Mason: Stimulated by your remarks about less perfect reproduction of the observed hydration structure for some particular ions I would like to comment on difficulties in representing with classical force fields the interactions of the group of ions whose electronic structures might be best represented as a superposition of resonance structures. An example would be the carboxylate ion which might be regarded as a superposition of O=C–O– and O–C=O, but might also apply to sulphate, carbonate *etc.* A classical force field will represent the charge distribution with an equal fractional charge on each O, this would be appropriate for the isolated molecule (in vacuum). However, if a cation approaches the group it is likely to "bind" to one oxygen and "lock" the group into one resonance structure – *e.g.* Na + O–C=O where the coulombic interaction is stronger than would be anticipated from the equal charge model. The consequences were examined in a paper by E. Leenders *et al.* (*J. Chem. Theory Comput.*, 2008, **4**, 898) where *ab initio* and force field simulations on the interactions of a proton with the glutamate anion in water were compared. Although the average structures of the *ab initio* simulations (*e.g.* the atom–atom rdfs) could be reproduced with a classical, equal charge model (though, as I recall, with a tendency to overestimate the bifurcated hydrogen bond) such a model would give a residence time for the bound proton which was much shorter than found in the *ab initio* calculations. The classical model lacks a crucial element required for transferability between different states of interaction which is sufficient flexibility to allow an appropriate description of the electron distribution for different states of coordination. The resonance structure effect is not a classical polarization effect, to be picked up by a normal polarisable simulation model.

Professor Bain addressed Professor Madden and Professor Mason: Could Prof. Madden comment on the extent to which resonance in the carboxylate ions could be treated through an anisotropic polarisability of CO_2^- head group, with enhanced polarisability associated with charge transfer between the two oxygens?

Professor Mason answered: It's an interesting question that needs to be addressed. While I have little to offer in the way of hard data, it's my impression that, say for instance with the hydration of the acetate ion, inclusion of polarizibility would lead to a significantly better modeling of reality, whereas the anisotropy is a lesser effect. This is of course excluding examples such as short strong hydrogen bonds found in proteins where two CO_2^- side chains essentially share one proton. There the asymmetry will be very important.

Professor Madden answered: I don't believe that simply introducing a group anisotropic polarizability could capture this effect whereby, because one of the oxygen atoms is tightly coordinated to a Lewis acidic site, the carboxylate group would be effectively locked into one of its resonance structures. A polarizable centre would respond in the same way to a charge placed at the same distance from it in the plane of the carboxylate group whereas the effect here is very much associated with the particular molecular and electronic structure of the group. In principle, I think that the best way to address the representation of the carboxylate ion would be through an empirical valence bond model (see *e.g.*: Y. Wu, H. Chen, F. Wang, F. Paesani and G. A. Voth, An improved multistate empirical valence bond model for aqueous proton transport, *J. Phys. Chem. B*, 2008, **112**, 467; *J. Phys. Chem. B*, 2008, **112**, 7146) in which the underlying electronic structure is a superposition of two quantum states which responds to the potential exerted by the neighbouring molecules. In a computational (MD) scheme, doing this for a single or isolated carboxylate ion is straightforward, but it would become demanding for a set of interacting carboxylate ions.

Professor Lynden-Bell opened the discussion of the paper by Liem Dang: You have shown us free energy calculations based on classical point charge models, *ab initio*, high level quantum and classical simulations with a modified interionic potential. Would you like to comment on the large difference between the quantum chemistry results and the *ab initio* results, in particular the presence or absence of a contact ion pair.

Dr Dang answered: This is a very interesting question and I may not have a good answer for it. I am not aware of any comparison between quantum chemistry results and *ab initio* results on the ion pair interaction energy. I will look into it because it is important and the presence or absence of a contact ion pair is very sensitive to the ion–ion interactions.

Miss Pluharova said: Based on averages and variances of mean forces calculated for different time spans, it seems that 5 ps per window as reported in J. Timko, D. Bucher and S. Kuyucak, *J. Chem. Phys.*, 2010, **132**, 114510 is not enough to obtain a converged result. At least 50 ps are needed.

Dr Dang replied: I do not have problems with your convergent criteria. Pair potential, the polarizable model and *ab initio* methods, each have their own convergent problems. I would do 25 ps and 50 ps per window using *ab initio* MD to see how they are different/the same? I do not speak for the authors of the paper you mentioned, but it seems to me that they are very competent scientists and may have more to say in this problem.

Professor Jungwirth commented: Since *ab initio* molecular dynamics is extremely time-consuming one can use classical MD for the same system to estimate the computational needs to obtain results with the desired statistical error.

Dr Dang responded: I do not have problems with this approach.

Miss Pluharova remarked: Estimates from classical molecular dynamics, that were mentioned in my previous comment, can be used for error estimation, if dynamics in classical and *ab initio* MD are comparable.

Dr Dang responded: I would be very surprised to see that dynamics in classical and *ab initio* MD are comparable!

Professor Bakker said: Fig. 2 of your paper shows a small minimum at 5 Å of the interionic potential of mean force for Sr–Cl which according to you implies the presence of solvent-separated ion-pairs (SSIP). Does this minimum really imply the presence of a longer-living SSIP, or is this minimum rather due to the structuring of the liquid in which there is a slight preference for having the ions separated by a discrete number of water molecules? Are there indications from the MD simulations that the cases in which the Sr and the Cl ion are separated by 1 H_2O constitute relatively long-living structures?

Dr Dang answered: This is a very interesting question and I will do my best to answer it. The SSIP is stabilized due to the solvent molecules around the ion pair and bridged water molecules between the ions (see Fig. 1). The complex you mentioned $Sr_2 + H_2OCl^-$ may not be stable at finite temperature (*i.e.*, 300 K), but will be stable at 0 K.

Dr Schwaab opened the discussion of the paper by Huib Bakker: You showed in Fig. 6 of your paper that the anisotropy decay of the OD groups is cation dependent. How can you determine experimentally if this effect is due to a cation effect of OD in the solvation shell of an anion *versus* an alternative explanation: cation hydration water that is (partly) immobilized? The latter effect is also expected to be stronger for Na compared to Cs. In principle both mechanisms should be possible.

Professor Bakker responded: The results shown in Fig. 6 from the paper are derived from an analysis in which we explicitly distinguish the water molecules that are hydrogen bonded with one OH group to Cl^- from the other water molecules (for which the OH groups are hydrogen bonded to the oxygen atoms of other water molecules). Fig. 6 exclusively shows the dynamics of the OH groups that are hydrogen bonded to the Cl^- anion. It could be that there is a contribution of $Na^+\cdots O\text{–}H\cdots Cl^-$ groups to the observed signals, *i.e.* of water molecules that are both in the first hydration shell of the anion and in that of the cation, but there cannot be a general contribution of the water molecules surrounding the cations, as the hydrogen bond to Cl^- is a prerequisite to show up in the signals shown in Fig. 6.

We think that the effect of the cation resulting from the electric field exerted by the cation limits the wobbling motion of the OH groups that are hydrogen bonded to Cl^-.

Professor Ben-Amotz remarked: Can you distinguish the hydrophobically induced slowing down of water rotational motions in the hydration shell of an isolated TMGndC molecule from that due to the crowding of water molecules that are pinned between neighboring TMGndC molecules?

Professor Bakker replied: We can get some information on these contributions by varying the concentration.

Fig. 1 Snapshot showing water molecules around the SSIP. If you zoom in close enough, you will see the bridged water molecules.

The strong slowing down of the water reorientation around hydrophobic groups (~10 ps) is an effect that already arises at low concentrations. The amplitude of this latter component increases with solute concentration and with the size of the hydrophobic part of the dissolved amphiphilic molecule or ion. At high concentrations (typically >2 M), we also observe a further increase of the reorientation time constant, and this further increase is most likely due to a crowding effect, *i.e.* an overlap of the hydration shells that further limits the orientational mobility of the water molecules.

Professor Lopes asked: How are the structure and dynamics of water affected by temperature in the different systems that you have presented, namely in the case of systems with hydrophobic effects?

Professor Bakker responded: Temperature has a very pronounced effect on the orientational dynamics of water molecules surrounding hydrophobic molecular groups. For all water molecules the reorientation rate increases when the temperature is increased, but for the water molecules in the hydrophobic hydration shell this effect is much stronger than for water molecules in bulk liquid water. As a result, the large difference in the reorientation time constant at 300 K (~10 ps *versus* 2.5 ps) becomes much smaller at higher temperatures (*e.g.* ~1.2 ps *versus* ~0.9 ps at 390 K). The stronger acceleration of the reorientation of the hydrophobic hydration shells expresses itself in a higher activation energy of the reorientation. For bulk liquid the activation energy is ~17 kJ mol^{-1}, for hydrophobic hydration shells it is ~30 kJ mol^{-1}. This means that the relatively rigid hydration shell of a hydrophobic solute at room temperature "melts" when the temperature is increased.

Professor van der Vegt asked: You discuss cooperative effects on the dynamics of water in solvent separated states of $MgSO_4$. Is your technique sensitive enough to probe differences in water dynamics comparing Li^+ and K^+ in solvent-shared pairing states with carboxylate salts? These states determine differences in ion specific osmotic activities of carboxylate salts in water (B. Hess and N. van der Vegt, *Proc. Natl. Acad. Sci. U. S. A.*, 2009, **106**, 13296; P. Ganguly *et al.*, *J. Phys. Chem. B*, 2011, **115**, 3734) and there are recent indications from simulations that the motion of OH groups is quite different for Li^+ and K^+ (T. Hajari *et al.*, *J. Chem. Theory Comp.*, 2012, **8**, 3804).

Professor Bakker replied: The femtosecond infrared technique is indeed sensitive enough to measure the differences in the dynamics of water molecules hydrating different carboxylate salts. In fact we recently performed such measurements for solutions of different formate salts: CsHCOO, NH4HCOO, LiHCOO, and NaHCOO. We observed a fraction of slowly reorienting water (t_{or} > 10 ps) of which the amplitude increases proportional with the salt concentration. The amplitude of the slow fraction also strongly depends on the nature of the cation. For CsHCOO the fraction of slow water is quite small.

Professor Goodall commented: Hydrophobes limit the phase volume of water, causing entropy decrease. We normally think about the effects of Hofmeister ions on proteins: is there merit in also considering the situation the other way round, *i.e.* how proteins affect the Hofmeister ions, through limiting the phase volume available to the ions?

Professor Bakker answered: This is an interesting idea, indeed it is to be expected that ions and molecular groups like that of proteins can have a mutual effect on each other *via* their influence on the water dynamics. Obviously this will most likely occur in highly concentrated (water-poor) solutions. It is even conceivable that the denaturation effect of an ion on a particular protein can be changed/controlled by the presence of other proteins or osmolytes: the repulsion of water from the ion's hydration shell by these other proteins/osmolytes will change the denaturation effect of the ion.

Professor Lynden-Bell presented Fig. 2: Participants may be interested in results that Jay Rasaiah and I obtained for the variation of solvation entropy as a function of charge for Lennard-Jones ions of various sizes. (R. M. Lynden-Bell and J. C. Rasaiah, *J. Chem. Phys.*, 1997, **107**, 1981). The lowest curve in the upper panel of the Figure shows results for an ion of the size of Na^+. Near zero charge there is a minimum solvation entropy corresponding to the hydration of a hydrophobic particle. As the magnitude of the charge is increased the solvation shell becomes more disordered and the ion could be described as chaotropic. At higher magnitudes of the charge the ordering effect of the electrostatic field of the ion dominates and the solvation entropy decreases. Similar behaviour is seen for larger ions, but with shifted entropy maxima. The lower part of Fig. 2 shows the variation of solvation entropies with ion size.

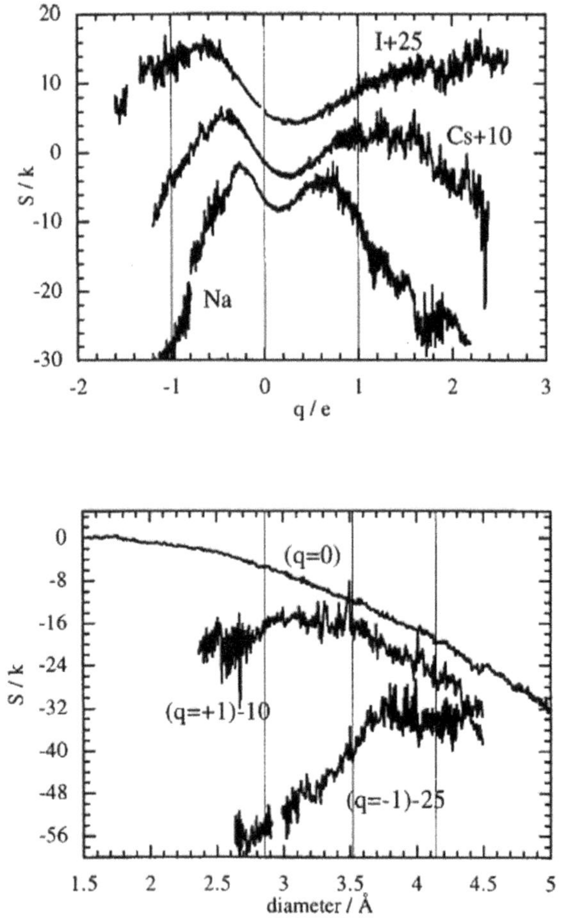

Fig. 2 Variation of solvation entropy with charge (above) and size (below). Note that the curves are displaced by the amounts shown for clarity. (Reproduced with permission from R. M. Lynden-Bell and J. C. Rasaiah, *J. Chem. Phys.*, 1997, **107**, 1981; copyright AIP 1997).

Professor Ben-Amotz replied: I am familiar with these results, which are indeed very interesting and thought provoking. I also find it noteworthy that these results, as well as experimental measurements, suggest that the solvation entropies of ionic solutes in water have remarkably similar magnitudes to those of hydrophobic solutes, while the corresponding solvation free energies and enthalpies differ by well over a factor of 10 (and in some cases near a factor of 100). Thus, one might expect that ions would also produce a much larger influence on the entropy of water than hydrophobic solutes. A plausible resolution to this dilemma is suggested if linear response theory is assumed to represent accurately electrostatic contributions to the solvation of ions in water.[1] This implies that charging a non-polar solute should induce two cancelling entropic contributions of nearly equal magnitude and opposite sign; one is an entropy associated with the reorganization of the water induced by charging the solute and the other is an entropy associated with fluctuations in the ion–water interaction energy. Linear response theory further implies that the magnitudes of the latter two terms are comparable to the experimental hydration free energy.

1. D. Ben-Amotz and R. Underwood, *Acc. Chem. Res.*, 2008, **41**, 957–967.

Mr Sharma returned the discussion of the paper by Huib Bakker: What is the pronounced difference in the average hydration dynamics of water molecules due to presence of distinct anions (chloride and iodide) with a common counter ion of sodium? How can such effects can get identified with change in concentration as well as temperature of the solutions?

Professor Bakker answered: The orientational dynamics of the water molecules making a hydrogen bond to the chloride and iodide are not very different. For both anions, the orientational dynamics consist of a fast component with a time constant of ~2 ps that can be assigned to a wobbling motion of the hydrogen-bonded OH group that keeps the hydrogen bond to the anion intact. The fast wobbling component has a larger amplitude for iodide than for chloride which can be explained from the larger size of the I^- anion, as this allows for a larger angular spread of the hydrogen bond without breaking the hydrogen bond. In addition, the hydrogen bond of water to I^- is weaker than the hydrogen bond of water to Cl^- which also makes the angular (wobbling) motion less restricted. The orientational dynamics of water in the hydration shells of Cl^- and I^- also show a distinct slow component with a time constant of ~9 ps for Cl^- and ~7 ps for I^-. In the past, we have studied this slow component in detail and found that the temperature dependence of its time constant closely follows a Debye–Stokes–Einstein expression for the orientational diffusion of a sphere in a viscous liquid (M. F. Kropman, H.-K. Nienhuys and H. J. Bakker, Real-time measurement of the orientational dynamics of aqueous solvation shells in bulk liquid water, *Phys. Rev. Lett.*, 2002, **88**, 077601).

The slow component shows a significant acceleration with temperature due to the decrease of the viscosity of water and the thermal activation of the orientational motion. This finding indicates that the slow component is due to the orientational diffusion of the intact hydration shell. We also observe that the I^- hydration shell shows a somewhat faster orientational diffusion than the Cl^- hydration shell at all temperatures, which indicates that the total effective size of the I^- ion plus its hydration shell is somewhat smaller than the total effective size of the Cl^- ion plus its hydration shell. This result forms an interesting contrast with the results for the fast component of the reorientation (the wobbling motion), for which the difference between I^- and Cl^- was governed by the sizes of the bare anions.

Dr Cremer asked: Please explain the effects of the size of the hydrophobic moiety on the slowing of water (methyl *vs.* ethyl). Would different behavior be expected at a flat hydrophobic surface? Would you expect this relative slowing to be different at different temperatures?

Professor Bakker replied: I do expect that the shape (curvature) of the hydrophobic surface plays a big role in the effects on the structure and dynamics of the hydration water. Such a dependence on the shape is also predicted by theory: for a small hydrophobic ball with a size of ~0.5 nm there is hardly any break-up of the hydrogen-bond network, and for a large hydrophobic ball there is a large concentration of defected hydrogen-bonds. Recent experiments by Dor Ben-Amotz indicate that the hydrogen-bond network surrounding methyl and methylene groups is even more ordered than in the bulk liquid; methyl and methylene groups seem to form ideal templates around which the water network can fold. This also agrees with our observation that the reorientation dynamics of the hydrating water molecules slow down significantly in comparison to bulk water. Indeed, for a more extended hydrophobic surface, the opposite could be expected, *i.e.* a weakening of the hydrogen bonding of water near the hydrophobe and an acceleration of the dynamics. In practice this is difficult to test, because large hydrophobic molecular solutes usually also possess structure on small length scales (corresponding to methyl and methylene groups). The same is true for a "flat" interface of water with a hydrophobic liquid. The interface is not flat at all, but will be highly corrugated. With regard to the temperature dependence, the experiments of Dor Ben-Amotz show that the enhanced hydrogen-bond structure surrounding methyl and methylene groups vanishes at higher temperatures, again in agreement with our observation that the reorientation strongly accelerates with increasing temperature.

Professor Jungwirth asked: Your combination of time-resolved IR and THz spectroscopy is beautiful and provides a rich amount of extremely useful data. But I am less enthusiastic about your "cartoons" used to explain the underlying molecular mechanisms, which are oversimplified and can even be misleading. *E.g.*, locking of the water dipole is indeed realistic for strongly hydrated cations like Mg^{2+} but it is not realistic for say K^+ (and Cs^+ even speeds up water dynamics compared to pure water). Likewise, only strongly hydrated F^- can lock the H–O bond; I^- certainly cannot. Concerning molecular ions like $SO_4(^{2-})$ the situation is even more complex and the ions cannot be faithfully represented as spheres. While MD simulations are certainly not perfect, I would argue that they provide a far more realistic molecular picture than the "cartoons" you employ.

Professor Bakker answered: The cartoon I showed was certainly not intended to suggest that locking would occur for any combination of cation and anion! I completely agree with you that locking of the water reorientation (of dipoles and OH groups) can only occur for strongly hydrating cations, as we indeed observe, both in the time-resolved IR experiments and in the dielectric relaxation studies. With both techniques we only observe an excess amount of slow water molecules for salt solutions containing Mg^{2+}, and even then only in case this cation is combined with a strongly hydrating anion like SO_4^{2-}. We do not observe any locking for salt solutions containing K^+ or Cs^+. For those cations the effects on the dynamics of water are really small.

I also agree with you that MD simulations, provided that they can account for the experimental observations, will provide a far more accurate molecular picture of the structure of water around ions than my cartoons! I would be happy to see MD simulations that can explain our experimental observations.

Professor Tiddy asked: I have three questions. First, how accurately can you deconvolute the curves into two exponential components for the anion-bound water, rather than (say) three curves or a distribution of curves? Second, how much of the slow motion observed with tetramethyl guanidinium is simply due to the fact that this ion is much larger? Third, when carbon is directly attached to a cationic nitrogen, some of the charge is transferred to the carbon (*e.g.* tetramethylammonium ions). Hence the interaction of this carbon with water is different from the interaction of water with CH_2 or CH_3 groups on a paraffin chain. Is this really "hydrophobic"?

Professor Bakker responded: (1) We can distinguish the anion-bound water from the other water molecules thanks to their difference in absorption spectrum and vibrational lifetime. This decomposition only works for water bound to anions with a sufficiently different absorption spectrum and/or vibrational lifetime. Examples for which this decomposition works are the halide anions. For these ions we can obtain a deconvoluted anisotropy decay that exclusively represents the orientational correlation function of the anion-bound water molecules. We observe a very pronounced non-exponential decay of this anisotropy that can be fitted well to a bi-exponential function with time constants of ~2 ps and ~10 ps. We cannot exclude that these time constants in fact represent distributions of time constants, the time scale over which we can follow the orientational dynamics does not allow us to discern such a distribution. Hence, we cannot exclude that the observed anisotropy decays in fact represent three or more components. However, we can conclude that the behavior of the reorientation is bimodal, showing fast components probably due to the wobbling motion of the hydrogen-bonded OH groups, and slow components probably due to more global rearrangements.

(2) The effect on the mobility of water appears to scale primarily with the size of the hydrophobic part of the molecule/ion, and not so much with the overall size of the molecule/ion. It thus follows that guanidinium has practically no effect on the dynamics of water while tetramethylguanidinium shows a very strong effect. Ammonium and tetraalkylammonium ions show the same behavior.

(3) Indeed one would expect that a CH_2 group directly adjacent to a cationic nitrogen atom will have a different effect on the dynamics of water than a CH_2 group embedded in an aliphatic chain. Unfortunately, it is difficult to quantify this difference. For different tetra-alkylammonium salts we observe that the effect on water scales quite well with the size of the aliphatic chains (methyl, ethyl, propyl, butyl), which suggests that the effect of the CH_2 group next to the cationic nitrogen is not that different from that of the more distant CH_2 groups.

Dr Vila Verde asked: Can you elaborate on the final statement of your paper?

Professor Bakker replied: The final statement of the conclusions relates the similar effect of tetramethyl-guanidinium-chloride (TMGndCl) and tetramethylurea (TMU) on the conformation of Photoactive Yellow Protein (PYP) to their similar influence on the dynamics of water. This statement is somewhat misleading and indeed needs to be elaborated.

We find that guanidinium-chloride (GndCl) has a much stronger denaturation effect on proteins than urea. However, tetramethyl-guanidinium-chloride (TMGndCl) and tetramethylurea (TMU) show a very similar effect on the denaturation of proteins, suggesting that this denaturion effect is governed by the four hydrophobic methyl groups that these substances have in common. TMGndCl and TMU also show a very similar slowing-down effect on the orientational dynamics of water, and this effect is also related to their four methyl groups. Hence, the denaturation and the slowing down of the reorientation find a common origin in the influence of the methyl groups on water. This common origin is that hydrophobes limit the configurational space of their surrounding water molecules, with the result that the entropy decreases and that the reorientation slows down. The effect on the entropy is likely the main driving force for the denaturation of proteins by TMGndCl and TMU. The denaturation is thus not a direct consequence of the slowing down of the orientational dynamics of water, but both effects (denaturation and slowing down) find a common origin in the influence of hydrophobic methyl groups on the hydrogen-bond structure of their surrounding water molecules.

Dr Vila Verde addressed Liem Dang: What are your thoughts on the best way of improving current force-fields for ions at high concentrations?

Dr Dang replied: This is a very good question and there is no easy answer for it. But the good news is that many groups are working on improving force fields. You should look at the recent work by Theresa Head-Gordon (UC Berkeley) or Roland Netz (Berlin). We need to go beyond simple combining rules. We also need to make use of experimental data and quantum chemical calculations.

Professor Madden addressed Phil Mason: In any classical simulation study, reliability depends on the quality of the force field. Many of the used force fields have been developed on an empirical basis, by fitting to an observable, with this information perhaps supplemented by distributed charges derived from *ab initio* calculations on isolated molecules. As these force field are employed in an expanding range of applications, their limitations become more apparent, especially when ions are involved. This comment is simply to point out that it is now routine to perform condensed-phase *ab initio* calculations and sample the electronic structure in a wide variety of coordination environments (many thousands generated along an MD trajectory, for example). By requiring that a force field reproduces well the forces on each nucleus obtained from the *ab initio* calculation (for exactly the same nuclear configuration) it is possible to refine a force field. This is a process of "force-fitting" (see *e.g.* P. A. Madden *et al.*, *J. Mol. Struct.: THEOCHEM*, 2006, **771**, 33–41). The process can be extended to polarisable models by simultaneously fitting the dipoles, and even to more complex potential forms if this proves necessary to represent the *ab initio* data. It gives a systematic approach to potential development.

The reason for stressing "condensed phase" is that the properties of molecules like water and ions vary enormously, depending upon their configuration environment – so that the F^- ion in crystalline LiF is very different from that in aqueous solution or when isolated in the gas-phase, and the effective dipole of water in the condensed phase is much larger than that of the gas-phase molecule. The remark is particular apposite for the polarizability of the ion (see *e.g.* M. Salanne, R. Vuilleumier, P. A. Madden, C. Simon, P. Turq and B. Guillot, *J. Phys.: Condens. Matter*, 2008, **20**, 492407; or J. J. Molina, S. Lectez, S. Tazi, M. Salanne, J.-F. Dufrêche, J. Roques, E. Simoni, P. A. Madden, and P. Turq, *J. Chem. Phys.*, 2011, **134**, 014511) but applies to other properties too.

Professor Mason replied: I find this comment right on the mark! This is undoubtedly the way forward.

Professor Lopes addressed the meeting: Modeling ions in electrolyte solutions (in the context of the Hofmeister series or not) and modeling ionic liquids have many points in common. In fact many of the problems related to the anisotropy of the liquid–air interface, surface potentials, polarization effects, *etc.*, are just exacerbated in the case of ionic liquids, which can be regarded as "electrolyte solutions without the solvent". In my opinion a large part of the "evolution" of different simulation techniques is somewhat determined by the available computing power. Some years ago when we developed and introduced a systematic, fixed-charge, and atomistic force-field for ionic liquids (*J. Phys. Chem. B*, 2004, **108**, 2038) we were quite pleased that we could predict some static properties and the structural characteristics of different homologous series of ionic liquids. These days we would like to extend the scope of the force field to dynamic properties and interfaces. One of the ways attempted by different authors was to try charge-transfer schemes, but such an approach has raised issues of internal consistency when applied to different series of ionic liquids or to the properties one can compute at the liquid–air interface. I think the use of polarizable models is one of the most promising ways to go at this moment, given the actual performance and availability of computer clusters. *Ab initio* methods are also becoming more and more competitive, at least as a way to complement MD simulations using classical models.

Dr Henchman addressed Huib Bakker: You discuss how non-polar solutes could inhibit hydrogen-bond rearrangements by occupying the interstitial positions around water molecules. This is not the only mechanism of hydrogen-bond switching. Fig. 3 shows three possible mechanisms by which donors form new hydrogen bonds (adapted from R. H. Henchman and S. J. Irudayam, *J. Phys. Chem. B*, 2010, **114**, 16792–16810). Mechanism (a) is the same as the jump mechanism of Laage and Hynes whereby a donor switches to an interstitial water *via* a bifurcated-hydrogen transition state. In mechanism (b) the donor switches to an existing donor and proceeds *via* a bifurcated-oxygen transition state. In mechanism (c) the donor switches to an existing acceptor and proceeds *via* a cyclic-dimer transition state. The frequency of these mechanisms for TIP4P/2005 water under ambient conditions is found to be 78%, 8% and 14%. Mechanisms (b) and (c) have higher energy barriers of 8.5 kJ mol^{-1} compared to 5.5 kJ mol^{-1} for mechanism (a) and therefore proceed less often in bulk water. However, in the first hydration shell of non-polar solutes, we find a more than doubling in the proportion of high-energy bifurcated-oxygen and cyclic-dimer intermediates found in mechanisms (b) and (c), as shown by their combined radial distribution function $g(r)$ around argon in Fig. 4 (adapted from S. J. Irudayam and R. H. Henchman, *J. Chem. Phys.*, 2012, **137**, 034508). Unlike mechanism (a), these two mechanisms would not be as inhibited by the presence of an interstitial non-polar solute.

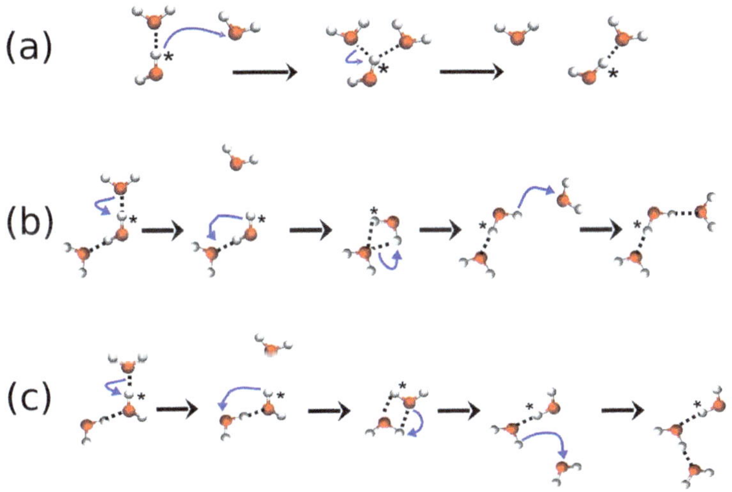

Fig. 3 The three possible mechanisms for hydrogen-bond rearrangement in water: switch of a donor (marked by an asterisk) to (a) an interstitial water *via* the bifurcated-hydrogen transition state, (b) an acceptor *via* the bifurcated-oxygen transition state, and (c) a donor via the cyclic-dimer transition state.

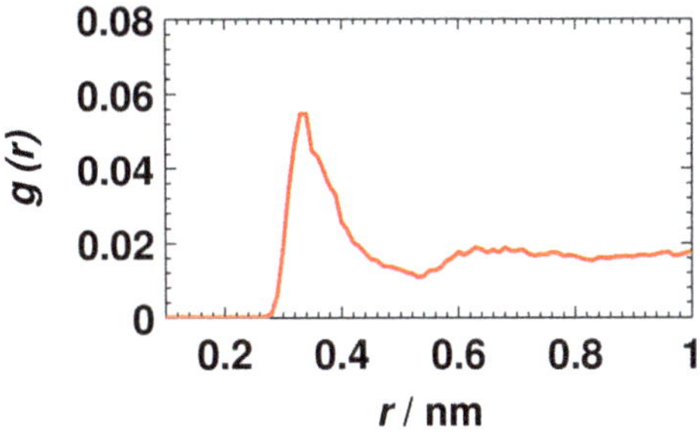

Fig. 4 Radial distribution function $g(r)$ for the single-donor hydrogen-bond species around water, normalised to the density of bulk water.

Professor Bakker responded: Thank you for this very interesting and important note! Indeed it makes sense that mechanism (a) is much more inhibited by the hydrophobic groups than mechanisms (b) and (c). In your simulations you find a much higher proportion of mechanisms (b) and (c) around hydrophobic groups than in bulk liquid water. I could even imagine that the reorientation of water around hydrophobic groups may become dominated by mechanisms (b) and (c) (which is not exactly what you find). A dominance of mechanisms (b) and (c) could explain why we observe a substantially higher activation energy for water reorientation around hydrophobic groups than for water reorientation in the bulk liquid.

Dr Reif spoke: Paul Madden discussed the electric potential obtained from integrating the charge density ("mean electrical potential") and the electric potential felt by an ion with excluded volume ("Madelung potential"). In this context, I would like to draw the connection to the ambiguities in calculating the electric potential at the air/water interface, which were hinted at already yesterday by Yan Levin. Calculations of the air/water interfacial potential *via* computer simulation give different results depending on whether one integrates the surface polarization or the surface charge density. The difference is due to the solvent quadrupole-moment trace (*J. Chem. Phys.*, 2006, **124**, 124106). Species of atomic size (excluded volume) probe the electric potential outside the volume occupied by water molecules and are thus sensitive to the electric potential obtained from integration of dipole moment rather than charge density. Just to provide an impression of the magnitude of the difference, using *ab initio* simulations, a value of 3.63 V was obtained for the air/water interfacial potential from integration of the charge density, whereas a significantly smaller value, 0.012 V, resulted after dissecting the dipolar contribution (*J. Phys. Chem. Lett.*, 2010, **1**, 496).

Mr Sharma added: How much do the cation electronic structure effects determine the solvation properties of trivalent cations, exclusively for the case of europium and gadolinium ions?

Professor Madden replied: I mentioned, in my response to a query by Ruth Lynden-Bell towards the end of the first session, the relative energetics of divalent and trivalent europium in some simulations of molten salts. I would agree that to fully understand the solvation of these ions one would need to include the influence of the f-electrons; this was not done in these simulations where the objective was to understand the role of the electrode potential on the ionic energetics in a generic sense, rather than to realistically model these particular species.

Professor Bakker addressed Phil Mason: You find that the MD simulations strongly overestimate the binding of pyridine to Gdm_2SO_4 system. You explain this from the fact that the MD simulations overestimate the ion pairing of Gdm_2SO_4, and the relatively strong binding of the thus formed ion pairs to pyridine. Why does the (overestimated) ion pairing of Gnd and SO_4 lead to a stronger binding of these ions to pyridine? What do the resulting structures of pyridine–Gdm_2SO_4 look like?

Professor Mason replied: I would agree that the expectation is that if the ion–ion interaction is over-emphasized then it would be expected that there would be less pyridine association to these ions. This however was not observed in these simulations. It was found that the ions clustered together and created clefts, typically with two or more Gdm ions protruding, which made for some relatively strong Gdm–pyridine interactions. Fig. 5 shows what these structures actually look like, with the ions colored in red and the pyridines in white.

Professor Madden followed on from this reply: You have commented on the efficacy of scaling down ionic charges in classical, unpolarizable force fields as a way to recapture, in an inexpensive way, some of the effects of the electronic polarization. Scaling down the overall charge of an ion might be justified as a representation of the weakening of the coulombic interactions between ions because they are, in effect, embedded in a dielectric medium whereas the vacuum is the dielectric background in unpolarizable simulations. Such a scaling has often been used in modelling ionic liquids, for example, as a way to obtain simulation models which predict diffusion coefficients, *etc.*, comparable to experimental values. However, it is not usually helpful to scale down uniformly the partial charges which are used to represent the internal charge distribution within a molecule. In fact, the effective dipole moments of molecules in condensed media are usually enhanced relative to their gas-phase values. The effect is well known in water simulation models, and for water the enhancement of the average dipole has been well documented in *ab initio* simulations. A consideration of the reaction-field induced enhancement of the effective dipole of a dipolar, polarisable molecule embedded in a polarisable medium provides a simple model which explains the origin of the effect (see *e.g.*: *Adv. Chem. Phys.*, 1984, **56**, 467).

Mr Marslaek remarked: There is a distinction between the effects the environment has on the molecular dipole and on the effective (partial) charges. While the change in the effective charge that is used in the simulation with "scaled charges" is only a technically convenient way of changing the dielectric constant, the molecular dipole really does change in the condensed phase; the charge density within the molecule is redistributed.

Professor Madden responded: I agree with this remark. The practical implications are, as I have said above, that the observation that reducing the total charge on an ion from its formal valence value may improve the properties predicted by some effective interaction models should not be taken to imply that uniformly scaling down the partial charges in a model of a molecular ion is a good idea. In the absence of direct guidance from a condensed phase electronic structure calculation on how the partial charges should be readjusted, I would suggest performing a simple calculation of the expected reaction-field enhancement of the molecular dipole by imagining the molecule placed in a continuum dielectric with the appropriate high

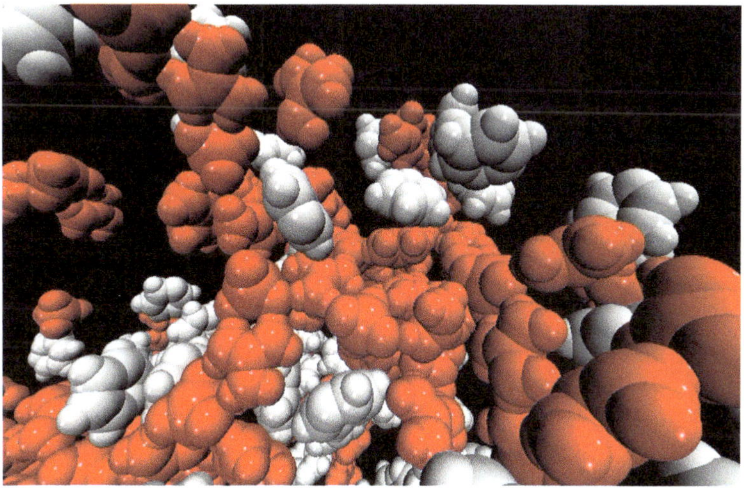

Fig. 5 Structures of pyridine–Gdm_2SO_4.

frequency dielectric permittivity and using this as a guide to the redistribution of charge within the molecule.

Professor Jungwirth commented: Accounting for polarization by charge scaling indeed improves agreement with experiment significantly for aqueous salts of polyvalent ions such as Na_2SO_4. We showed that, *e.g.*, for sodium sulfate it is not possible to achieve such a good agreement within strictly non-polarizable models unless unrealistic ionic sizes are assumed (see Wernersson *et al.*, *J. Chem. Theory Comput.*, 2010, **6**, 3233).

Professor Mason replied: I think this is symptomatic of a wider issue concerning the accurate parameterization of molecular dynamics force fields for ions. A similar problem was encountered with attempting to accurately model neutron scattering data of K_2CO_3 and KNO_3 where it was found that reducing the ions' charges (to account for the polarizibility of the media) was necessary in order to obtain a good fit to the experimental data (*J. Phys. Chem. B*, 2012, **116**(28), 8145–8153).

Professor Peter Halling asked a question on the paper by Professor Yi Qin Gao by communicating: Do I understand correctly that your simulations show Na^+ ions interacting more strongly with peptide backbone carbonyl groups than with charged carboxylates? This would be interesting as we have crystallographic evidence (shown on my poster) that cations are detected more commonly interacting with clusters of carbonyls, despite the presence elsewhere of formally negative carboxylates.

Professor Gao communicated in reply: In our simulations, we found that the relative bindings of Na^+ with the carbonyl and carboxylates are different, and depend heavily on the counterions used. In sodium sulphite aqueous solution, the binding of the cation with the carbonyl and carboxylates is very similar. However, in the iodide solution, the RDF has a much higher peak around the carboxylate oxygen than that found around the carbonyl oxygens. This observation again shows the importance of ion cooperativity. Therefore, the binding of cations is very strongly dependent on the environment; and we would expect variations between solution and crystal structures as well.

Diversity in the mechanisms of cosolute action on biomolecular processes

Shahar Sukenik, Liel Sapir, Regina Gilman-Politi and Daniel Harries*

Received 6th May 2012, Accepted 24th May 2012
DOI: 10.1039/c2fd20101a

Numerous cellular cosolutes significantly impact the way that proteins and other biomacromolecules act and interact. We have followed the thermodynamic effect of several cosolute classes, including polymers, cellular osmolytes, and inorganic salts, on the stability of biomolecular folding and complexation. By comparing changes in free energy, enthalpy, and entropy upon cosolutes addition for these processes, we identify several thermodynamically distinct mechanisms. Surprisingly, even while many cosolutes display similar scaling of the change in stabilizing free energy with their concentration, a breakdown of this free energy into enthalpic and entropic contributions distinguishes different families of cosolutes. We discuss how these "thermodynamic fingerprints" can direct towards possible underlying mechanisms that govern the cosolute effect.

1 Introduction

Salts and other cellular solutes profoundly affect the interactions between and within biomacromolecules.[1–11] Moreover, it is well appreciated that these effects are largely specific to the type of solute. As early as 1888, Hofmeister showed that some ions effectively precipitate or "salt out" many proteins from solution, while others help to solvate them.[12] This effect of salts is described by a common Hofmeister ranking of ions that has since been shown to extend to many other physical properties, including surface tension at the air–water interface, interactions and phase transitions in lipid membranes, and protein folding and complexation.[13–17] Over the past couple of decades, much effort has been devoted to gaining a better molecular-level understanding of these ion-specific effects at interfaces, as well as to developing quantitative predictive theories to account for the experimentally observed consequences of salt and other cosolute addition to macromolecular interactions.[18–26]

As our understanding of the cosolute-specific effects deepens, we seek further vindications and appropriate tests, while also attempting to extend our insight beyond the phenomenology and into the underlying molecular and thermodynamic mechanisms. In this effort, an important goal is to dissect the change in free energy upon cosolute addition into the associated contributions of enthalpy and entropy, forming a "thermodynamic fingerprint" for biomolecular processes in the presence of cosolutes. Through links to current theories that describe the forces driving macromolecular association or protein folding in the presence of cosolutes, this fingerprint should allow us to support or challenge different molecular mechanisms for different classes of cosolutes. A persistently lingering question remains: are there common unifying features in the mechanisms by which seemingly disparate

Institute of Chemistry and The Fritz Haber Research Center, The Hebrew University, Jerusalem, Israel. E-mail: daniel@fh.huji.ac.il

cosolutes, such as salts and neutral polyol osmolytes, exert their action on macromolecules?

In recent years there have been significant advances in linking the thermodynamics of cosolute action to specific mechanisms. It is now realized that while the changes in free energy may follow similar trends for different cosolutes, the underlying molecular driving forces can differ dramatically and, as a result, the breakdown into entropic and enthalpic contributions should vary. Macromolecular association that is driven by ion release, for example, shows an entropic gain upon binding due to the release of previously associated counterions from the macromolecular proximity.[28] For this mechanism, the addition of salt would, therefore, mainly impact the association entropy. Macromolecular crowding has also been shown to be entropically driven; the primary effect of molecularly large "crowders" is often due to their exclusion from a volume around proteins that drives burial of the macromolecular surfaces to allow subsequent favorable gains in free volume.[1] By contrast, we have previously shown that not all excluded cosolutes act in an entropy-driven mechanism. For example, small polyols, important "stabilizing osmolytes", were found to induce peptide folding primarily through enthalpic gains that are offset by a large unfavorable entropic contribution.[29] These cosolutes, used by cells to regulate cytoplasmic solution conditions with respect to their often deleterious environment, typically stabilize proteins in their folded state.[5,30] We have linked this favorable enthalpic contribution to folding due to cosolutes with the corresponding changes upon osmolyte addition to the hydrogen bond network of the waters surrounding the protein.[31] From these examples it already becomes clear that matching observed enthalpy–entropy thermodynamic fingerprints can help to support or rule out some of the possible underlying molecular mechanisms, and point to the potential applicability of others.

We have been using model biomolecules to help dissect the different contributing interactions that, in concert, modulate the response of proteins to the addition of cosolutes in aqueous solution. We start here by describing the effect of various salts on the association of β-cyclodextrin with adamantane, Fig. 1a. This guest–host complexation reaction involves the burial of uncharged, mostly nonpolar "hydrophobic" surfaces.[32,33] In the absence of direct electrostatic interactions between the cosolute and the interacting surfaces, we find that salt-specific stabilization of the complexed state follows the Hofmeister ranking of ions. The added complex

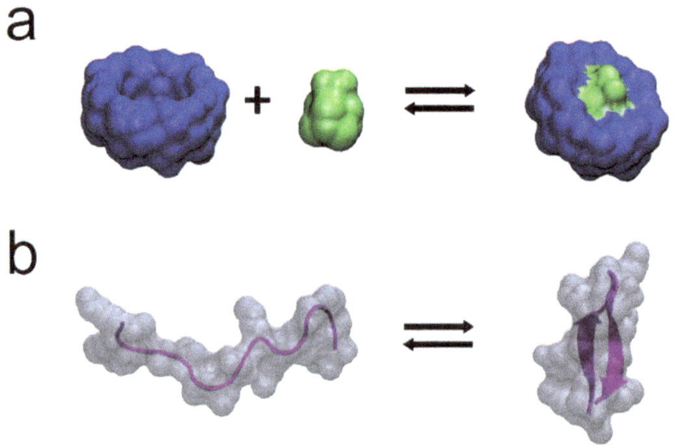

Fig. 1 Schematics of β-cyclodextrin association with adamantane carboxylate (**a**), and of the folding of a 16-amino acid long peptide (**b**). Representations demonstrate the decrease in solvent accessible surface area (SASA) in the complexation or folding process. Molecules rendered using VMD.[27]

stabilization can be well fit by integrated preferential interactions of salts with the surfaces of cyclodextrin and adamantane. This fit supports the Tanford–Bolen approach and that developed by Record and coworkers,[34,35,21,36] which ascribes additive transfer free energies to various surface groups in proportion to exposed surface area. The thermodynamic fingerprint of this complexation process reveals that complex stabilization by cosolute salt is dominated by a favorable entropic contribution with an added favorable enthalpic contribution at larger concentrations.

We next turn to follow the effect of cosolute addition on the folding of a model peptide into a β-hairpin, Fig. 1b. We have previously used naturally occurring "stabilizing osmolytes" to drive this folding process, and compared these osmolytes with the effect of molecularly larger PEG and dextran polymers, serving as "macromolecular crowders". Despite exhibiting a similar, size dependant stabilization effect on the peptide, the mechanisms observed for these two classes are disparate. While osmolytes stabilize folding through an enthalpic mechanism,[29] the larger PEGs and dextrans predominantly show entropic stabilization at low concentrations, which gradually takes on a larger enthalpic contribution as concentrations rise. This change with concentration seen in the thermodynamics may reflect a balance of entropically-driven excluded volume effects (or "depletion interactions") *versus* direct "chemical" interactions that can involve significant enthalpic contributions.

Finally, we address the effect of cosolute ions on the folding of the same peptide. As is often found for proteins, the charged peptide's stability responds to salts non-monotonically.[37,14,38] At low salt concentrations, we find peptide unfolding, while at higher concentrations there is regain of folding stability in an ion-specific manner that tracks the Hofmeister series. The changes at low salt concentrations can be assigned to electrostatic effects that can be reasonably described by Poisson–Boltzmann (PB) theory. The ion-specific stabilization effect seen at higher salt concentrations can be traced to the preferential interaction of salt with uncharged residues. The thermodynamic fingerprint for salts shows that, for the concentrations tested, they all act in an entropy-driven mechanism. Modeling the net effect of salts on folding as a sum of salt interactions with charged and uncharged surfaces suggests that the lower enthalpic contribution results from a cancelation of interaction of salts with the two types of surfaces.

By presenting this series of molecular models, we demonstrate how tracing the entropic and enthalpic contributions of cosolute action allows further insights into plausible underlying molecular mechanisms. Moreover, this strategy highlights remaining challenges in formulating quantitative theories to predict the cosolute effect on biomacromolecular processes.

2 Neutral interacting interfaces in the presence of cosolutes

To explore how different cosolutes influence biomacromolecular interactions, it is instructive to first consider interactions between neutral macromolecules, or study processes that involve burial or exposure of neutral, nonpolar, surfaces. Cosolute effects are simplified here since direct electrostatic interactions between the solute and cosolute become less significant. A useful example is the association of β-cyclodextrin (CD) and adamantane carboxylate and the impact on this complexation by various cosolutes, both neutral and charged.[39] Soluble CD is shaped as a truncated-cone with a hydrophobic cavity. Adamantane (AD) is highly insoluble in water, but when mixed with CD can associate to form an inclusion complex with increased solubility, Fig. 1a.

The CD/AD association is accompanied by a decrease in solvent accessible surface area (SASA) of about 400 Å^2, mainly associated with burial of nonpolar hydrophobic moieties. The preferential interaction of added cosolutes with CD and AD is generally different for the unassociated molecules and for the complex. Preferential exclusion of cosolute from the nonpolar macromolecular surfaces should lead to net stabilization of the complex, whereas preferential inclusion would result in destabilization.[6,42,2,3] The extent of stabilization is quantified by the free energy of complexation, $\Delta\Delta G_{com}$, defined as $\Delta\Delta G_{com} = \Delta G_{com}(c) - \Delta G_{com}(0)$, where $\Delta G_{com}(c)$ and $\Delta G_{com}(0)$ are the free energies of complexation in cosolute solution of concentration c and in pure water, respectively. Isothermal titration calorimetry and vapor pressure osmometry experiments[43,39] have demonstrated that, for many cosolutes, $\Delta\Delta G_{com}$ is linear with respect to added cosolute concentration, Fig. 2a. Generally, salts stabilize the complex, whereas other solutes, such as betaine, destabilize it. The strength of the salt effect follows the Hofmeister ranking of anions, $SO_4^{2-} > F^- \approx Cl^- > Br^- > SCN^-$, seen as the differences in the slopes of the lines in Fig. 2a.

It is possible to predict $\Delta\Delta G_{com}$ by using the Tanford–Bolen and the Record approaches towards preferential solvation.[34,35,44] Such predictions are based on calculated changes in exposed surface areas for different chemical moieties, weighted by corresponding effective preferential interaction coefficients per unit area that are calibrated from transfer energies of model compounds.[21,44,36] To derive changes in exposed surface areas in CD/AD association, we performed all-atom molecular dynamics (MD) simulations of solutions of AD, CD, and the CD/AD complex in pure water using NAMD.[45] Water molecules were represented by the TIP3P potential and CD by the CHARMM-type CSFF model,[46] using MD simulation parameters as previously described.[47] The forcefield was appropriately extended to account for adamantane. The resulting molecular ensemble was used for SASA calculations of each species, where we have assumed that the conformational landscape of the complexed and unassociated molecules is unaltered by the presence of salt (while the ratio of complexed to uncomplexed molecules changes). Each configuration was assigned a SASA using the Richards set of atomic radii[35,44,48,49] and a 1.4 Å water probe radius, and these molecular surfaces were further decomposed into different classes (aliphatic, hydroxyl, or "other"). The contribution of each surface type to the change in free energy of transfer of the entire molecule (or complex) into each cosolute was then calculated using the available group transfer free energies (Table 1), while assuming group additivity.[21,44,36] Calculating the average transfer free energy for each separate molecule and for the complex allows calculation of $\Delta\Delta G_{com}$ due to the presence of salt, where $\Delta\Delta G_{com} = \Delta G_{CD+AD} - (\Delta G_{AD} + \Delta G_{CD})$, and terms on the right hand side represent the transfer free energy of the molecules from water into a solution containing cosolute at a specific concentration.

Assuming $\Delta\Delta G_{com}$ to be linear with cosolute osmolality (as is often found for weakly interacting or excluded cosolutes), we have predicted m-values, where the

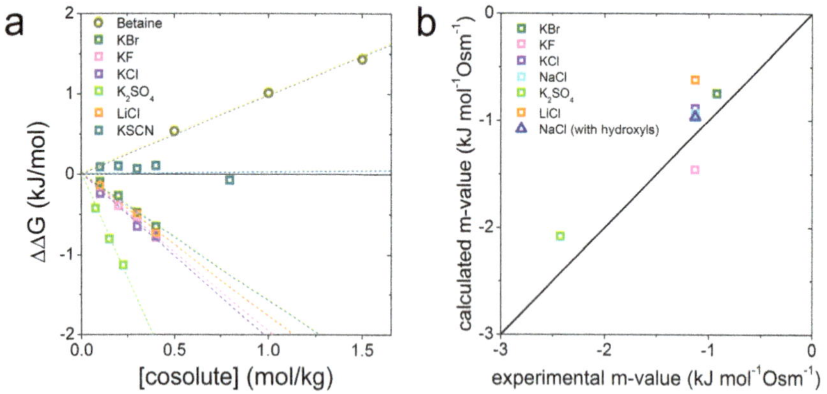

Fig. 2 (**a**) Changes in the free energy of β-cyclodextrin association with adamantane *versus* cosolute concentration. Lines are linear fits of the data. (**b**) Calculated *versus* experimentally derived m-values. Data represented by the triangle is based on an alternate set of group transfer free energies that in addition ascribes a value for hydroxyl groups (Table 1) based on measurements of alcohol model compounds.[40,41]

Table 1 Group transfer free energies for various cosolutes used in this study

Cosolute	Group transfer free energies, J mol^{-1} m^{-1} Å^{-2}				
	Aliphatic	Aromatic	Amide	Hydroxyl	Carboxyl
KBr^a	4.2 ± 0.2	3.0 ± 0.5	−10 ± 1		
KF^a	8.4 ± 0.2	5.9 ± 0.2	−11.4 ± 0.7		
KCl^a	5.0 ± 0.2	4.2 ± 0.2	−10.4 ± 0.5		
$NaCl^a$	5.5 ± 0.2	5.0 ± 0.2	−10.4 ± 0.5		
$NaCl^b$	6.3 ± 0.5			−5 ± 2	
$K_2SO_4{}^a$	13.9 ± 0.2	13.1 ± 0.5	−17 ± 1		
$LiCl^c$	4.0 ± 0.5	4.0 ± 0.5	−9 ± 1		
$GuHCl^a$	0.7 ± 0.2	−2.0 ± 0.5	−9 ± 1		
NH_4Cl^a	2.0 ± 0.7		−3 ± 2		
$Betaine^d$	0.87 ± 0.40	−5 ± 1	6 ± 1 (O) −4.2 ± 0.7 (N)	0.2 ± 0.3	6.9 ± 0.5

^{a} Data by Pegram *et al.*44 The values for K_2SO_4 were derived from the available data on NaCl, KCl and Na_2SO_4, assuming ion-additivity. ^{b} Values were determined by analyzing salting out constants for various alcohols in NaCl41,40 according to the procedure described in ref. 21. ^{c} Derived from ref. 21 and 44 by globally fitting the solubility data. ^{d} Data by Capp *et al.*36

m-value is defined here by $\Delta\Delta G_{com} = mc_{Osm}$, for the complexation reaction, and c_{Osm} is the cosolute concentration on the osmolal scale.[50–53] These values are compared with experiment in Fig. 2b. The *m*-values show quantitatively excellent fits for KCl, KBr, and K_2SO_4, and fits for the other salts are close to the experimental and calculation errors. By contrast, the effect predicted for betaine is stabilizing ($m = -0.36$ kJ mol^{-1} Osm^{-1}), while experiments show a destabilizing effect on the complex ($m = 0.98$ kJ mol^{-1} Osm^{-1}) that suggests preferential inclusion of betaine in the solute vicinity. This discrepancy is probably due to cooperative effects of the segments of the convex inner interface of CD on the favorable association with betaine, a relatively large molecule with respect to simple inorganic ions. This may suggest that the group transfer free energies per surface area are non-additive in special cases where interacting surfaces are structured very differently from expanded, well-exposed, or flat interfaces. For such convex or corrugated interfaces (relative to the cosolute's size) the transfer free energy may scale nonlinearly with surface area.

We conclude that the free energy change upon addition of cosolute as determined by the Record approach allows fair estimates of the experimental *m*-values. In addition, decomposition of SASA to the different moieties to which transfer free energies are assigned works best when the different SASA elements are decoupled. However, when such coupling does exist, surface convexity and cosolute size become important, and could result in non-additive free energy contributions. In addition, we expect some discrepancies may result when not all group types are accounted for in the overall molecular SASA. Indeed, we find that by including the hydroxyl group transfer free energy into NaCl solutions, the fit of predicted and experimental *m*-value was improved, though only very slightly (compare the blue square and triangle in Fig. 2b). We note, finally, that taking group transfer free energy to be linear with either osmolal or molal concentration result in almost indistinguishable predictions for the concentration range used here.

3 Charged peptide in the presence of neutral cosolute

Protein folding poses additional challenges for analysis. Beyond non-polar hydrophobic residues, the macromolecular protein–water interface itself may carry charges, so that subtle changes to the solvating environment can strongly impact the stability of the folded and unfolded protein conformational ensembles. For simplicity, we first describe the effect of non-electrolytes on peptide folding. The model system we used in these studies is a 16 amino-acid peptide previously described[54] that carries a net-total of +3 elementary charges at neutral pH. The peptide was determined in circular dichroism and NMR experiments to fit a two-state folding process from the unfolded to folded, β-hairpin states, Fig. 1b. We have followed the effect of two electrostatically neutral cosolute classes on the stability of this peptide. The first includes stabilizing osmolytes: sugars, such as the dissacharide trehalose, as well as polyols, including sorbitol, glycerol, erythriol, manitol, and xylitol. The second cosolute group consists of "macromolecular crowding" polymers, including various molecular weights of polyethylene glycol (PEG 100, 400, and 4000) and dextran (20 k and 40 k).

Molecular crowding theory has been proposed to rationalize the observed discrepancies between macromolecular interactions in the crowded milieu of the cell and *in vitro* experiments performed in dilute aqueous solutions. Briefly, by taking up volume in solution, non-interacting or "excluded" cosolutes promote any process that results in freeing up of space excluded by the macromolecules, or in liberating translational and conformational degrees-of-freedom for cosolutes. Such crowding has been shown to drive protein folding,[55,56] complexation,[57] and aggregation.[56,58] The magnitude of this stabilization scales with the molecular size of the cosolute. Indeed, the changes in free energy have been accurately predicted by using scaled particle theory for protein association in the presence of various macromolecular crowders including dextran.[58]

Using circular dichroism spectroscopy to monitor the change in free energy of peptide folding with increasing cosolute concentrations, we find that all cosolutes tested here stabilize the folded state. In addition, the change in free energy for the folding process, $\Delta\Delta G$, is linear with cosolute concentration, Fig. 3. This linearity is characteristic of stabilization through non-specific interactions, and similar trends have been shown for other folding and interacting proteins upon addition of these and other cosolutes.[59–61,44] The change in free energy is proportional to the size of the perturbing cosolute, as might also be expected for macromolecular crowders. Interestingly, however, our recent experiments[29] showed that for polyol osmolytes, the stabilizing mechanism is dominated by a favorable enthalpy, which cannot be accounted for by the simple molecular crowding model alone. As we further discuss

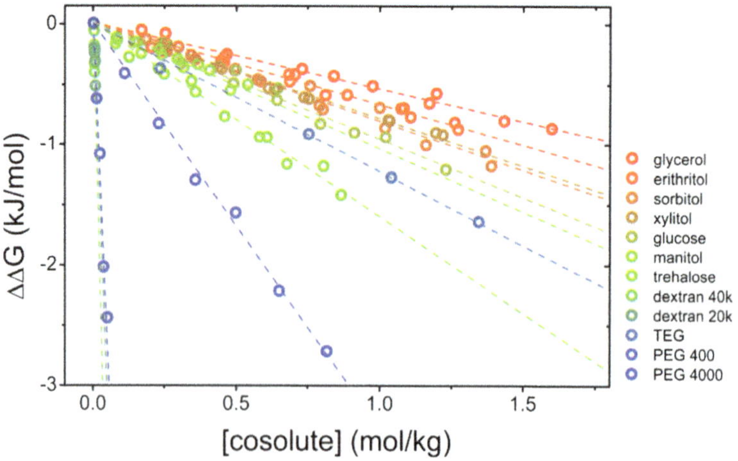

Fig. 3 Change in free energy of peptide folding with cosolute concentration. Dashed lines are linear fits of the data.

in section 5, these findings may indicate a fundamental mechanistic difference between the effects of macromolecular crowders and stabilizing osmolytes.

4 Charged peptide in the presence of salt

Because both electrolytes and protein are charged, their interactions can show the most complex cosolute effects, so that both electrostatic and "chemical" or ion-specific preferential interactions must be considered. We used circular dichroism spectroscopy to follow folding of the model peptide presented in section 3 in various salts at several concentrations. These data allow us to derive the change in free energy of folding, $\Delta\Delta G$, in the presence of the cosolute salts, Fig. 4. At low concentrations (less than ~0.25 M) all salts destabilize the native conformation of the peptide. As salt concentration increases, however, a stabilizing effect becomes prominent for most salts at concentrations higher than ~0.5 M, leading to a maximum in $\Delta\Delta G$. This type of non-monotonic behavior is often encountered for proteins.[37,38] For the more strongly destabilizing salts, GuHCl and to a lesser extent NH_4Cl, $\Delta\Delta G$ continues to rise with salt concentration, albeit with a reduced slope.

The Hofmeister ranking of ions dictates that certain cations and anions more strongly confer stability to proteins than others.[14,12] The classical Hofmeister ranking for the ions we have used, listed from the most stabilizing to the most destabilizing, is: for anions $SO_4^{2-} > Cl^-$, and for cations $Na^+ > K^+ > Li^+ > NH_4^+ > Gu^+$. Notably, our results follow this ranking, showing that the peptide stabilizing effects are strongest (when comparing anions) for Na_2SO_4, followed by NaCl and (comparing cations) are largest for NaCl, followed by KCl, LiCl, NH_4Cl, and finally GuHCl, which is most destabilizing towards the peptide.

Pegram and Record[44] were able to successfully account for the unfolding process of a different protein, the *lac*-protein DNA binding domain, by fitting the energetic contributions of the unfolding process to a functional form that accounts for electrostatic and ion-specific terms due to salt addition. The latter terms were predicted quite accurately by summing contributions to polar and non-polar peptide groups, derived by calculating the changes in SASA in the folded and unfolded states with calibrated group transfer free energies, as described in section 2. Using a similar approach, we have tried to fit the data in Fig. 4 with a sum of electrostatic and ion-specific contributions. In addition, we directly calculated the electrostatic free energy by numerically solving the nonlinear Poisson–Boltzmann (PB) equation. These electrostatic contributions were summed with the preferential interaction estimates, determined as described in section 2.

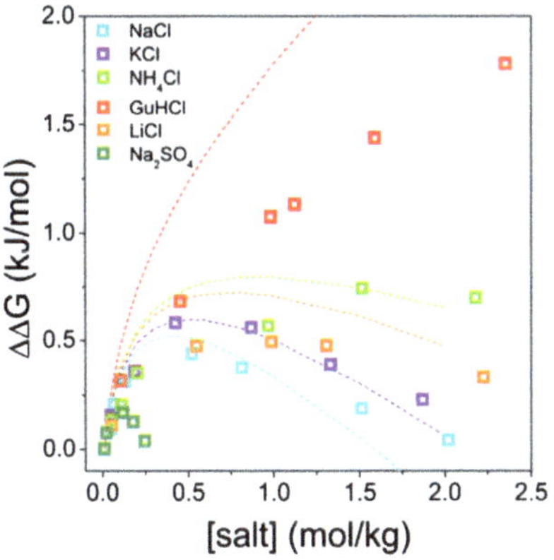

Fig. 4 Free energy of peptide folding as a function of salt concentration. Calculated values derived from the addition of PB and preferential interaction energies are shown as dashed lines.

To account for the electrostatic contribution to the folding free energy, folded and unfolded conformations were chosen from a peptide ensemble derived from all-atom MD simulations in explicit water.[31] The unfolded ensemble was obtained by sampling over 800 conformations that fulfilled a criteria based on SASA (>2300 Å^2). All conformations were assigned charges according to the CHARMM potential to the net-charged moieties only. This was done to avoid double-counting of ions around net-neutral surfaces that are separately accounted for in the preferential interaction energies; this procedure significantly improved the subsequent fits with experiments. The PB equation was solved numerically using APBS[62] for each conformation using the same grid dimensions, at different uni-univalent salt concentrations in the concentration range 0.01–2.0 M, The change in PB free energy for folding ($\Delta\Delta G_{\mathrm{PB}}$) is the difference in the charging free energy G_{PB} values for the unfolded and the folded conformations at a specific salt concentration.

The predictions of $\Delta\Delta G$, evaluated as the sum of calculated transfer free energy for uncharged surfaces and PB electrostatics for charged groups, are shown as dashed lines in Fig. 4. While the trends and ordering of all salts from experiments are faithfully reproduced, we do not find a full quantitative match. This may be a result of ion-specific effects that are absent from the PB calculations, yet could become important, particularly at higher salt content. We and others have previously shown that modified PB equations can be used to properly account for this added salt-contribution.[63,64] However, these formulations require additional knowledge of the non-electrostatic preferential interaction of ions with the *charged* moieties, which is at present lacking for peptides.[65,66]

In an attempt to more faithfully reproduce the experimental values, we tested four different ensembles for both folded and unfolded conformations. These included trajectories derived from replica-exchange MD, implicit solvation models and additional MD simulations (for descriptions of these simulation methodologies see for example ref. 67 and 68). Because all ensembles were derived under pure water conditions, we also tested reassigning probabilities to conformations in the ensemble according to predicted shifts in free energy upon salt addition, which also contributed to changes in the average differences in free energies. While the energies for preferential interactions of ions with the non-charged peptide surface were relatively insensitive to the choice of ensemble, the energies derived from the electrostatic PB calculations showed large variations with the choice of ensemble. In contrast with the folded ensemble that had relatively low variability (standard deviation) for all the ensembles we tested, the energies derived for the unfolded conformations were more variable. Some forcefields tended to yield more extended, linear conformations, while others show more compact forms. The sensitivity we find underscores a significant challenge for protein stability predictions, because proper sampling is clearly essential for quantitative estimates. Importantly, our prediction estimates of $\Delta\Delta G$ can only be considered to be semi-quantitative. Fig. 4 shows the ensemble that best fit the experimental results for NaCl (the salt for which PB is usually most accurate). The unfolded ensemble is composed of roughly 800 conformations, taken from a 50 ns MD ensemble in explicit TIP3P waters. In addition, in line with the approach that takes folded proteins using a single or few experimentally determined states, we represent the folded state by a single conformation. While the theory seems to properly reproduce the trends in the ranking of ions and their effect on peptide stability, it is hard to give quantitatively exact predictions. The most sensitive element of modeling remains the electrostatic contributions related to ion–charged residue interactions.

5 Thermodynamic fingerprints characterizing mechanistic differences in cosolute effect

The breakdown of folding free energy into its enthalpic and entropic contributions affords new insight into the mechanisms by which different cosolute families exert their stabilizing or destabilizing actions. We have measured the peptide folding free energy in the presence of various cosolutes, at different concentrations and at different temperatures, to determine the respective changes in enthalpy and entropy using a van 't Hoff analysis. These values are then subtracted from the enthalpy–entropy values for the peptide folding in buffered aqueous solution to obtain $\Delta\Delta H$ and $T\Delta\Delta S$ values, where $T = 298.15$ K (for additional information see ref. 29). A similar decomposition was performed for CD/AD association using isothermal titration calorimetry.[39] We find a clear separation of different cosolute families in the resulting entropy–enthalpy plots, whereby the distinct groups tend to occupy different regimes of the plot, Fig. 5.

The diagonal line in Fig. 5, representing full enthalpic–entropic compensation, separates stabilizing cosolutes that lie above it (blue area highlighted in the inset of Fig. 5a) from those causing destabilization lying beneath it (pink area). The stabilizing region can be further divided into two specific regions corresponding to different underlying thermodynamic effects. Region I includes solutes showing an enthalpically driven stabilization, while cosolutes that fall within region II induce increased stability that is primarily entropically driven. Similarly, regions III and IV indicate an enthalpic and entropic destabilizing effect, respectively. For the CD/AD complexation process (discussed in section 2), all salts tested stabilize the complex *via* an entropically dominated process (region II) at low concentrations, while at higher concentrations the effect becomes more enthalpically driven (region I), Fig. 5a.

For the peptide folding process, the stabilizing cosolutes tested include polyol osmolytes as well as the larger, polymeric crowders PEG and dextran, Fig. 5b. Polymers span much of the upper left area of the plot, moving from an entropically dominated mechanism at lower concentrations (region II) to a larger enthalpic contribution to their stabilizing effect (region I) at higher concentrations. All polyols, by contrast, fall close to a single line, which represents an enthalpic stabilizing effect (region I). This effect is generally weaker than for the polymers at a given concentration, seen as a shorter distance of data points from the diagonal. This difference in thermodynamic fingerprint is surprising, as there is no hint to this apparent difference in mechanism in the data shown in Fig. 3, where we find the same trend and scaling of $\Delta\Delta G$ with cosolute molecular size for both cosolute types,

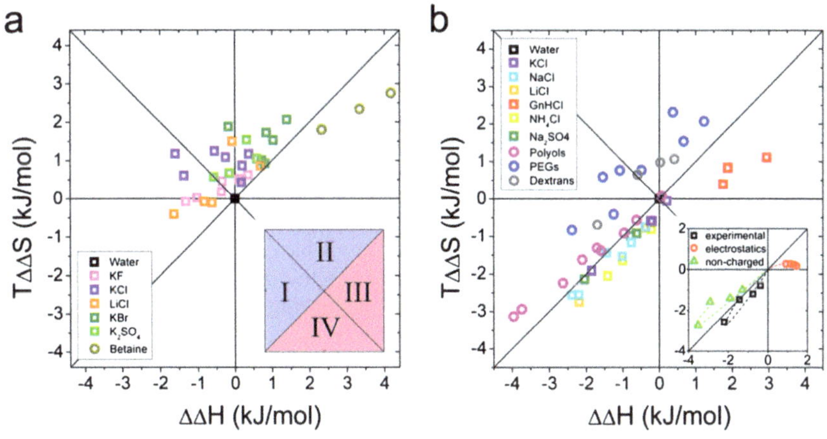

Fig. 5 Entropy *versus* enthalpy plots for changes in macromolecular processes upon cosolute addition. (**a**) CD/AD complexation. Inset shows the different corresponding regimes on the plot, and the blue and red colors signify stabilizing and destabilizing regions, respectively; see text for details. (**b**) Folding of a peptide. Inset shows the model enthalpic and entropic contributions to charged (PB) and non-charged terms of electrolytes on peptide folding.

The disparities seen in the thermodynamic fingerprint bespeak mechanistic differences in cosolute stabilizing effect for the polyol osmolytes *versus* the polymeric crowders. We have previously traced the enthalpically driven stabilization induced by osmolytes to the reordering of the hydrogen bond network in aqueous solution in the presence of the polyols.[31,69] Polymeric crowders, on the other hand, have long been shown to act in an entropically driven fashion through steric, excluded volume interactions,[59] as also described in section 1. Interestingly, recent evidence shows that at high concentration the mechanism of action may change for these crowder molecules. Record and coworkers have found that while PEG induces a stabilizing effect on DNA duplexation at low cosolute concentration, as predicted by molecular crowding theory, high concentrations show a destabilizing trend that has been attributed to the formation of a "mesh" of polymer molecules which comes into direct interaction with the DNA surface.[70] We propose that a similar balance of forces may explain the shift we find from entropically to enthalpically driven peptide folding at intermediate polymer concentrations.

The denaturing cosolute regime occupies the area below the full-compensation diagonal (regions III and IV). Here we find salts that tend to have an overall destabilizing effect on the model peptide (discussed in section 4). Over the entire concentration range tested, all salts, except GuHCl, promote entropically driven destabilization (region IV). GuHCl represents a special electrolytic cosolute since it bears a charge but also shows strong preferential inclusion to peptides and proteins that are not purely electrostatic.[71] We find that GuHCl shows a strong enthalpic denaturing effect, probably due to appreciable binding to the peptide, and accordingly its thermodynamic fingerprint lies in region III.

The model calculations we have performed (detailed in section 4) dissect the effect of salts into contributions from electrostatics and preferential (or ion-specific) interactions, $\Delta\Delta G_{\mathrm{exp}} = \Delta\Delta G_{\mathrm{PB}} + \Delta\Delta G_{\mathrm{m}}$. It is, therefore, possible to further decompose these variations in free energy into enthalpic and entropic contributions, by requiring that the contributions too must sum up to the experimental total changes in entropies and enthalpies. Thus, we have derived the enthalpic and entropic contributions to the PB free energy. We then derived the corresponding contributions to $\Delta\Delta G_{\mathrm{m}}$, the transfer free energy of the uncharged surfaces for NaCl. Because the sum of PB and preferential interaction contributions (Fig. 4, dashed lines) does not completely overlap the experimental data, the values obtained from this breakdown should only be considered as semi-quantitative. Nonetheless, the trends that appear using this analysis are rather insensitive to the quality of the fit seen in Fig. 4.

We find that the free energy contribution to the interaction of ions with the charged surfaces, $\Delta\Delta G_{\mathrm{PB}}$, is enthalpically destabilizing (region III), while the interaction with the non-charged surface $\Delta\Delta G_{\mathrm{m}}$ falls in the same region as the polyol osmolytes, and is enthalpically stabilizing (region I), Fig. 5 inset. While the interaction of salts with both the charged and non-charged surfaces of the protein is enthalpically driven, the two contributions cancel each other to a large extent, and the sum of these interactions creates an overall destabilizing, yet entropically dominated change in folding free energy, as seen in Fig. 5b.

Comparing Fig. 5b with its inset suggests a possible common mechanism for the non-electrostatic salt–peptide interaction and for the polyol–peptide interactions. For both, the thermodynamic fingerprint is stabilizing and localized in region I. This resonates with the "kosmotropic" nature of some salts – a term used to describe the order-inducing tendencies of salts in water.[14,13] Indeed, our simulations of peptide folding have pointed to the strengthening of water's hydrogen-bonds in the vicinity of polyol cosolutes as the source for peptide stabilization.[31] Possibly, both polyol osmolytes and salts that interact with non-polar surfaces act through similar cosolute-induced "water order" or water structuring.

Interestingly, while the salt induced CD/AD complex stabilization is entropically driven, Fig. 5a, the calculated non-PB contribution of the same salts to the peptide folding process is enthalpically driven, Fig. 5b (inset). This difference could be reconciled in at least two possible ways. First, the processes of peptide folding and CD/AD complexation involve changes in exposed surface area carrying different (chemical) propensities that could lead to different mechanisms. Alternatively, simple electrostatic PB calculations cannot account for ion-specific effects, so that more accurate approximations are possibly required. These could change the estimated interaction free energy with charged residues in the presence of different salts.

6 Concluding remarks

Control of cosolute content and concentrations is an integral means of cellular regulation and adaptation. While typically lacking strong, specific interactions, these cosolutes are nonetheless able to induce a measurable effect on the stability of many biological molecules and impact the way they interact. Here we have compared the action of several families of cosolutes: sugars and polyols, polymers, and salts, by measuring their effects on distinct macromolecular processes such as association and folding. Following the unique enthalpic–entropic fingerprint of each cosolute, we have found that certain cosolutes act through thermodynamically similar mechanisms that correlate with their chemical and physical characteristics. This fingerprinting allows to naturally separate cosolutes into families according to their common action, and sheds new light on the possible underlying mechanisms of cosolute effect. We suggest that similar analysis could be helpful not only in resolving the behavior of additional cosolutes, but also in formulating and testing predictive theories of their impact on macromolecular processes. Beyond thermodynamic changes in stability, we have recently shown that polyols and PEG crowders have disparate effects also on the kinetics of peptide amyloid aggregation.[56,72] Different salts, too, can differently influence kinetic steps (such as association *versus* dissociation) in molecular complexation.[43,73] We propose that there could be a possible mechanistic link between these kinetic changes and the corresponding thermodynamic mechanism we find for the different cosolute families. These types of links may eventually reshape our understanding of the way solvating environments are actively involved in numerous biochemical processes.

Acknowledgements

We thank D. Avnir for use of the circular dichroism spectrometer. DH acknowledges support from the Israel Science Foundation (ISF grants 1011/07, 1012/07). The Fritz Haber Center is supported by the Minerva Foundation, Munich, Germany.

References

1 R. J. Ellis and A. P. Minton, *Nature*, 2003, **425**, 27–28.
2 S. N. Timasheff, *Proc. Natl. Acad. Sci. U. S. A.*, 2002, **99**, 9721–9726.
3 V. A. Parsegian, *Int. Rev. Cytol.*, 2002, **215**, 1–31.
4 P. H. Yancey, M. E. Clark, S. C. Hand, R. D. Bowlus and G. N. Somero, *Science*, 1982, **217**, 1214–1222.
5 G. N. Somero, C. B. Osmond, and C. L. Bolis, *Water and life: comparative analysis of water relationships at the organismic, cellular, and molecular levels*, Springer-Verlag, 1992.
6 D. Harries and J. Rösgen, *Methods Cell Biol.*, 2008, **84**, 679–735.
7 Z. Ignatova and L. M. Gierasch, *Proc. Natl. Acad. Sci. U. S. A.*, 2006, **103**, 13357–13361.
8 B. W. Berger, C. M. Gendron, A. M. Lenhoff and E. W. Kaler, *Protein Sci.*, 2006, **15**, 2682–2696.
9 A. Nayak, C.-C. Lee, G. J. McRae and G. Belfort, *Biotechnol. Prog.*, 2009, **25**, 1508–1514.
10 R. Sousa, *Acta Crystallogr., Sect. D: Biol. Crystallogr.*, 1995, **51**, 271–277.
11 A. Kumar, P. Attri and P. Venkatesu, *Thermochim. Acta*, 2012, **536**, 55–62.
12 F. Hofmeister, *Arch. Pharmacol.*, 1888, **24**, 247–260.
13 K. D. Collins and M. W. Washabaugh, *Q. Rev. Biophys.*, 1985, **18**, 323–422.
14 W. Kunz, *Specific Ion Effects*, 1st edn, 2009.

15 H. I. Petrache, S. Tristram-Nagle, D. Harries, N. Kucerka, J. F. Nagle and V. A. Parsegian, *J. Lipid Res.*, 2006, **47**, 302–309.
16 H. I. Petrache, *Proc. Natl. Acad. Sci. U. S. A.*, 2006, **103**, 7982–7987.
17 L. M. Pegram and M. T. Record, *J. Phys. Chem. B*, 2007, **111**, 5411–5417.
18 R. L. Baldwin, *Biophys. J.*, 1996, **71**, 2056–2063.
19 P. Jungwirth and D. J. Tobias, *Chem. Rev.*, 2005, **106**, 1259–1281.
20 W. Kunz, P. Lo Nostro and B. W. Ninham, *Curr. Opin. Colloid Interface Sci.*, 2004, **9**, 1–18.
21 L. M. Pegram and M. T. Record, *J. Phys. Chem. B*, 2008, **112**, 9428–9436.
22 J. Rösgen, B. M. Pettitt, D. W. Bolen and J. Rosgen, *Biophys. J.*, 2005, **89**, 2988–2997.
23 D. Shukla, C. P. Schneider and B. L. Trout, *Adv. Drug Delivery Rev.*, 2011, **63**, 1074–1085.
24 M. H. Priya, H. S. Ashbaugh and M. E. Paulaitis, *J. Phys. Chem. B*, 2011, **115**, 13633–13642.
25 R. Chitra and P. E. Smith, *J. Phys. Chem. B*, 2001, **105**, 11513–11522.
26 T. Ghosh, A. Kalra and S. Garde, *J. Phys. Chem. B*, 2005, **109**, 642–651.
27 W. Humphrey, A. Dalke and K. Schulten, *J. Mol. Graphics*, 1996, **14**, 33–38.
28 M. T. Record, C. F. Anderson and T. M. Lohman, *Q. Rev. Biophys.*, 1978, **11**, 103–178.
29 R. Politi and D. Harries, *Chem. Commun.*, 2010, **46**, 6449–6451.
30 P. H. Yancey, *Integr. Comp. Biol.*, 2001, **41**, 699–709.
31 R. Gilman-Politi and D. Harries, *J. Chem. Theory Comput.*, 2011, 3816–3828.
32 B. C. Gibb, *Isr. J. Chem.*, 2011, **51**, 798–806.
33 M. V. Rekharsky and Y. Inoue, *Chem. Rev.*, 1998, **98**, 1875–1918.
34 K. C. K. C. Aune and C. Tanford, *Biochemistry*, 1969, **8**, 4586–4590.
35 M. Auton and D. W. Bolen, *Proc. Natl. Acad. Sci. U. S. A.*, 2005, **102**, 15065–15068.
36 M. W. Capp, L. M. Pegram, R. M. Saecker, M. Kratz, D. Riccardi, T. Wendorff, J. G. Cannon and M. T. Record, *Biochemistry*, 2009, **48**, 10372–10379.
37 R. Komsa-Penkova, R. Koynova, G. Kostov and B. G. Tenchov, *Biochim. Biophys. Acta, Protein Struct. Mol. Enzymol.*, 1996, **1297**, 171–181.
38 W. Kunz, *Pure Appl. Chem.*, 2006, **78**, 1611–1617.
39 D. Harries, D. C. Rau and V. A. Parsegian, *J. Am. Chem. Soc.*, 2005, 2184–2190.
40 Z. Sir, L. Strnadova and V. Rod, *Collect. Czech. Chem. Commun.*, 1980, **45**, 679.
41 G. Perron, D. Joly, J. E. Desnoyers, L. Avédikian and J.-P. Morel, *Can. J. Chem.*, 1978, **56**, 552–559.
42 J. Gibbs, in *The Scientific Papers of J. Willard Gibbs, Vol.1*, ed. H. Bumstead and R. van Name, Ox Bow, Woodbridge, CT, 1993.
43 P. A. Gurnev, D. Harries, V. A. Parsegian and S. M. Bezrukov, *ChemPhysChem*, 2009, **10**, 1445–1449.
44 L. M. Pegram, T. Wendorff, R. Erdmann, I. Shkel, D. Bellissimo, D. J. Felitsky and M. T. Record, *Proc. Natl. Acad. Sci. U. S. A.*, 2010, **107**, 7716–7721.
45 J. C. Phillips, R. Braun, W. Wang, J. Gumbart, E. Tajkhorshid, E. Villa, C. Chipot, R. D. Skeel, L. Kalé and K. Schulten, *J. Comput. Chem.*, 2005, **26**, 1781–1802.
46 M. Kuttel, J. W. Brady and K. J. Naidoo, *J. Comput. Chem.*, 2002, **23**, 1236–1243.
47 L. Sapir and D. Harries, *J. Phys. Chem. B*, 2011, **115**, 624–634.
48 G. D. Rose, A. R. Geselowitz, G. J. Lesser, R. H. Lee and M. H. Zehfus, *Science*, 1985, **229**, 834–838.
49 B. Lee and F. M. Richards, *J. Mol. Biol.*, 1971, **55**, 379–IN4.
50 R. A. Robinson and R. H. Stokes, *Electrolyte solutions*, Dover Publications, 2002.
51 B. Wishaw and R. Stokes, *Trans. Faraday Soc.*, 1953, **49**, 27–31.
52 D. J. Davis, C. Burlak and N. P. Money, *Mycol. Res.*, 2000, **104**, 800–804.
53 E. Courtenay, M. Capp, R. Saecker and M. Record Jr., *Proteins: Struct., Funct., Genet.*, 2000, **4**, 72–85.
54 A. J. Maynard, G. J. Sharman and M. S. Searle, *J. Am. Chem. Soc.*, 1998, **120**, 1996–2007.
55 M. S. Cheung, D. Klimov and D. Thirumalai, *Proc. Natl. Acad. Sci. U. S. A.*, 2005, **102**, 4753–4758.
56 S. Sukenik, R. Politi, L. Ziserman, D. Danino, A. Friedler and D. Harries, *PLoS One*, 2011, **6**, e15608.
57 N. Kozer and G. Schreiber, *J. Mol. Biol.*, 2004, **336**, 763–774.
58 D. M. Hatters, A. P. Minton and G. J. Howlett, *J. Biol. Chem.*, 2002, **277**, 7824–7830.
59 D. Hall and A. P. Minton, *Biochim. Biophys. Acta, Proteins Proteomics*, 2003, **1649**, 127–139.
60 M. Auton and D. W. Bolen, *Biochemistry*, 2004, **43**, 1329–1342.
61 J. C. Lee and S. N. Timasheff, *Biochemistry*, 1974, **13**, 257–265.
62 N. A. Baker, D. Sept, S. Joseph, M. J. Holst and J. a. McCammon, *Proc. Natl. Acad. Sci. U. S. A.*, 2001, **98**, 10037–10041.
63 D. Ben-Yaakov, D. Andelman, D. Harries and R. Podgornik, *J. Phys.: Condens. Matter*, 2009, **21**, 424106.

64 D. Ben-Yaakov, D. Andelman, R. Podgornik and D. Harries, *Curr. Opin. Colloid Interface Sci.*, 2011, **16**, 542–550.
65 A. P. dos Santos and Y. Levin, *Phys. Rev. Lett.*, 2011, **106**, 167801.
66 M. Lund, R. Vacha and P. Jungwirth, *Langmuir*, 2008, **24**, 3387–3391.
67 A. E. Garcia, H. Herce, and D. Paschek, *Simulations of Temperature and Pressure Unfolding of Peptides and Proteins with Replica Exchange Molecular Dynamics*, Elsevier, 2006, Volume 2.
68 T. Lazaridis and M. Karplus, *Proteins: Struct., Funct., Genet.*, 1999, **35**, 133–152.
69 R. Politi, L. Sapir and D. Harries, *J. Phys. Chem. A*, 2009, **113**, 7548–7555.
70 D. B. Knowles, A. S. Lacroix, N. F. Deines, I. Shkel and M. T. Record, *Proc. Natl. Acad. Sci. U. S. A.*, 2011, **108**, 12699–12704.
71 P. E. Mason, J. W. Brady, G. W. Neilson and C. E. Dempsey, *Biophys. J.*, 2007, **93**, L04–6.
72 S. Sukenik and D. Harries, *Prion*, 2012, **6**, 26–31.
73 P. A. Gurnev, D. Harries, V. A. Parsegian and S. M. Bezrukov, *J. Phys.: Condens. Matter*, 2010, **22**, 454110.

Ion interactions with non-polar and amphiphilic solutes in water

D. T. Bowron*[a] and J. L. Finney[b]

Received 11th April 2012, Accepted 27th April 2012
DOI: 10.1039/c2fd20071f

Despite its extensive use in, for example, the concentration and crystallization of proteins, the salting out phenomenon remains poorly understood, with many – sometimes contradictory – explanations found in the literature. Following our earlier work using isotope substitution neutron scattering on an aqueous tertiary butyl alcohol (TBA) with added NaCl that examined in detail the molecular-level interactions in the system, and suggested that attempts to understand salting out through ion perturbation of the non-polar hydration shell may not be appropriate, we report here on two further sets of high resolution structural experiments that detail the structural interactions between ion, solvent and amphiphile in a wider range of relevant systems. First, a set of X-ray absorption spectroscopy experiments probed the effects on the hydrophobic hydration shell of Kr of adding a range of different salts (specifically Na_2SO_4, $NaClO_4$, NaCl, $NaNO_3$ and $Mg(ClO_4)_2$). The bottom line from these experiments is that the hydration shell is essentially unaffected by the added ions, underlining further the stability of the shell to such perturbations. The second set of experiments was a further isotope substitution neutron scattering study of the molecular-level solution structures of aqueous TBA, this time for a range of different added ions (CsF, NaBr, and NaCl at three concentrations), for which we have in addition extracted equilibrium constants to quantify the relative strengths of the various interactions. From this second set of results, we can conclude again that the solvent structure is essentially unperturbed by both the various ions and the amphiphile, and also identify a number of ion-specific effects. Comparing our results with those obtained from simulations that are not constrained by the experimental data underlines how sensitive structural conclusions are to the assumed potential functions, and that drawing conclusions from simulations not constrained to fit experimental data can lead to seriously erroneous conclusions.

1 Introduction

The interaction of amphiphilic molecules with ions in solution is a fundamentally important issue for many areas of chemistry and biochemistry. For example, the addition of salts is commonly used for the concentration or crystallization of biomolecules from solution. The classical view found in many textbooks[1] explains the effect in terms of competition between the salt and protein for water of solvation. Arakawa and Timasheff's work[2] proposes that proteins are preferentially hydrated in the presence of salting out agents, while that of Chick and Martin[3] suggests that

[a]*STFC ISIS Facility, Rutherford Appleton Laboratory, Harwell-Oxford, OX11 0QX, United Kingdom. E-mail: daniel.bowron@stfc.ac.uk; Fax: +44 (0)1235 445720; Tel: +44 (0)1235 446381*

[b]*Department of Physics and Astronomy and London Centre for Nanotechnology, University College London, Gower Street, London, WC1E 6BT, United Kingdom*

the salt is excluded from the protein precipitate. In a detailed neutron scattering and modelling study[4,5] we showed that the addition of a small quantity of sodium chloride to a 0.02 mole fraction aqueous solution of tertiary butanol resulted in the formation of a simple anion bridging interaction between the hydroxyl groups of tertiary butanol molecules that inverted the preference seen in the salt free system for non-polar interactions between the amphiphilic alcohol molecules. The addition of salt thus had the tendency of exposing more of the non-polar surface of the amphiphilic alcohol to the solvent water and pointed to a possible mechanism for salting out in such systems.

Our aim in this work is to extend our earlier neutron scattering study of salt effects upon the solution structure of tertiary butanol–water solutions[4,5] to cover a wider range of monovalent ions and amphiphile concentrations. This should enable us to investigate underlying common features and behaviours in a more systematic fashion than permitted by the earlier single salt, single concentration approach.

2 Ion interactions with pure hydrophobes

Before one can properly appreciate the interaction of ions with amphiphiles in aqueous solution it is necessary to understand the structural interactions that the ions have with pure non-polar solute moieties and with other charged species. Although there is extensive information in the literature[6] on the interactions of ions with water, the former issue is arguably one of the most poorly understood mechanisms in modern liquid state physics, chemistry and biology.[7]

To establish a baseline for an understanding of ion interactions with non-polar solutes in water we have undertaken an extensive X-ray absorption spectroscopy investigation of the structure of the hydrophobic hydration shell of krypton in water,[8] as a function of a wide range of ionic co-solutes and ion concentrations. This experimental technique is a chemically specific local probe of inter-atomic structure and the measured signal contains information relating to inter-atomic distances, coordination numbers and local structural disorder from a viewpoint centred on the noble gas atoms.

Following the methodology established in our earlier investigation,[8] aqueous solutions of krypton gas were prepared in a pressure cell suitable for X-ray absorption measurements at a gas pressure of 110 bar and at room temperature, 293 K.[9] As before, a measurement of krypton in pure water (H_2O) was performed as a control. This was then followed by measurements in 0.5 M and 1 M Na_2SO_4, 0.25 M and 0.5 M $NaClO_4$, 0.5 M NaCl, 0.25 M and 0.5 M $NaNO_3$ and lastly in 0.25 M and 0.5 M $Mg(ClO_4)_2$ aqueous solutions. At the 110 bar gas pressure investigated, the concentration of Kr in water is 0.004 mole fraction;[10] this corresponds to a ratio of 1 : 250 Kr atoms to water molecules. The salt concentrations correspond to ion concentrations ranging between 1 : 111 and 1 : 28 ions to water molecules.

All measurements were performed at the ESRF general purpose X-ray absorption spectroscopy beamline BM29.[11] The storage ring was operating in 2/3 fill mode with typical ring currents in the 150 mA to 200 mA range and at an electron energy of 6 GeV. The spectrometer was equipped with a pair of Si(311) monochromator crystals operating with 40% detuning for harmonic rejection. Spectra were collected with an energy resolution of ≈ 1 eV as compared with the natural lifetime broadening of 2.7 eV for the Kr K absorption edge (14.324 keV).

Fig. 1 shows the measured X-ray absorption spectra for all systems, including that of pure water. Immediately it is clear that there is no significant difference induced by the presence of salts when compared with the signal measured for the salt free control. The insensitivity of system to ionic perturbation is further reflected in the energy stability of the measured signals when an analysis of the position of the X-ray absorption threshold is made. This value was found to be stable around a mean of 14326.4 eV, with a ± 0.25 eV variation over the complete set of experimental data – pure water and salts. The intrinsic sensitivity of the XAS technique

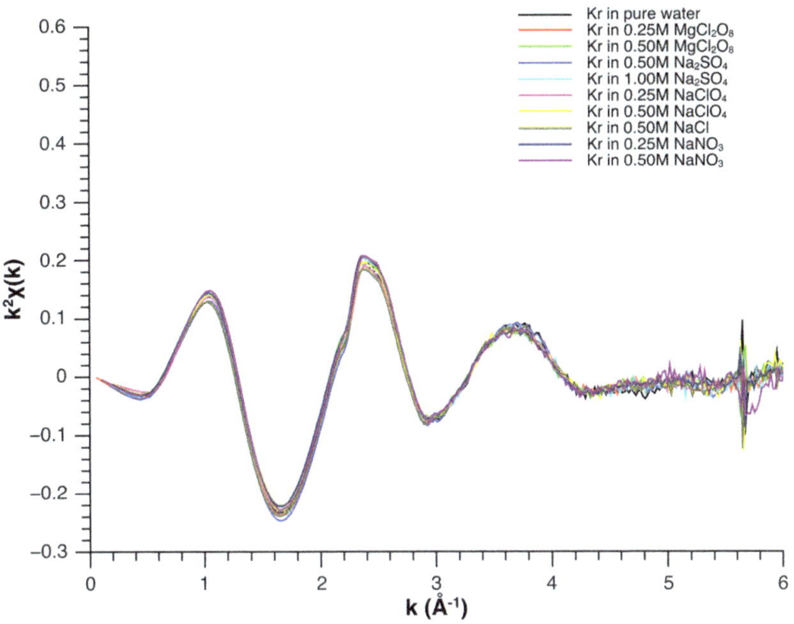

Fig. 1 Kr K-edge $k^2\chi(k)$ X-ray absorption spectra measured for solutions of Kr in water, and in 0.25 M and 0.5 M $Mg(ClO_4)_2$, 0.5 M and 1 M Na_2SO_4, 0.25 M and 0.5 M $NaClO_4$, 0.5 M NaCl, and lastly in 0.25 M and 0.5 M $NaNO_3$ aqueous solutions.

to both electronic and structural components provides considerable weight to the conclusion that the similarity between all the measured spectra arises from Kr atoms occupying identical aqueous environments regardless of salt concentration or species. The oscillatory component of the measured signals pins the structural similarities, whilst the absorption energy threshold pins the electronic environment *i.e.* the chemical composition.

At the lowest ion concentrations, it is not unreasonable to expect that the hydration shell of the krypton atoms is unaffected by the presence of ionic species as there is sufficient solvent water to fully hydrate both charged and non-polar species independently. However, at the highest ion concentrations, the resilience of the hydrophobic hydration shell of Kr to structural perturbation becomes increasingly interesting. It has been shown that a Kr atom requires twenty water molecules to form a complete hydration shell to a radial distance of 5 Å;[8] hence it could be expected that at the higher concentrations, ionic species might well become involved in this structure. The experimental evidence does not support such an interaction and suggests that structural perturbations induced by ions are restricted to the bulk solvent beyond the immediate hydrophobic hydration shell. Further, it is not inconceivable that the ionic perturbation is restricted to the region beyond the second hydration shell – the presence of ions at the second shell distance would be likely to modify the structure found in the first hydrophobic hydration shell. The latter effect may be rather subtle and difficult to detect by this technique, particularly if the perturbation is primarily linked to the water hydrogen correlations. However, the experimental results clearly indicate that the direct hydrophobic hydration shell defined by the water oxygen atoms remains essentially unperturbed by these ions at these concentrations.

3 Ion interactions with amphiphiles

Having established that structural measurements show no evidence for direct ion to pure non-polar solute interactions in aqueous solution, we turn our attention to the more complex system of ions, amphiphilic solutes and water. Aqueous solutions of tertiary butanol are an excellent model system for these investigations since numerous experimental studies of the binary mixtures have established a detailed picture of their underlying structure and behaviour at a wide range of concentrations and temperatures.[12–17]

3.1 Experiment

Neutron scattering methods are an ideal structural probe of the intermolecular correlations in these solutions since selective hydrogen/deuterium isotopic substitution can be effectively used to pick out selected aspects of the amphiphilic solute and solvent interactions.[14,15] To investigate the interaction of ions with the alcohol and water molecules in these complex mixtures, five systems have been studied. Working with molecular ratios that are both relevant to the importance of the interactions being studied and that also facilitate the isotopic substitution method, the systems were:

1. 1 tertiary butanol : 50 water : 0.5 Cs^+ : 0.5 F^-
2. 1 tertiary butanol : 50 water : 0.5 Na^+ : 0.5 Br^-
3. 1 tertiary butanol : 50 water : 0.5 Na^+ : 0.5 Cl^-
4. 2 tertiary butanol : 50 water : 0.5 Na^+ : 0.5 Cl^-
5. 3 tertiary butanol : 50 water : 0.5 Na^+ : 0.5 Cl^-

These molecular ratios correspond to ≈0.02, 0.04 and 0.06 mole fraction solutions of alcohol in water with ≈0.01 mole fraction of salt. Solution 3 is a repeat of the previously studied system.[4,5]

To maximise the structural information that can be extracted from the neutron scattering measurements, five isotopic contrasts were used for each of these systems, namely:

1. $(CD_3)_3OD$ in D_2O + salt
2. $(CD_3)_3OH$ in H_2O + salt
3. 1 : 1 $(CD_3)_3OD$–$(CD_3)_3OH$ in 1 : 1 D_2O–H_2O + salt
4. $(CH_3)_3OH$ in H_2O + salt
5. 1 : 1 $(CD_3)_3OD$–$(CH_3)_3OH$ in 1 : 1 D_2O–H_2O + salt

where solutions 1,2 and 3 weight the structural correlations to the water–water interactions, and solutions 1,4 and 5 weight the structural correlations to the water and alcohol interactions with each other.

Each sample was prepared by weight and transferred to null-scattering alloy cells of composition $Ti_{0.68}Zr_{0.32}$, that makes no structural contribution to the coherent neutron scattering signal. The cells were sealed to vacuum with a poly(tetrafluoroethylene) (PTFE) O-ring. The sample dimensions, delimited by the cell walls of 1 mm thickness, were 35 mm height × 35 mm width × 1 mm thickness. Neutron scattering measurements were performed by use of the small-angle neutron diffractometer for amorphous and liquid samples (SANDALS) at the ISIS Facility, Rutherford Appleton Laboratory, U.K. Time-of-flight neutron scattering data were collected over an angular range from 3° to 40° by use of arrays of lithium-doped ZnS scintillation detectors over neutron wavelengths in the range from 0.05 to 3.5 Å. This instrument configuration gives a corresponding Q-range of 0.2–50 $Å^{-1}$ that can be used in the structural analysis. As all the cells were completely filled and sealed at room temperature, that is, leaving no expansion volume, the atomic density used throughout the analysis was 0.1 atom $Å^{-3}$.

The data were corrected for instrument and cell background scattering, multiple scattering, and absorption by use of the Gudrun codes.[18] The data were placed on an absolute scale by normalization to the scattering of a known vanadium standard, and corrections for self-scattering and inelastic scattering were performed by the methods outlined by Soper and Luzar.[19]

For each isotopic composition the experiment measures the interference differential scattering cross section $F(Q)$, shown for the 0.02 mole fraction NaBr system in Fig. 2. $F(Q)$ is defined as

$$F(Q) = \sum_{\alpha \leq \beta} (2 - \delta_{\alpha\beta}) c_\alpha c_\beta b_\alpha b_\beta [S_{\alpha\beta}(Q) - 1] \tag{1}$$

Q is the magnitude of the momentum transfer vector of the scattering process, defined as $Q = 4\pi sin\theta/\lambda$ where λ is the wavelength of the incident neutron and θ is half the scattering angle, c_α and c_β are the concentrations of the atomic species α and, β whilst b_α and b_β are their neutron scattering lengths.[20] $S_{\alpha\beta}(Q)$ are the partial structure factors representing pairwise structural correlations between the atoms and

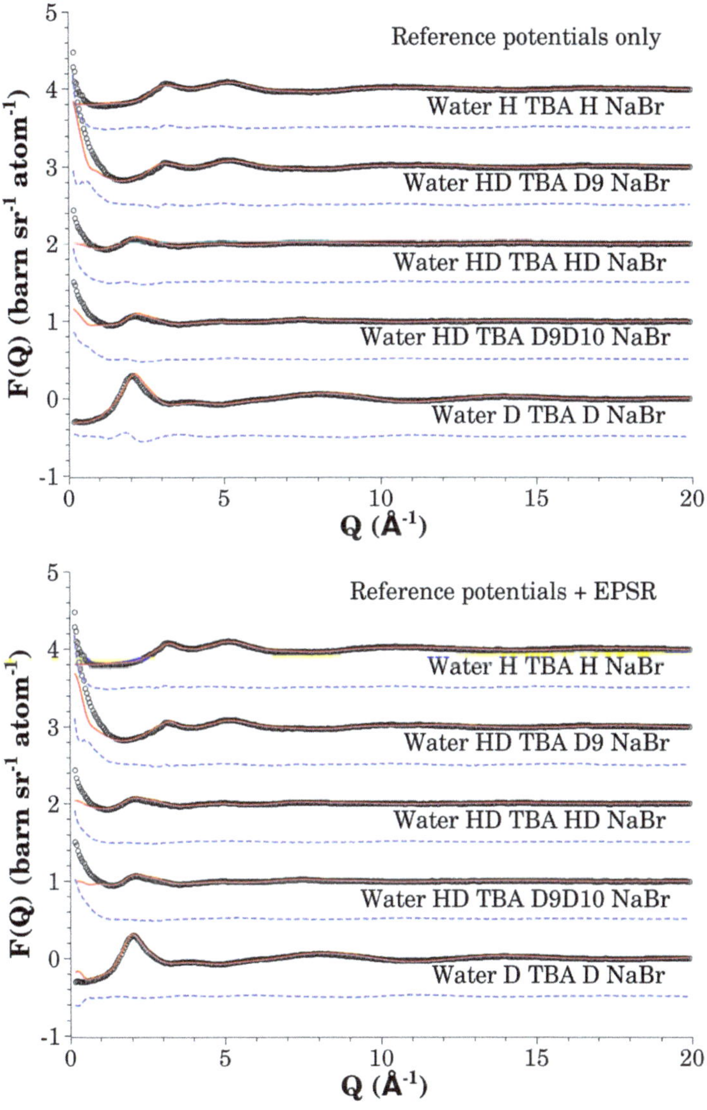

Fig. 2 Experimental data and model structure factors for the 1 tertiary butanol : 50 water : 0.5 Na^+ : 0.5 Br^- system. The top panel shows the experimental data, the model structure factors and fit residuals produced using just the reference potentials that seed the Empirical Potential Structure Refinement. The lower panel shows the experimental data, the model structure factors that are obtained following the EPSR procedure, and the final fit residuals. Note that the quality of the fits for the other investigated systems and solution concentrations (CsF and NaCl) are similar.

$\delta_{\alpha\beta}$ is the Kronecker delta function to avoid double counting the correlations between like atomic species. $S_{\alpha\beta}(Q)$ is related to the corresponding real space partial pair distribution functions, $g_{\alpha\beta}(r)$, by a Fourier transform weighted by the atomic density, ρ of the system under study.

$$[g_{\alpha\beta}(r) - 1] = \frac{1}{2\pi^2\rho}\int_0^\infty Q^2[S_{\alpha\beta}(Q) - 1]\,\frac{\sin Qr}{Qr}\,dQ \qquad (2)$$

3.2 Structure refinement

To maximise the amount of structural information that can be extracted from the complementary sets of experimental data, we have applied the technique of Empirical Potential Structure Refinement developed by Soper.[21,22] This method uses the available experimental scattering data to generate a set of interatomic perturbation potentials that are used to drive a classical Monte Carlo simulation of the system into structural configurations that agree with the driving $F(Q)$s. The modelling framework constrains the resulting three dimensional atomistic structural model to be consistent with the basic physico-chemical constraints of the system's bulk atomic density and the known conformations of the molecular units from which it is built.

Once an acceptable model that reproduces the driving experimental structure factors has been generated, it is then possible to interrogate the structure to obtain an estimate of any of the atomic partial pair distribution functions, despite the fact that only a subset are fully constrained by the experimental data. However, it is important to recognise that any pair distribution functions not heavily constrained by the scattering data will primarily reflect the prior assumptions made for the underlying interatomic interaction potentials, the packing requirements that result from the imposed basic geometries of the molecules and the requirement that the model conforms to the known bulk density of the system under study. In spite of this, provided that these three primary constraints are reasonable, the models can provide a useful guide for improving our general understanding of how the structure of the system gives rise to its observed physical and chemical properties.

The EPSR procedure begins by first using a set of Lennard-Jones plus Coulomb charge reference potentials to equilibrate a Monte Carlo simulation of the solution. These potentials are summarised in Table 1, and the quality of the reference model can be judged from the calculated structure factors shown in the top panel of Fig. 2. The water potentials are the SPC/E parameters of Berendsen *et al.*[23]

The atom labels used in the structural model are CC for the tertiary butanol central carbon, C for the tertiary butanol methyl group carbons and M for the methyl group hydrogens, O for the tertiary butanol hydroxyl group oxygen and H for the hydroxyl group hydrogen sites, OW for water molecule oxygens and HW for water molecule hydrogens, and Na, Cs, F, Cl and Br for the ionic components of the salts.

The structural simulations were performed in a cubic box under periodic boundary conditions. The 0.02 mole fraction systems were modelled using 1000 water molecules, 20 tertiary butanol molecules, 10 cations and 10 anions, in a box of side length 32.1 Å. The 0.04 and 0.06 mole fraction systems were modelled in boxes containing the proportionate increase in tertiary butanol molecules whilst maintaining the concentration of water molecules and ions as used in the 0.02 mole fraction models.

Once the simulation has equilibrated, the EPSR algorithm is turned on and the experimental data are used to derive the set of perturbation functions that will then drive the structural configurations into consistency with the scattering data. The comparison of the residuals shown in the lower panel of Fig. 2 with those in the upper panel shows the improvement obtained in the refined model. Once this

Table 1 Atom names and types, Lennard-Jones and charge parameters used in the simulation of tertiary butanol + water + salt solutions. Within the EPSR model, these are combined using the Lorenz–Berthelot mixing rules $\sigma_{\alpha\beta} = \frac{1}{2}[\sigma_\alpha + \sigma_\beta]$ and $\varepsilon_{\alpha\beta} = \sqrt{[\varepsilon_\alpha \varepsilon_\beta]}$

Atom label	Element	Neutron scattering length/fm	ε/kJ mol^{-1}	σ/Å	q/e
CC	C	6.646	0.2093	3.800	0.2650
C	C	6.646	0.6069	3.960	0.0000
M	H (or D)	−3.739 (or 6.671)	0.0000	1.350	0.0000
O	O	5.803	0.7116	3.070	−0.7000
H	H (or D)	−3.739 (or 6.671)	0.0000	1.350	0.4350
OW	O	5.803	0.6500	3.166	−0.8476
HW	H (or D)	−3.739 (or 6.671)	0.0000	0.000	0.4238
Na	Na	3.63	0.1250	2.500	1.0000
Cs	Cs	5.42	0.1250	4.150	1.0000
F	F	5.654	0.5660	3.080	−1.0000
Cl	Cl	9.577	0.5660	4.191	−1.0000
Br	Br	6.795	0.5660	4.538	−1.0000

is achieved the simulation is continued and ensemble average information is accumulated from periodic snapshots of the structural configurations of the atoms and molecules. These data can include the partial distribution functions, bond angle distributions, coordination number histograms *etc.*

3.3 Structural results

Water is the major component of all the investigated solutions and consequently it is worth investigating the structure of the solvent medium that supports the ionic and amphiphilic components. Fig. 3 shows the OW–OW, OW–HW and HW–HW partial pair distribution functions that result from the EPSR refined structural configurations, whilst Fig. 4 shows the corresponding running OW–OW coordination numbers. The similarity in this function shows that regardless of salt type or amphiphile concentration, the structure of the solvent water is very similar in all these cases.

For completeness Fig. 5 shows how the EPSR model has enhanced the structure of the solvent medium beyond that delivered by the SPC/E reference potential, and this element plays a critical role in the improvement of the fit of the model structure factors to the experimental data. The lack of the characteristic second OW–OW peak in the reference model is particularly notable.

Ion hydration is a critical aspect of all electrolyte solutions and Fig. 6 shows the ion–water partial pair distributions and running coordination numbers for the three salt systems investigated. The difference in the positions of the first peak in the OW and HW partial pair distribution functions for each cation and anion reflects the strong orientational correlation that the ion charge imposes on the water molecules in the first hydration shell. The negatively polarised water oxygen atom appears to prefer to sit close the cations, whilst the anions preferentially hydrate *via* water hydrogen atoms. The running coordination numbers show that in the dilute amphiphile solutions, the sodium cation is hydrated by 5.5 water molecules on average, whilst the larger cesium cation has closer to 7 water molecules in its first hydration shell. The hydration numbers for the anions are found to be between 6 and 7, with the fluoride anion having the most cleanly defined hydration shell consisting of 6 water molecules.

As the concentration of tertiary butanol is increased in the NaCl system, there is a concomitant fall in the average hydration number of the cations and anions. This reflects the competition for hydration water between the amphiphilic alcohol molecule and the ionic species.

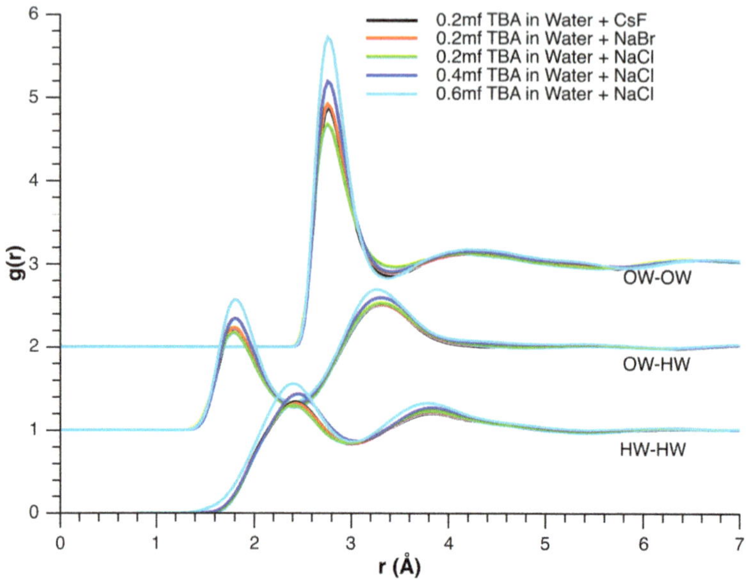

Fig. 3 EPSR refined water partial pair distribution functions ($g_{OWOW}(r)$, g_{OWHW} and $g_{HWHW}(r)$) for the five tertiary butanol + salt solutions that have been investigated. The largest differences in these functions reflect the concentration variation in the NaCl series data.

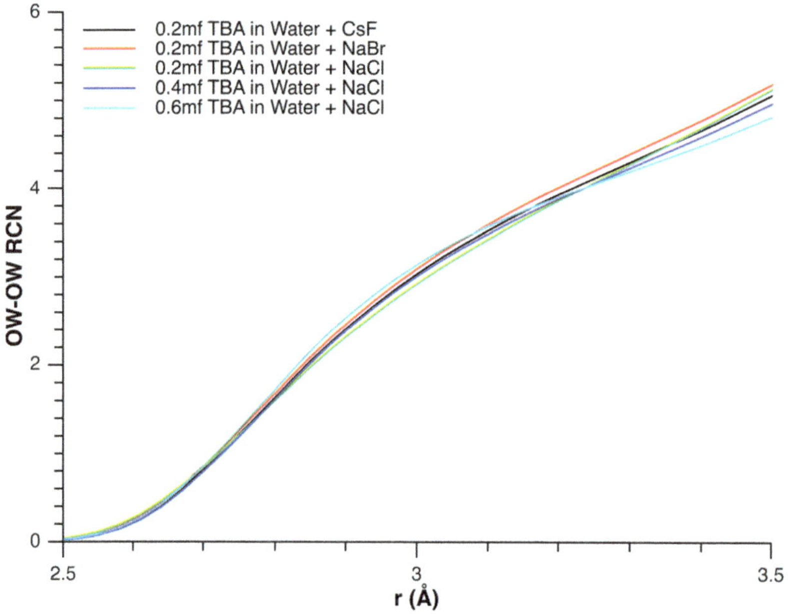

Fig. 4 EPSR refined running coordination number for $g_{OWOW}(r)$ for the five tertiary butanol + salt solutions that have been investigated.

Fig. 7 shows the partial pair distribution functions that correspond to interactions between the ions and the polar hydroxyl group on the tertiary butanol molecules. It is interesting to note that in the NaBr and NaCl systems, both the cations and anions are found in close proximity to the polar group on the alcohol, whilst the CsF system shows a distinctly different behaviour. In this system the interaction between the

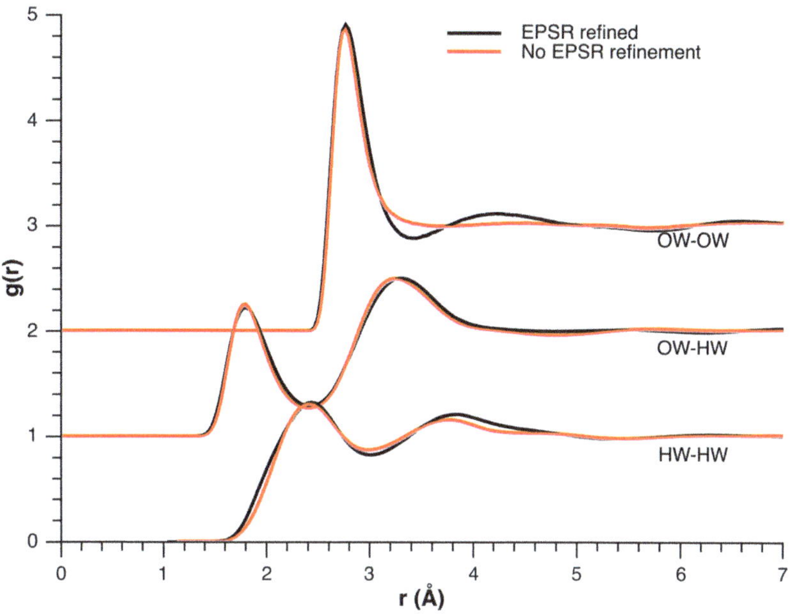

Fig. 5 Comparison of the water partial pair distribution functions ($g_{OWOW}(r)$, g_{OWHW} and $g_{HWHW}(r)$) obtained using the reference potentials alone (Table 1), and following structure refinement using the EPSR procedure.

alcohol hydroxyl group and the fluoride anion is always found to be mediated by the fluoride ion hydration shell, as reflected in the more distant (≈4 Å) correlation between the anion and the hydroxyl hydrogen sites. The cesium cation in contrast coordinates the hydroxyl oxygen atoms directly at ≈3 Å, slightly further out than found for the sodium cations due to the former's larger ionic radius.

To complete the picture of solute hydration Fig. 8 shows the alcohol centre to water oxygen partial pair distribution function $g_{CCOW}(r)$ for the five salt systems that have been investigated. The peak at 3.6 Å corresponds to the water molecules that hydrate the alcohol's hydroxyl group, whilst the feature at ≈5 Å corresponds to the water molecules that hydrate the non-polar methyl groups on the alcohol molecule.[14] The similarity between these functions (and those already in the literature for the salt-free systems, *e.g.* ref. 5 & 14) tells us that in all the investigated systems, the presence of the investigated ionic co-solutes does not significantly disrupt the hydration shells of the amphiphiles.

Fig. 9 shows the alcohol centres partial pair distribution function $g_{CCCC}(r)$. This shows the principal alcohol–alcohol interaction peak between 5.8 Å and 6.0 Å, that corresponds to direct contacts between the non-polar ends of the alcohol molecules.[14] Interestingly, only the 0.04 and 0.06 mole fraction tertiary butanol solutions show strong evidence of ion- or water-bridged interactions as reflected in the feature between 9 Å and 10 Å in this function,[4,5] although there is slight evidence for this type of correlation in the NaBr system.

3.4 Ion-amphiphile association

To gain further insight into the magnitude of the ion interactions with the amphiphilic alcohol molecules it is possible to estimate the equilibrium constants for the various ion complexes from a knowledge of the concentrations of the various species in solution.[24] For example, for the equilibrium between ions and the alcohol, we can write

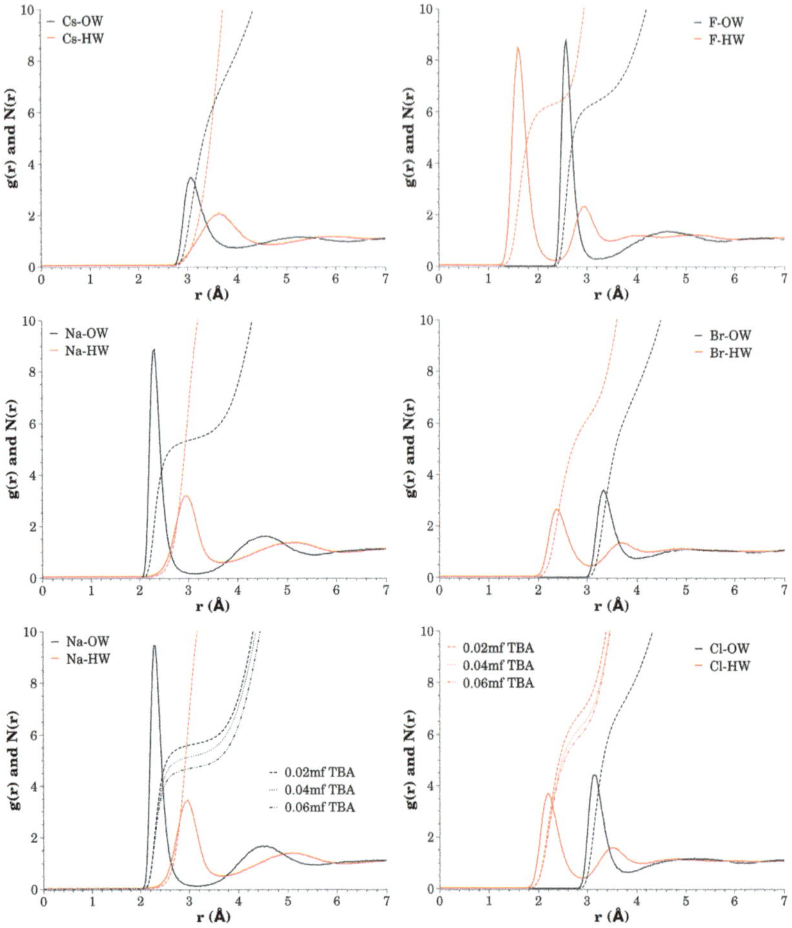

Fig. 6 Ion hydration partial pair distribution functions and running coordination numbers for the five tertiary butanol + water + salt systems investigated. For the concentration series of NaCl samples, three sets of running coordination numbers are given to highlight the structural differences. These differences are not immediately evident from the similar features found in the corresponding radial distribution functions.

$$(CH_3)_3COH_{(aq)} + \text{cation}^+_{(aq)} \rightleftharpoons \{(CH_3)_3OH\cdot\text{cation}\}^+_{(aq)} \tag{3}$$

$$(CH_3)_3COH_{(aq)} + \text{anion}^-_{(aq)} \rightleftharpoons \{(CH_3)_3OH\cdot\text{anion}\}^-_{(aq)} \tag{4}$$

and calculate the equilibrium constants as

$$k_{\text{ion},(CH_3)_3COH} = \frac{\left[\{(CH_3)_3COH\cdot\text{ion}\}^{+/-}\right]}{\left[\text{ion}^{+/-}\right]\left[(CH_3)_3COH\right]} \tag{5}$$

The square brackets indicate the equilibrium concentrations of the paired and unpaired species. These concentrations can be calculated from the known concentrations of the molecular and ionic species in the model and using the coordination histograms generated from the atomic configurations, for example see Fig. 10, which examples the data for the NaBr system. These coordination histograms use a simple distance criterion to decide whether two components are paired or not. Due to the

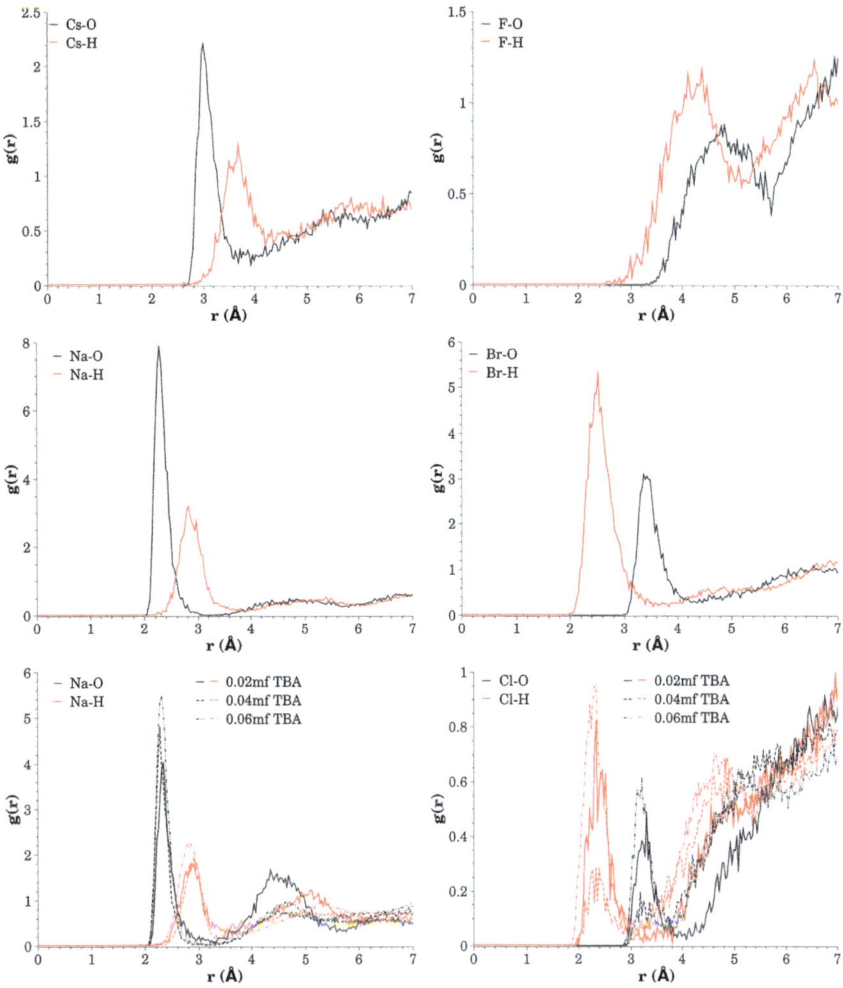

Fig. 7 Ion to tertiary butanol hydroxyl group partial pair distribution functions for the five tertiary butanol + water + salt systems investigated.

molecular nature of the alcohol in the paired configuration, two choices of pairing criterion were evaluated; the first was the distance between the cation or anion and the oxygen or hydrogen site on the alcohol hydroxyl group, and the second was the distance of the cation or anion from the tertiary-carbon located at the nominal centre of the alcohol. Differences in the equilibrium constants determined from the two sets of pairing criteria should indicate the existence of alcohol–ion pairs coordinating through the non-polar methyl groups of the amphiphile as opposed to ions coordinating through polar interactions with the alcohol hydroxyl group. The results are summarized in Table 2.

The calculations of the equilibrium constants have been performed with and without refinement against the neutron scattering data, to investigate the effect of the experimental information on the equilibrium conditions generated by the chosen sets of atomic reference potentials. The comparison shows that in the tertiary butanol solution with added CsF and the 0.02 and 0.04 mole fraction solutions with added NaCl, the effect of the empirical potentials is to generally reduce the extent of the alcohol–ion pairing present in the unrefined models. This is

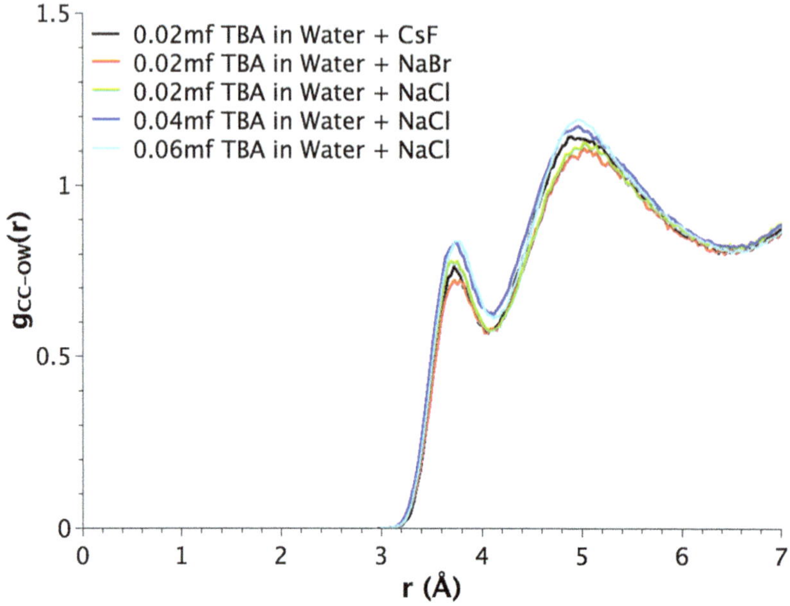

Fig. 8 Tertiary butanol central carbon to water oxygen partial pair distribution functions $g_{CCOW}(r)$ for the five tertiary butanol + water + salt systems investigated.

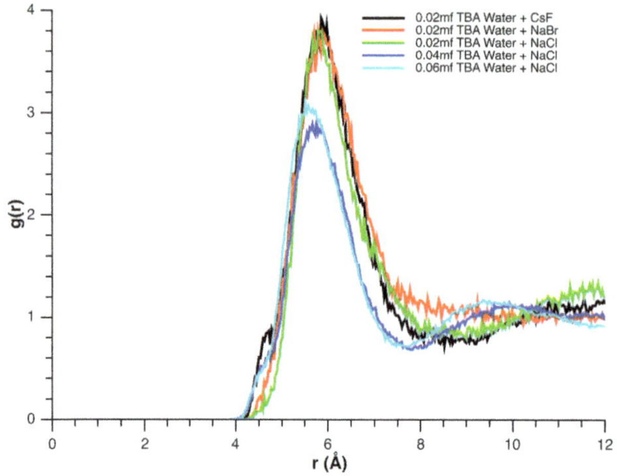

Fig. 9 Tertiary butanol to tertiary butanol partial pair distribution functions ($g_{CCCC}(r)$) for the five tertiary butanol + water + salt systems investigated.

demonstrated by the reduction of the magnitude of the equilibrium constants. The situation in the NaBr solution is similar for the cation, but inverted for the interactions between the bromide anion and the alcohol molecules where the interaction constant increases by a factor of approximately five, whilst in the highest concentration tertiary butanol solution it is the cation that shows a slightly increased preference for interacting with the alcohol molecules in the refined system and *vice-versa* for the anion.

The similarity in NaBr and NaCl systems between the tertiary-carbon centred and hydroxyl group centred equilibrium constants, tells us that the ion interactions with the alcohol molecules are effectively mediated through the polar group on the

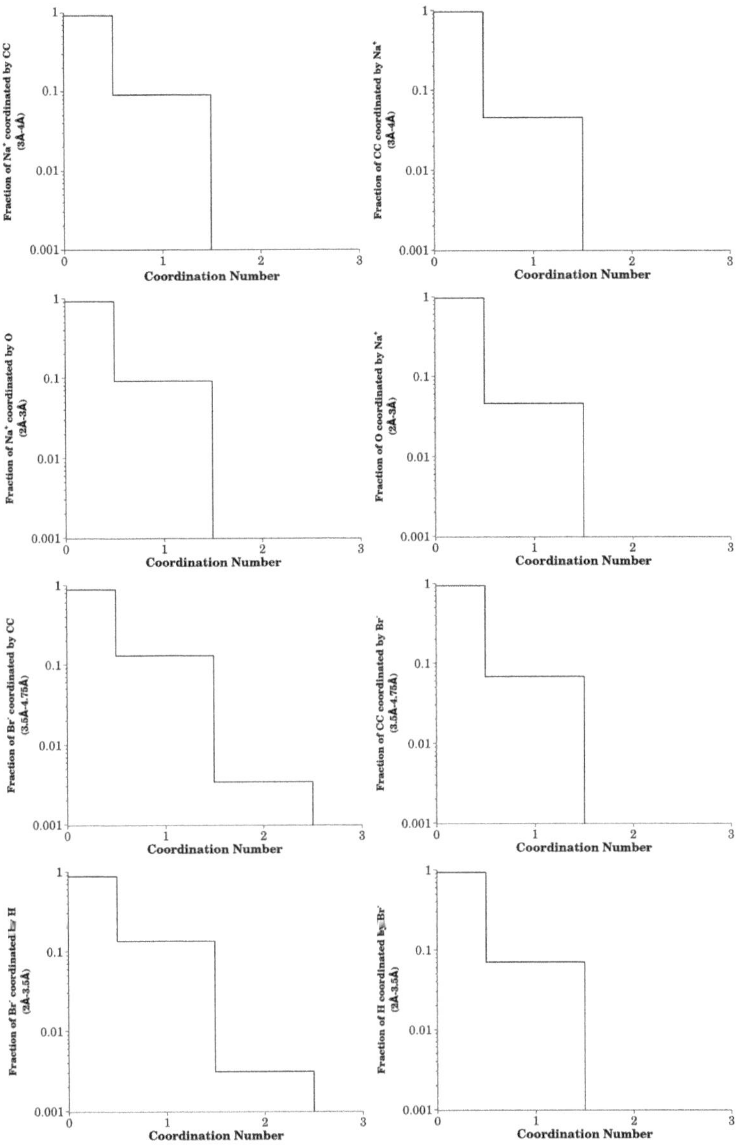

Fig. 10 Coordination number histograms calculated for the ion-alcohol interactions in the tertiary butanol + water + NaBr system. Note the logarithmic scale on the ordinate axis. The distance ranges used to determine the existence of ion–alcohol interactions are given in the axis label. Three interaction sites on the alcohol molecules are considered: CC the centre of the alcohol molecule for either cation or anion, and the O or H atoms of the alcohol hydroxyl group for the cation or anion respectively.

amphiphile. This is also the case for the cation–alcohol interactions in the CsF system; however the fluoride–alcohol equilibrium constants indicate that there is a measurable interaction for the fluoride ion with the non-polar groups of the alcohol molecules. This is most likely because this ion is very strongly hydrated (see for example Fig. 6) and so its interaction with the other components of the solution will be mediated by hydration water molecules, regardless of whether these interactions are with polar or non-polar groups.

Table 2 Equilibrium association constants calculated for ion–alcohol interactions in the models produced using the reference potentials alone (fourth column), and following refinement using the EPSR procedure (final column). The minimum and maximum separation distances used to determine whether the ion and alcohol site are coordinated, taken from the relevant minima in the partial RDFs, are given in the second and third columns

tBuOH–ion pair	r_{min}/Å	r_{max}/Å	$k_{ion,alcohol}$ no-EPSR/ mol^{-1} dm^{-3}	$k_{ion,alcohol}$ EPSR/ mol^{-1} dm^{-3}
0.02 mf TBA in water with 1 : 50 Cs^+ or F^-–water (*NB*. F^- is always found hydrated)				
Cs–CC	3.00	4.70	0.33(1)	0.10(1)
Cs–O	2.50	4.00	0.35(1)	0.09(1)
F–CC	4.25	4.60	0.34(1)	0.25(1)
F–H	3.00	4.60	0.28(1)	0.16(1)
0.02 mf TBA in water with 1 : 50 Na^+ or Br^-–water				
Na–CC	3.00	4.00	0.10(1)	0.10(1)
Na–O	2.00	3.00	0.10(1)	0.10(1)
Br–CC	3.50	4.75	0.03(1)	0.16(1)
Br–H	2.00	3.50	0.03(1)	0.17(1)
0.02 mf TBA in water with 1 : 50 Na^+ or Cl^-–water				
Na–CC	2.80	4.10	0.07(1)	0.06(1)
Na–O	2.00	3.00	0.07(1)	0.06(1)
Cl–CC	3.50	4.50	0.07(1)	0.01(1)
Cl–H	1.80	3.20	0.08(1)	0.01(1)
0.04 mf TBA in water with 1 : 50 Na^+ or Cl^-:water				
Na–CC	2.80	4.10	0.12(1)	0.06(1)
Na–O	2.00	3.00	0.12(1)	0.06(1)
Cl–CC	3.50	4.50	0.02(1)	0.01(1)
Cl–H	1.80	3.20	0.02(1)	0.01(1)
0.06 mf TBA in water with 1 : 50 Na^+ or Cl^-–water				
Na–CC	2.80	4.10	0.02(1)	0.04(1)
Na–O	2.00	3.00	0.02(1)	0.03(1)
Cl–CC	3.50	4.50	0.05(1)	0.02(1)
Cl–H	1.80	3.20	0.05(1)	0.02(1)

4 Discussion and conclusions

The results presented in the previous two sections are consistent with a picture where under the reasonably dilute solution conditions studied here, there is no direct interaction between ions and non-polar species in solution, whether they be perfect non-polar molecules such as the Kr system, or the non-polar groups on amphiphilic molecules. Moreover, none of the ions studied perturb the direct hydration shell of either non-polar moiety studied, and if there are ions present in the second hydration shell, these are likely to be anions rather than cations. The structure of the solvent is also similar in all the solutions studied, suggesting that explanations of salting out that involve significant solvent structure perturbation look unlikely. As expected, ions tend to favour interaction with polar entities, be they the solvent water molecules themselves or the polar groups on the amphiphiles. Not surprisingly, the results confirm chemical intuition that cations prefer to sit close to water or alcohol oxygens while anions preferentially interact *via* hydroxyl hydrogens. Average ion hydration shells themselves vary from 5.5 to 7 for the cations, and from 6 to 7 for the anions. There are differences in "sharpness" of the various ion hydration shells, with those of F^- and Na^+ defined most clearly. Reduction in the average hydration number of both the cations and anions as the concentration of tertiary butanol is increased in the NaCl system reflects the competition for hydration water between the amphiphilic alcohol molecule and the ionic species

Beyond these general comments, conclusions can be drawn about the interactions of specific ions in the systems examined. These are discussed below, but before looking at these in more detail we should note the significant issue raised by the comparison of association constants between the structural models obtained directly from the reference potentials (a) without and (b) with EPSR refinement against the experimental data.

These comparisons highlight a strong sensitivity of structural conclusions to small variations in the underlying assumptions that are made about atomic site interactions. Although the reference potentials lead to structures that seem to be in reasonably close agreement with experimental data (upper panel of Fig. 2 and Fig. 5), the ion–amphiphile association constants given in Table 2 highlight quite different interaction behaviours for the ionic and amphiphilic solute components when the structural models are refined to fit the constraints of the experimental data. In all cases except for the bromide anion, the reference potential models display enhanced levels of ion–amphiphile association. The addition of structural information from the experimental neutron scattering data generally drives the models to reduce these interactions.

Similar structural sensitivity to refinement against experimental data has been noted before in even simpler systems such as pure liquid amphiphiles such as TBA itself,[25,26] where the chemical nature of the dominant TBA–TBA interaction was incorrectly predicted by the unrefined reference model. Another example is that of liquid benzene, where the reference potential simulations predicted T-configurations while refining against the experimental neutron scattering data led to a clear shift to Y-configurations.[27] Such sensitivities are likely to be even more subtle and exquisite in ternary systems of weakly interacting systems such as the ion–amphiphile–water systems discussed here, underlining the need to constrain models by high resolution experimental data if viable conclusions are to be drawn about interactions and mechanisms.

In contrast to our earlier findings[4,5] this systematic study of a wider range of solutions does not support the suggestion of an enhanced interaction between anions and amphiphiles in solution. These later results demonstrate that both anions and cations can be found to preferentially coordinate with the polar hydroxyl group on the alcohol molecules, with cation interactions favoured in the NaCl system. The explanation of this (perhaps subtle) difference in conclusions concerning the ion–polar group interaction is perhaps an object lesson relating to the sensitivity of the structural model to refinement against experimental data discussed above. It is clear from Table 2 that in the NaCl salted system, models derived solely from the reference potentials have a markedly greater propensity to display anion–TBA interactions than does the refined system. This preference appears to not have been fully "refined out" in the earlier work; the fact that it has been in the current work most likely reflects the enhanced computational abilities of modern processors for the study of dilute systems. The enhanced processing resources now available allow larger structural configurations to be used and these more rapidly reach equilibrium with the underlying potential schemes. The studies performed a decade ago appear not to have reached full equilibration with the EPSR refined interaction potentials and thus would be biased in favour of enhanced anion–alcohol interactions prescribed by the reference potentials.

Returning to specific ion effects revealed in this work, the behaviour of the fluoride ions in the alcohol–water solutions appears to be particularly interesting. Their greater distance correlations with the hydroxyl group atoms (Fig. 7) tell us that this ion remains fully hydrated in solution and that its hydration water shell mediates the interactions it has with the amphiphile. This propensity for hydration means that this anion does appear to be able to interact with the non-polar end of the alcohol, *via* its hydration shell, as well as the polar hydroxyl group (as indicated by the difference between the CC and H association constants given in Table 2). The cesium cation, being much less strongly hydrated, can interact directly with the polar alcohol end of the amphiphile.

In the sodium salt systems, the chloride and bromide ions are also more weakly hydrated than the fluoride ions and can directly interact with the amphiphiles. However, in contrast to its smaller cousin, the bromide ion displays a greater preference for interaction with the alcohol molecules and this dominates over the interaction between the alcohol and the sodium ions.

With respect to interactions between the amphiphiles, as in the earlier NaCl study we see direct interactions through the non-polar headgroups. With respect to amphiphile association through the polar tails, the higher concentration NaCl solutions display the strongest evidence for the formation of ion- or water-bridged alcohol–alcohol interactions (Fig. 9), suggesting that solution crowding may become an important issue at concentrations of 0.04 mole fraction and above.

The experimentally-consistent models obtained in these experiments can be mined further for specific structural interaction information that could throw further light on the molecular-level processes relevant to amphiphile association (salting out) in aqueous solvent and the influence of specific salts in enhancing or suppressing such association in salted systems (the Hofmeister mechanism). The selection of results made here show the quality of information that can be obtained by careful experiment, and underlines the need to constrain simulations of such systems by high quality experimental data if the conclusions drawn are to be robust. Further experiments along similar lines are needed to explore a wider range of ionic species, concentrations (of both ions and amphiphile) and temperature.

Acknowledgements

We would like to thank the European Synchrotron Radiation Facility and the ISIS pulsed neutron facility for X-ray and neutron beam time and support facilities, under proposals CH-554 and RB620377 respectively.

References

1 H. R. Mahler and E. H. Cordes, *Biological Chemistry*, Harper and Row, New York, 1971.
2 T. Arakawa and S. N. Timasheff, *Biochemistry*, 1982, **21**, 6545–6552.
3 H. Chick and C. J. Martin, *Biochem. J.*, 1913, **7**, 380–398.
4 D. T. Bowron and J. L. Finney, *Phys. Rev. Lett.*, 2002, **89**, 215508.
5 D. T. Bowron and J. L. Finney, *J. Chem. Phys.*, 2003, **118**, 8357–8372.
6 H. Ohtaki and T. Radnai, *Chem. Rev.*, 1993, **93**, 1157–1204.
7 W. Blokzijl and J. B. F. N. Engberts, *Angew. Chem., Int. Ed. Engl.*, 1993, **32**, 1545–1579.
8 A. Filipponi, D. T. Bowron, C. Lobban and J. L. Finney, *Phys. Rev. Lett.*, 1997, **79**, 1293–1296.
9 D. T. Bowron, R. Weigel, A. Filipponi, M. A. Roberts and J. L. Finney, *Mol. Phys.*, 2001, **99**, 761–765.
10 R. P. Kennan and G. L. Pollack, *J. Chem. Phys.*, 1990, **93**, 2724–2735.
11 A. Filipponi, M. Borowski, D. T. Bowron, S. Ansell, A. D. Cicco and S. D. Panfilis, *Rev. Sci. Instrum.*, 2000, **71**, 2422–2432.
12 K. Nishikawa, Y. Kodera and T. Iijima, *J. Phys. Chem.*, 1987, **91**, 3694–3699.
13 K. Nishikawa, H. Hayashi and T. Iijima, *J. Phys. Chem.*, 1989, **93**, 6559–6565.
14 D. T. Bowron, J. L. Finney and A. K. Soper, *J. Phys. Chem. B*, 1998, **102**, 3551–3563.
15 D. T. Bowron, A. K. Soper and J. L. Finney, *J. Chem. Phys.*, 2001, **114**, 6203–6219.
16 D. T. Bowron and S. D. Moreno, *J. Chem. Phys.*, 2002, **117**, 3753–3762.
17 D. T. Bowron and J. L. Finney, *J. Phys. Chem.*, 2007, **111**, 9838–3552.
18 A. K. Soper, *GudrunN and GudrunX: programs for correcting raw neutron and X-ray diffraction data to differential scattering cross section*, Rutherford Appleton Laboratory Report, Oxfordshire, UK, 2011, vol. RAL-TR-2011-013.
19 A. K. Soper and A. Luzar, *J. Chem. Phys.*, 1992, **97**, 1320–1331.
20 V. F. Sears, *Neutron News*, 1992, **3**, 26–37.
21 A. K. Soper, *Chem. Phys.*, 1996, **202**, 295–306.
22 A. K. Soper, *Phys. Rev. B: Condens. Matter Mater. Phys.*, 2005, **72**, 104204.
23 H. J. C. Berendsen, J. R. Grigera and T. P. Straatsma, *J. Phys. Chem.*, 1987, **91**, 6269–6271.
24 F. Meersman, D. T. Bowron, A. K. Soper and M. H. J. Koch, *Phys. Chem. Chem. Phys.*, 2011, **13**, 13765–13771.
25 D. T. Bowron, J. L. Finney and A. K. Soper, *Mol. Phys.*, 1998, **93**, 531–543.
26 D. T. Bowron, J. L. Finney and A. K. Soper, *Mol. Phys.*, 1998, **94**, 249–251.
27 T. F. Headen, C. A. Howard, N. T. Skipper, M. A. Wilkinson, D. T. Bowron and A. K. Soper, *J. Am. Chem. Soc.*, 2010, **132**, 5735–5742.

Interactions between halide anions and a molecular hydrophobic interface

Blake M. Rankin, Michael D. Hands, David S. Wilcox, K. Rebecca Fega, Lyudmila V. Slipchenko and Dor Ben-Amotz*

Received 24th April 2012, Accepted 7th June 2012
DOI: 10.1039/c2fd20082a

Interactions between halide ions (fluoride and iodide) and t-butyl alcohol (TBA) dissolved in water are probed using a recently developed hydration-shell spectroscopic technique and theoretical cluster and liquid calculations. High signal-to-noise Raman spectroscopic measurements are combined with multivariate curve resolution (Raman-MCR) to reveal that while there is little interaction between aqueous fluoride ions and TBA, iodide ions break down the tetrahedral hydration-shell structure of TBA and produce a red-shift in its CH stretch frequency, in good agreement with the theoretical effective fragment potential (EFP) molecular dynamics simulations and hybrid quantum/EFP frequency calculations. The results imply that there is a significantly larger probability of finding iodide than fluoride in the first hydration shell of TBA, although the local iodide concentration is apparently not as high as in the surrounding bulk aqueous NaI solution.

1 Introduction

The affinity of ions for aqueous interfaces is at the heart of Hofmeister's original observations of the influence of salts on protein stability.[1,2] However, many fundamental questions remain regarding the interactions between various ions and proteins as well as macroscopic and molecular hydrophobic interfaces. Although it is widely accepted that ions such as Na^+ and F^- (which "salt-out" non-polar solutes) are repelled from hydrophobic interfaces, some experimental and theoretical results suggest that the opposite is the case for large polarizable anions such as I^- and SCN^- (which "salt-in" non-polar solutes). However, substantial questions remain regarding the magnitude of the latter affinity and the distribution of ions with respect to a hydrophobic interface, particularly when the interface is the hydration-shell of a dissolved hydrophobic group. Here we address the latter issue using a recently developed hydration-shell spectroscopic technique,[3–7] combined with effective fragment potential (EFP) molecular dynamics (MD) simulations and hybrid QM/EFP calculations.[8–11]

Previous experiments, including mass spectrometry,[12] X-ray photoelectron and fluorescence spectroscopies,[13,14] and nonlinear optical surface spectroscopies,[15] all suggest that large anions such as I^- have an affinity for air–water interfaces, while small ions such as F^- do not (although some surface tension[16] and gas bubble electrophoretic mobility[17] experiments are apparently inconsistent with that latter conclusion). However, none of the above experiments are able to determine definitively the degree to which I^- has an enhanced likelihood of residing precisely at the interface, as opposed to one or more hydration layers below the surface. Theoretical studies further suggest that the surface affinity of anions in liquid water and

Department of Chemistry, Purdue University, West Lafayette, Indiana, United States

hydrate clusters can be quite sensitive to both ion size[1,12,18–20] and polarizibility,[21,22] although the relative importance of ion size and polarizability remains a subject of debate.[23]

Here we focus on experimentally probing the interactions between halide anions and molecular hydrophobic hydration-shells, whose high curvature and structure might influence ion surface affinities. Previous theoretical and experimental studies relevant to this issue have reached mixed conclusions. For example, classical polarizable molecular dynamics simulations have suggested that[22a] "...heavier halides exhibit affinity to the amide group and to non-polar protein patches...resembling their behavior at the air/water interface" while a more recent combined NMR and simulation study of an uncharged elastin-like polypeptide, $(VPGVG)_{120}$, in aqueous NaSCN has refined this notion by showing that,[22b] although SCN^- interacts with backbone CH_n groups, "...hydrophobic side chains do not contribute significantly to anion binding or the corresponding salting-in behavior...".

In order to further investigate the interaction between salts and molecular hydrophobic interfaces we have performed hydration-shell spectroscopic measurements of aqueous salt solutions containing an alcohol, $(CH_3)_3COH$ (t-butyl alcohol, TBA), whose three methyl groups provide a molecular hydrophobic interfaces which we use to probe the interfacial affinity of F^- and I^- ions. Our results reveal clear differences between the influence of F^- and I^- on vibrational spectral features arising from both the CH stretch of TBA and its hydration shell. Although some of our results speak for themselves, we have attempted to further elucidate their structural significance with the aid of both *ab initio* cluster calculations, liquid MD simulations, and hybrid quantum-classical calculations performed using the effective fragment potential (EFP) methodology.[8–11] More specifically, the key question which we aim to address is whether either F^- or I^- significantly populate the first hydration-shell of TBA, and if so, how does the local ion concentration (and angular distribution) compare with that in the surrounding bulk aqueous salt solution?

The remainder of this manuscript is organized as follows. In Section 2 we describe the experimental procedures used to investigate the affinities of ions for the molecular hydrophobic hydration-shell of TBA, as well as the cluster and liquid calculations used to aid and extend the interpretation of our experimental measurements. The results are described and discussed in Section 3, and then summarized in Section 4.

2 Experimental and theoretical methods

A Raman-MCR experiments

The hydration-shell spectroscopic experimental technique employed in the present work combines high signal-to-noise Raman spectra of variable concentration solutions with multivariate curve resolution (Raman-MCR). We have previously demonstrated the capability of this method in detecting H-bonds and pi-H-bonds between water and polar solutes, ions, and aromatics, as well as the presence of dangling OH bonds in the hydration shells of hydrophobic groups.[3–7] In essence, Raman-MCR facilitates the decomposition of two-component solution spectra into pure solvent and solute-correlated (SC) contributions. The resulting SC spectra may contain features arising from both solute intramolecular vibrations and solute-induced perturbations of solvent molecules. Most of the results described in this work have been obtained using an MCR algorithm called self-modeling curve resolution (SMCR),[24] which is an analytical (as opposed to iterative) method for performing two-component MCR decompositions.

Since the present studies are concerned with interactions between species in three component mixtures containing water, ions, and TBA, it might seem that a three-component MCR algorithm would be required to decompose the measured spectra into contributions arising from each component. We have previously demonstrated that such three-component analyses (performed using MCR-ALS) can be used to extract SC spectra from reactive mixtures (such as acid–base equilibria).[7] However, in the present studies we have found that it is difficult to obtain reliable three-component MCR-ALS results, presumably because the OH stretch features that are correlated with each of the three components are all highly overlapped. Instead, we have found that more reliable results are produced when using two-component SMCR to analyze spectra obtained by varying one component concentration at a time. For example, if only the alcohol concentration is varied, while holding the salt concentration constant, then the three component mixture is reduced to an effective two-component mixture in which the solvent is a salt solution, and thus the resulting SC spectrum contains features arising from the hydration-shell of the alcohol in the corresponding salt solution. On the other hand, if the alcohol concentration is held constant and the salt concentration is varied, then the resulting SC spectra contain features arising from the hydration-shell of the ions, and thus may include features arising from both water and alcohol molecules which are perturbed by the ions. Both of the latter two-component analysis strategies have proved to be useful in the present studies, as they each provide different information pertaining the interactions between ions and TBA.

All of the Raman spectroscopic measurements described in this work were performed as described previously,[6] with the following exceptions. The power of the 514.5 nm Ar-ion laser was ~50 mW and the integration time for each spectrum was 5 min. Two replicate spectra were collected from each sample and both were used in the MCR analysis of similar pairs of spectra obtained from solutions in which the concentration of one of the components is varied. The measurements were carried out using 1200 gr/mm grating, which produces a dispersion of approximately 1 cm^{-1} per pixel in the OH Raman band spectral region. Raman peak positions were obtained with an accuracy of better than ± 0.5 cm^{-1} by fitting the region near the top of the peak of interest to a Gaussian function.

Fig. 1 contains results which illustrate how Raman-MCR works when applied to salt-free mixtures containing only TBA and water, with either unpolarized (V+H) or depolarized (H) input Raman spectra. Fig. 1A shows the unpolarized Raman spectra of pure water and a 0.5 M aqueous TBA solution. The most obvious difference between these spectra is the presence of the CH stretch bands between 2850 cm^{-1} and 3000 cm^{-1}. These spectra (including replicate measurements) were analyzed using SMCR to obtain the pure water and SC components corresponding to the solid curves in Fig. 1B. Essentially identical results are obtained using MCR-ALS, as well as when using input solution spectra with several different TBA concentrations between 0.1 M and 0.5 M.[7] The dashed curve in Fig. 1B shows depolarized SC spectra (obtained from depolarized input Raman spectra of water and 0.5 M TBA). The difference between the solid and dashed SC curves in Fig. 1B indicates that the lower frequency SC OH stretch (near 3200 cm^{-1}) is more highly polarized than the remaining (higher frequency) SC OH features. Previous theoretical studies of the polarized Raman spectra of pure water have found that highly tetrahedral water structures give rise to strongly polarized low frequency OH stretch Raman scattering.[25–27] Thus, the appearance of a highly polarized low frequency band in the SC spectrum of the hydration-shell of TBA strongly suggests that there is greater tetrahedral order in the hydration-shell of TBA than in bulk water (at 20 °C). In other words, if the hydration-shell of TBA had the same degree of tetrahedral order as the surrounding bulk water, then this polarized low frequency OH band would not appear in the SC spectrum of TBA.

The only pre-processing which was done to the input spectra in order to obtain the results shown in Fig. 1 (as well as the other Raman-MCR results described in this work) is to introduce a slight vertical offset to the solution spectra (but not the pure water spectra), so as to obtain properly converged minimum area SC spectra, which do not dip below the baseline at any point, but reach the baseline at some point within the spectral range within which the solvent spectrum has a finite intensity (above the baseline).

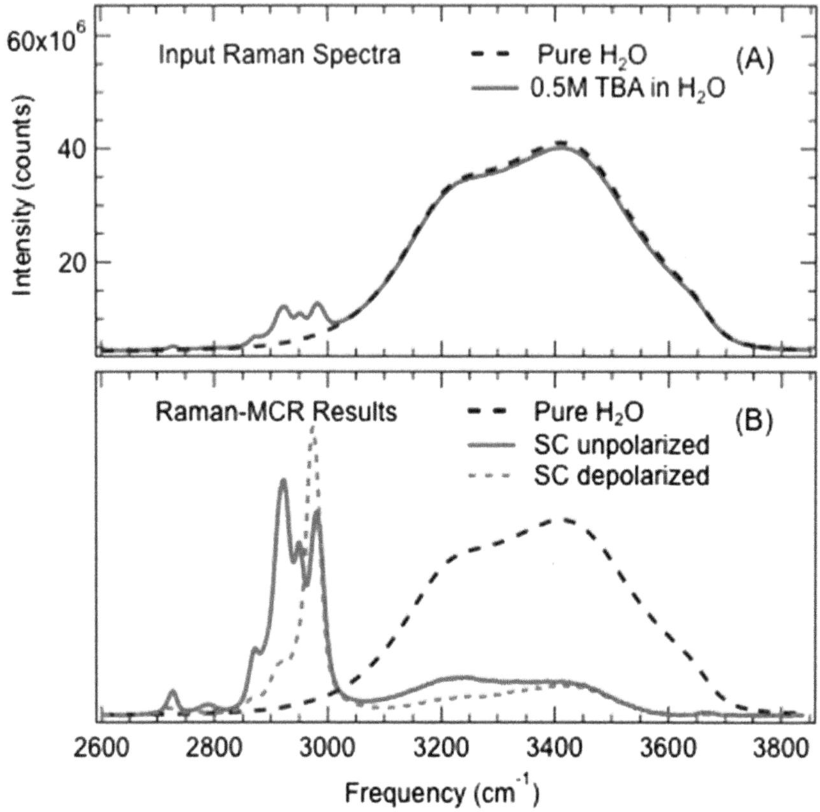

Fig. 1 (A) The input Raman spectra of pure water and 0.5 M TBA in water are compared with (B) the pure water and both the unpolarized and depolarized TBA solute-correlated (SC) spectra obtained using SMCR. The intensities of the spectra in (A) represent the number of counts obtained at each CCD pixel in 5 min of signal averaging, using a 50 mW 514.5 nm excitation laser. The two SC curves in (B) are scaled to the same amplitude in the high frequency region (and the pure water component is scaled to approximately the same height as the SC spectra).

B Theoretical methods and computational details

Water–TBA–ion mixtures were simulated with the general effective fragment potential (EFP) method[8–11] and with the hybrid QM/EFP1 approach. EFP is a quantum mechanics based potential that provides a computationally inexpensive way of modeling intermolecular interactions in non-covalently bound systems. General EFP (originally called EFP2) has four interaction terms, each of which may be thought of as a truncated expansion: Coulomb (electrostatic), induction (polarization), exchange repulsion, and dispersion. An optional charge-transfer term has also been developed.[28] The original implementation of EFP (called EFP1) was specifically designed for describing interactions in water.[8,9,29]

Previously, EFP has been successfully applied for investigation of the non-covalent interactions in small molecular clusters for which EFP was shown to provide accurate results as compared to *ab initio* methods.[30–36] Recently, we employed general EFP to characterize structure and hydrogen bonding in water–TBA solutions.[37] EFP has also been a very useful tool in hybrid quantum mechanics QM/EFP investigations of chemical reactions and solvatochromic effects.[38–46]

EFP implementation has been described elsewhere;[8–11] here we provide computational details specific to the present study. Molecular dynamics (MD) simulations of water–TBA–ion mixtures were performed with general EFP composed of Coulomb, polarization, dispersion, and exchange-repulsion terms. To account for short-range charge-penetration effects and avoid polarization collapse, Coulomb, polarization, and dispersion interactions were moderated by screening terms.[47–49] Overlap-based electrostatic and dispersion screenings and Gaussian-type polarization screening functions were employed. The polarization screening parameter for Na^+ and I^- ions was set to 0.1; it was kept at a default value of 0.6 for F^-, water, and TBA. Effective potentials for water, TBA, and ions were constructed using the mixed-basis approach, as previously described.[37] Water and TBA potentials were the same as those used previously;[37] multipoles for the Coulomb term were obtained in the 6-31+G* basis,[50–52] while the 6-311++G(3df,2p) basis[53,54] was used for the other terms in water and TBA potentials. 6-31+G* and aug-cc-pVQZ[55,56] basis sets were used for preparing ion potentials.

MD simulations on systems containing 1 TBA molecule, 5 halide and 5 Na^+ ions, and 100 water molecules were performed in a periodic cubic box with dimensions of 14.5 Å and 15.5 Å for F^- and I^- containing mixtures, respectively (corresponding to ~0.55 M TBA added to ~2.7 M aqueous salt solutions). After 10 ps equilibration, five independent 60 ps production runs for NaI solutions and five 100 ps runs for NaF solutions were performed. The results of the independent simulations were averaged. At long range, Coulomb interactions were treated with Ewald summations for charge–charge, charge–dipole, charge–quadrupole, and dipole–dipole terms. Switching functions were employed for other EFP terms.[57] The equations of motion were integrated using the velocity Verlet algorithm with a time step of 0.5 fs. Simulations were performed in the NVT ensemble at 300 K; the temperature was maintained constant using the Nosé–Hoover chain algorithm.

Radial distribution functions (RDF) and angular distributions of ions near TBA and statistics of the water orientations around TBA were gathered from these MD simulations. Additionally, 100 snapshots (collected at random) were sampled for QM/EFP frequency calculations.

The following strategy was employed for evaluating frequency shifts of the CH stretching vibrations of TBA due to the presence of iodide. Sampled structures from EFP/MD simulations were prepared such that TBA was placed in the middle of the simulation box, surrounded by water and ions. Periodic boundary conditions were switched off. The TBA frequencies were calculated using QM/EFP1 and the partial Hessian approach. That is, TBA and ions were comprised in the quantum part and described with the second-order Møller–Plesset perturbation theory using the 6-311++G**[53,54] basis set for TBA and the Hay–Wadt effective core potential for ions.[58] Waters were described by EFP1 water potential. Since only the first derivatives (gradients) are available analytically for QM/EFP1, the second derivatives were calculated numerically. Namely, the Hessian matrix for TBA normal modes was constructed through a set of independent QM/EFP1 gradient calculations with TBA coordinates shifted in both positive and negative directions by 0.005 Å along each x, y, z direction. Positions of ions and waters were kept fixed in these calculations. Vibrational frequencies of TBA were obtained though diagonalization of this partial Hessian matrix. The accuracy of this procedure was tested on small water–TBA clusters by comparing full quantum and partial Hessian QM/EFP1 approaches. Discrepancies in TBA frequencies did not exceed 1 cm^{-1}. To estimate TBA frequency shifts due to a presence of iodide, the averaged TBA frequencies in a pure (salt-free) TBA–water solution were compared to averaged TBA frequencies with iodide ions present. Calculations on 10 structures were performed for averaging the pure-water TBA frequencies and 100 snapshots were used for averaging the frequencies in salt solutions. For salt solutions, separate statistics were gathered for randomly selected snapshots (evenly distributed along MD trajectories) and for the snapshots that contained I^- in the first hydration shell of TBA.

Additionally to liquid calculations, full quantum computations on TBA–water and TBA–I^- dimers were performed. Optimized structures and vibrational frequencies of these clusters were obtained at the MP2/6-311++G** level of theory.

All calculations were performed in the GAMESS electronic structure package.[59,60]

3 Results and discussion

A Aqueous TBA in NaF control experiments

We have performed measurements of TBA in NaF solutions as control experiments, to see if our results confirm that these ions do not significantly interact with TBA. Note that NaF has a maximum aqueous solubility near 1 M, and so that is the highest salt concentration which could be used in these experiments.

Fig. 2 shows the Raman spectra collected from three aqueous solutions, each of which contain 0.5 M TBA but different concentrations of NaF (ranging from 0 M to 1 M). A separate set of spectra was collected from the same solution with no TBA. The latter spectra look very similar to those in Fig. 2, except that they do not contain the CH stretch features between 2850 cm^{-1} and 3000 cm^{-1}. In order to see whether either Na^+ or F^- have any significant interaction with TBA we have performed an SMCR analysis of the spectra obtained at each salt concentration (with and without TBA), just as was done to obtain the salt-free results in Fig. 1.

The resulting SC spectra of TBA, which are shown in Fig. 3, clearly indicate that there is little interaction between NaF and TBA. More specifically, SC spectra in Fig. 3A show no evidence of any significant change in the H-bonded OH hydration-shell features, except for a small (but reproducible) change in the water dangling OH band area (as further discussed in section 3C, below). Similarly, the CH stretch shift results in Fig. 3B show that NaF has no measurable effect on the CH stretch frequency of TBA up to ~1 M NaF. Thus, these results are consistent with the expectation that neither Na^+ nor F^- have any significant affinity for TBA. However, the small change in the dangling OH intensity in the hydration-shell of TBA suggests that the ions in the second or third hydration shell of TBA may slightly influence the structure of TBA's first hydration-shell (as further discussed in Section 3C).

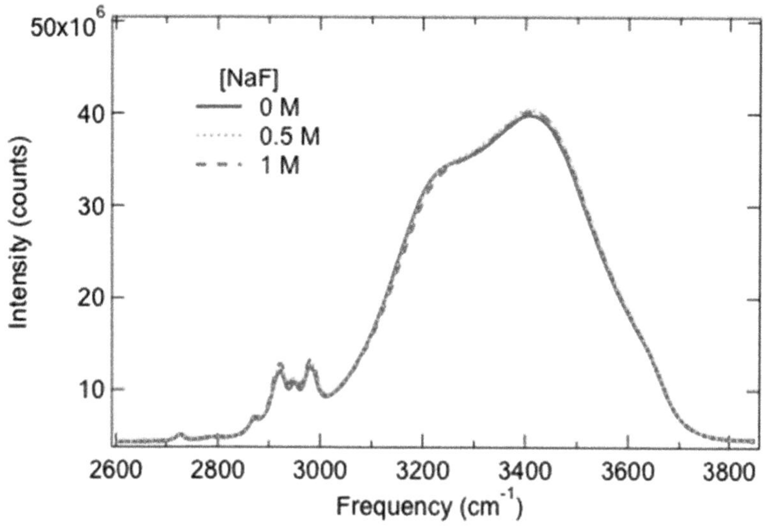

Fig. 2 Raman spectra of water containing 0.5 M TBA and NaF concentrations of either 0 M, 0.5 M or 1 M. The spectra are obtained as described in Fig. 1.

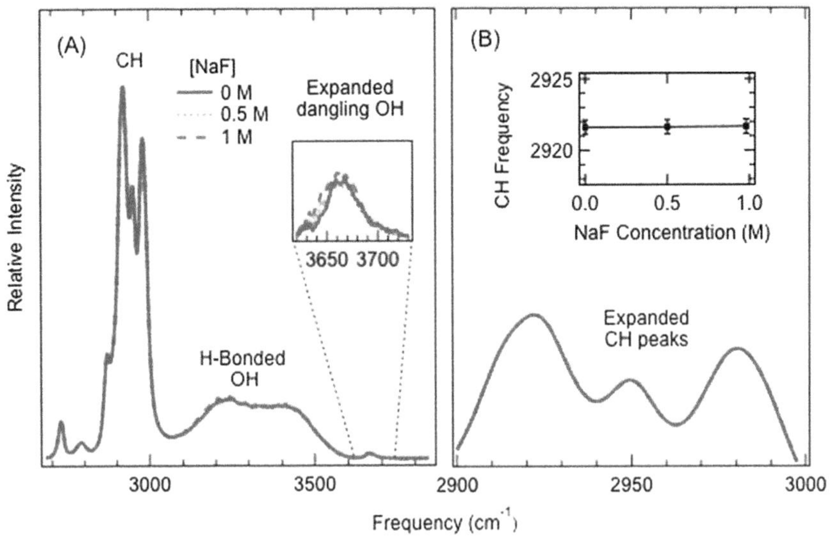

Fig. 3 TBA solute-correlated (SC) spectra obtained from the input spectra shown in Fig. 2 (as well as replicate measurements) in solutions with three different NaF concentrations (0 M, 0.5 M, and 1 M). These SC spectra were each obtained using an independent two-component SMCR analysis of the spectra measured at each salt concentration (with and without TBA).

B Influence of ions on the TBA hydration-shell tetrahedral order

Fig. 4 shows Raman spectra collected from five aqueous solutions, each with ~0.5 M TBA but different concentrations of NaI (ranging from 0 M to 3 M). A separate set of five spectra was collected from the same salt solutions with no TBA. The latter spectra again look quite similar to those in Fig. 4, except that they do not contain the CH stretch features between 2850 cm^{-1} and 3000 cm^{-1}. The change in shape (and increase in intensity) of the OH stretch band with increasing salt concentration is duc primarily to H-bonding between water and I^- (and the associated increase in the water OH Raman cross-section) as previously discussed.[4,61] Our present interest

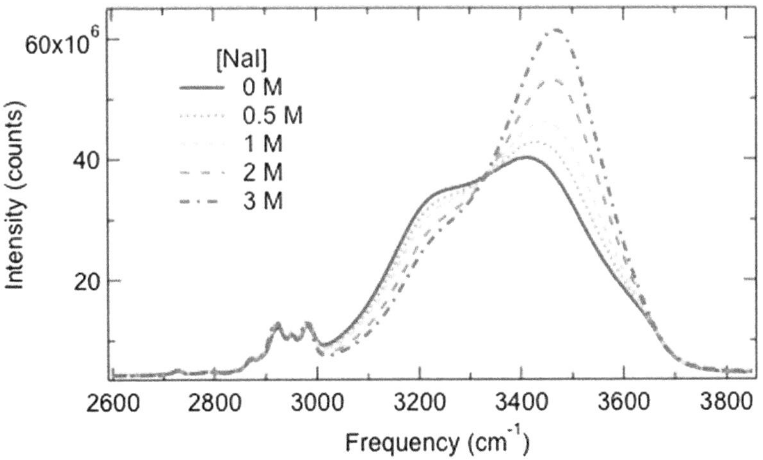

Fig. 4 Raman spectra of aqueous solutions containing 0.5 M TBA and various concentrations of NaI (as indicated in the legend) at 20 °C. The spectra are obtained as described in Fig. 1.

is in elucidating how I$^-$ influences the hydration-shell of TBA, using SC obtained in the same way as was done to get the results in Fig. 1 and 3, by independently analyzing the Raman spectra obtained at each salt concentration (with and without TBA).

The resulting SC spectra of TBA in NaI solutions of various concentrations are shown in Fig. 5A. Note that all of the features in these spectra are correlated with the concentration of TBA and thus include the TBA CH stretch band as well as OH features arising from water molecules in the hydration-shell of TBA. The latter OH features would only appear in the SC spectrum if they differed from the spectra of water molecules in the corresponding salt solution (or pure water). Note that the NaF results shown in Fig. 3 imply that Na^+ cannot be the source of the perturbations that are evidenced in Fig. 5. Thus, any salt concentration dependent changes in the SC spectra shown in Fig. 5 reveal the influence of I$^-$ ions on TBA or its hydration-shell.

Our previous studies have confirmed that alcohol OH head groups contribute very little to hydration-shell OH spectra as, for example, benzene and phenol have virtually identical SC spectra.[6] Thus, the H-bonded OH features in the SC spectra in Fig. 5A (as well as Fig. 1 and 3) arise primarily from water molecules in the hydrophobic hydration shell of TBA.

The most evident effect of NaI on the SC spectrum of TBA is the change in the H-bonded OH stretch features shown in Fig. 5A. Increasing the concentration of NaI clearly leads to a decrease in the intensity of the lower frequency H-bonded OH band, near 3200 cm^{-1}. Since the latter highly polarized Raman band is assigned to water molecules with a high degree of tetrahedral order (as discussed in Section 2), the observed decrease in the area of that low frequency band implies the I$^-$ disrupts the tetrahedral order of the TBA hydration shell (relative to TBA in pure water).

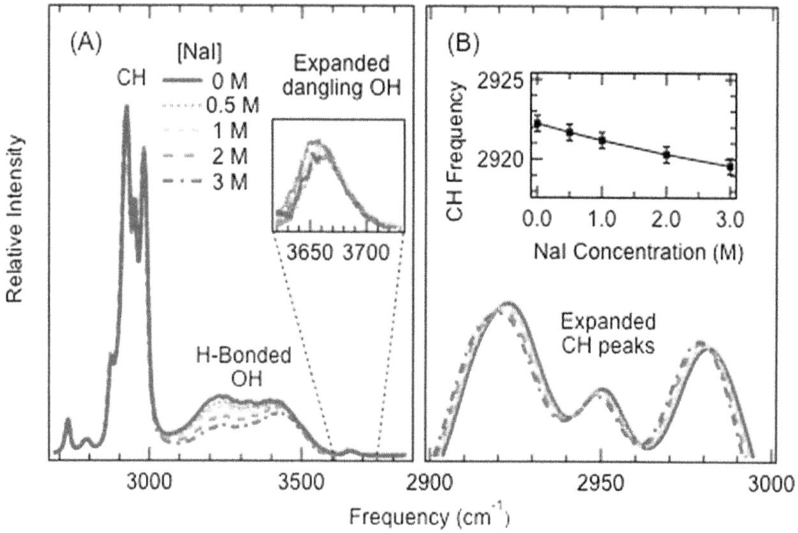

Fig. 5 Solute-correlated (SC) spectra of TBA obtained in solvents with different NaI concentrations. The SC spectra have all been normalized to the same TBA CH band area, so as to compare properly the relative intensities of the corresponding hydration-shell features. Panel (A) shows the full TBA SC spectrum in the CH/OH stretch region, which reveals salt concentration dependent changes in the H-bonded OH region, as well as small salt concentration dependent changes in the intensity of the water dangling OH in the TBA hydration-shell. Panel (B) shows an expanded view of the TBA CH stretch peaks, to reveal that NaI produces a small red-shift in the TBA CH stretch, with a nearly linear NaI concentration dependence (and a slope of ~1 cm^{-1} M^{-1}).

C Influence of ions on the TBA hydration-shell dangling OH bonds

The inserted panel in Fig. 5A shows an expanded view of the small high frequency dangling OH peak that appears in TBA's hydration-shell. Such water dangling OH features have previously been observed both at macroscopic air–water and oil–water interfaces,[62,63] as well as in the hydration shells of hydrocarbon groups dissolved in water.[5,64] In this work, we are interested in using the water dangling OH band to probe interactions between ions and molecular hydrophobic interfaces. If there were a high probability of finding an I^- ion in contact with the CH_3 groups of TBA, then one might expect that increasing the concentration of I^- would *decrease* the dangling OH population in TBA's hydration-shell. However, our results show no such decrease but rather a small, but reproducible, salt-induced increase in the dangling OH intensity.

The magnitude of the dangling OH increase produced by 1 M NaI is in fact very similar to that which was found in 1 M NaF (as shown in Fig. 3A), suggesting that both I^- and F^- produce a slight increase in the number of dangling bonds in the hydration shell of TBA. However, an intensity increase can in general either arise from an increase in population or from an increase in Raman scattering cross-section. Thus, an alternative explanation for the observed dangling OH increase might be that both F^- and I^- increase the Raman cross section of the dangling OH. However, the following argument suggest that the observed dangling bond intensity increase is not due to ion-induced changes in the dangling OH Raman cross section.

Raman spectral results such as those shown in Fig. 4 indicate I^- produces an increase in the Raman cross section of the OH groups to which it is H-bonded,[4,61] and so the other OH group on such water molecules might perhaps also have an enhanced Raman cross section. If that second OH also happened to form one of the dangling OH groups in TBA's hydration shell, then it could contribute the observed intensity increase. However, since F^- *decreases* rather than increases the Raman cross-section of water molecules to which it is H-bonded,[4,61] the above argument would imply that the dangling OH intensity should *decrease* in NaF solutions. Thus, the fact that a similar dangling OH *increase* is found in both NaF and NaI, suggests that the observed intensity increase does not result from anion-induced OH Raman cross section changes, but rather arises from a dangling OH population change.

D Influence of ions on hydrophobic CH vibrational frequencies

The vibrational frequencies of solute CH groups provide another potentially valuable probe of the affinity of ions for molecular hydrophobic hydration-shells. When TBA is dissolved in NaF solutions the resulting SC spectra show essentially no change in the CH stretch frequency of TBA (see Fig. 3B). However, when TBA is dissolved in NaI solutions a small but readily measureable CH red-shift is observed, as shown in Fig. 5B. The approximately linear concentration dependent slope of the latter red-shift has a magnitude of only $\sim$1 cm^{-1} M^{-1}, corresponding to a total shift of $\sim$3 cm^{-1} in a 3 M NaI solution (and all the CH sub-peaks shown in Fig. 5B have comparable shifts). The small magnitude of this red-shift is remarkable given that in a 3 M NaI solution there are fewer than 10 water molecules per ion, and so even random mixing would assure that there would be approximately one I^- in the first hydration shell of every TBA molecule (since there are about $\sim$22 water molecules in the first hydration-shell of TBA).

In order to gauge whether the observed CH frequency shift is consistent with the presence of I^- ions in the first hydration-shell of TBA, we have performed calculations on TBA–water and TBA-I^- dimers as well as TBA dissolved in both liquid water and aqueous NaI. Our dimer MP2/6-311++G** calculations were performed on TBA–water and TBA-I^- dimers in which either water or I^- are situated on the hydrophobic side of TBA, approximately equally distant from its three CH_3 groups, as shown in Fig. 6. Interestingly, I^- in this hydrophobic hydration-shell location produces a red-shift in the CH frequency of about 20 cm^{-1}, relative to the CH frequency obtained when I^- is replaced by a water molecule. (More specifically, placing a water molecule on the hydrophobic side of TBA produces a *blue*-shift of about 5 cm^{-1} relative to an isolated TBA, while replacing the water by I^- produces a CH *red*-shift of $\sim$15 cm^{-1} relative to an isolated TBA.) The fact that the latter shift is significantly larger that the experimentally observed CH shift (see Fig. 5) might suggest that I^- ions do not significantly occupy the first hydration-shell of TBA. However, the above cluster result may not accurately reflect the predicted shift induced by I^- in hydration-shell of TBA dissolved in aqueous NaI, both because I^- may not be located in the same place as in the above dimers and because the negative charge of I^- may be screened and/or redistributed among surrounding water molecules. Thus, the following liquid simulations were performed in order to more accurately predict the CH frequency shift induced by I^- ions in the first hydration shell of TBA.

Hybrid QM/EFP1 simulations of TBA in pure water and aqueous NaI were performed as described in Section 2B. The resulting TBA CH frequency shifts are shown in Table 1. These TBA frequency shift predictions were obtained using the same level of theory (MP2/6-311++G**) which was used to obtain the above dimer results, while representing water molecules by semi-classical EFP1 potentials. The frequency shifts listed in the fourth column of Table 1 correspond to those obtained for TBA molecules whose first hydration-shell contains an I^- ion. The average value of the red-shift of all the strongly Raman active CH stretch modes is $\sim 4 \pm 1$ cm^{-1}, which is significantly smaller than the $\sim$20 cm^{-1} red-shift predicted for the gas phase dimers (containing TBA and either water or I^-). Thus, our liquid QM/EFP1 calculations imply that the relatively small magnitude of experimentally observed frequency shift (as shown in Fig. 5) is consistent with the presence of I^- ions in the first hydration shell of TBA (as further discussed below).

The experimental results in Fig. 5 pertain to the CH frequency shift of the total population of TBA molecules in a given salt solution (compared to TBA in pure water). Our simulation results for the total TBA population in aqueous NaI are not sufficiently accurate to determine reliably the corresponding average CH frequency shift, relative to TBA in pure water. More specifically, the average CH frequency of TBA in $\sim$2.7 M NaI was found to be within less than 1 cm^{-1} from that of TBA in pure water.

In order to more directly compare the predictions in Table 1 with experimental measurements, as well as to experimentally estimate the concentration of I^- in the first hydration-shell of TBA, we have used the following alternative two-component SMCR analysis strategy. This was done by collecting Raman spectra of 0.5 M TBA solutions to which various amounts of NaI were added. The resulting I^- SC spectra,

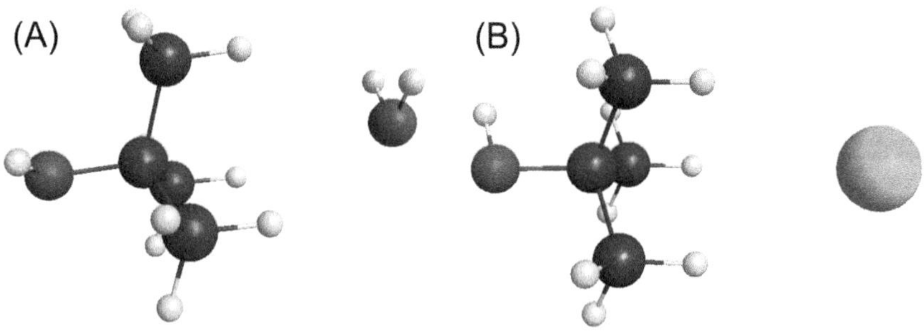

Fig. 6 Local minimum energy structures of TBA–water and TBA-I^- clusters.

Table 1 Predicted TBA CH stretch frequencies (cm^{-1}) and shifts in a ~2.7 M aqueous NaI solution (with respect to the TBA frequencies in pure water, so negative values correspond to red shifts)

TBA in pure water			I^- in the first hydration shell	
CH Frequency	σ^a	Strongly Raman active[b]	Shift[c]	σ^a
3110.4	±1.9	no	−5.71	±1.06
3114.9	±2.2	yes	−5.01	±0.96
3124.1	±2.4	yes	−4.00	±1.40
3199.4	±2.3	no	−7.88	±1.46
3205.6	±2.8	yes	−6.11	±1.12
3216.6	±1.6	yes	−2.20	±1.23
3221.3	±1.8	no	−2.50	±1.01
3224.9	±2.1	yes	−2.18	±0.96
3230.9	±4.8	yes	−2.14	±2.77
Mean frequency shift[c]			−4	±1

[a] Standard deviation of the average frequency of a mode in pure water. [b] Raman activity based on gas phase and TBA–water dimer calculations. [c] CH frequency shift, relative to TBA in pure water. The mean frequency shift (on the bottom row) is obtained from the average of the strongly Raman active CH stretch frequency shifts.

obtained using SMCR, necessarily contain features arising from the hydration-shell of the I^- ions. Previous Raman-MCR studies of aqueous solutions of various alkali-halides have shown that the OH stretch spectra of water are primarily influenced by the anion to which water is directly H-bonded. In the present three component systems, we expect the I^- SC spectrum to contain both features arising from water molecules that are H-bonded to I^-, as well as features arising from the sub-population of TBA molecules that are strongly perturbed by I^-.

Fig. 7 shows Raman-MCR results obtained by adding various concentrations of NaI to a "solvent" composed of 0.5 M aqueous TBA. The spectra shown in Fig. 7 pertain to a mixture containing 3 M NaI (and 0.5 M TBA); the blue curve is the aqueous TBA "solvent" component, while the red curve is the I^- SC spectrum. The higher frequency H-bonded OH stretch band in the red SC spectrum is virtually identical to that previously obtained from solutions of NaI in water (with no TBA), and thus pertains to water OH groups that are H-bonded to I^-. The more interesting feature in the spectra shown in Fig. 7 is the CH band which appears in both the aqueous TBA solvent and I^- SC spectra. The fact that this band appears in the I^- SC spectrum implies that it arises from TBA molecules which are strongly perturbed by I^-. The frequency of the SC CH stretch is red-shifted by $\sim 5 \pm 0.5$ cm^{-1} relative to the CH stretch of TBA in pure water. The latter frequency shift is very close to the $\sim 4 \pm 1$ cm^{-1} red-shift shown in Table 1, which was obtained from the liquid QM/EFP1 simulations for TBA molecules whose first hydration-shells contain an I^- ion.

The relative areas of the CH bands in Fig. 7 may further be used to estimate the number of TBA molecules which are strongly perturbed by I^-. More specifically, the relative area of the CH band in the SC (solid) and total (dashed) curves implies that about 64% of the TBA molecules in a 3 M NaI solution are strongly perturbed by I^-. However, a fundamental "rotational ambiguity" associated with MCR[7] implies that minimum area SC spectra (such as those obtained in this work) have areas which represent a *lower bound* to the true SC spectrum. Thus, although it is possible that only 64% of the TBA molecules are perturbed by I^- it is also possible that some larger fraction of TBA molecules are (less strongly) perturbed by I^-. If we assume that 64% is the actual number of TBA molecules whose first hydration shells contain

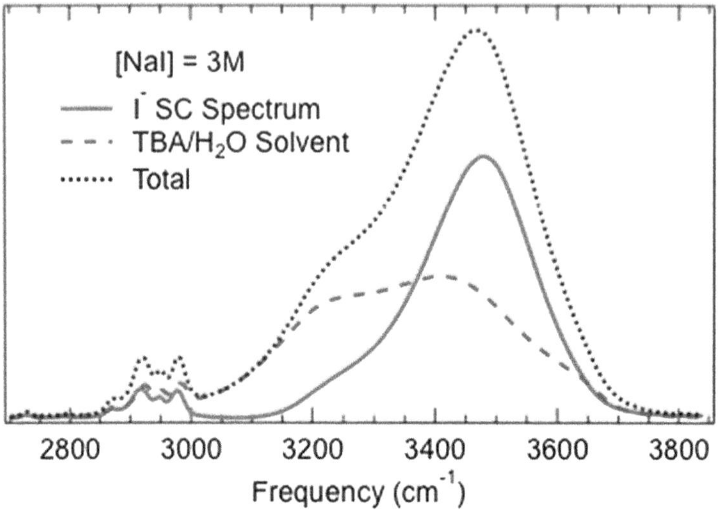

Fig. 7 Raman-MCR is used to decompose the measured spectrum of a solution containing 0.5 M TBA and 3 M NaI (dotted curve) into aqueous TBA (dashed curve) and I^- SC (solid curve) components. The frequency and area differences between the CH bands in the red and blue spectra may be used to quantify the average I^- induced CH frequency shift of the TBA molecules that are strongly perturbed by I^- as well as the number of such strongly perturbed TBA molecules.

one iodide, that would imply that the local concentration I^- is less than 2 M (and thus is below the 3 M concentration of the surrounding NaI solution). This conclusion is consistent with our EFP/MD liquid simulations which suggest that about 89% of the TBA molecules in a ~2.7 M NaI solution have one I^- ion in their first hydration shell. The latter number includes I^- ions that are in the proximity of the hydroxyl group of TBA (as further explained below).

The radial distribution functions shown in Fig. 8 provide additional information that supports and extends the conclusions reached based on our experimental Raman-MCR measurements. For example, our liquid simulations of TBA in ~2.7 M NaF confirm that F^- is essentially never found in the first hydration shell of TBA, which is consistent with the results shown in Fig. 3, which indicate that NaF has virtually no influence on the hydration shell of TBA or its CH frequency. More importantly, our simulation results are consistent with the experimental finding that there is a significant probability of finding an I^- ion in the first hydration shell of TBA, although the local concentration of I^- is lower than it is in the surrounding bulk, and so I^- is not attracted to the hydration shell of TBA, although it is apparently not repelled from the hydration-shell as much as Na^+ and F^-

Our EFP/MD simulations also provide information pertaining to the angular distribution of I^- ions within TBA's first hydration-shell, as shown in Fig. 9. This angular distribution is normalized such that an isotropic (perfectly spherical) distribution would have a value of 1 at all angles. The distribution shown in the lower curve of Fig. 9, which corresponds to ions found within 6.5 Å from the central carbon atom of TBA (near the minimum in TBA–water RDF, see Fig. 8), implies that the I^- ions are most likely to be located in the regions of ~0°–20°, ~30°–70°, ~80°–140°, or near 180°. Angles greater that ~80° correspond to I^- ions located around the periphery of TBA's CH_3 groups, while smaller angles correspond to ions interacting with the OH head group of TBA. The peak at ~30°–70° is consistent with I^- ions H-bonded to the OH head group of TBA. The peak at ~0°–20° completely disappears in the angular distribution pertaining to the shorter (5.3 Å) C_{TBA}-I^- cut-off distance (see Fig. 9, upper curve). This means that I^- ions in this angular range are located between 5.3 Å and 6.5 Å from central TBA carbon; a

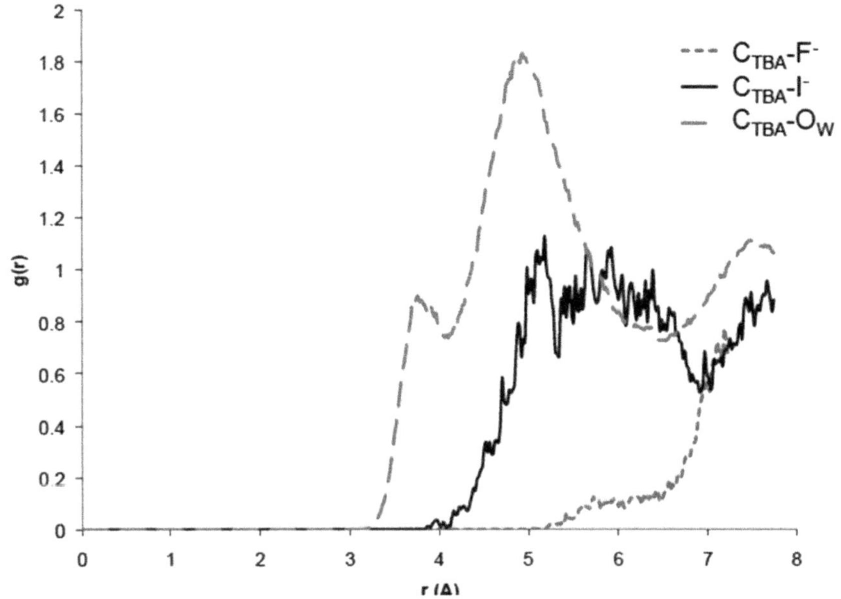

Fig. 8 Radial distribution functions between the central carbon in TBA and F^- (dotted curve), I^- (solid curve), and water oxygen (dashed curve). The first maximum in the TBA-I^- RDF at ~5.1 Å corresponds to ions directly in contact with TBA; the second broader peak extending out to ~6.5 Å includes ions that are interacting with a water molecule that is H-bonded to the TBA OH head group.

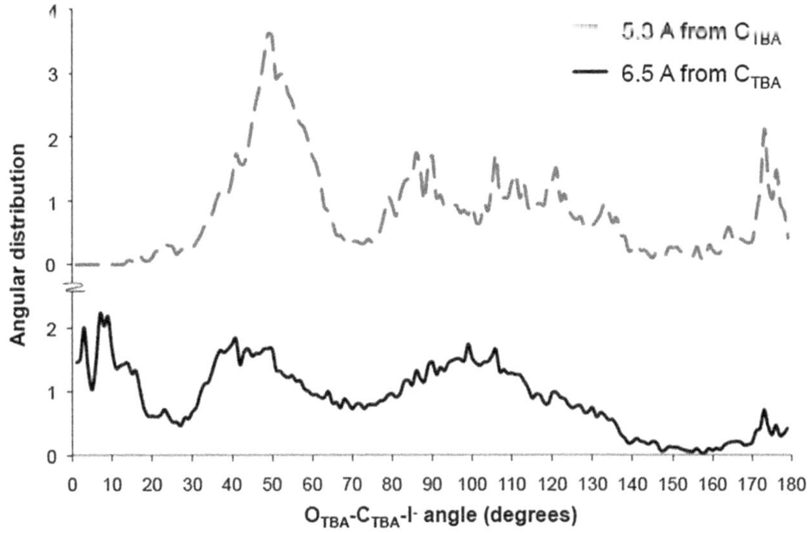

Fig. 9 The angular distribution of I^- ions within TBA's first hydration shell is normalized such that a perfectly isotropic distribution would have a value of 1 at all angles. At angles greater than 80° the ion is on the hydrophobic side of TBA. The upper (dashed) curve corresponds to ions that are within 5.3 Å of the central carbon atom of TBA. The lower (solid) curve corresponds to ions that are within 6.5 Å of the central carbon atom of TBA.

detailed investigation of the corresponding snapshots shows that these ions are bridged by a water molecule H-bonded to the OH head group of TBA.

4 Conclusions

We have combined experimental (Raman-MCR) measurements with theoretical cluster and liquid (EFP/MD) calculations to probe quantitatively and compare the affinities of F^- and I^- for the first hydration-shell of t-butyl alcohol (TBA). Our results indicate that while F^- (as well as Na^+) is strongly excluded from the first hydration shell of TBA, the local concentration of I^- around TBA is nearly as large as the surrounding bulk NaI concentration. The latter conclusion is supported by our experimental findings that in ~3 M NaI approximately two thirds of the TBA molecules are strongly perturbed by I^-, and those strongly perturbed TBA molecules have a CH stretch Raman frequency which is red-shifted by ~5 cm^{-1} relative to TBA in pure water. Both of the latter experimental observations are in good agreement with our theoretical EFP/MD and QM/EFP1 predictions. Our liquid simulation results further indicate that I^- ions are most likely to be located either near TBA's head group or around the periphery of its CH_3 groups, rather than up against the hydrophobic end of TBA. We have further found experimental evidence that I^- disrupts the tetrahedral order in the hydration shell of TBA and that both F^- and I^- slightly increase the intensity of the narrow high-frequency dangling OH peak in the hydration-shell spectrum of TBA. The latter dangling OH intensity increase is not likely to be due to an anion-induced Raman cross section change (as F^- and I^- have opposite effects on the OH cross section of water), but may indicate that the probability that water will form a dangling OH bond around TBA is increased by ions in the second (or higher) hydration shell(s) of TBA.

Acknowledgements

The work done by D. B. A., B. M. R., and D. W. was supported by a grant from the National Science foundation (CHE-0847928), while that done by L. V. S. and M. D. H. was supported by a CAREER grant from the National Science foundation (grant CHE-0955419) and Purdue University. We thank Ben Nebgen for help in preparing a script for performing semi-numerical frequency calculations, and Joe Thomaz for performing MCR-ALS three component analyses to compare with the SMCR results reported in this work.

References

1 Y. Levin, A. P. dos Santos and A. Diehl, *Phys. Rev. Lett.*, 2009, **103**(257802), 257801–257804.
2 W. Kunz, P. Lo Nostro and B. W. Ninham, *Curr. Opin. Colloid Interface Sci.*, 2004, **9**, 1–18.
3 P. Perera, M. Wyche, Y. Loethen and D. Ben-Amotz, *J. Am. Chem. Soc.*, 2008, **130**, 4576–4579.
4 P. N. Perera, B. Browder and D. Ben-Amotz, *J. Phys. Chem. B*, 2009, **113**, 1805–1809.
5 P. N. Perera, K. R. Fega, C. Lawrence, E. J. Sundstrom, J. Tomlinson-Phillips and D. Ben-Amotz, *Proc. Natl. Acad. Sci. U. S. A.*, 2009, **106**, 12230–12234.
6 K. P. Gierszal, J. G. Davis, M. D. Hands, D. S. Wilcox, L. V. Slipchenko and D. Ben-Amotz, *J. Phys. Chem. Lett.*, 2011, **2**, 2930–2933.
7 K. R. Fega, D. S. Wilcox and D. Ben-Amotz, *Appl. Spectrosc.*, 2011, **66**, 282–288.
8 P. N. Day, J. H. Jensen, M. S. Gordon, S. P. Webb, W. J. Stevens, M. Krauss, D. Garmer, H. Basch and D. Cohen, *J. Chem. Phys.*, 1996, **105**, 1968–1986.
9 M. S. Gordon, M. A. Freitag, P. Bandyopadhyay, J. H. Jensen, V. Kairys and W. J. Stevens, *J. Phys. Chem. A*, 2001, **105**, 293–307.
10 M. S. Gordon, L. V. Slipchenko, H. Li and J. H. Jensen, *Annu. Rep. Comput. Chem.*, 2007, **3**, 177–193.

11 D. Ghosh, D. Kosenkov, V. Vanovschi, C. F. Williams, J. M. Herbert, M. S. Gordon, M. W. Schmidt, L. V. Slipchenko and A. I. Krylov, *J. Phys. Chem. A*, 2010, **114**, 12739–12754.
12 J. Cheng, M. R. Hoffmann and A. J. Colussi, *J. Phys. Chem. B*, 2008, **112**, 7157–7161.
13 S. Ghosal, J. C. Hemminger, H. Bluhm, B. S. Mun, E. L. D. Hebenstreit, G. Ketteler, D. F. Ogletree, F. G. Requejo and M. Salmeron, *Science*, 2005, **307**, 563–566.
14 V. Padmanabhan, J. Daillant, L. Belloni, S. Mora, M. Alba and O. Konovalov, *Phys. Rev. Lett.*, 2007, **99**(086105), 086101–086104.
15 H. C. Allen, N. N. Casillas-Ituarte, M. R. Sierra-Hernandez, X. K. Chen and C. Y. Tang, *Phys. Chem. Chem. Phys.*, 2009, **11**, 5538–5549.
16 P. B. Petersen and R. J. Saykally, *J. Am. Chem. Soc.*, 2005, **127**, 15446–15452.
17 J. K. Beattie, P. Creux and A. Gray-Weale, *Aust. J. Chem.*, 2011, **64**, 1580–1582.
18 B. L. Eggimann and J. I. Siepmann, *J. Phys. Chem. C*, 2008, **112**, 210–218.
19 D. Horinek, A. Herz, L. Vrbka, F. Sedlmeier, S. I. Mamatkulov and R. R. Netz, *Chem. Phys. Lett.*, 2009, **479**, 173–183.
20 B. A. Bauer, S. C. Ou and S. Patel, *Chem. Phys. Lett.*, 2012, **527**, 22–26.
21 P. Jungwirth and D. J. Tobias, *Chem. Rev.*, 2006, **106**, 1259–1281.
22 (*a*) P. Jungwirth, *Faraday Discuss.*, 2009, **141**, 9–30; (*b*) K. B. Rembert, J. Paterová, J. Heyda, C. Hilty, P Jungwirth and P. S. Cremer, *J. Am. Chem. Soc.*, 2012, **134**, 10039–10046.
23 M. D. Baer and C. J. Mundy, *J. Phys. Chem. Lett.*, 2011, **2**, 1088–1093.
24 W. H. Lawton and E. A. Sylvestre, *Technometrics*, 1971, **13**, 617–633.
25 H. Torii, *J. Phys. Chem. A*, 2006, **110**, 9469–9477.
26 M. Yang and J. L. Skinner, *Phys. Chem. Chem. Phys.*, 2010, **12**, 982–991.
27 B. M. Auer and J. L. Skinner, *Chem. Phys. Lett.*, 2009, **470**, 13–20.
28 H. Li, M. S. Gordon and J. H. Jensen, *J. Chem. Phys.*, 2006, **124**, 214108.
29 I. Adamovic, M. A. Freitag and M. S. Gordon, *J. Chem. Phys.*, 2003, **118**, 6725–6732.
30 P. N. Day, R. Pachter, M. S. Gordon and G. N. Merrill, *J. Chem. Phys.*, 2000, **112**, 2063–2073.
31 I. Adamovic and M. S. Gordon, *J. Phys. Chem. A*, 2006, **110**, 10267–10273.
32 I. Adamovic, H. Li, M. H. Lamm and M. S. Gordon, *J. Phys. Chem. A*, 2006, **110**, 519–525.
33 T. Smith, L. V. Slipchenko and M. S. Gordon, *J. Phys. Chem. A*, 2008, **112**, 5286–5294.
34 Q. A. Smith, M. S. Gordon and L. V. Slipchenko, *J. Phys. Chem. A*, 2011, **115**, 4598–4609.
35 Q. A. Smith, M. S. Gordon and L. V. Slipchenko, *J. Phys. Chem. A*, 2011, **115**, 11269–11276.
36 L. V. Slipchenko and M. S. Gordon, *J. Phys. Chem. A*, 2009, **113**, 2092–2102.
37 M. D. Hands and L. V. Slipchenko, *J. Phys. Chem. B*, 2012, **116**, 2775–2786.
38 J. M. Mullin and M. S. Gordon, *J. Phys. Chem. B*, 2009, **113**, 8657–8669.
39 J. M. Mullin and M. S. Gordon, *J. Phys. Chem. B*, 2009, **113**, 14413–14420.
40 D. Kosenkov and L. V. Slipchenko, *J. Phys. Chem. A*, 2011, **115**, 392–401.
41 L. V. Slipchenko, *J. Phys. Chem. A*, 2010, **114**, 8824–8830.
42 A. DeFusco, N. Minezawa, L. V. Slipchenko, F. Zahariev and M. S. Gordon, *J. Phys. Chem. Lett.*, 2011, **2**, 2184–2192.
43 D. Ghosh, O. Isayev, L. V. Slipchenko and A. I. Krylov, *J. Phys. Chem. A*, 2011, **115**, 6028–6038.
44 S. P. Webb and M. S. Gordon, *J. Phys. Chem. A*, 1999, **103**, 1265–1273.
45 I. Adamovic and M. S. Gordon, *J. Phys. Chem. A*, 2005, **109**, 1629–1636.
46 S. Yoo, F. Zahariev, S. Sok and M. S. Gordon, *J. Chem. Phys.*, 2008, **129**, 144112–144118.
47 M. A. Freitag, M. S. Gordon, J. H. Jensen and W. J. Stevens, *J. Chem. Phys.*, 2000, **112**, 7300–7306.
48 L. V. Slipchenko and M. S. Gordon, *J. Comput. Chem.*, 2007, **28**, 276–291.
49 L. V. Slipchenko and M. S. Gordon, *Mol. Phys.*, 2009, **107**, 999–1016.
50 W. J. Hehre, R. Ditchfield and J. A. Pople, *J. Chem. Phys.*, 1972, **56**, 2257.
51 P. C. Hariharan and J. A. Pople, *Theor. Chim. Acta*, 1973, **28**, 213–222.
52 T. Clark, J. Chandrasekhar, G. W. Spitznagel and P. V. Schleyer, *J. Comput. Chem.*, 1983, **4**, 294–301.
53 R. Krishnan, J. S. Binkley, R. Seeger and J. A. Pople, *J. Chem. Phys.*, 1980, **72**, 650–654.
54 M. J. Frisch, J. A. Pople and J. S. Binkley, *J. Chem. Phys.*, 1984, **80**, 3265–3269.
55 T. H. Dunning, *J. Chem. Phys.*, 1989, **90**, 1007–1023.
56 R. A. Kendall, T. H. Dunning and R. J. Harrison, *J. Chem. Phys.*, 1992, **96**, 6796–6806.
57 H. Li, H. M. Netzloff and M. S. Gordon, *J. Chem. Phys.*, 2006, **125**, 194103.
58 W. R. Wadt and P. J. Hay, *J. Chem. Phys.*, 1985, **82**, 284–298.
59 M. W. Schmidt, K. K. Baldridge, J. A. Boatz, S. T. Elbert, M. S. Gordon, J. H. Jensen, S. Koseki, N. Matsunaga, K. A. Nguyen, S. J. Su, T. L. Windus, M. Dupuis and J. A. Montgomery, *J. Comput. Chem.*, 1993, **14**, 1347–1363.

60 M. S. Gordon and M. W. Schmidt, in *Theory and Applications of Computational Chemistry*, ed. C. E. Dykstra, G. Frenking, K. S. Kim and G. E. Scuseria, Elsevier, 2005, ch. 41.
61 J. D. Smith, R. J. Saykally and P. L. Geissler, *J. Am. Chem. Soc.*, 2007, **129**, 13847–13856.
62 Q. Du, E. Freysz and Y. R. Shen, *Science*, 1994, **264**, 826–828.
63 F. G. Moore and G. L. Richmond, *Acc. Chem. Res.*, 2008, **41**, 739–748.
64 J. Tomlinson-Phillips, J. Davis, D. Ben-Amotz, D. Spangberg, L. Pejov and K. Hermansson, *J. Phys. Chem. A*, 2011, **115**, 6177–6183.

Solution electrostatics beyond pH: a coarse grained approach to ion specific interactions between macromolecules

Anıl Kurut* and Mikael Lund*

Received 13th April 2012, Accepted 10th May 2012
DOI: 10.1039/c2fd20073b

Oblivious to ion specificity, pH has been a key parameter for macromolecular solutions for little more than a century. We here widen the concept by describing the ionization of macromolecules not only *via* pH, but also pX where X are other binding species. Using binding constants, measured by NMR, of chloride and thiocyanate to amino acid motifs on γ-crystallin, we calculate i) titration curves as a function of pH and pX and ii) estimate second virial coefficients using both approximate theory and computer simulations. In agreement with experiment, a Hofmeister reversal for protein–protein interactions is observed when crossing iso-electric conditions. Thiocyanate binding further leads to large charge fluctuations that may trigger intermolecular charge regulation interactions.

1 Introduction

For nearly a century, chemists and other natural scientists have been brought up with concepts such as pH, Brønsted acids and bases, auto-protolysis, and pK_a-values. This terminology focuses on thermodynamic equilibrium processes of merely a single ionic species, namely *protons*. pH is hence a key parameter throughout a number of scientific disciplines and is commonly used as a convenient handle to control inter- and intra-molecular electrostatic interactions. Protons are, however, not alone. Plenty of other ions may be present, yet our standard terminology is unable to account for those. During the last two decades there has been a large development in the molecular understanding of ion specific effects; the reader may refer to recent reviews for a full account.[1–3] It has for example long been known that proteins can be titrated with anions and that this changes the thermodynamic properties,[4–6] yet only recently have the underlying molecular mechanisms started to unravel. Experimental and theoretical works both show that small and large anions distinctly distribute on molecular surfaces; large anions bind to non-polar motifs[7–11] and thiocyanate, for example, binds to the backbone of model peptides.[12] The binding affinities to molecular motifs are currently being systematically quantified[12,13] and an obvious question is if specific ion binding can be treated at the same level as protons have been for so long. That is, through the chemical potential (*cf.* pH) and binding equilibrium constant to molecular motifs (*cf.* pK_a values) we may reduce the microscopic ion-motif potential of mean force to a two state binding model:

$$MX \rightleftharpoons M + X \quad (1)$$

where M is the motif and X is the ion in question. Arguably, protons distinguish themselves from other ions in that they are always present in water and their binding

Department of Theoretical Chemistry, Lund University, P.O.B. 124, SE-22100 Lund, Sweden. E-mail: anil.kurut@teokem.lu.se; mikael.lund@teokem.lu.se; Tel: +46 462 223167

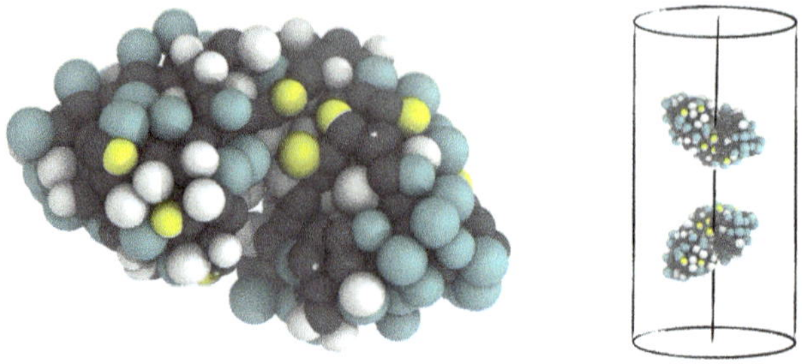

Fig. 1 Left: protein model used in the Monte Carlo simulations. Protons can bind to acidic and basic sites (turquoise); thiocyanate may bind to the backbone (black) as well as to hydrophobic side chains (yellow). Right: two-body simulation model in a cylindrical container – solvent and salt are handled implicitly.

is of covalent nature. This is fitting for a two-state model as the configurational space may readily be divided into a well defined bound and unbound state. This (arbitrary) division becomes more difficult for weakly bound ions, yet formally a thermodynamic binding constant can always be defined, provided that the ion-motif potential of mean force has a distinct minimum and decays with distance, r, as $1/r^3$ of faster.[14] The mere fact that a binding constant can be reliably measured, preferably with different experimental techniques, of course also supports a two-state model.

In this work we present a coarse grained model for macromolecules in solution, taking into account specific ion binding to dedicated molecular motifs. The model is based on continuum electrostatics, combined with two-state binding of protons and an arbitrary number of other ions. Ultimately this allows for studies of ion specific effects at large length scales hitherto unattainable with existing models.

2 Model and theory

2.1 Simulation model

Solvent and salt particles are treated at the Debye–Hückel level while the rigid proteins are described in mesoscopic detail where each amino acid is represented by two soft spheres: one for the backbone and another for the side chain – see Fig. 1. These spheres are placed at their respective mass centers as found in the atomistic NMR or crystal structure and their radii are determined from the atomic weight, assuming a density of 1 g ml^{-1}. To capture specific ion binding to the macromolecular surface we specify binding sites (Table 1) where the free energy difference between the bound and free state is $k_BT\ln10(\text{pX} - \text{p}K_d)$. Analogous to pH, pX is the minus logarithm of the activity† of the binding ion, X, and K_d is the dissociation constant for the process. In addition to electrostatic and van der Waals pair interactions, hydrophobic side-chains interact with a square-well potential and the final system Hamiltonian becomes:

$$\beta U = \sum_{i\neq j}^{N} \frac{\lambda_B z_i z_j e^{-\kappa r_{ij}}}{r_{ij}} + \sum_{i\neq j}^{N} 4\beta\varepsilon_{LJ}\left[\left(\frac{\sigma_{ij}}{r_{ij}}\right)^{12} - \left(\frac{\sigma_{ij}}{r_{ij}}\right)^{6}\right] + \sum_{i\neq j}^{N_{\text{hydroph.}}} \beta u_{sw}$$

$$+ \sum_{i\neq j}^{N_{\text{bound}}} (\text{pX}_i - \text{p}K_{d,i})\ln 10 \quad (2)$$

where the first two sums run over all particles, the third over hydrophobic side chains and the last term over sites with bound ions, only. $\lambda_B = 7.1$ Å is the Bjerrum length

† In this work we assume that the activity is equal to the concentration.

Table 1 Amino acid properties and binding sites of γD-crystallin (PDB: 2KLJ[16]). Only particles with solvent accessible surface area (SASA) larger than 30 Å^2 are included as ion binding sites; this is indicated by parenthesis in the N column. When ions bind to a hydrophobic site, this is rendered non-hydrophobic. A probe with radius 1.4 Å is used in SASA measurements

Atom type	N	σ/Å	pK_{d,H^+}	pK_{d,SCN^-} [12]	Hydrophobic
Asp	12 (10)	4.8	4.0		no
Glu	10 (9)	5.2	4.4		no
His	6 (4)	5.4	6.3		no
Tyr	14 (7)	5.9	9.6		no
Lys	1 (1)	5.2	10.4		no
Cys	6 (0)	4.5	10.8		no
Arg	21 (21)	5.8	12.0		no
Ctr	1 (1)	4.7	2.6		no
Ntr	1 (1)	4.7	7.5		no
Ala	4 (0)	3.1		0.82	yes/no
Ile	6 (0)	4.8		0.82	-//-
Leu	17 (3)	4.8		0.82	-//-
Met	4 (3)	5.2		0.82	-//-
Phe	6 (1)	5.6		0.82	-//-
Pro	4 (4)	4.3		0.82	-//-
Trp	4 (0)	6.3		0.82	-//-
Val	6 (0)	4.3		0.82	-//-
Ser	17	3.9			no
Thr	4	4.4			no
Asn	7	5.6			no
Gln	11	5.2			no
Gly	14	1.2			no
backbone	156 (5)	4.7		0.60	no
backbone (Pro, Gly)	18 (3)	4.7		1.30	no

for water, z are particle charge numbers, r_{ij} the distance between the i'th and j'th particle, $1/\kappa = 3.04/\sqrt{C_s}$ Å is the Debye screening length,[15] C_s the molar 1 : 1 salt concentration, $\beta = 1/k_BT$ is the inverse thermal energy, and $\beta u_{sw} = -0.8$ for surface separations less than 1.4 Å – zero otherwise. Finally, $\sigma_{ij} = (\sigma_i + \sigma_j)/2$ is the Lennard–Jones diameter and $\beta\varepsilon_{LJ} = 0.088$.

Using Metropolis Monte Carlo (MC) simulations[17] we simulate either one or two of the proteins in a cylindrical container and average over i) mass center positions (fixed along the main axis of the cylinder), ii) orientations, and iii) ion binding states. Translations and rotations for two-body simulations, only, are performed in a combined MC move where a random displacement vector and angle is used to generate configurations in the canonical ensemble. Averaging over ion binding states is done by swap moves where the state of randomly chosen binding sites – see Table 1 – is alternated between bound and unbound. "Bound" simply means that the site charge is changed by $+1e$ or $-1e$ for proton and anion binding, respectively. Ion binding to a hydrophobic group renders the group non-hydrophobic, thus excluding it from hydrophobic interaction. All simulations were performed using the Faunus project.[18]

2.2 Charge regulation

Charge binding sites on molecular surfaces give rise to a fluctuating molecular charge distribution that can be perturbed by an external electric potential. This leads to a correlation effect, known as charge regulation or "fluctuation forces", that lowers the system free energy.[19,20] The mechanism is well known for proton binding sites, yet for other ions it is largely unexplored and we therefore give a brief overview.

Consider an unperturbed macromolecule with a fluctuating charge distribution with a charge valency, Z, that occurs with the intrinsic probability $P(Z)$. Exposing the molecule to an external electric potential, φ, due to the surrounding chemical environment, the charge ensemble average becomes,

$$\langle Z\rangle = \frac{\int P(Z)Z\exp(-\beta\varphi eZ)\mathrm{d}Z}{\int P(Z)\exp(-\beta\varphi eZ)\mathrm{d}Z}. \tag{3}$$

It then follows that,

$$-\frac{\partial\langle Z\rangle}{\beta e\partial\varphi} = \langle Z^2\rangle - \langle Z\rangle^2 \equiv C. \tag{4}$$

where we have introduced the charge capacitance, C, which is simply the variance of the mean charge.[21] If φ stems from another fluctuating charge distribution, $P(Z')$, the two distributions have the interaction free energy,

$$\beta w(R) = -\ln\int_{-\infty}^{\infty}\int_{-\infty}^{\infty} P(Z)P(Z')\exp(-\lambda_B ZZ'/R)\mathrm{d}Z\mathrm{d}Z' \approx \frac{\lambda_B\langle Z\rangle\langle Z'\rangle}{R} - \frac{\lambda_B^2}{2R^2}\left(C\langle Z'\rangle^2 + C'\langle Z\rangle^2 + CC'\right) \tag{5}$$

where R is the center–center distance between the two distributions and in the last step we have assumed that P, P' are normal distributions. The same result can be obtained using a multipole expansion[20–22] and the first term is clearly the monopole–monopole interaction term, while the remaining attractive terms are monopole-induced and induced-induced monopole interactions. For the sake of simplicity we do not discuss angular dependent terms and further note that these decay faster than the above terms – $1/R^4$ and $1/R^6$ for monopole–dipole and dipole–dipole interactions, respectively.[23]

Thus, in the intermolecular interaction energy given by eqn (5), electrostatics are condensed to two molecular parameters, i) the average net charge, $\langle Z\rangle$ and ii) the charge capacitance, C. The latter property is central for describing charge regulation phenomena[21] and entails the overall fluctuation effect of all binding ions and thus depends on solution conditions, structure and protein sequence.

2.3 Osmotic second virial coefficient

Through eqn (5), the protein net charge and capacitance can be used to estimate the osmotic second virial coefficient of protein solutions using,

$$B_2 = \frac{2\pi\sigma^3}{3} - 2\pi\int_{\sigma}^{\infty}\left(e^{-\beta w(R)} - 1\right)R^2\mathrm{d}R \tag{6}$$

where the first term, B_2^{HS}, is a simple hard sphere contribution for σ smaller than protein–protein contact, $\sigma = 25$ Å. To take into account the effect of salt screening, λ_B in eqn (5) is replaced with[15]

$$\lambda_B\exp[-\kappa(R-\sigma)]/(1+\kappa\sigma) \tag{7}$$

where we note that the quadratic charge regulation terms are subject to stronger salt screening than the direct monopole–monopole term.

3 Results and discussion

3.1 Binding of protons, chloride, and thiocyanate

The overall electrostatic properties of bio-molecules are traditionally classified from the pH titration curve, *i.e.* the net charge as a function of the proton activity (pH). To describe how these properties are influenced by additional ions, further dimensions must be added to the titration curve: one for each additional binding species. In the following we assume that the salt cation does not bind to the protein surface, although this could easily be included by expanding Table 1; the computational cost would be negligible.

Fig. 2, top, shows the net charge of γ-crystallin in the presence of chloride and thiocyanate, respectively. Due to weak binding at moderate salt concentrations[5,12,24] we shall assume that chloride interacts only through double layer forces, that is through the Debye screening length. The iso-electric point, pI where the net-charge is zero, thus remains constant over a large span of salt concentrations and the classical one-dimensional pH picture thus gives a good description of the electrostatic properties. For pH $\neq$ pI the net charge does however change with salt concentration due to screening of intramolecular electrostatic interactions. At high ionic strengths the protein can hence accommodate a higher net-charge, while if $\langle Z \rangle = 0$, screening has little effect since the internal electrostatic energy, roughly proportional to $\langle Z \rangle^2$, is near a minimum.

In contrast, the strongly bound thiocyanate ion behaves very differently. Changing the salt concentration from 30 mM to 300 mM causes a drop in the iso-electric point from nine to five because the increased salt activity drives anion adsorption to the backbone and side-chain binding sites. This process is further aided by the positive net-charge of the protein at low pH, although at sufficiently high salt concentrations the binding sites saturate while at the same time the favorable electrostatic interactions diminish due to screening.

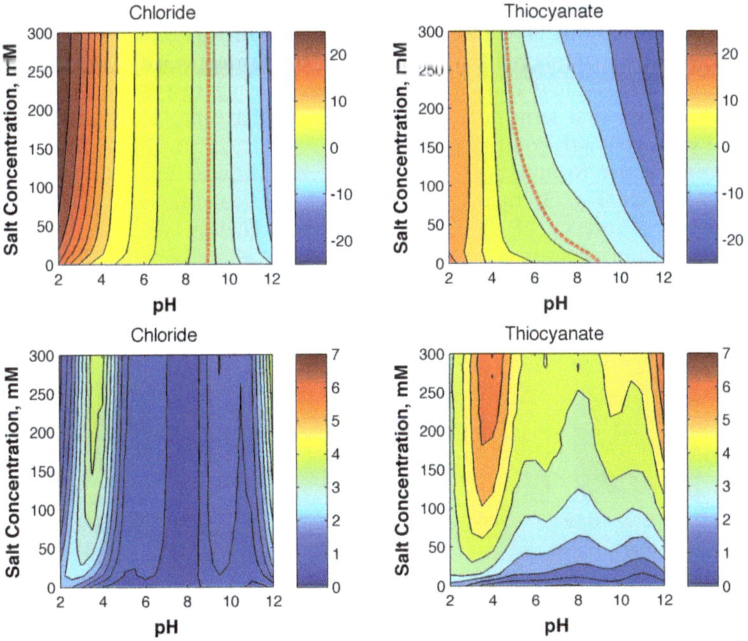

Fig. 2 Protein net charge (top) and charge capacitance (bottom) for γ-crystallin in NaCl (left) and NaSCN (right). The dashed lines on the charge plots represent the iso-electric conditions.

The charge capacitance, C, describes the protein's ability for charge regulation. In the chloride solution, the capacitance is influenced by salt only through the ionic strength; the proton concentration causes maxima at high and low pH, corresponding to the pK_d of acidic and basic sites. As the salt concentration increases the electrostatic interactions between sites are screened and whereby they become independent of each other. In the limit of infinite salt, the capacitance approaches an ideal value, given by[21]

$$C^{\text{ideal}} = \sum_i \frac{10^{\text{pX}_i - \text{p}K_{\text{d},i}}}{(1 + 10^{\text{pX}_i - \text{p}K_{\text{d},i}})^2} \tag{8}$$

where the sum runs over all binding sites. Compared to the capacitance where sites interact, the ideal curve tends to be larger in magnitude and hence less broad.

Thiocyanate also here displays a remarkably different picture compared to chloride and, overall, the capacitance is higher in magnitude for all salt and pH conditions. This is trivially due to the added number of binding sites that allows for more charge fluctuations. Increasing the thiocyanate concentration leads to higher capacitances that approach the ideal value as given by eqn (8) and may potentially result in strong charge regulation interactions with other molecular matter. However, as already mentioned, the quadratic charge regulation terms in eqn (5) are screened with $\sim e^{-2\kappa r}$ compared to $\sim e^{-\kappa r}$ for the direct charge–charge term and hence diminish rapidly with increasing salt. Still, thiocyanate causes capacitances several times higher than chloride even at low salt concentrations and the addition of binding ions may hence be a useful tool to tune intermolecular interaction mechanisms, charge regulation in particular.

3.2 Protein–protein interactions

3.2.1 Virial coefficients. In the previous section we discussed the electric properties of γ-crystallin using the average charge, $\langle Z \rangle$, and capacitance, C, while considering both protons and binding anions. Using these isotropic terms we now estimate the osmotic second virial coefficient, B_2, using eqn (5) and eqn (6). As shown in Fig. 3, B_2 for both SCN^- and Cl^- reaches minima for pH close to the iso-electric points, while at extreme pH the net charge of the protein causes a high (repulsive) virial coefficient. Note that B_2 here does not become negative due to the dominance of the repulsive first term in eqn (5) and that we have neglected attractive short ranged van der Waals contributions.

It is instructive to investigate the B_2 *difference* between the NaSCN and NaCl solutions – see Fig. 3, right. At near iso-electric conditions, ΔB_2 is zero and thus marks a reversal of the Hofmeister series. That is, when the protein has

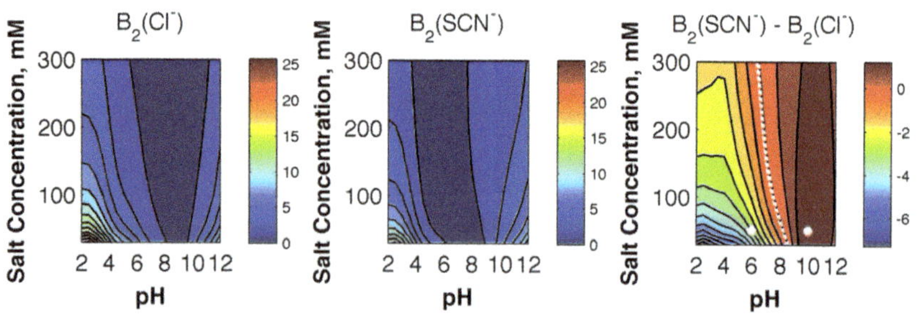

Fig. 3 Left and middle: estimated virial coefficients for γ-crystallin in NaCl and NaSCN, respectively using eqn (5) and eqn (6). Right: difference in second virial coefficient between NaSCN and NaCl. The dashed white line represents zero (Hofmeister reversal), while the two white dots mark the conditions used for the two-body simulations in Fig. 4. Note that the B_2 values are normalized with B_2^{HS}.

a net positive charge, bound thiocyanate anions lower the protein–protein repulsion more than chloride; the *reverse* Hofmeister series.[25] For a net negatively charged protein, binding of thiocyanate causes more repulsion than chloride, *i.e.* the *direct* Hofmeister series. This is exactly what is found in SAXS measurements for γ-crystallin as well as other proteins.[26] Experimentally, the inversion for γ-crystallin occurs at pH 4.5 and at 500 mM salt[26] which is lower than the iso-electric point in dilute solution (pI 6.7–8.9)[26,27] and thus fits the picture that addition of binding salts down shifts pI, *cf.* Fig. 2. While the low salt Hofmeister reversal predicted here also agrees with previous theoretical observations[8,28,30] the current work is distinguished in that ion binding adheres to the complex surface topology of the protein, while at the same time no explicit ions are required.

3.2.2 Two-body simulations. In the previous section we used perturbation theory to estimate the second virial coefficient of γ-crystallin solutions and crudely neglected anisotropy due to electrostatics, van der Waals and excluded volume. This can be remedied by instead explicitly simulating two coarse grained proteins (Fig. 1, right) and from the radial distribution function, $g(R)$, extracting the potential of mean force, $\beta w(R) = -\ln g(R)$. As shown in Fig. 4 the protein–protein attraction is larger for thiocyanate anions than chloride anions at pH 6, while the reverse is true at pH 10. This result is in qualitative agreement with the simplistic multipole calculations in Fig. 3, and more elaborate computer simulations with explicit ions,[8] as well as with experimental data.[26] Note also, that there are now distinct minima in the interaction free energies, stemming mostly from van der Waals interactions and, to a lesser extent, hydrophobic interactions *cf.* eqn (2).

4 Conclusion

Using a simple theoretical model that combines ion binding equilibria to macromolecular motifs with continuum electrostatics, we have investigated ion specific charge properties of the protein γ-crystallin. In particular we note that the iso-electric "point" of the protein in thiocyanate solution varies strongly with the salt concentration and is thus a function of both pH and pSCN. Osmotic second virial coefficients – calculated both from a multipole expansion and from two body computer simulations – show a reverse salting-in Hofmeister series for the cationic protein while the direct Hofmeister series is followed under anionic conditions. This is in agreement with SAXS measurements as well as with other theoretical work. While our model accounts for salting in effects, salting out due to highly solvated ions is

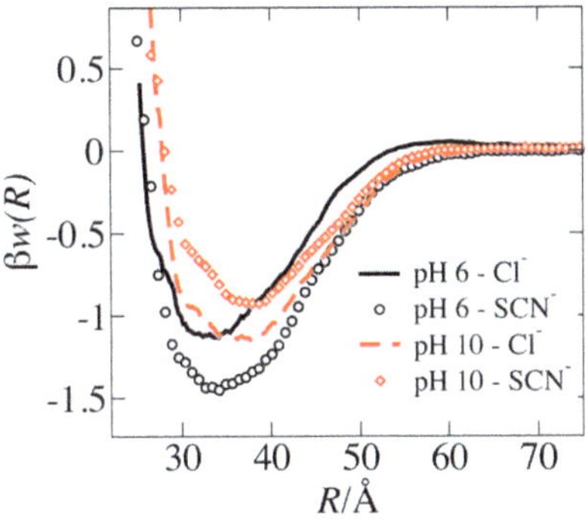

Fig. 4 Simulated potential of mean force, $w(R)$, along the protein–protein mass center coordinate, R, at different pH and anionic species (50 mM). Note the Hofmeister reversal going from low to high pH.

neglected. This may be remedied by introducing salt sensitive hydrophobic interactions;[29] a simple surface tension argument could be a viable option.

With multiple binding sites on the protein surface, thiocyanate brings about large fluctuations in the molecular charge distribution. This may lead to significant charge regulation phenomena when interacting with other molecular matter and specific ion effects may thus provide yet another handle to control intermolecular interactions.

Acknowledgements

We thank Paul Cremer and Pavel Jungwirth for useful discussions. For financial support we thank the eSSENCE@LU strategic program and the Linneaus Center of Excellence "Organizing Molecular Matter", Lund, Sweden.

References

1 P. Jungwirth and B. Winter, *Annu. Rev. Phys. Chem.*, 2008, **59**, 343–66.
2 Y. Zhang and P. S. Cremer, *Annu. Rev. Phys. Chem.*, 2010, **61**, 63–83.
3 P. Lo Nostro and B. W. Ninham, *Chem. Rev.*, 2012, **112**, 2286–2322.
4 E. Volkin, *Journal of Biological Chemistry*, 1948, **175**, 675–681.
5 P. Retailleau, M. Riès-Kautt and A. Ducruix, *Biophys. J.*, 1997, **73**, 2156–63.
6 K. D. Collins, *Methods (San Diego, Calif.)*, 2004, **34**, 300–11.
7 R. L. Baldwin, *Biophys. J.*, 1996, **71**, 2056–63.
8 M. Lund and P. Jungwirth, *J. Phys.: Condens. Matter*, 2008, **20**, 494218.
9 H. Krienke, V. Vlachy, G. Ahn-Ercan and I. Bako, *J. Phys. Chem. B*, 2009, **113**, 4360–4371.
10 C. Gibb, *J. Am. Chem. Soc.*, 2011, **133**, 7344–7347.
11 E. A. Algaer and N. F. A. van der Vegt, *J. Chem. Phys. B*, 2011, 13781–13787.
12 K. Rembert, J. Paterova, J. Heyda, C. Hilty, P. Jungwirth and P. Cremer, *J. Am. Chem. Soc.*, 2012, **134**, 10039–10046.
13 L. M. Pegram and M. T. Record, *J. Phys. Chem. B*, 2008, **112**, 9428–36.
14 D. F. Evans and H. Wennerström, *The Colloidal Domain - Where Physics, Chemistry, Biology and Technology Meet*, VCH Publishers, New York, 1994.
15 T. L. Hill, *An Introduction to Statistical Thermodynamics*, Dover Publications Inc., New York, 1986.
16 J. Wang, X. Zuo, P. Yu, I.-J. L. Byeon, J. Jung, X. Wang, M. Dyba, S. Seifert, C. D. Schwieters, J. Qin, A. M. Gronenborn and Y.-X. Wang, *J. Am. Chem. Soc.*, 2009, **131**, 10507–10515.
17 N. Metropolis, A. Rosenbluth, M. Rosenbluth, A. Teller, E. Teller and Others, *J. Chem. Phys.*, 1953, **21**, 1087.
18 M. Lund, M. Trulsson and B. Persson, *Source Code Biol. Med.*, 2008, **3**, 1.
19 K. Linderstrøm-Lang, *Compt. Rend. trav. lab. Carlsberg*, 1924, **15**, 1–29.
20 J. G. Kirkwood and J. B. Shumaker, *Proc. Natl. Acad. Sci. U. S. A.*, 1952, **38**, 855–62.
21 M. Lund and B. Jönsson, *Biochemistry*, 2005, **44**, 5722–7.
22 M. L. Grant, *J. Phys. Chem. B*, 2001, **105**, 2858–2863.
23 D. Bratko, A. Striolo, J. Z. Wu, H. W. Blanch and J. M. Prausnitz, *J. Phys. Chem. B*, 2002, **106**, 2714–2720.
24 X. Chen, T. Yang, S. Kataoka and P. Cremer, *J. Am. Chem. Soc.*, 2007, **129**, 12272–12279.
25 Y. Zhang and P. S. Cremer, *Proc. Natl. Acad. Sci. U. S. A.*, 2009, **106**, 15249–53.
26 S. Finet, F. Skouri-Panet, M. Casselyn, F. Bonneté and A. Tardieu, *Curr. Opin. Colloid Interface Sci.*, 2004, **9**, 112–116.
27 S. Zigman, J. Schultz and T. Yulo, *Comp. Biochem. Physiol., Part B: Biochem. Mol. Biol.*, 1983, **75**, 425427.
28 M. Boström, F. W. Tavares, S. Finet, F. Skouri-Panet, A. Tardieu and B. W. Ninham, *Biophys. Chem.*, 2005, **117**, 217–24.
29 M. Jönsson, M. Skepö and P. Linse, *J. Phys. Chem. B*, 2006, **110**, 8782–8.
30 M. Boström, D. F. Parsons, A. Salis, B. W. Ninham and M. Monduzzi, *Langmuir*, 2011, **27**, 9504–9511.

Ionic interactions in biological and physical systems: a variational treatment

Bob Eisenberg

Received 8th April 2012, Accepted 15th May 2012
DOI: 10.1039/c2fd20066j

Chemistry is about chemical reactions. Chemistry is about electrons changing their configurations as atoms and molecules react. Chemistry has for more than a century studied reactions as if they occurred in ideal conditions of infinitely dilute solutions. But most reactions occur in salt solutions that are not ideal. In those solutions everything (charged) interacts with everything else (charged) through the electric field, which is short and long range extending to the boundaries of the system. Mathematics has recently been developed to deal with interacting systems of this sort. The variational theory of complex fluids has spawned the theory of liquid crystals (or *vice versa*). In my view, ionic solutions should be viewed as complex fluids, particularly in the biological and engineering context. In both biology and electrochemistry ionic solutions are mixtures highly concentrated ($\sim$10 M) where they are most important, near electrodes, nucleic acids, proteins, active sites of enzymes, and ionic channels. Ca^{2+} is always involved in biological solutions because the concentration (really free energy per mole) of Ca^{2+} in a particular location is the signal that controls many biological functions. Such interacting systems are not simple fluids, and it is no wonder that analysis of interactions, such as the Hofmeister series, rooted in that tradition has not succeeded as one would hope. Here, we present a variational treatment of hard spheres in a frictional dielectric with the hope that such a treatment of an electrolyte as a complex fluid will be productive. The theory automatically extends to spatially nonuniform boundary conditions and the nonequilibrium systems and flows they produce. The theory is unavoidably self-consistent since differential equations are derived (not assumed) from models of (Helmholtz free) energy and dissipation of the electrolyte. The origin of the Hofmeister series is (in my view) an inverse problem that becomes well posed when enough data from disjoint experimental traditions are interpreted with a self-consistent theory.

Introduction

Science flows in certain directions because scientists, like all people, are creatures of our training. Theories taught in our youth, before we learn critical skills, are hard for us to see objectively, and even harder to modify. But modify we must, as disciplines overlap, questions intermingle, and methods evolve. Modify we must if important questions remain unsolved despite the diligent work of generations of scientists.

Chemistry

Chemistry is first of all about chemical structures and reactions on the atomic scale. Chemists were excited by molecules and flows of electrons within those structures.

Department of Molecular Biophysics and Physiology, Rush University, 1750 West Harrison Street, Chicago IL 60612, USA. E-mail: beisenbe@rush.edu; Fax: +312 942 8711; Tel: +312 942 6467

Chemistry evolved without much thinking about macroscopic flow of molecules in space and time. Chemistry made remarkable advances because of this tight focus on chemical reactions. Physical chemists studied the tiny part of phase space that describes systems without macroscopic flow. The immense power of equilibrium thermodynamics and statistical mechanics[1–3] culminated in the theory of simple liquids, with its cathedrals[4–8] of analysis. The theory of simple liquids provides elegant insights of power and complexity. The founders of thermodynamics would surely have shared my view that the theory of simple liquids is an aesthetic and scientific masterpiece.

Simulations

In the last decades, aesthetics have often been replaced by calculations. Simulations have replaced much of the theory of simple liquids, judging by the number of papers published, but simulations have not yet dealt successfully with most macroscopic phenomena in my opinion,[9] because in most cases simulations have not yet been able to confront macroscopic phenomena in the form found in the laboratory[10–47] and in chemical engineering every day.[48]

As the simulation profession adopts the standards of validation of the experimental community,[48–50] it may well replace analysis as the best connector between atomic scale descriptions of chemical reactions and macroscopic measurements of the output of these reactions. But that has not happened yet, in my view and the consequences are serious.

Simulations are not mathematics, as much as I wish they were. Simulations cannot rely on the theorems that buttress and underlie numerical analysis. In my view, simulations need to be continually checked by comparison with test cases (*e.g.*, conservation laws, model systems with known properties) just as do experiments. Until they fit experiments in conditions similar to those of interest, they have limited use. In the case of biological systems, those conditions are troublesome. All of life occurs in salt solutions derived from oceans. Without the proper ionic environment, cells burst, proteins coagulate, life stops. If simulations are not performed in solutions resembling those of life, they clearly cannot fit the properties of those solutions. Thus, simulations must be done in mixtures of Na^+, K^+, Ca^{2+} and Cl^- ions of substantial concentration (~0.5 M). Biological solutions are mixtures always involving Ca^{2+} ions over a wide range of concentrations (10^{-7} M as intracellular signals; ~10^1 M near and in active sites, nucleic acids, and ion channels). I am unaware of simulations done in these conditions, although surely some will be done soon enough. Until simulations reproduce the properties of life's solutions that are essential to maintain life, they are in a never-never land, in which their awesome detail and power can actually mislead. As any laboratory biologist knows, uncalibrated experiments, or solutions of unknown composition can produce a bewildering array of results, that take more than a lifetime to sort out, most of which are artifacts, having little to do with biological function. One cannot expect uncalibrated simulations to do better than uncalibrated experiments.

Mathematics

Applied mathematics has a history quite different from chemistry. Applied math has always described macroscopic movement and flow. Indeed, mathematics of flow developed rapidly in the 1800's before molecules were thought to be real by many physicists.

Mathematics developed so powerfully that measurements of infinitesimal changes in fluids could be transformed into successful predictions of macroscopic flows. Infinitesimal measurements could be describe by infinitesimal (partial differential) equations (*e.g.*, Navier Stokes) describing matter within boundaries. The meaning of the symbols in the equations could be forgotten, and the differential equations treated as abstractions. The abstractions could be integrated analytically and numerically (as mathematical objects without additional physical assumption).

An infinitesimal treatment could tell us what large masses of fluids do after the infinitesimals were integrated into macroscopic reality by mathematicians who knew nothing of what they were integrating.[51,52] Water flowing in showers, water waves in oceans, even the action of wind on the water could be determined by mathematics. Indeed, computational fluid dynamics can predict complex flows (with shock waves, over complex structures) like those of air around a supersonic airplane, with a power and accuracy that leaves outsiders like me breathless.

Mathematics is different

Mathematics learned to deal with certain classes of systems (conservative systems without dissipation) in a more abstract way, using the variational treatment of Hamilton. This variational approach had the important advantage of dealing with interactions automatically by algebra. The differential equations of motion were derived from the Least Action Principle. Components could be added to the Hamiltonian and the interactions of the components and the differential equations of motions re-derived for the composite system. The interactions of the new and old components appeared automatically in the differential equations without further physical argument. In this way, complex systems could be studied without *ad hoc* physical study of interactions, if the original and composite systems involved the same physics described by similar Hamiltonians.

The complex systems of the mathematicians did not involve chemicals, sad to say. The chemicals of the fluids and the atoms of the Hamiltonian were ignored by mathematicians, just as the interactions and macroscopic flow of the chemicals were ignored by the chemists, for many years. Chemicals were mysterious shadows to most mathematicians, as partial differential equations were mysterious moving shadows, a cinema of darkness, to most chemists.

An interdisciplinary attempt to fill in the shadows

I argue here that problems like the physics and physical chemistry of the Hofmeister series should be viewed in a new light, the additional light of mathematics, with a good measure of engineering insight thrown in. The light of just one discipline casts shadows too deep to see into. Light from another direction can fill those shadows substantially. Of course, the light of the other discipline creates its own shadows. So the trick is to use each discipline's light and ignore its shadows, melding this vision into a coherent self-consistent view of the entire phenomena. It should go without saying that I do not know how to do that, but I do think I know where to aim the light.

The essential difficulty in dealing with Hofmeister effects is, as I see it, that everything depends on interactions. To put it crudely, in the ideal noninteracting world, there is no Hofmeister series.

Inverse problems: reverse engineering the Hofmeister series

Of course, saying everything depends on interactions says too much, and says it too vaguely. There are many types of interactions and keeping them separate is essential if we are to avoid confusion. We need to keep the interactions if we wish to calculate what a solution would do from first principles.

The problem of interactions is much worse if we face it backwards, as I think we actually do. We are actually trying to infer the atomic basis of the Hofmeister series from macroscopic data taken under a wide range of conditions. This is an inverse problem, a problem in reverse engineering. Such problems are notoriously tricky because they are both ill-posed and over-determined, in the typical case. Sorting out the class of physical interactions is crucial if we are to make progress with our inverse problem.[53,54] As difficult as the problems seem, there is real hope of progress, at least in the biological case, when the natural selection of evolution has created a reasonably robust device described by reduced equations,[55] as it apparently has when dealing with the selectivity of calcium,[56,57] sodium,[58] and ryanodine receptor[59] channels.

Let's start with the simplest case. Let's talk about an implicit solvent model of ionic solutions, in which ions are hard spheres, of different charge and diameter, in a dielectric which also provides friction for flow. We consider the possibility that the Hofmeister effect might arise in even this oversimplified system[60] because of interactions that have not been dealt with self-consistently by previous methods. Of course, it is much more likely (from the chemists' perspective) that additional physics, beyond that of the implicit solvent primitive model will be needed to produce the effect. From the mathematicians' perspective, one cannot tell what a model is unless it is computed self-consistently, so the mathematician has no opinion of earlier work.

Consider a homogeneous system of say NaCl. We have spheres of two diameters and two charges. Monte Carlo simulations of this system do surprisingly well compared to real experimental data (see[18] for a recent treatment and references to the enormous literature). Monte Carlo simulations also do well for another monovalent salt say KCl. If we mix NaCl and KCl over a range of concentrations, difficulties arise because of interactions of the hard spheres within the physical model I just described. If divalents are involved—$CaCl_2$, for example—we start with pure simulations of pure $CaCl_2$, and then mix $CaCl_2$ with NaCl and/or KCl.

Mixed solutions have many interaction terms. Even if the physics in these mixtures is the same as in the pure solution (spatially uniform dielectric coefficient, *etc.*), the interactions are difficult to deal with. So many terms are involved, and theories and measurements have enough uncertainties, that it is very difficult to evaluate the coefficients of interactions.

In fact, mixed solutions of this type are very important to chemical engineers and have received a great deal of attention by them, summarized in.[40,43,61] Nonetheless, it seems clear that the many attempts to create a coherent self-consistent model of the activity (free energy per mole) of ionic mixtures has been a difficult one. A generation of physical chemists tried[38,42,43,45–47,62–69] culminating in the compendium of reference 43 that summarized the work of a taskforce of chemical engineers funded for many years for this purpose. Equations of state were awkwardly complex (summarized in ref. 70, see also[71–78]). Apparently promising simplifications[40,69,79] seem not to have taken hold. Modern treatments and compendia of results for chemical engineers[29,61] seem to lack physical basis or 'transferability'. That is to say, models and parameters developed for one type of mixture do not apply to other mixtures and other cases.

Enormous efforts of many workers (only some of which I know about[10–27,30,32–34,59,80–108]) have not produced a coherent view of ionic mixtures, containing divalents, as far as I can tell. This is not just a biologist's view, looking from the outside. A recent volume[29] described progress in dealing with specific properties of ionic solutions of many groups. An overview article by Kunz and Neueder[28] concluded

"It is still a fact that over the last decades, it was easier to fly to the moon than to describe the free energy of even the simplest salt solutions beyond a concentration of 0.1 M or so."

This situation is rather upsetting, particularly to me as a biologist, mostly interested in biological solutions, because our solutions are those that Kunz and Neueder say are not well described, let alone understood.

Biology occurs in salt solutions evolved from the oceans of the world. Biological salt solutions are invariably mixtures of NaCl and KCl with $CaCl_2$. Ca^{2+} plays a particularly important role because the concentration of Ca^{2+} acts as a specific control signal for many of life's processes. Indeed, Ca^{2+} signals are as important and diverse in biology as electrical voltage is in computers! The quotation of Kunz and Neueder implied what I feared:[109–111] the physical chemistry of life's solutions is not understood.

Here I suggest a different approach. I present a new variational treatment of ionic solutions[112–115] that produces (nearly) unique results for mixtures. I propose that this treatment can be used to systematically refine treatments of mixtures allowing a (hopefully) rapid convergence to a full understanding. But this is an approach not a result. The computations to compare with specific experiments have not been done. The feasibility of those computations has been shown, but the detail necessary to compare with results has not been computed.

The variational approach

The approach is to start with a simple model of a pure solution (say the primitive model of NaCl). Write it in a variational form. Derive the appropriate differential equations by the Euler Lagrange process. Learn to integrate those equations efficiently in a setup that allows evaluation of activities. Compare with experiment.

The approach repeats the same procedure with another pure solution, say $CaCl_2$ to be definite.

Then a variational model of a mixture of NaCl and $CaCl_2$ is made (by combining energy and dissipation) and converted into differential equations with boundary conditions appropriate for the experimental setup. The resulting differential equation has minimal adjustable parameters. Most coupling terms are determined by the Euler Lagrange process itself. The differential equation for the mixture is then integrated and applied in conditions that determine activity coefficients of the components of the mixture. Comparison is made with experiments.

When conditions are found in which the model fails to fit experiments, additional physics is included in the variational model. For example, a specific model of electron interaction between Ca^{2+} and water is made, or a specific model of the spatial variation of dielectric coefficient is made, and included in the variational model. This model is then converted into a set of differential equations for mixtures just as was the primitive model, in the last few paragraphs.

Ionic solutions as complex fluids

Ionic solutions are complex fluids because they contain charged particles. Ions in solution interact so strongly through the electric field that they always come 'in pairs' (within ~1 part in 10^{15}). Ionic solutions are necessarily neutral or they explode.[116] Most ions also interact through their finite size.[23,29,35,62,63,111] At biological concentrations, the shape of the ion dramatically distorts the shape of the electric field. The electric field of 1 M spheres (approximately 8 Å apart) of diameter 2 Å is not well approximated by the electric field of 1 M points. Finite size ions cannot occupy the same space, so ionic solutions are particularly complex in crowded confines near electrodes, active sites, and ion channels. Devices that use ions often depend on the nonideal properties of crowded ions. Ionic solutions are not simple fluids, particularly where they are most important, *i.e.* where they are crowded. Ionic solutions need to be treated as the complex fluids that they are, in my opinion.

It is best now to add some specifics to these abstractions. To do that, we consider a specific variational model of ionic solutions,[112–115,117] that has grown from the theory of complex fluids[118–126] developed in the context of liquid crystals,[120,122,127–129] more than anything else.

The approach uses the dissipation principle as described by Liu[118] and co-workers built on the analysis of dissipative systems of Rayleigh [Strutt],[130–132] Onsager,[133–136] and Doi, explicitly[121,137] and implicitly,[129] among many others.[11,119,120,122–126,138–142]

We start with the key fact that the ions in water satisfy the dissipation law

$$\frac{\mathrm{d}}{\mathrm{d}t}\int\overbrace{\left\{k_{\mathrm{B}}T\sum_{i=1}^{N}c_i\log c_i+\frac{1}{2}\left(\rho_0+\sum_{i=1}^{N}z_iec_i\right)\phi\right\}}^{\text{Helmholtz Free Energy}}\mathrm{d}\vec{x}$$

$$=-\underbrace{\int\frac{\mathrm{d}}{\mathrm{d}t}\int\left\{k_{\mathrm{B}}T\sum_{i=1}^{N}c_i\log c_i+\frac{1}{2}\left(\rho_0+\sum_{i=1}^{N}z_iec_i\right)\phi\right\}\mathrm{d}\vec{x}}_{\text{Raleigh [Strutt] Dissipation}}. \tag{1}$$

This ensures that ions follow both an Energetic Variational Principle and also a set of simply derived differential equations. Either can be derived from the other.

$$\frac{\mathrm{d}E}{\mathrm{d}t}=-\Delta\;;\text{ or }\frac{\delta E}{\delta\vec{x}}=\frac{1}{2}\frac{\delta\Delta}{\delta\vec{u}}\;;$$

or

$$\underbrace{\text{Conservative Force}}_{\substack{\text{from variation with respect to}\\ \textbf{Position }\vec{x}}}=\underbrace{\text{Dissipative Force}}_{\substack{\text{from variation with respect to}\\ \textbf{Velocity }\vec{u}}} \tag{2}$$

E is the Helmholtz free energy; Δ is the Rayleigh[Strutt]/Onsager dissipation discussed at length in the literature.[112,118,121,130–136,138,139]

I think it better physically to start with the variational principle because additional fields or components are best handled that way. The variational form of this expression allows easy derivation of the Euler Lagrange differential equations, which can also be written down directly in this case.

$$\begin{aligned}\frac{\partial c_i}{\partial t}&=\nabla\cdot\left(D_i\left(\nabla c_i+\frac{z_ie}{k_{\mathrm{B}}T}c_i\nabla\phi\right)\right),\\ \nabla\cdot(\varepsilon\nabla\phi)&=-\left(\rho_0+\sum_{i=1}^{N}z_iec_i\right).\end{aligned} \tag{3}$$

These are the usual Poisson Nernst Planck equations, called the Vlasov equations in plasma physics,[143–145] drift diffusion in semiconductor physics[146–151] and PNP first[152] in biophysics[11,55,106,111,117,153–212] and then in electrochemistry[158,200,213–215] and applied mathematics,[11,106,117,146,148,149,158,159,165,186,202–204,206–209,213,214,216] in which c_i are the number densities of ions of species i, with diffusion coefficient D_i, valence z_i, charge z_ie with e the unit charge of a proton, diffusion (free) energy of $k_{\mathrm{B}}T$, at electrical potential ϕ, in permittivity ε, with protein permanent charge density ρ_0 as it represents the charge density of doping in semiconductors. ρ_0 depends on location and is zero in bulk solutions.

The coupling of these equations is tricky because the net charge $\sum_{i=1}^{N}z_iec_i$ is well determined by the Laplacian of the electrical potential but is difficult to compute by summing the ***individual*** concentrations c_i because the charge weighted sum is so tiny ($\sim10^{-15}$) compared to individual concentrations. The system must be (nearly) electroneutral or potentials would strip electrons from atoms.

This treatment so far deals with point ions making a solution that is nonideal in a simple way. Interactions are only electrostatic, as in classical Debye–Hückel and Poisson Boltzmann equilibrium theories. This is not the primitive model.

Spatially nonuniform boundary conditions

The variational method and the resulting Poisson Nernst Planck (PNP)[11,55,106,109,117,146–148,151–153,155,159,163,164,169,170,172–175,179,183,184,188,190,193,194,197,201,202,204,206,207,209–212,214–219] equations have substantial advantages, however, over the traditional equilibrium approach. The PNP equations allow spatially nonuniform boundary conditions such as those needed to drive flux. In our electronic technology those spatially nonuniform boundary conditions are the power supply potentials. In biology, those spatially nonuniform boundary conditions are the electrical potential and concentration gradients across the cell membrane. In chemistry, those spatially nonuniform boundary conditions are the concentrations of the classical Nernst equation of electrochemistry and polarography, and perhaps an applied electrical potential as well.

A great deal is gained by using PNP as a nonideal theory allowing flow: all of the components of a computer can be built using circuit elements that satisfy PNP.[148–151,220–231] Resistors, capacitors, and transistor switches, transistor amplifiers, *etc.* are actually mathematical solutions of the PNP equations. Thus, moving to PNP allows one to write "all human knowledge and logic" (that can be performed by a computer) as a solution of a partial differential equation in a spatially complex domain with spatially nonuniform (but otherwise simple) boundary conditions.

Finite size of ions

To deal with the finite size of ions, and extend PNP into a theory of ionic solutions, we need to decide how to describe the finite diameter of the ions. This decision is surely not unique,[10–12,15,16,19,22–25,27,100,103,104,106,232] and the choice of description depends on the feasibility of calculations and the accuracy of fit of the resulting model to a broad range of data, from different types of experiments, in a broad range of solutions and concentrations. This work has not yet been done. The purpose of this paper is to facilitate and motivate such work.

We have included finite size at the length scale of PNP by adding terms to the above equations. The resulting model is variational, so it is a model in the spirit of the theory of complex fluids, but the fluid is the primitive model of electrolytes, leading to peculiar oxymoronic names for our model, like 'complex primitive theory' of electrolytes! Given the present state of checking of our models, (oxy)moronic names are perhaps appropriate.

To be specific, our variational theory describes the (Helmholtz free) energy and dissipation of ionic solutions by

$$\overbrace{\frac{\mathrm{d}}{\mathrm{d}t}\int\left\{k_\mathrm{B}T\sum_{i=n,p}c_i\log c_i+\frac{1}{2}\left(\rho_0+\sum_{i=n,p}z_iec_i\right)\phi+\sum_{i,j=n,p}\frac{c_i}{2}\int\underset{i,j}{\widetilde{\Psi}}\,c_j\,d\vec{y}\right\}d\vec{x}}^{\text{Dissipative}}$$

$$=\underbrace{-\int\left\{\sum_{i=n,p}\frac{D_ic_i}{k_\mathrm{B}T}\left|k_\mathrm{B}T\frac{\nabla c_i}{c_i}+z_ie\nabla\phi-\sum_{j=n,p}\nabla\int\underset{i,j}{\widetilde{\Psi}}\,c_j\,d\vec{y}\right|^2\right\}d\vec{x}}_{\text{Conservative}}\qquad(4)$$

$\underset{i,j}{\widetilde{\Psi}}$ represents the Lennard Jones crowded charges terms defined in references 112,118. The variational formulation and variational derivatives are computed and discussed in those papers.

If ions are modeled as Lennard Jones spheres, the variational principle produces 'Euler Lagrange' equations of a drift-diffusion theory with finite sized solutes that are a generalization and correction of PNP.

$$\frac{\partial c_n}{\partial t} = \nabla \cdot \left[D_n \left\{ \nabla c_n + \frac{c_n}{k_B T} \left(z_n e \nabla \phi - \int \frac{12\varepsilon_{n,n}(a_n + a_n)^{12}(\vec{x} - \vec{y})}{|\vec{x} - \vec{y}|^{14}} c_n(\vec{y})\, d\vec{y} \right.\right.\right.$$
$$\left.\left.\left. - \int \frac{6\varepsilon_{n,p}(a_n + a_p)^{12}(\vec{x} - \vec{y})}{|\vec{x} - \vec{y}|^{14}} c_p(\vec{y})\, d\vec{y} \right) \right\} \right] \quad (5)$$

always combined with the Poisson equation

$$\nabla \cdot (\varepsilon \nabla \phi) = -\left(\rho_0 + \sum_{i=1}^{N} z_i e c_i \right) \; i = n \text{ or } p \quad (6)$$

We write the equation only for negative monovalent ions with valence $z = -1$ to keep the formulas reasonably compact. Programs have been written for all valences. $c_{p,n}(\vec{y})$ is the number density of positive p or negative n ions at location $\vec{y}$. $\varepsilon_{n,n}$ and $\varepsilon_{n,p}$ are coupling coefficients. $a_{p \text{ or } n}$ are the radii of ions.

Field theory of ionic solutions

These equations form a field theory of electrolytic solutions and so pretend to explain all phenomena of those solutions, at equilibrium and in flow. They automatically are self-consistent and change form as new ions are added into the system. Thus, these equations form a necessarily self-consistent model of the flow of mixtures of ionic solutions. Up to now, most such models have been rather *ad hoc*, and one could not be sure if they were self-consistent or not.[44,233–253] Additional fields[254] (*e.g.*, heat flow[138] or convection[112,115]) can be added to our variational model by including their (Helmholtz free) energy in the energy functional E and the dissipation in the dissipation functional Δ of equation (2).

Equations of this generality can be misleading, however, because difficulties often appear in the numerical implementation. Work to date shows that the calculations needed to describe electrochemical cells and ion channels are feasible and not particularly irksome, but it took some years to reach that conclusion. Each experimental setup requires translation into an approximate model, and then numerical procedures must be designed that focus work on the important features of the model these difficulties can be overcome one by one, as they have been in other models of the experimental setups.

Applications to specific problems

The field theory of ionic solutions has been applied to a number of problems of biochemical and chemical interest. The original publications[112,113] computed curves of binding selectivity in two classical channels of considerable biological interest, the calcium channel of cardiac muscle[57,255–268] and the voltage activated sodium channel of nerve.[269–277] Both were represented by the implicit solvent model with side chains represented as spheres free to move within the channel selectivity filter, but not able to move out of that region.[56–58,111,267,278,279] This model has proven remarkably successful in dealing with the important selectivity phenomena of these channels. A single model, with one set of parameters that are never changed, using crystal radii of ions, and one dielectric coefficient and one channel diameter, is able to account for selectivity data in a wide range of solutions (over 4 orders of magnitude of calcium concentration, and in solutions of varying K^+, Na^+, Rb^+ and Cs^+ concentration, for example). The calcium channel is represented by side chains Glu Glu Glu Glu and the sodium channel by Asp Glu Lys Ala. This work is reported in some 35 papers using various numerical methods including the variational model described here. It has been possible to invert the procedure. The inverse problem of determining the distribution of side chains inside the channel from current voltage relations in a range of solutions has actually been explicitly solved, using established methods of inverse problems, including the effects of noise and systematic error.[55]

The variational method allows calculations of current *vs.* time and current *vs.* voltage not previously performed on this reduced model of calcium and sodium channels. Such calculations are of some significance since they confront the measurements performed by hundreds of membrane biologists following the paradigms introduced by Hodgkin and Huxley, reviewed in reference 280.

The variational method was extended[115] to allow computation of the combined effects of water flow and electrodiffusion that underlie the function of many tissues and cells, from kidney, to epithelia, to cerebrospinal fluid and so. Indeed, classical physiology from Harvey to Pappenheimer[281,282] might be considered the study of the flow of water and ions. As far as we know, this paper represents the first selfconsistent treatment of this classical subject, allowing cell volume changes, convection, and ionic electrodiffusion.

The application of the variational method to classical problems of physical chemistry has started with the classical charged wall problem. There we have shown[283] direct comparisons with the Metropolis Monte Carlo calculations of Henderson and Boda,[86,92,284–289] which form a significant fraction of our understanding of ionic solutions in homogeneous and inhomogeneous situations.

The most striking success of these reduced models has been the description of the main calcium channel that controls contraction in muscle (and appears in neurons and many other cell types as well, so far with unknown function), the Ryanodine Receptor Ryr.[290–297] Gillespie has shown that an extension of the reduced model called PNP-DFT[185,298,299] does remarkably well[59,268,300–303] in predicting experiments of some complexity and subtlety (*i.e.*, anomalous mole fraction effects and three cation mixtures) as well as drastic mutations changing charge densities from some 13 molar to zero. The theory used was developed[185,297,298] before selfconsistent variational methods were known to me, and is not selfconsistent. It leaves out the effects of changes in shape of the ionic atmosphere thought to be important 'forever',[44,235–240,247,249,304] and now found in textbooks.[35,36,305] These effects may in fact be unimportant in applications to channels, where the shape of the ionic atmosphere may be mostly determined by a (rather) unchanging protein. But the lack of selfconsistency in PNP-DFT prevents it from being a good model of ionic solutions in physical chemistry, where changes in shape of the ionic atmosphere are thought to be dominant.[44] The variational treatment of ionic conductance[112,113,115,118,283] makes no simplifying assumptions about the shape of the ionic atmosphere: everything (that is in the model) interacts with everything else, to minimize the dissipation and Helmholtz free energy of the model.

This work is more practical science than vague theory. It is meant to connect the mathematical results of the theory of complex fluids with the practical problems of experimental biophysicists and physical chemists. Sadly, the authors know much less of those problems in physical chemistry than in biophysics, in which they have worked for many years. Thus, the utility of this approach for classical physical chemistry remains to be seen. Applications to the Hoffmeister series, and the experiments that define the Hoffmeister series are one important part of our approach to classical physical chemistry.

Experimental setups

The experimental setups used to measure the Hofmeister series are varied and in my view must be understood in some detail before they can be well modelled by this new approach. The subtleties are nontrivial, as can be seen in the magnificent treatise of Hünenberger and Reif[16] which describes thousands of papers using hundreds of methods (I estimate) to determine the free energy per mole of individual ions. Many different setups are used and each requires a separate idealization, model, and mathematical description. In my view, theories should be tested against idealizations of experimental setups and not against metaphorical abstractions like thermodynamic 'theorems'. Too often, the theorems of thermodynamics and statistical mechanics have hidden assumptions in my personal opinion. Fitting data to metaphors is likely to cause even more confusion in that case, compounding the felony, as the legal saying goes. Fitting to real data avoids this problem, at the cost of some complexity in analysis.

Application of our variational model to the Hofmeister series requires models and calculations of the free energy per mole of individual ions. Those calculations will need the help of workers familiar with the experimental setups. As a primitive model for the Hofmeister series[60]—presented here in variational clothing—fails to fit some data, as it surely will, additional physics will no doubt need to be added, and the less primitive models will then have to be compared to experiments. It is not at all clear how much improvement a fully self-consistent model will provide at equilibrium in solutions of one salt. One may expect more improvement when interactions are larger and more complex, in concentrated environments, in mixtures, and in divalents. Of course, those are just the characteristics of biological solutions, so one can expect variational models to be most important as they are applied to living systems.

Along the way, the variational method will also allow introduction of boundary conditions and nonequilibrium properties that arise from spatially nonuniform concentrations and voltages. In my view, this is an important social and scientific advantage. Socially, this unites the classical thermodynamic approach (which had been isolated in its thermodynamic limit) with the classical field theories of the rest of physics. Scientifically, the nonequilibrium systems have properties of enormous importance that provide, for example, our modern electronic technology, and also allow voltage signals ('action potentials') to propagate along our nerve fibres.

Specific plan to implement a variational approach

I present a specific plan that might implement these ideas. I am very aware that it is dangerous and presumptuous to present any plan for science. Science proceeds best when it is initiated and energized by the curiosity of individuals. The following is meant to provoke discussion to test the utility of the variational approach.

1) identify a specific set of experimental measurements that operationally define 'the Hofmeister Effect'.

2) make a physical model of the setup of those experiment.

3) make a model of how estimates of parameters are made from the measurements of that experiment.

4) make a mathematical model of the physical model, in the spirit of the variational theory of complex fluids.

5) derive the field theory (Euler Lagrange) equations of the model.

6) solve the equations of the model for the experimental boundary conditions.

7) compute estimates of parameters from the mathematical solutions of the equations, treating the output of the equations as if they were outputs from the experimental setup.

8) compare with experiments in a range of solutions and concentrations.

9) revise the model to include more and different physics to see if the fit to experiments is improved.

Chemical conclusions

Chemistry has focused on atoms and molecules with less concern about the surrounding environment, understandably enough, since mathematics was not available to deal with systems in which 'everything' interacts with 'everything' else. The solutions containing these atoms and molecules were treated as ideal fluids, in which nothing interacts with nothing, if my lapse in grammar may be excused. The variational theory of complex fluids has successfully dealt with systems in which everything interacts with everything else over many scales.

Chemical reactions in the liquid state

In my admittedly biased view, chemical reactions in the liquid state should be analysed in the framework of complex fluids. Chemical reactions are interactions of reactants (microelements in the language of the theory of complex fluids) involving rearrangements of the internal (electronic) structure of reactants, according to the wave equation of the electron, Schrödinger's equation. Daunting interactions of electrons, microelements, and the macroscopic world can be handled automatically and self-consistently in the theory of complex fluids that cannot easily be handled by any other method I know of.

Future history of physical chemistry

The future history of physical chemistry will deal with ionic solutions as complex fluids, I believe. When mixtures and divalents and flow are involved, the variational approach to complex fluids is likely to lead to substantial progress. The crucial step is to force the theory to confront the actual properties of such solutions as measured with many techniques under many conditions. Determining an appropriate model of ionic solutions is an inverse problem that requires masses of disjoint data, and a variational theory, to provide useful well-posed results.

Acknowledgements

This paper is possible because of the support of others. Chun Liu, more than anyone else, has developed the energetic variational approach to the theory of complex fluids (and liquid crystals) and spared no effort to help me understand how we can apply it to ions in liquids and channels. The physical chemistry community in general, represented in the present instance by Pavel Jungwirth, has welcomed my view[9,109–111,266,306] of long standing approaches in a most generous way, in the best of the academic tradition, with criticism and scepticism to be sure, but always listening. The computational electronics community[195,307–317] has been similarly generous. Biophysicists have not always been like that.[9,109–111,163,164,171,177,266,306,318–321] Support was from Rush Medical College and I am particularly grateful for the administrative tasks the Dean's Office has not asked me to do.

References

1 L. Boltzmann, *Lectures on Gas Theory*, University of California, Berkeley CA, 1964.
2 S. G. Brush, *The Kind of Motion We Call Heat*, North Holland, New York, 1986.
3 E. Garber, S. G. Brush and C. W. F. Everitt, *Maxwell on Molecules and Gases*, MIT Press, Cambridge MA, 1986.
4 J. Barker and D. Henderson, *Rev. Mod. Phys.*, 1976, **48**, 587–671.
5 J.-L. Barratt and J.-P. Hansen, *Basic concepts for simple and complex liquids*, Cambridge University Press, 2003.
6 J.-P. Hansen and I. R. McDonald, *Theory of Simple Liquids*, First Edition, Academic Press, New York, 1986.
7 J.-P. Hansen and I. R. McDonald, *Theory of Simple Liquids*, Third Edition, Academic Press, New York, 2006.
8 S. A. Rice and P. Gray, *Statistical Mechanics of Simple Fluids*, Interscience (Wiley), New York, 1965.
9 B. Eisenberg, *J. Phys. Chem. C*, 2010, **114**, 20719–20733.
10 S. Tazi, J. J. Molina, B. Rotenberg, P. Turq, R. Vuilleumier and M. Salanne, *J. Chem. Phys.*, 2012, **136**, 114507–114512.
11 Q. Zheng, D. Chen and G.-W. Wei, *J. Comput. Phys.*, 2011, **230**, 5239–5262.
12 T. Xiao and X. Song, *J. Chem. Phys.*, 2011, **135**, 104104–104114.

13 M. M. Reif and P. H. Hunenberger, *J. Chem. Phys.*, 2011, **134**, 144103–144130.
14 M. M. Reif and P. H. Hunenberger, *J. Chem. Phys.*, 2011, **134**, 144104–144125.
15 J. J. Molina, J.-F. Dufreche, M. Salanne, O. Bernard and P. Turq, *J. Chem. Phys.*, 2011, **135**, 234509–234520.
16 P. H. Hünenberger and M. Reif, *Single-Ion Solvation*, RSC Publishing, Cambridge UK, 2011.
17 M. B. Gee, N. R. Cox, Y. Jiao, N. Bentenitis, S. Weerasinghe and P. E. Smith, *Journal of Chemical Theory and Computation*, 2011.
18 J. Vincze, M. Valisko and D. Boda, *J. Chem. Phys.*, 2010, **133**, 154507–154506.
19 Y. V. Kalyuzhnyi, V. Vlachy and K. A. Dill, *Phys. Chem. Chem. Phys.*, 2010, **12**, 6260–6266.
20 I. Kalcher, J. C. F. Schulz and J. Dzubiella, *Phys. Rev. Lett.*, 2010, **104**, 097802.
21 J. J. Howard, J. S. Perkyns and B. M. Pettitt, *J. Phys. Chem. B*, 2010, **114**, 6074–6083.
22 D. Fraenkel, *J. Phys. Chem. B*, 2010, **115**, 557–568.
23 D. Fraenkel, *Mol. Phys.*, 2010, **108**, 1435–1466.
24 L. Vrbka, M. Lund, I. Kalcher, J. Dzubiella, R. R. Netz and W. Kunz, *J. Chem. Phys.*, 2009, **131**, 154109–154112.
25 Y. Luo and B. t. Roux, *J. Phys. Chem. Lett.*, 2009, **1**, 183–189.
26 B. Li, *SIAM J. Math. Anal.*, 2009, **40**, 2536–2566.
27 B. Li, *Nonlinearity*, 2009, **22**, 811.
28 W. Kunz and R. Neueder, in *Specific Ion Effects*, ed. W. Kunz, World Scientific Singapore, 2009, pp. 11–54.
29 W. Kunz, *Specific Ion Effects*, World Scientific Singapore, 2009.
30 Z. Abbas, E. Ahlberg and S. Nordholm, *J. Phys. Chem. B*, 2009, **113**, 5905–5916.
31 L. L. Lee, *Molecular Thermodynamics of Electrolyte Solutions*, World Scientific Singapore, 2008.
32 P. Jungwirth and B. Winter, *Annu. Rev. Phys. Chem.*, 2008, **59**, 343–366.
33 P. Grochowski and J. Trylska, *Biopolymers*, 2008, **89**, 93–113.
34 A. Grattoni, M. Merlo and M. Ferrari, *J. Phys. Chem. B*, 2007, **111**, 11770–11775.
35 W. R. Fawcett, *Liquids, Solutions, and Interfaces: From Classical Macroscopic Descriptions to Modern Microscopic Details*, Oxford University Press, New York, 2004.
36 K. J. Laidler, J. H. Meiser and B. C. Sanctuary, *Physical Chemistry*, Fourth edn., BrooksCole, Belmont CA, 2003.
37 T. N. Heinz, W. F. van Gunsteren and P. H. Hunenberger, *J. Chem. Phys.*, 2001, **115**, 1125–1136.
38 K. S. Pitzer, *Thermodynamics*, 3rd edn., McGraw Hill, New York, 1995.
39 A. Chhih, O. Bernard, J. M. G. Barthel and L. Blum, *Ber. Bunsenges. Phys. Chem.*, 1994, **98**, 1516–1525.
40 V. S. Patwardhan and A. Kumar, *AIChE J.*, 1993, **39**, 711–714.
41 H. Cabezas and J. P. O'Connell, *Ind. Eng. Chem. Res.*, 1993, **32**, 2892–2904.
42 K. S. Pitzer, *Activity Coefficients in Electrolyte Solutions*, CRC Press, Boca Raton FL USA, 1991.
43 J. F. Zemaitis, Jr., D. M. Clark, M. Rafal and N. C. Scrivner, *Handbook of Aqueous Electrolyte Thermodynamics*, Design Institute for Physical Property Data, American Institute of Chemical Engineers, New York, 1986.
44 J.-C. Justice, in *Comprehensive Treatise of Electrochemistry Volume 5 Thermondynbamic and Transport Properties of Aqueous and Molten Electrolytes*, eds. B. E. Conway, J. O. M. Bockris and E. Yaeger, Plenum, New York, 1983, pp. 223–338.
45 R. M. Pytkowicz, *Activity Coefficients in Electrolyte Solutions*, CRC, Boca Raton FL USA, 1979.
46 R. A. Robinson and R. H. Stokes, *Electrolyte Solutions*, Second edn., Butterworths Scientific Publications, London, 1959.
47 H. S. Harned and B. B. Owen, *The Physical Chemistry of Electrolytic Solutions*, Third edn., Reinhold Publishing Corporation, New York, 1958.
48 E. J. Maginn, *AIChE J.*, 2009, **55**, 1304–1310.
49 D. E. Post and L. G. Votta, *Phys. Today*, 2005, **58**, 35–41.
50 B. Eisenberg, *Physics Today*, 2005, **58**, p.13 (Letter to the Editor).
51 O. Zikanov, *Essential Computational Fluid Dynamics*, Wiley, New York, 2010.
52 J. Anderson, *Computational Fluid Dynamics*, McGraw-Hill Science/Engineering/Math, 1995.
53 H. W. Engl, M. Hanke and A. Neubauer, *Regularization of Inverse Problems* Kluwer, Dordrecht, The Netherlands, 2000.
54 J. Kaipio and E. Somersalo, *Statistical and Computational Inverse Problems* Springer, New York, 2005.
55 M. Burger, R. S. Eisenberg and H. Engl, *SIAM J. Appl. Math.*, 2007, **67**, 960–989.

56 D. Boda, W. Nonner, D. Henderson, B. Eisenberg and D. Gillespie, *Biophys. J.*, 2008, **94**, 3486–3496.
57 D. Boda, M. Valisko, D. Henderson, B. Eisenberg, D. Gillespie and W. Nonner, *J. Gen. Physiol.*, 2009, **133**, 497–509.
58 D. Boda, W. Nonner, M. Valisko, D. Henderson, B. Eisenberg and D. Gillespie, *Biophys. J.*, 2007, **93**, 1960–1980.
59 D. Gillespie, *Biophys. J.*, 2008, **94**, 1169–1184.
60 M. Lund, B. Jagoda-Cwiklik, C. E. Woodward, R. Vacha and P. Jungwirth, *J. Phys. Chem. Lett.*, 2010, **1**, 300–303.
61 G. M. Kontogeorgis and G. K. Folas, *Thermodynamic Models for Industrial Applications: From Classical and Advanced Mixing Rules to Association Theories*, John Wiley & Sons, Ltd, 2009.
62 C. A. Kraus, *Bull. Am. Math. Soc.*, 1938, **44**, 361–383.
63 H. L. Friedman, *Ionic Solution Theory*, Interscience Publishers, New York, 1962.
64 P. M. V. Resibois, *Electrolyte Theory*, Harper & Row, New York, 1968.
65 B. E. Conway, *Electrochemical Data*, Greenwood Press Publishers, Westport CT USA, 1969.
66 K. S. Pitzer and J. J. Kim, Journal of the American Chemical Society, 1974, **96**, 5701–5707.
67 H. L. Friedman and W. D. T. Dale, in *Statistical Mechanics, Part A: Equilibrium Techniques*, ed. B. J. Berne, Plenum Press, Nwe York, 1977, vol. 1, pp. 85–135: Ch. 133.
68 A. L. Hovarth, *Handbook of aqueous electrolyte solutions: physical properties, estimation, and correlation methods*, Ellis Horwood, New York, 1985.
69 V. S. Patwardhan and A. Kumar, *AIChE J.*, 1986, **32**, 1429–1438.
70 J. V. Sengers, R. F. Kayser, C. J. Peters and H. J. White, Jr., *Equations of State for Fluids and Fluid Mixtures (Experimental Thermodynamics)* Elsevier, New York, 2000.
71 R. Inchekel, J.-C. de Hemptinne and W. Fürst, *Fluid Phase Equilib.*, 2008, **271**, 19–27.
72 R. T. Jacobsen, S. G. Penoncello, E. W. Lemmon and R. Span, in *Equations of State for Fluids and Fluid Mixtures*, eds. J. V. Sengers, R. F. Kayser, C. J. Peters and H. J. White, Jr., Elsevier, New York, 2000, pp. 849–882.
73 B. Larsen, *J. Chem. Phys.*, 1976, **65**, 3431–3438.
74 Y. Lin, K. Thomen and J.-C. d. Hemptinne, *AIChE J.*, 2007, **53**, 989–1005.
75 N. Metropolis, A. W. Rosenbluth, M. N. Rosenbluth, A. H. Teller and E. Teller, *J. Chem. Phys.*, 1953, **21**, 1087–1092.
76 J. A. Myers, S. I. Sandler and R. H. Wood, *Ind. Eng. Chem. Res.*, 2002, **41**.
77 G. Odriozola and L. Berthier, *J. Chem. Phys.*, 2011, **134**, 054504–054509.
78 Y. Rosenfeld, *Phys. Rev. E: Stat. Phys., Plasmas, Fluids, Relat. Interdiscip. Top.*, 1996, **54**, 2827–2838.
79 A. Kumar and V. S. Patwardhan, *AIChE J.*, 1992, **38**, 793–796.
80 M. A. Kastenholz and P. H. Hunenberger, *J. Chem. Phys.*, 2006, **124**, 224501–224520.
81 M. A. Kastenholz and P. H. Hunenberger, *J. Chem. Phys.*, 2006, **124**, 124108–124112.
82 M. A. Kastenholz and P. H. Hunenberger, *J. Chem. Phys.*, 2006, **124**, 124106–124127.
83 P. B. Petersen and R. J. Saykally, *Annu. Rev. Phys. Chem.*, 2006, **57**, 333–364.
84 H. Kokubo, J. Rosgen, D. W. Bolen and B. M. Pettitt, *Biophys. J.*, 2007, **93**, 3392–3407.
85 A. A. Kornyshev, *J. Phys. Chem. B*, 2007, **111**, 5545–5557.
86 D. Boda and D. Henderson, *Mol. Phys.*, 2008, **106**, 2367–2370.
87 J. Che, J. Dzubiella, B. Li and J. A. McCammon, *J. Phys. Chem. B*, 2008, **112**, 3058–3069.
88 M. V. Fedorov and A. A. Kornyshev, *J. Phys. Chem. B*, 2008, **112**, 11868–11872.
89 I. S. Joung and T. E. Cheatham, *J. Phys. Chem. B*, 2008, **112**, 9020–9041.
90 I. V. Khavrutskii, J. Dzubiella and J. A. McCammon, *J. Chem. Phys.*, 2008, **128**, 044106.
91 B.-Y. Dan, D. Andelman, D. Harries and R. Podgornik, *J. Phys.: Condens. Matter*, 2009, **21**, 424106.
92 D. Henderson and D. Boda, *Phys. Chem. Chem. Phys.*, 2009, **11**, 3822–3830.
93 D. Horinek, S. Mamatkulov and R. Netz, *J. Chem. Phys.*, 2009, **130**, 124507–124521.
94 J. G. Ibarra-Armenta, A. Martin-Molina and M. Quesada-Perez, *Phys. Chem. Chem. Phys.*, 2009, **11**, 309–316.
95 J. JaneCek and R. R. Netz, *J. Chem. Phys.*, 2009, **130**, 074502–074515.
96 I. Kalcher and J. Dzubiella, *J. Chem. Phys.*, 2009, **130**, 134507.
97 S. Buyukdagli, M. Manghi and J. Palmeri, *Phys. Rev. Lett.*, 2010, **105**, 158103.
98 I. Kalcher, J. C. F. Schulz and J. Dzubiella, *J. Chem. Phys.*, 2010, **133**, 164511–164515.
99 F. Lipparini, G. Scalmani, B. Mennucci, E. Cances, M. Caricato and M. J. Frisch, *J. Chem. Phys.*, 2010, **133**, 014106–014111.
100 H. Paul and S. Matthias, *J. Phys.: Condens. Matter*, 2010, **22**, 325108.
101 G. b. Rutkai, D. Boda and T. s. Kristóf, *J. Phys. Chem. Lett.*, 2010, **1**, 2179–2184.
102 H. Saidani, N. Ben Amar, J. Palmeri and A. Deratani, *Langmuir*, 2010, **26**, 2574–2583.
103 J. Sala, E. Guardia and J. Marti, *J. Chem. Phys.*, 2010, **132**, 214505–214511.

104 H. Yu, T. W. Whitfield, E. Harder, G. Lamoureux, I. Vorobyov, V. M. Anisimov, A. D. MacKerell and B. t. Roux, *J. Chem. Theory Comput.*, 2010, **6**, 774–786.
105 C. Zhang, S. Raugei, B. Eisenberg and P. Carloni, *J. Chem. Theory Comput.*, 2010, **6**, 2167–2175.
106 B. Li, X. Cheng and Z. Zhang, *SIAM J. Appl. Math.*, 2011, **71**, 2093–2111.
107 S. Zhao, Z. Jin and J. Wu, *The Journal of Physical Chemistry B*, 2011.
108 N. Schwierz, D. Horinek and R. R. Netz, *Langmuir*, 2010, **26**, 7370–7379.
109 B. Eisenberg, *Chem. Phys. Lett.*, 2011, **511**, 1–6.
110 B. Eisenberg, *Posted on arXiv.org with Paper ID arXiv:1105.0184v1*, 2011.
111 B. Eisenberg, in *Advances in Chemical Physics*, John Wiley & Sons, Inc., 2011, pp. 77–223 also available at http:\\arix.org as arXiv 1009.1786v1001.
112 B. Eisenberg, Y. Hyon and C. Liu, *J. Chem. Phys.*, 2010, **133**, 104104.
113 Y. Hyon, B. Eisenberg and C. Liu, *Communications in Mathematical Sciences*, 2011, **9**, 459–475 also available as preprint# 2318 (IMA, University of Minnesota, Minneapolis) http://www.ima.umn.edu/preprints/jun2010/jun2010.html, 2010.
114 Y. Hyon, J. E. Fonseca, B. Eisenberg and C. Liu, *Biophys. J.*, 2011, **100**, 578a.
115 Y. Mori, C. Liu and R. S. Eisenberg, *Phys. D*, 2011, **240**, 1835–1852.
116 A. Singer, Z. Schuss and R. S. Eisenberg, *J. Stat. Phys.*, 2005, **119**, 1397–1418.
117 Q. Zheng and G.-W. Wei, *J. Chem. Phys.*, 2011, **134**, 194101–194117.
118 Y. Hyon, D. Y. Kwak and C. Liu, available at URL: http://www.ima.umn.edu as IMA Preprint Series # 2228, 2010, **26**, 1291–1304, available at URL: http://www.ima.umn.edu as IMA Preprint Series # 2228.
119 Y. Hyon, Q. Du and C. Liu, *J. Comput. Theor. Nanosci.*, 2010, **7**, 756–765.
120 C. Liu, in *Multi-scale Phenomena in Complex Fluids: Modeling, Analysis and Numerical Simulations*, ed. T. Y. Hou, Liu, C., Liu, J.-g, World Scientific Publishing Company, Singapore, 2009.
121 M. Doi, *J. Phys. Soc. Jpn.*, 2009, **78**, 052001.
122 J. Zhang, X. Gong, C. Liu, W. Wen and P. Sheng, *Phys. Rev. Lett.*, 2008, **101**, 194503.
123 P. Sheng, J. Zhang and C. Liu, *Prog. Theor. Phys., Suppl.*, 2008, **175**, 131–143.
124 Y. Hyon, J. A. Carrillo, Q. Du and C. Liu, *Kinetic and Related Models*, 2008, **1**, 171–184.
125 F.-H. Lin, C. Liu and P. Zhang, *Commun. Pure Appl. Math.*, 2007, **60**, 838–866.
126 F.-H. Lin, C. Liu and P. Zhang, *Commun. Pure Appl. Math.*, 2005, **58**, 1437–1471.
127 R. G. Larson, *The Structure and Rheology of Complex Fluids*, Oxford, New York, 1995.
128 P.-G. d. Gennes and J. Prost, *The Physics of Liquid Crystals*, Oxford University Press, New York, 1993.
129 M. Doi and S. F. Edwards, *The Theory of Polymer Dynamics*, Oxford University Press, New York, 1988.
130 L. Rayleigh and previously John Strutt, *Proceedings of the London Mathematical Society*, 1873, **IV**, 357–368.
131 L. Rayleigh and previously John Strutt, *Phil. Mag.*, 1892, **33**, 209.
132 L. Rayleigh and previously John Strutt, *Theory of Sound*, 2 edn., Dover Reprint 1976, New York, 1896.
133 S. Machlup and L. Onsager, *Phys. Rev.*, 1953, **91**, 1512.
134 L. Onsager, *Phys. Rev.*, 1931, **38**, 2265–2279.
135 L. Onsager, *Phys. Rev.*, 1931, **37**, 405–109.
136 L. Onsager and S. Machlup, *Phys. Rev.*, 1953, **91**, 1505–1512.
137 M. Doi, *J. Phys.: Condens. Matter*, 2011, **23**, 284118.
138 M. A. Biot, *Variational Principles in Heat Transfer: A Unified Lagrangian Analysis of Dissipative Phenomena*, Oxford University Press, New York, 1970.
139 E. Sciubba, *International Journal of Thermodynamics*, 2004, **7**, 115–122.
140 T. Y. Hou, C. Liu and J.-g. Liu, *Multi-scale Phenomena in Complex Fluids: Modeling, Analysis and Numerical Simulations*, World Scientific Publishing Company, Singapore, 2009.
141 S. Buyukdagli, M. Manghi and J. Palmeri, *Phys. Rev. E: Stat., Nonlinear, Soft Matter Phys.*, 2010, **81**, 041601.
142 G.-W. Wei, *Bull. Math. Biol.*, 2010, **72**, 1562–1622.
143 R. J. Goldston and P. H. Rutherford, *Introduction to Plasma Physics*, Institute of Physics Publishing, London, 1995.
144 R. M. Kulsrud, *Plasma Physics for Astrophysics*, Princeton, NJ, USA, 2005.
145 E. Mason and E. McDaniel, *Transport Properties of Ions in Gases*, John Wiley and Sons, NY, 1988.
146 W. Van Roosbroeck, *Bell System Technical Journal*, 1950, **29**, 560–607.
147 J. R. Macdonald, *Phys. Rev.*, 1953, **92**, 4–17.
148 P. A. Markowich, C. A. Ringhofer and C. Schmeiser, *Semiconductor Equations*, Springer-Verlag, New York, 1990.

149 J. W. Jerome, *Analysis of Charge Transport. Mathematical Theory and Approximation of Semiconductor Models*, Springer-Verlag, New York, 1995.
150 M. Lundstrom, *Fundamentals of Carrier Transport*, Second edn., Addison-Wesley, NY, 2000.
151 S. Selberherr, *Analysis and Simulation of Semiconductor Devices*, Springer-Verlag, New York, 1984.
152 R. Eisenberg and D. Chen, *Biophysical Journal*, 1993, **64**, A22.
153 R. de Levie and H. Moreira, *J. Membr. Biol.*, 1972, **9**, 241–260.
154 D. G. Levitt, *Biophys. J.*, 1975, **15**, 533–551.
155 D. G. Levitt, *Biophys. J*, 1982, **37**, 575–587.
156 K. Cooper, E. Jakobsson and P. Wolynes, *Prog. Biophys. Mol. Biol.*, 1985, **46**, 51–96.
157 D. G. Levitt, *Biophys. J.*, 1985, **52**, 575–587.
158 I. Rubinstein, *Electro-diffusion of ions*, SIAM, Philadelphia, 1990.
159 V. Barcilon, D. P. Chen and R. S. Eisenberg, *SIAM J. Appl. Math.*, 1992, **52**, 1405–1425.
160 D. P. Chen, V. Barcilon and R. S. Eisenberg, *Biophys. J.*, 1992, **61**, A11.
161 D. P. Chen, V. Barcilon and R. S. Eisenberg, *Biophys. J.*, 1992, **61**, 1372–1393.
162 D. P. Chen and R. S. Eisenberg, *Biophys. J.*, 1993, **64**, 1405–1421.
163 R. S. Eisenberg, *J. Membr. Biol.*, 1996, **150**, 1–25. Also available on http:\\arxiv.org as arXiv 1009.2857.
164 R. S. Eisenberg, in *New Developments and Theoretical Studies of Proteins*, ed. R. Elber, World Scientific, Philadelphia, 1996, vol. 7, pp. 269–357. Published in the Physics ArXiv as arXiv:0807.0715.
165 V. Barcilon, D. P. Chen, R. S. Eisenberg and J. W. Jerome, *SIAM J. Appl. Math.*, 1997, **57**, 631–648.
166 D. P. Chen, J. Lear and R. S. Eisenberg, *Biophys. J.*, 1997, **72**, 97–116.
167 S.-H. Chung, M. Hoyles, T. Allen and S. Kuyucak, *Biophys. J.*, 1998, **75**, 793–809.
168 W. Nonner and B. Eisenberg, *Biophys. J.*, 1998, **75**, 1287–1305.
169 B. Corry, S. Kuyucak and S. H. Chung, *J. Gen. Physiol.*, 1999, **114**, 597–599.
170 G. R. Dieckmann, J. D. Lear, Q. Zhong, M. L. Klein, W. F. DeGrado and K. A. Sharp, *Biophys. J.*, 1999, **76**, 618–630.
171 R. S. Eisenberg, *J. Membr. Biol.*, 1999, **171**, 1–24.
172 D. Gillespie, Rush, 1999.
173 U. Hollerbach, D. Chen, W. Nonner and B. Eisenberg, *Biophysical Journal*, 1999, **76**, A205.
174 M. G. Kurnikova, R. D. Coalson, P. Graf and A. Nitzan, *Biophys. J.*, 1999, **76**, 642–656.
175 A. E. Cardenas, R. D. Coalson and M. G. Kurnikova, *Biophys. J.*, 2000, **79**.
176 B. Corry, S. Kuyucak and S. H. Chung, *Biophys. J.*, 2000, **78**, 2364–2381.
177 B. Eisenberg, in *Biophysics Textbook On Line "Channels, Receptors, and Transporters"* http://www.biophysics.org/btol/channel.html#5, ed. L. J. DeFelice, Published in ArXiv as arXiv:0807.0721, 2000.
178 P. Graf, A. Nitzan, M. G. Kurnikova and R. D. Coalson, *J. Phys. Chem. B*, 2000, **104**, 12324–12338.
179 U. Hollerbach, D. P. Chen, D. D. Busath and B. Eisenberg, *Langmuir*, 2000, **16**, 5509–5514.
180 W. Im, S. Seefeld and B. Roux, *Biophys. J.*, 2000, **79**, 788–801.
181 G. Moy, B. Corry, S. Kuyucak and S. H. Chung, *Biophys. J.*, 2000, **78**, 2349–2363.
182 W. Im and B. Roux, *Biophysical Journal*, 2001, **115**, 4850–4861.
183 Z. Schuss, B. Nadler and R. S. Eisenberg, *Phys. Rev. E: Stat. Phys., Plasmas, Fluids, Relat. Interdiscip. Top.*, 2001, **64**, 036116.
184 T. A. van der Straaten, R. S. Eisenberg, J. M. Tang, U. Ravaioli and N. Aluru, *Biophysical Journal*, 2001, **80**, 115a.
185 D. Gillespie, W. Nonner and R. S. Eisenberg, *J. Phys.: Condens. Matter*, 2002, **14**, 12129–12145.
186 U. Hollerbach, D.-P. Chen and R. S. Eisenberg, *Journal of Computational Science*, 2002, **16**, 373–409.
187 U. Hollerbach and R. Eisenberg, *Langmuir*, 2002, **18**, 3262–3631.
188 W. Im and B. Roux, *J. Mol. Biol.*, 2002, **322**, 851–869.
189 W. Im and B. Roux, *J. Mol. Biol.*, 2002, **319**, 1177–1197.
190 Z. Schuss, B. Nadler, A. Singer and R. Eisenberg, AIP Conference Proceedings, 3–6 September 2002: Unsolved Problems Of Noise And Fluctuations, UPoN 2002, 3rd International Conference on Unsolved Problems of Noise and Fluctuations in Physics, Biology, and High Technology Washington, DC, 2002.
191 B. Corry, S. Kuyucak and S. H. Chung, *Biophys. J.*, 2003, **84**, 3594–3606.
192 B. Eisenberg, *Journal of Computational Electronics*, 2003, **2**, 245–249.

193 A. B. Mamonov, R. D. Coalson, A. Nitzan and M. G. Kurnikova, *Biophys. J.*, 2003, **84**, 3646–3661.
194 B. Nadler, U. Hollerbach and R. S. Eisenberg, *Phys. Rev. E: Stat. Phys., Plasmas, Fluids, Relat. Interdiscip. Top.*, 2003, **68**, 021905.
195 T. A. van der Straaten, J. M. Tang, U. Ravaioli, R. S. Eisenberg and N. R. Aluru, *Journal of Computational Electronics*, 2003, **2**, 29–47.
196 M. Aguilella-Arzo, V. Aguilella and R. S. Eisenberg, *Eur. Biophys. J.*, 2005, **34**, 314–322.
197 R. D. Coalson and M. G. Kurnikova, *IEEE Trans. NanoBiosci.*, 2005, **4**, 81–93.
198 B. Corry and S.-H. Chung, *Eur. Biophys. J.*, 2005, **34**, 208–216.
199 B. Lu, D. Zhang and J. A. McCammon, *J. Chem. Phys.*, 2005, **122**, 214102–214107.
200 M. Saraniti, S. Aboud and R. S. Eisenberg, *Rev. Comput. Chem.*, 2006, **22**, 229–294.
201 B. Eisenberg and W. Liu, *SIAM J. Math. Anal.*, 2007, **38**, 1932–1966.
202 N. Abaid, R. S. Eisenberg and W. Liu, *SIAM J. Appl. Dyn. Syst.*, 2008, **7**, 1507.
203 A. Singer, D. Gillespie, J. Norbury and R. S. Eisenberg, *Eur. J. Appl. Math.*, 2008, **19**, 541–560.
204 K. Arning, Johannes Kepler University Linz, 2009.
205 D. G. Luchinsky, R. Tindjong, I. Kaufman, P. V. E. McClintock and R. S. Eisenberg, *Phys. Rev. E: Stat., Nonlinear, Soft Matter Phys.*, 2009, **80**, 021925–021912.
206 A. Singer and J. Norbury, *SIAM J. Appl. Math.*, 2009, **70**, 949–968.
207 W. Liu and B. Wang, *Journal of Dynamics and Differential Equations*, 2010, **22**, 413–437.
208 C. Hsiao-Bing, C. Li-Tien and L. Bo, *Nonlinearity*, 2011, **24**, 3215.
209 S. Zhou, Z. Wang and B. Li, *Physical Review E*, 2011, **84**.
210 E. Csanyi, D. Boda, D. Gillespie and T. Kristof, *Biochim. Biophys. Acta, Biomembr.*, 2012, **1818**, 592–600.
211 B. Eisenberg, Ions in Fluctuating Channels: Transistors Alive, *Fluctuations and Noise Letters*, 2012, **11**(2), 76–96.
212 R. Tindjong, I. Kaufman, P. V. E. Mcclintock, D. G. Luchinsky and R. S. Eisenberg, *Fluctuation Noise Lett.*, 2012, **11**, 83–93.
213 M. Z. Bazant, M. S. Kilic, B. D. Storey and A. Ajdari, *Adv. Colloid Interface Sci.*, 2009, **152**, 48–88.
214 M. Z. Bazant, K. Thornton and A. Ajdari, *Phys. Rev. E: Stat., Nonlinear, Soft Matter Phys.*, 2004, **70**, 021506.
215 B. P. Richard, *J. Membr. Sci.*, 1984, **17**, 1–62.
216 K. Arning, M. Burger, R. S. Eisenberg, H. W. Engl and L. He, *Proc. Appl. Math. Mech.*, 2007, **7**, 1120801–1120802.
217 D. G. Levitt, *J. Gen. Physiol.*, 1999, **113**, 789–794.
218 A. B. Mamonov, M. G. Kurnikova and R. D. Coalson, *Biophys. Chem.*, 2006, **124**, 268–278.
219 Damocles, in http://www.research.ibm.com/DAMOCLES/home.html, 2007.
220 W. Shockley, *Electrons and Holes in Semiconductors to applications in transistor electronics*, van Nostrand, New York, 1950.
221 B. G. Streetman, *Solid State Electronic Devices*, 4th edn., Prentice Hall, Englewood Cliffs, NJ, 1972.
222 S. M. Sze, *Physics of Semiconductor Devices*, John Wiley & Sons, New York, 1981.
223 D. J. Rouston, *Bipolar Semiconductor Devices*, McGraw-Hill Publishing Company, New York, 1990.
224 M. Shur, *Physics of Semiconductor Devices*, Prentice Hall, New York, 1990.
225 R. F. Pierret, *Semiconductor Device Fundamentals*, Addison Wesley, New York, 1996.
226 R. T. Howe and C. G. Sodini, *Microelectronics: an integrated approach*, Prentice Hall, Upper Saddle River, NJ USA, 1997.
227 D. L. Critchlow, *Proc. IEEE*, 1999, **87**, 659–667.
228 P. R. Gray, P. J. Hurst, S. H. Lewis and R. G. Meyer, *Analysis and Design of Analog Integrated Circuits*, 4th edn., John Wiley, NewYork, 2001.
229 S. J. Fiedziuszko, I. C. Hunter, T. Itoh, Y. Kobayashi, T. Nishikawa, S. Stitzer and K. Wakino, *IEEE Trans. Microwave Theory Tech.*, 2002, **50**, 706–720.
230 J. D. Cressler, *Silicon Heterostructure Handbook: Materials, Fabrication, Devices, Circuits and Applications of SiGe and Si Strained-Layer Epitaxy*, CRC Boca Raton, FL, 2005.
231 K. Varsos, J. Luntz, M. Welsh and K. Sarabandi, *Proc. IEEE*, 2011, **99**, 2112–2124.
232 A. Ben-Naim, *Molecular Theory of Water and Aqueous Solutions Part II: The Role of Water in Protein Folding, Self-Assembly and Molecular Recognition*, World Scientific Publishing Company, 2011.
233 R. M. Fuoss and C. A. Kraus, *J. Am. Chem. Soc.*, 1933, **55**, 1019–1028.
234 R. M. Fuoss and C. A. Kraus, *J. Am. Chem. Soc.*, 1933, **55**, 2387–2399.
235 R. M. Fuoss, *Chem. Rev.*, 1935, **17**, 27–42.
236 R. M. Fuoss and L. Onsager, *Proc. Natl. Acad. Sci. U. S. A.*, 1955, **41**, 274–283.

237 R. M. Fuoss and L. Onsager, *J. Phys. Chem.*, 1958, **62**, 1339–1340.
238 R. M. Fuoss, *J. Phys. Chem.*, 1959, **63**, 633–636.
239 R. M. Fuoss and F. Accascina, *Electrolytic Conductance*, Interscience, New York, 1959.
240 R. M. Fuoss and L. Onsager, *J. Phys. Chem.*, 1962, **66**, 1722–1726.
241 R. M. Fuoss and L. Onsager, *J. Phys. Chem.*, 1962, **66**, 1722–1726.
242 R. M. Fuoss, *J. Am. Chem. Soc.*, 1963, **85**, 2536–2536.
243 R. M. Fuoss and L. Onsager, *J. Phys. Chem.*, 1963, **67**, 621–628.
244 R. M. Fuoss, *J. Phys. Chem.*, 1974, **78**, 1383–1386.
245 R. M. Fuoss, *J. Phys. Chem.*, 1975, **79**, 525–540.
246 J. E. Anderson, *J. Non-Cryst. Solids*, 1994, **172–174, Part 2**, 1190–1194.
247 S. Durand-Vidal, P. Turq, O. Bernard, C. Treiner and L. Blum, *Phys. A*, 1996, **231**, 123–143.
248 S. Van Damme and J. Deconinck, *J. Phys. Chem. B*, 2007, **111**, 5308–5315.
249 G. M. Roger, S. Durand-Vidal, O. Bernard and P. Turq, *J. Phys. Chem. B*, 2009, **113**, 8670–8674.
250 S. Van Damme, N. Smets, D. De Wilde, G. Weyns and J. Deconinck, *J. Phys. Chem. B*, 2009, **113**, 3105–3111.
251 S. Van Damme, N. Smets, D. De Wilde, G. Weyns and J. Deconinck, *J. Phys. Chem. A*, 2009, **113**, 4972–4975.
252 T. Yamaguchi, T. Matsuoka and S. Koda, *J. Chem. Phys.*, 2011, **135**, 164511–164510.
253 D. Boda and D. Gillespie, *J. Chem. Theory Comput.*, 2012, **8**, 824–829.
254 D. Chen, R. Eisenberg, J. Jerome and C. Shu, *Biophys. J.*, 1995, **69**, 2304–2322.
255 S. Hagiwara and S. Nakajima, *J. Gen. Physiol.*, 1966, **49**, 793–806.
256 P. G. Kostyuk, O. A. Krishtal and P. A. Doroshenko, *Pfluegers Arch.*, 1974, **348**, 83–93.
257 S. Hagiwara and L. Byerly, *Annu. Rev. Neurosci.*, 1981, **4**, 69–125.
258 W. Almers and E. W. McCleskey, *J. Physiol.*, 1984, **353**, 585–608.
259 P. Hess and R. W. Tsien, *Nature*, 1984, **309**, 453–456.
260 E. W. McCleskey and W. Almers, *Proc. Natl. Acad. Sci. U. S. A.*, 1985, **82**, 7149–7153.
261 S. H. Heinemann, H. Terlau, W. Stuhmer, K. Imoto and S. Numa, *Nature*, 1992, **356**, 441–443.
262 E. W. McCleskey, M. D. Womack and L. A. Fieber, *International review of cytology*, 1993, **137C**, 39–54.
263 P. T. Ellinor, J. Yang, W. A. Sather, J.-F. Zhang and R. Tsien, *Neuron*, 1995, **15**, 1121–1132.
264 E. W. McCleskey, *Biophys. J.*, 2000, **79**, 1691–1692.
265 W. A. Sather and E. W. McCleskey, *Annu. Rev. Physiol.*, 2003, **65**, 133–159.
266 B. Eisenberg, *Institute of Mathematics and its Applications*, 2009, IMA University of Minnesota http://www.ima.umn.edu/2008–2009/W2012.2008–2012.2008/abstracts.html and also http://arxiv.org/0906.5173.
267 D. Boda, J. Giri, D. Henderson, B. Eisenberg and D. Gillespie, *J. Chem. Phys.*, 2011, **134**, 055102–055114.
268 D. Krauss, B. Eisenberg and D. Gillespie, *Eur. Biophys. J.*, 2011, **40**, 775–782.
269 A. L. Hodgkin, *Biol. Rev.*, 1951, **26**, 339–409.
270 A. L. Hodgkin and A. F. Huxley, *J. Physiol.*, 1952, **116**, 449–472.
271 I. Atwater, F. Bezanilla and E. Rojas, *J Physiol*, 1969, **201**, 657–664.
272 F. Bezanilla, J. Vergara and R. E. Taylor, in *Methods of Experimental Physics Volume 20*, eds. G. Ehrenstein and H. Lecar, Elsevier Inc., New York, 1982, vol. 20, pp. 445–511.
273 F. Bezanilla, *J. Membr. Biol.*, 1985, **88**, 97–111.
274 F. Bezanilla, *Biophys. J.*, 1987, **52**, 1087–1090.
275 A. M. Correa, R. Latorre and F. Bezanilla, *J. Gen. Physiol.*, 1991, **97**, 605–625.
276 F. Bezanilla, *Neuron*, 2008, **60**, 456–468.
277 J. Payandeh, T. Scheuer, N. Zheng and W. A. Catterall, *Nature*, 2011, **475**, 353–358.
278 J. Giri, J. E. Fonseca, D. Boda, D. Henderson and B. Eisenberg, *Phys. Biol.*, 2011, **8**, 026004.
279 B. Eisenberg, *Biophys. Chem.*, 2003, **100**, 507–517.
280 B. Hille, *Ionic Channels of Excitable Membranes*, 3rd edn., Sinauer Associates Inc., Sunderland, 2001.
281 J. R. Pappenheimer, *Physiol. Rev.*, 1953, **33**, 387–423.
282 J. R. Pappenheimer, *Annu. Rev. Physiol.*, 1987, **49**, 1–16.
283 Y. Hyon, J. E. Fonseca, B. Eisenberg and C. Liu, Energy variational approach to study charge inversion (layering) near charged walls, *Discrete and Continuous Dynamical Systems - Series B (DCDS-B)*, 2012, **17**(8), 2725–2743.
284 D. Boda, K.-Y. Chan and D. Henderson, *J. Chem. Phys.*, 1998, **109**, 7362–7371.
285 S. Varga, D. Boda, D. Henderson and S. Sokolowski, *J. Colloid Interface Sci.*, 2000, **227**, 223–226.

286 J. Reszko-Zygmunt, S. Sokolowski, D. Henderson and D. Boda, *J. Chem. Phys.*, 2005, **122**, 84504.
287 D. Boda, K. Y. Chan and I. Szalai, *Molecular Physics*, 1997, **92**, 1067–1072.
288 A. Malasics, D. Gillespie and D. Boda, *J. Chem. Phys.*, 2008, **128**, 124102.
289 J. Vincze, M. Valisko and D. Boda, *J. Chem. Phys.*, 2010, **133**, 154507.
290 A. S. Fairhurst and D. J. Jenden, *Proc. Natl. Acad. Sci. U. S. A.*, 1962, **48**, 807–813.
291 D. J. Jenden and A. S. Fairhurst, *Pharmacol Rev*, 1969, **21**, 1–25.
292 G. Meissner, *The Journal of biological chemistry*, 1986, **261**, 6300–6306.
293 M. Fill and J. A. Copello, *Physiological reviews*, 2002, **82**, 893–922.
294 G. Meissner, *Front. Biosci.*, 2002, **7**, d2072–2080.
295 D. Gillespie, L. Xu, Y. Wang and G. Meissner, *J. Phys. Chem. B*, 2005, **109**, 15598–15610.
296 Y. Wang, L. Xu, D. Pasek, D. Gillespie and G. Meissner, *Biophys. J.*, 2005, **89**, 256–265.
297 T. Nagrani, M. Siyamwala, G. Vahid and S. Bekheit, *Neurologist*, 2011, **17**, 91–94.
298 D. Gillespie, W. Nonner and R. S. Eisenberg, *Phys. Rev. E: Stat. Phys., Plasmas, Fluids, Relat. Interdiscip. Top.*, 2003, **68**, 0313503.
299 D. Gillespie, M. Valisko and D. Boda, *J. Phys.: Condens. Matter*, 2005, **17**, 6609–6626.
300 D. Gillespie and D. Boda, *Biophys. J.*, 2008, **95**, 2658–2672.
301 D. Gillespie and M. Fill, *Biophys. J.*, 2008, **95**, 3706–3714.
302 D. Gillespie, J. Giri and M. Fill, *Biophys. J.*, 2009, **97**, 2212–2221.
303 D. Krauss and D. Gillespie, *Eur. Biophys. J.*, 2010, **39**, 1513–1521.
304 R. M. Fuoss and L. Onsager, *Proc. Natl. Acad. Sci. U. S. A.*, 1961, **47**, 818–825.
305 J. Barthel, H. Krienke and W. Kunz, *Physical Chemistry of Electrolyte Solutions: Modern Aspects*, Springer, New York, 1998.
306 B. Eisenberg, *Fluctuations and Noise Letters*, 2012, **11**, 76–96. Earlier version 'Living Transistors: a Physicist's View of Ion Channels' available on http://arxiv.org/as q-bio/0506016v0506012.
307 S. Aboud, D. Marreiro, M. Saraniti and R. S. Eisenberg, *J. Comput. Electron.*, 2004, **3**, 117–133.
308 J. R. Barker and D. K. Ferry, *Solid-State Electron.*, 1980, **23**, 531–544.
309 D. K. Ferry, *Semiconductor Transport*, Taylor and Francis, New York, 2000.
310 M. Goryll, S. Wilk, G. M. Laws, T. J. Thornton, S. M. Goodnick, M. Saraniti, J. M. Tang and R. S. Eisenberg, *Superlattices Microstruct.*, 2003, **34**, 451–457.
311 K. Hess, *Monte Carlo Device Simulation: Full Band and Beyond*, Kluwer, Boston, MA USA, 1991.
312 K. Hess, *Advanced Theory of Semiconductor Devices*, IEEE Press, New York, 2000.
313 K. Hess, J. P. Leburton and U. Ravaioli, *Computational Electronics: Semiconductor Transport and Device Simulation*, Kluwer, Boston, MA USA, 1991.
314 M. Saraniti, S. Hu, S. Goodnick and S. J. Wigger, *Physica B*, 2002, in the press.
315 D. Vasileska, S. M. Goodnick, G. Klimeck, *Computational Electronics: Semiclassical and Quantum Device Modelling and Simulation*, CRC Press, New York, 2012, 764.
316 T. A. van der Straaten, G. Kathawala, R. S. Eisenberg and U. Ravaioli, *Mol. Simul.*, 2004, **31**, 151–171.
317 T. A. van der Straaten, J. Tang, R. S. Eisenberg, U. Ravaioli and N. R. Aluru, *Journal of Computational Electronics*, 2002, **1**, 335–340.
318 K. E. Cooper, P. Y. Gates and R. S. Eisenberg, *Q. Rev. Biophys.*, 1988, **21**, 331–364.
319 R. S. Eisenberg, *J. Membr. Biol.*, 1990, **115**, 1–12. Available on arXiv as http://arxiv.org/abs/1112.2363.
320 D. Chen, L. Xu, A. Tripathy, G. Meissner and R. Eisenberg, *Biophys. J.*, 1997, **73**, 1349–1354.
321 B. Eisenberg, available on http://arxiv.org/as q-bio/0506016v2, 24 pages, 2005.

Ionic specific effects on the structure, mechanics and interfacial softness of a polyelectrolyte brush

Francisco Rodríguez-Ropero and Nico F. A. van der Vegt*

Received 13th April 2012, Accepted 30th April 2012
DOI: 10.1039/c2fd20072d

Polymer brushes are assemblies of polymer chains grafted with one end group to a solid surface in a number high enough so that the chains are stretched away from the interface due to steric hindrance and/or electrostatic repulsion between neighbouring polymer units. Diblock copolymers have been used to build controlled switchable surfaces with a selective response to external stimuli specific to one of the blocks. Herein we study the structure, mechanical behaviour and interfacial softness of a diblock copolymer brush composed of *N*,*N′*-dimethylmethylacrylamide (DMMAA) and methacrylic acid (MAA) considering different counterions and Hofmeister salts by means of atomistic molecular dynamics simulations. Li^+, Na^+, K^+, Rb^+ and Cs^+ were used as counterions while NaF, NaCl, NaBr and NaI 1 molal aqueous salt solutions were used to study anion effects. Our results show that Li^+ and Na^+ cations form a strong salt bridge network with the carboxylic groups of MAA monomers belonging to neighbouring chains leading to a stretched monodisperse arrangement of the polymeric chains. On the other hand, the weaker salt bridge network upon addition of Rb^+ and Cs^+ cations promotes a partial collapse of the polyelectrolyte brushes. Addition of K^+ cations leads to an intermediate arrangement. These results have a direct impact on the mechanical behaviour of such soft surfaces and the packing of the polymer chains, *i.e.* the rigidity of the brush is ion specific and follows a Hofmeister series in the order $Li^+ > Na^+ \gg K^+ > Rb^+ > Cs^+$. We furthermore report that in those systems where a sodium halide salt is added, the same swelling behaviour is observed independently of the nature of the added salt. However, the degree of ionic absorption within the brush is again ion specific in the order NaCl > NaBr > NaI > NaF at 1 m salt concentration.

Introduction

Polymer brushes are a specific kind of macromolecular arrangement formed by polymer chains grafted by an end group to a solid substrate in a number high enough so that the chains are stretched away from the surface due to steric hindrance and/or electrostatic repulsion between neighbouring polymer units.[1] It is precisely this deformed configuration at equilibrium that differentiates polymer brushes from their unperturbed counterparts conferring them a unique set of properties.[2] More specifically the brush regime is achieved if the height of the grafted chains ($\langle h \rangle$) is higher than the end to end distance ($\langle r^2 \rangle^{1/2}$) of a free chain in the same solvent and the anchoring points between two chains are separated by less than $\langle r^2 \rangle^{1/2}$. Otherwise the grafted polymers are within the so called mushroom regime.[3] Neutral brushes and polyelectrolyte brushes differ basically in their swelling behaviour. The height of neutral brushes in good solvents is proportional to the chain length (N) and

Center of Smart Interfaces, Technische Universität Darmstadt, Petersenstrasse 17, 64287 Darmstadt, Germany. E-mail: vandervegt@csi.tu-darmstadt.de

the grafting density (σ) following the Alexander-de Gennes relation $h \sim N \cdot \sigma^{1/3}$,[4–6] while for weak and strong polyelectrolyte brushes <h> is independent of σ but depends on the salt concentration of the medium.[7–9]

The distinctive architecture of polymer brushes and the possibility of easily tailoring their properties by addition of different functional groups makes them a flexible tool to modify the surface properties of materials and modulate their sensitivity towards specific stimuli without affecting the bulk properties.[10] Therefore the properties of a given polymer brush and its structure are strongly dependent on the nature and molecular weight of the polymer(s) tethered to the surface, the grafting density and external stimuli such as temperature, solvent, light, magnetic or electric fields. Moreover polymer brushes exhibit a great chemical stability and a reversible extension/collapse switching response, which makes them attractive systems to be used in a broad range of fields.[11] One of the pioneering and more widespread applications of polymer brushes is the stabilization of colloidal dispersions.[12] Nonetheless, during the past years polymer brushes have been used to develop electrochemical biosensors,[13] mechanosensitive surfaces,[14] voltage-dependent nanofluidic channels,[15] coatings to suppress or enhance the adhesion of a specific cell type,[16] *etc.*

Several experimental studies have examined polymer brush structure and properties by the use of techniques such as Atomic Force Microscopy, Neutron Reflectivity, X-ray Photoelectron Spectroscopy or Ellipsometry.[10] Nevertheless some key aspects regarding the responsive mechanism and the behavior of the polymer brush under different conditions remain difficult or impossible to elucidate with only physical experiments. At this point, theoretical[1,4–7,17–21] and molecular simulation studies[22–32] are required to have a more accurate description of polymer brushes under different conditions.

In the framework of polymer brushes, block copolymers offer new possibilities to design controlled switchable surfaces with a selective response to external stimuli specific to one of the blocks. Thus the use of block copolymer brushes enables tuning the surface properties of a film from the properties of one of the blocks to the properties of the other block or a mix of the properties of all the blocks.[33,34] One interesting example of application of a block copolymer brush to modulate the hydrophilic/hydrophobic nature of a film is the use of a diblock copolymer constituting a hydrophobic block A grafted to the surface with a hydrophilic block B exposed to the environment. By using a polyelectrolyte as block B one could modulate the microstructure and mechanical properties of the brush by using different counterions and/or salts. Moreover the charged block B might be swollen or collapsed under an external electric field enabling the exposure of either block A or B to the bulk water phase depending on the magnitude and sign of the external field. The dynamics of such switching phenomena can be dramatically affected by the nature of the chosen counterion. This latter aspect can be modulated by considering specific ion effects through the so-called Hofmeister series,[35] which rank a set of ions according to the effect on target properties such as surface tension, protein stability or enzymatic activity.[36]

In this article, we study the structure, mechanical behaviour and interfacial softness of a diblock copolymer brush composed of *N*,*N*′-dimethylmethylacrylamide (DMMAA) and methacrylic acid (MAA) of the same block length $(PDMMAA)_{40}$-*b*-$(PMAA)_{40}$ considering different alkali metals as counterions and sodium halide salts by means of atomistic molecular dynamics simulations. The work is organized as follows. In the next section we give a description of all the systems under study and the simulation details. Thereafter in the results and discussion section we give a detailed comparative analysis of the influence of the considered counterion and/or added salt on the structure and mechanics of the diblock copolymer brush. Finally, the conclusions and the outlook of this work are presented in the last section.

Methodology

System description

25 atactic $(PDMMAA)_{40}$-*b*-$(PMAA)_{40}$ diblock copolymer chains were homogeneously grafted in a square grid to a hydroxylated (7.746×7.567) nm^2 Si(1 0 0) surface using a silane linker (Scheme 1) leading to a grafting density $\sigma = 0.4265$ chains nm^{-2}. The distance between grafting points was ~1.5 nm. Polymer chains were initially constructed as fully stretched rod-like chains perpendicular to the 1.5 nm thick silicon surface. 1000 alkali counterions (Li^+, Na^+, K^+, Rb^+ or Cs^+) were added to neutralize the system, which also contained 34532 water molecules. The initial box length on the direction perpendicular to the surface was set to 24 nm. To build those systems containing 1 m (molal) of a sodium halide salt (NaF, NaCl, NaBr or NaI) we used the equilibrated brush system in which Na^+ was considered as counterion and exchanged 1194 water molecules belonging to the bulk water phase to add 597 salt molecules.

The diblock copolymer was simulated using the OPLS-AA force field[37] while the SPC/E[38] model was used to represent water molecules. The force field parameters for the silicon surface were taken from van der Spoel *et al.*[39] The ion force field parameters were taken from Gee *et al.*,[40] which were designed to reproduce the experimental Kirkwood–Buff integrals[41] in conjunction with the SPC/E water model as a function of salt concentration. This force field has been proven to reproduce not only the solution salt activities but also the ion diffusion constants, dielectric decrements and the excess of heat of mixing in a good agreement with experimental data.[40]

Simulation details

The simulations were performed using the GROMACS 4.0.7. package.[42,43]

Periodic boundary conditions were set in all dimensions, so that the surface and the polymer were infinitely extended in the *x*–*y* direction. Since the box size in the *z* direction is larger than the stretched chains plus the non-bonded interaction cut-off, interaction of the top of the polymer chains with the periodic image of the surface on the vertical direction is avoided. Atoms that constituted the surface remained completely fixed during our simulations.

Each system was equilibrated in three steps. Firstly we relaxed the initial structure by means of 5000 steps of the steepest descent minimization algorithm. Secondly the system was submitted to a simulated annealing protocol at constant volume from 0 K to 300 K using a simulation timestep of 1 fs for 5 ns. The target 300 K temperature was reached after 0.3 ns. In this stage the Berendsen thermostat[44] was used

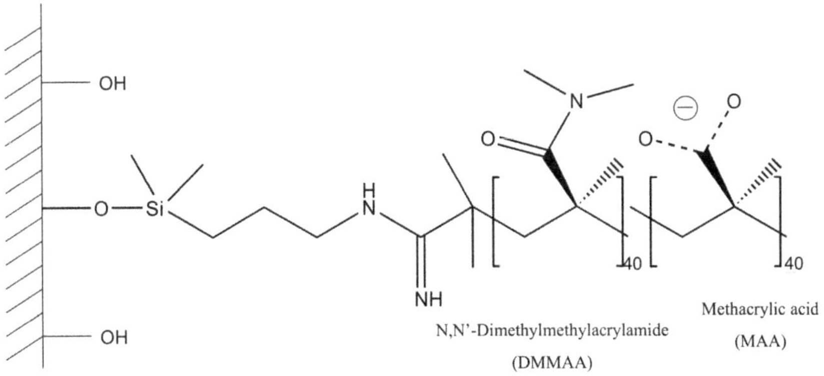

Scheme 1

with a coupling constant $\tau_T = 0.1$ ps. In the third and last equilibration step the systems were simulated for 10 ns using a simulation timestep of 2 fs at semi-isotropic NPT conditions fixing $P = 1$ bar and $T = 300$ K fixing the box length in the x and y direction and allowing it to fluctuate in the z direction. The Parrinello–Rahman barostat[45,46] was used with a $\tau_P = 0.5$ ps combined with the Nosé–Hoover thermostat using a coupling constant $\tau_T = 0.1$ ps.[47,48] The last snapshot of the equilibration process was used as initial configuration of the 100 ns NVT-MD production run at 300 K using the Nosé–Hoover thermostat and 2 fs of simulation timestep, the neighbour list being updated every 10 fs. Atomic coordinates were saved every 1 ps for subsequent analysis. Van der Waals interactions were treated using a 12-6 Lennard-Jones potential applying a cut-off at 1.40 nm. The electrostatic interactions have been calculated using Particle-Mesh Ewald method[49] with a real space cut-off of 1.40 nm. All the analyses presented in this article have been performed considering the last 50 ns of each of our simulations.

Results and discussion

Influence of alkali counterions on the diblock copolymer brush

The final configurations of all the systems under study are depicted in Fig. 1, while the monomer number density distributions for both blocks are plotted in Fig. 2. As expected, given the high grafting density considered in this work, in all the cases all the chains are fully extended perpendicular to the surface and no layering over the surface is observed in any of the systems. The presence of hydroxyl groups on the surface combined with the hydrophobic nature of the DMMAA block leads to the appearance of a water-rich polymer depletion layer near the surface, as it can be observed in the peak near the surface in the water density profiles (Fig. 3). It should be mentioned that all the water profiles can be divided in four regions moving from the surface to the bulk water phase. First the aforementioned water-rich layer near the surface, second a region within the DMMAA block (see below) where the water density is around 65–70% with respect to the water bulk density, third a region within the MAA block where density increases up to 80–85% and finally the bulk

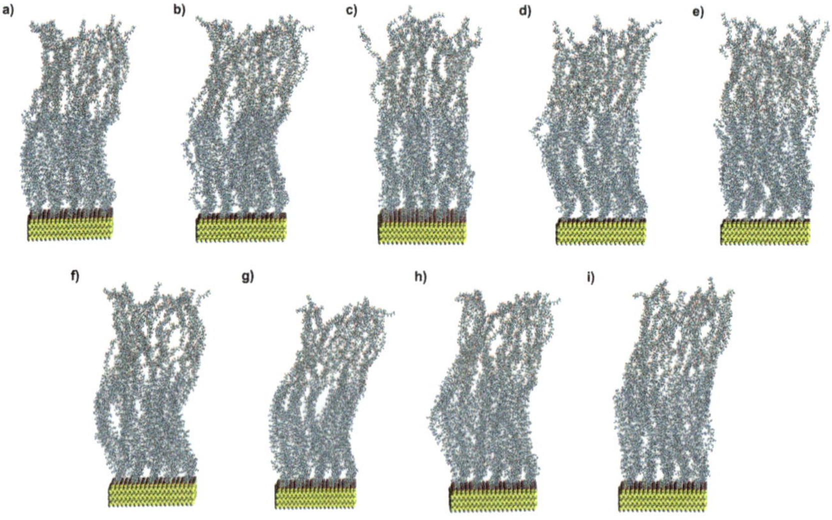

Fig. 1 Snapshots displaying the final configuration of the polymer brushes upon addition of (a) Li^+, (b) Na^+, (c) K^+, (d) Rb^+ and (e) Cs^+ alkali ions and 1 m of (f) NaF, (g) NaCl, (h) NaBr and (i) NaI salts. Water and ions are not displayed for clarity.

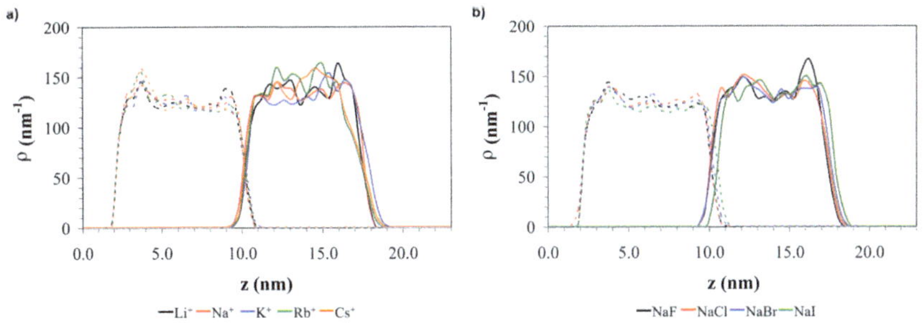

Fig. 2 Monomer distribution of the DMMAA (dotted line) and the MAA blocks (continuous line) for the polymer brushes upon addition of (a) alkali ions and (b) 1 m of halide salts. Profiles are normalized over the total number of each type of monomer (1000 monomers each) and the surface area.

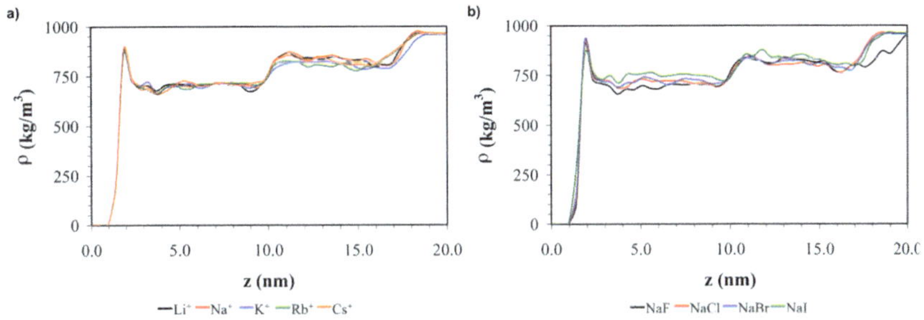

Fig. 3 Water density profiles upon addition of (a) alkali ions or (b) 1 m of halide salt.

water phase. The average brush height $\langle h\rangle$ values are reported in Table 1 and were calculated according to eqn (1), which is approximated to the doubled normalized first moment of the brush density distribution:[50]

$$\langle \mathrm{h} \rangle = \frac{2 \cdot \int_0^\infty z \cdot \rho(z) \mathrm{d}z}{\int_0^\infty \rho(z) \mathrm{d}z} \quad (1)$$

where $\rho(z)$ is the polymer density and z is the distance from the surface. The same equation has been used to compute the average block heights $\langle h_{\mathrm{DMMAA}}\rangle$ and

Table 1 Average brush height $\langle h\rangle$, counterion layer height $\langle h_+\rangle$, DMMAA block height $\langle h_{\mathrm{DMMAA}}\rangle$, MAA block height $\langle h_{\mathrm{MAA}}\rangle$ and osmotic compressibility for each block (κ) for all the systems under study

			DMMAA		MAA	
	$\langle h\rangle$/nm	$\langle h_+\rangle$/nm	$\langle h_{\mathrm{DMMAA}}\rangle$/nm	$\kappa \times 10^5$/bar^{-1}	$\langle h_{\mathrm{MAA}}\rangle$/nm	$\kappa \times 10^5$/bar^{-1}
Li^+	16.1	9.9	9.6	29.6	10.0	24.7
Na^+	16.1	9.8	9.5	27.8	9.9	25.7
K^+	16.4	10.4	9.6	31.9	10.5	130.1
Rb^+	16.0	9.8	9.4	38.3	9.8	132.4
Cs^+	16.0	9.8	9.3	57.2	9.9	146.3
NaF	16.0	—	9.5	30.6	10.1	22.1
NaCl	16.0	—	9.5	23.1	9.9	18.4
NaBr	16.3	—	9.7	25.7	10.0	28.0
NaI	16.5	—	9.8	22.1	10.5	26.9

$\langle h_{MAA}\rangle$, which are also summarized in Table 1. As it can be seen the height of the polymer brush is independent of the counterion. However interesting differences are observed in the polymer density profiles (Fig. 2), especially at the polymer/water interface. In this sense there are two remarked situations, *i.e.* a sharp MAA/water interface upon addition of Li^+ or Na^+ and a parabolic MAA/water interface upon addition of Rb^+ or Cs^+ cations. Addition of K^+ led to an intermediate situation. These results point towards a partial collapse when bigger alkali counterions are added. This feature is reinforced after examination of the density distribution of the MAA terminal monomers (Fig. 4a). Addition of Li^+ and Na^+ led to a narrow monodisperse distribution, while addition of Rb^+ or Cs^+ and, at some extent K^+, not only broadens the end monomer distributions but also promotes the appearance of, respectively, one and two extra peaks closer to the block–block interface. It should be noted that the counterion profiles (Fig. 5) are fully coincident with the MAA density profile, evidencing that counterions are fully confined within the charged block layer. This feature has been observed in other polyelectrolyte brushes at high grafting densities and high charged fraction.[51] The average counterion layer height $\langle h_+\rangle$ was calculated using eqn (1) considering the counterion density profile (Table 1). The similarity between $\langle h_+\rangle$ and $\langle h_{MAA}\rangle$ confirms that in all cases practically all the counterions are trapped inside the charged block leading to a local electroneutrality.

In order to check how different cations may affect the mechanical behaviour of such soft surfaces, we have calculated the osmotic compressibility κ of each block from the monomer number fluctuations as it is shown in eqn (2) and reported the results in Table 1:

$$\kappa = \frac{\langle N^2\rangle - \langle N\rangle^2}{\langle N\rangle^2}\frac{V}{k_B T} \quad (2)$$

where k_B is the Boltzmann constant, T is the working temperature and V is the considered volume. It should be pointed that eqn (2) is valid only in the thermodynamic limit of a grand-canonical system, which is satisfied as the number of N monomers approaches infinite. To satisfy this condition we have calculated the compressibility considering $(7.746 \times 7.567 \times 4.600)\ nm^3$ orthorhombic boxes centred in each of the blocks. Regarding the neutral DMMAA block it can be seen that its compressibility remains unaltered independently of whether Li^+, Na^+ or K^+ has been added as a counterion. However this block becomes much more compressible upon addition of Cs^+ and, to a lesser extent, Rb^+ cations. This difference in compressibility can be noticed in the appearance of a first high peak in the DMMAA density profile (Fig. 2a) closer to the surface. A more remarkable change in the compressibility is observed within the MAA block. Addition of bigger cations

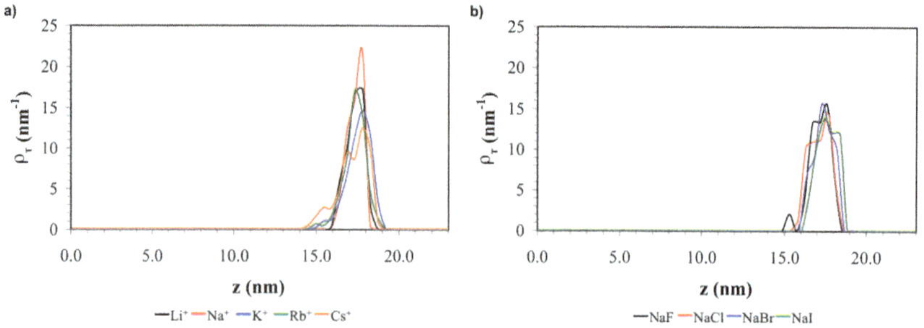

Fig. 4 Free terminal monomer distribution for the polymer brushes upon addition of (a) alkali ions and (b) 1 m of halide salts. Profiles are normalized over the total number of terminal monomer (25 monomers) and the surface area.

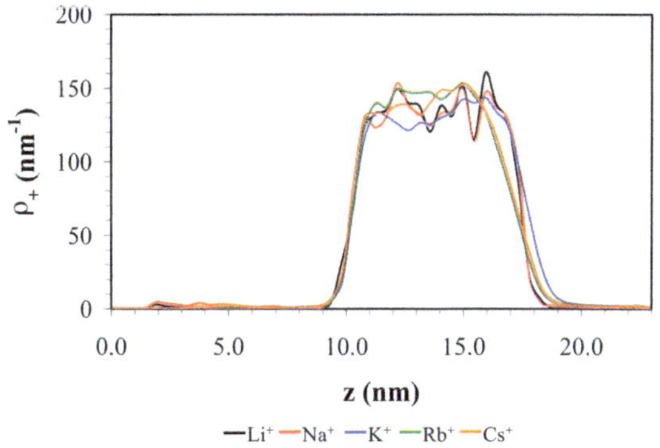

Fig. 5 Counterion distribution profiles for the polymer brushes upon addition of alkali ions. Profiles are normalized over the total number of counterions (1000) and the surface area.

K^+, Rb^+ and Cs^+ promotes an approximately five-fold increase in the compressibility of such block compared with the values obtained for Li^+ and Na^+. This increase in the compressibility explains the partial collapse observed in the MAA density profiles and the final monomer distributions when bigger alkali cations are considered.

This difference in the compressibility is related to the way neighbouring chains organize or pack within the brush architecture. Given the two dimensional nature of such packing we examined the lateral structure of the brushes by computing a two dimension radial distribution function (2D-rdf) considering only the carbon atoms in the backbone bearing a functional side chain. This expression $g_{2D}^{AA}(r,z)$ for two identical A particles has been defined according to eqn (3):

$$g_{2D}^{AA}(r,z) = \frac{1}{4\pi r^2 N_A^2} \sum_{ij}^{N_A \in z_n} \delta(r - r_{ij}) \; ; \qquad r = \sqrt{x^2 + y^2} \text{ and } z_n = n \cdot \Delta z \tag{3}$$

where n and Δz are respectively the total number of points and the grid thickness. In our calculations we have assumed $0 < n < 182$ and $\Delta z = 0.12$ nm. The resulting 2D-rdfs are depicted in Fig. 6. Focusing our attention in the lower part of each plot, which corresponds to the 2D-rdf of the neutral DMMAA, one can observe the presence of 5 bands of decreasing intensity which refer to 5 peaks in the rdf separated by ~0.65 nm. These bands, which are slightly more intense upon addition of Li^+ or Na^+, evidence a great degree of ordering in the packing on the x–y planes between neighbouring polymer chains. However a different situation is presented within the charged MAA block, which can be seen on the top of each plot. As it can be seen, the degree of lateral ordering within the polyelectrolyte block follows a Hofmeister series in the order $Li^+ \sim Na^+ \gg K^+ \sim Rb^+ \sim Cs^+$. More concretely, up to 4 bands can be distinguished when the two smaller cations are added while only 1 band can be distinguished when the bigger ones take part.

Hitherto we can briefly summarize that addition of Li^+ and Na^+ alkali cations increases not only the rigidity but also the degree of structured packing of the $(PDMMAA)_{40}$-b-$(PMAA)_{40}$ diblock copolymer brush under study. On the other hand Rb^+, Cs^+ and to a lesser extent K^+ do not affect dramatically the structure of the neutral block but promote the increase of the compressibility of the charged block and disrupt the degree of lateral ordering within such a block. On the other hand we have seen that counterions are mostly confined within the charged block. In order to understand what is the cause of this ionic specific effect on the structure and mechanics of the brush, we have studied the salt bridge network formed upon

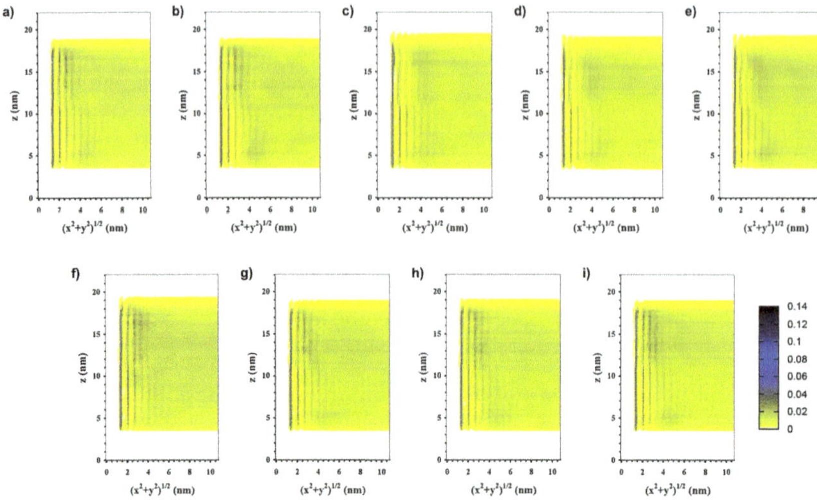

Fig. 6 2D-Radial distribution functions between the monomers belonging to different chains upon addition of (a) Li^+, (b) Na^+, (c) K^+, (d) Rb^+ and (e) Cs^+ alkali ions and 1 m of (f) NaF, (g) NaCl, (h) NaBr and (i) NaI salts.

coordination of the cationic species with the carboxylic groups in the MAA monomers. To that end we have considered that a given cation forms a salt bridge with one carboxylic group belonging to a given polymer chain if they are separated by less than a cut-off distance which has been previously determined from the first peak of the $COO^-\cdots$cation radial distribution function (not shown). Given this definition we have considered 4 different situations: (a) the cation is not coordinated to any carboxylic group; (b) the cation is only coordinated to one carboxylic group; (c) the cation is coordinated to two carboxylic groups belonging to two different chains; and (d) the cation is coordinated to three carboxylic groups belonging to three different chains. The averaged results over the last half of each trajectory are summarized in Table 2. As it can be seen the lowest averaged population of non-coordinated cations occurs when Li^+ and Na^+ are used. Moreover in such cases the number of dicoordinated cations is one order of magnitude higher than compared with the K^+, Rb^+ and Cs^+ cases. It should be noted that such $COO^-\cdots M^+\cdots {}^-OOC$ salt bridges play a crucial role in fixing neighbouring polymer chains and induce a global ordering. Additionally, for Li^+ and Na^+ tricoordinated salt bridges are observed throughout the trajectory. The salt bridge network was further characterized by calculating the structure relaxation time on basis of a time correlation function $C(t)$ accounting for the intermittent time correlation function similar to the one used to characterize the hydrogen bonding dynamics in liquid water[52] or in polyamide-66.[53]

$$C(t) = \frac{\langle \delta c(O)\delta c(t)\rangle}{\langle \delta c^2\rangle} = \frac{\langle c(O)c(t)\rangle - \langle c\rangle^2}{\langle c\rangle - \langle c\rangle^2} \approx \frac{\langle c(O)c(t)\rangle}{\langle c\rangle} \quad (4)$$

In eqn (4) $c(t)$ is a binary function that can be either equal to one when a target salt bridge is present at time t or zero otherwise. Fig. 7 shows the time correlation function $C(t)$ for both $COO^-\cdots M^+$ and $COO^-\cdots M^+\cdots {}^-OOC$ salt bridges in all the systems under study. As it can be seen we have a similar situation in both kinds of salt bridges: neither for Li^+ nor for Na^+ $C(t)$ functions relax within the first 30 ns, while for the rest of the cations both curves relax very rapidly in the order $K^+ > Rb^+ > Cs^+$. This inverse Hofmeister trend is related to the slower diffusivity

Table 2 Cut-off distance considered for the calculation of the different salt bridges, average number of non-coordinated cations, average number of mono-, di- and tri-coordinated salt bridges and characteristic structure relaxation times $\tau_{R,1}$ and $\tau_{R,2}$ deduced from the autocorrelation function $C(t)$ for respectively mono- and di-coordinated salt bridge networks

	Cut-off/nm	Free cation	Salt bridge #1	Salt bridge #2	Salt bridge #3	$\tau_{R,1}$/ns	$\tau_{R,2}$/ns
Li^+	0.286	21.3 ± 3.3	661.3 ± 7.3	316.2 ± 7.0	1.2 ± 0.4	16.6	18.6
Na^+	0.302	97.8 ± 8.8	538.7 ± 13.2	359.6 ± 12.6	4.0 ± 0.9	4.6	10.3
K^+	0.379	462.7 ± 13.4	497.6 ± 14.4	39.8 ± 6.3	0.9 ± 0.1	0.2	0.06
Rb^+	0.364	433.7 ± 13.9	509.8 ± 15.3	56.6 ± 6.1	0.8 ± 0.01	0.2	0.1
Cs^+	0.342	448.8 ± 14.0	484.2 ± 15.8	67.1 ± 8.6	0.6 ± 0.02	0.3	0.3
NaF	0.302	625.7 ± 6.6	586.3 ± 9.3	383.5 ± 5.4	1.6 ± 0.6	5.9	11.2
NaCl	0.302	470.8 ± 8.7	793.1 ± 10.4	331.1 ± 5.3	2.1 ± 0.3	4.6	9.9
NaBr	0.302	517.0 ± 10.6	670.2 ± 11.7	401.6 ± 7.4	2.2 ± 0.6	3.7	10.3
NaI	0.302	542.6 ± 10.2	653.8 ± 11.4	400.4 ± 8.0	0.9 ± 0.2	3.9	10.9

as the ion size increases. A characteristic structure relaxation time τ_R was defined as $\tau_R = \int_0^\infty C(t)dt$ (Table 2). Both relaxation times $\tau_{R,1}$ and $\tau_{R,2}$ follow the same trend $Li^+ > Na^+ \gg Cs^+ \sim Rb^+ \sim K^+$ that allows us to confirm that the salt bridge network when considering the smallest ions is not only stronger in *quantity* (higher number of salt bridges per frame) but also in *quality* (longer relaxation times).

Influence of added salt on the diblock copolymer brush

The impact of added salt on the structure and mechanics of the $(PDMMAA)_{40}$-*b*-$(PMAA)_{40}$ diblock copolymer brush was evaluated by adding 1 m of different sodium halide salts, *i.e.* NaF, NaCl, NaBr and NaI. It should be emphasized that carboxylic groups in the MAA block are expected to interact mostly with the Na^+

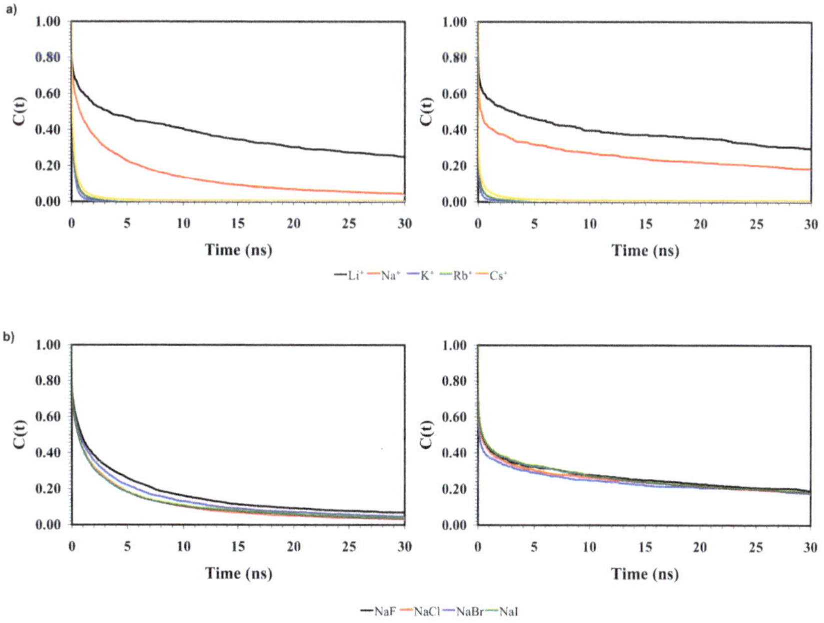

Fig. 7 Time correlation function C(t) for both $COO^- \cdots M^+$ (left) and $COO^- \cdots M^+ \cdots {}^-OOC$ (right) salt bridges upon addition of (a) alkali ions and (b) halide salts.

cation, thus we will compare the results in this section with those obtained in the previous section upon addition of Na^+ counterion.

The last snapshot of the polymer brushes are plotted in Fig. 1, showing a similar configuration to the brushes discussed in the last section. However the brush height $<h>$ reported in Table 2 follows a trend NaF ~ NaCl < NaBr < NaI which is mostly due to the longer $<h_{DMMAA}>$ in the last two cases plus a wider MAA phase when NaI is considered. Interestingly a reduction on the water content within the neutral DMMAA block (Fig. 3b) is observed in the decreasing order NaF < NaCl ~ NaBr < NaI. This different water content within the neutral block seems to have a clear impact on the compressibility of the neutral block, which is slightly lower upon addition of NaI salt. Block density profiles depicted in Fig. 2b show a sharp MAA/water interface in all the cases, the high peak displayed in the case of NaF being especially remarkable. Additionally it should be mentioned that a narrower DMMAA/MAA interface is observed upon addition of NaI salt. Examination of the density distribution of the MAA terminal monomers (Fig. 4b) shows an interesting situation. All these profiles are wider than those shown in the previous section and a knee can be observed in all of them. In the case of NaF one extra peak can be observed which is due to the partial collapse of some polymer chains. Independently of the added salt the lateral packing of the polymer chains, which was examined through the 2D-rdf presented in the previous section, follows the same pattern and it can be directly compared with that of the polymer brush upon addition of sodium counterion (*vide supra*) (Fig. 6).

A more interesting situation is given when analyzing the ion distributions (Fig. 8). If we focus our attention in the neutral block we can see that both cation and anion profiles practically overlap and that the ionic content follows a trend such as NaF < NaCl < NaBr < NaI, being the presence of NaF practically null. The distribution within the charged block follows a trend NaF ≪ NaI < NaBr < NaCl showing that the polymer brush under study is practically impenetrable for F^- anions. At this point it should be pointed that solubility of NaF at room temperature is approximately 1 mol L^{-1}. This concentration is clearly exceeded in the MAA/water interface (Fig. 8 at $z \sim 18.5$ nm), which leads to a local crystallization of the NaF salt near the brush/water interface. However if we consider lower NaF concentrations, where the salt remains soluble in water and such crystallization is not expected, a higher F^- concentration within the MAA block would be expected. For the rest of salts in the MAA/water interface (Fig. 8 at $z \sim 18.5$–19 nm) the local ionic concentration follows a trend such as NaCl < NaBr < NaI which resembles the situation given in water/air interfaces.[54–57] This different availability of Na^+ cations within

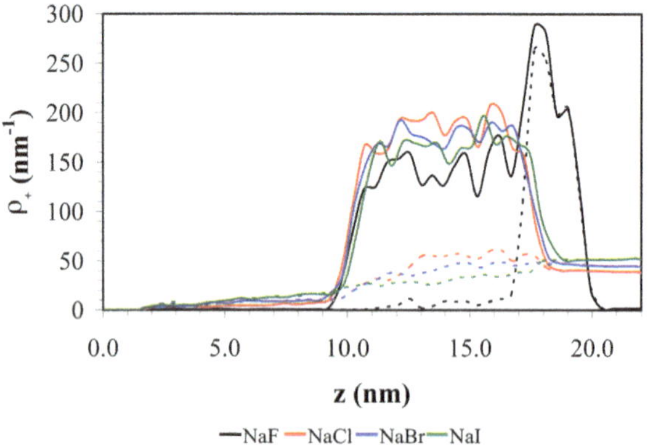

Fig. 8 Anionic (dotted line) and Na^+ (continuous) distribution profiles for the polymer brushes upon addition of 1 m sodium halide salt. Profiles are normalized over the total number of halide anions (597) and Na^+ (1597) and the surface area.

Table 3 Diffusion potentials for different halides in the water/MAA ($\Psi_{water\rightarrow MAA}$) and in MAA/DMMAA ($\Psi_{MAA\rightarrow DMMAA}$) interfaces

	$\Psi_{water\rightarrow MAA}$/mV	$\Psi_{MAA\rightarrow DMMAA}$/mV
F^-	81.1	78.3
Cl^-	−13.2	51.1
Br^-	−4.6	37.1
I^-	5.8	18.9

the charged block may influence the salt bridge network formed upon interaction of the cations with the carboxylic groups in the MAA monomers. Results included in Table 2 show a moderate increase in the number of $COO^- \cdots Na^+ \cdots {}^-OOC$ salt bridges (except for NaCl) and a great increase in the number of monocoordinated salt bridges compared with the salt-free case. Considering that the total number of carboxylic groups is 1000 by simply summation of the results in Table 2 it can be seen that some acidic groups are able to accommodate two Na^+ cations simultaneously. Autocorrelation functions $C(t)$ (Fig. 7b) show a similar behaviour in the salt bridge network compared to that of the polymer brush-Na^+ system. Structure relaxation times $\tau_{R,1}$ and $\tau_{R,2}$ for NaF were found to be slightly higher than in the other sodium halide systems.

Given the difference in ionic distribution within the brush and the water phases the polymer brush can be seen as an ionic specific selective semipermeable membrane. To characterize better the ionic distribution in our systems we have calculated the diffusion potential[58] of such a "membrane" by approximating the potential to the Nernst equation using the equilibrium anionic concentrations at both the MAA and water phases.[59] Additionally we have calculated in the same way the potential considering the DMMAA-MAA interface. Results are summarized in Table 3. As it can be seen the polymer brush has especial affinity to allow permeation of Cl^- and, in a lesser extent, Br^- anions. On the other hand permeation from the MAA block to the DMMAA block follows a Hofmeister trend in the order $F^- < Cl^- < Br^- < I^-$. In a recent study it was shown that among all the halide anions only I^- seems to have an appreciable attractive interaction with the isopropyl CH_3 of NiPAM (*N*-isopropylacrylamide) while the rest of anions are strongly repelled in the order $F^- > Cl^- > Br^-$.[60] This explains the observed trend on basis of attractive hydrophobic interactions between the DMMAA monomers and the big polarisable iodide anion.

Conclusions and outlook

In this article we have examined by means of Molecular Dynamics simulations the impact of different alkali counterions and the addition of 1 m sodium halide salts on the interfacial softness and the structure at the nanoscopic level of a $(PDMMAA)_{40}$-*b*-$(PMAA)_{40}$ diblock copolymer brush.

Regarding the effect of the alkali counterion results can be divided in two extreme groups, *i.e.* the effect of smaller ions Li^+ and Na^+ and the effect of bigger ions Rb^+, Cs^+ and in a lesser extent K^+. Addition of small cations led to rigid and homogeneously packed polymer brushes displaying a sharp brush/bulk water interface. On the other hand addition of bigger counterions results in a larger compressibility of the brush due to a more disordered interchain packing and results in a parabolic brush/bulk water interface. Moreover in such systems the polymer brush underwent a partial collapse. This different impact has been related to the different strength of the salt bridge network formed upon coordination of the cationic species with one, two or three carboxylic groups in different MAA blocks. On the basis of the average number of mono-, di- and tri-coordinated salt bridges and their characteristic structure relaxation times the effectiveness of the network can be classified according to $Li^+ > Na^+ \gg Cs^+ > Rb^+ > K^+$.

Addition of 1 m of sodium halides did not dramatically affect the structure of the polymer brush, being practically comparable to that observed in the polymer brush neutralized with Na^+ cations. However the polymer brush showed a different degree of ionic absorption at the working 1 m concentration in the order NaCl > NaBr > NaI ≫ NaF, the being F^- anions poorly permeated. In the last case it should be emphasized that 1 m exceeds the solubility concentration of NaF salt in water. Addition of a much smaller amount of salt would lead to a higher concentration of F^- within the MAA block. Absorption within the neutral block followed the series NaI > NaBr > NaCl > NaF, which is related to the attractive hydrophobic interactions between the DMMAA monomers and I^- anion.

In this work we have mostly focused our attention on the specific ionic effects on the $(PDMMMAA)_{40}$-*b*-$(PMAA)_{40}$ diblock copolymer brush. However, important questions remain to be answered. In this sense we are currently working on the examination of the mushroom/brush regime transition by studying the system at different coverage densities. The present diblock copolymer brush may be considered as a model system to design voltage-modulated wettability switchable surfaces. To that end we are currently studying the extension/collapse response of this polymer brush under external electric fields.

Acknowledgements

This work has been supported by the Deutsche Forschungsgemeinschaft (DFG) Cluster of Excellence 259 "Smart Interfaces – Understanding and Designing Fluid Boundaries". The authors would like to express their thanks to Markus Biesalski and Sander Klemme for fruitful discussions. We also thank Pim Schravendijk for assistance in writing the code to compute the 2D-rdf shown in this work.

References

1 S. T. Milner, *Science*, 1991, **251**, 905–914.
2 A. Halperin, M. Tirrell and T. P. Lodge, *Adv. Polym. Sci.*, 1992, **100**, 31–71.
3 S. Minko, *Polym. Rev.*, 2006, **46**, 397–420.
4 S. Alexander, *J. Phys.*, 1977, **38**, 977–981.
5 P. G. De Gennes, *J. Phys.*, 1976, **37**, 1445–1452.
6 P. G. De Gennes, *Macromolecules*, 1980, **13**, 1069–1075.
7 P. Pincus, *Macromolecules*, 1991, **24**, 2912–2919.
8 M. Biesalski and J. Rühe, *Macromolecules*, 2002, **35**, 499–507.
9 M. Biesalski, D. Johannsmann and J. Rühe, *J. Chem. Phys.*, 2002, **117**, 4988–4994.
10 R. C. Advincula, W. J. Britain, K. C. Caster and J. Rühe, *Polymer Brushes: Synthesis, Characterization, Applications*, Wiley-VCH, Weinheim, 2004.
11 P. Uhlmann, H. Merlitz, J.-U. Sommer and M. Stamm, *Macromol. Rapid Commun.*, 2009, **30**, 732–740.
12 D. H. Napper, *Polymeric Stabilization of Colloidal Dispersions*, Academic, London, 1983.
13 M. Welch, A. Rastogi and C. Ober, *Soft Matter*, 2011, **7**, 297–302.
14 J. Bünsow, T. S. Kelby and W. T. S. Huck, *Acc. Chem. Res.*, 2010, **43**, 466–474.
15 B. Yameen, M. Ali, R. Neumann, W. Ensinger, W. Knoll and O. Azzaroni, *Chem. Commun.*, 2010, **46**, 1908–1910.
16 N. Ayres, *Polym. Chem.*, 2010, **1**, 769–777.
17 S. T. Milner, T. A. Witten and M. E. Cates, *Macromolecules*, 1988, **21**, 2610–2619.
18 S. T. Milner, T. A. Witten and M. E. Cates, *Europhys. Lett.*, 1988, **5**, 413–418.
19 A. Halperin and E. B. Zhulina, *Macromolecules*, 1991, **24**, 5393–5397.
20 C. M. Wijmans, J. M. H. M. Scheutjens and E. B. Zhulina, *Macromolecules*, 1992, **25**, 2657–2665.
21 P. M. Biesheuvel, W. M. De Vos and V. M. Amoskov, *Macromolecules*, 2008, **41**, 6254–6259.
22 I. Coluzza and J.-P. Hansen, *Phys. Rev. Lett.*, 2008, **100**, 016104.
23 M. Murat and G. S. Grest, *Macromolecules*, 1989, **22**, 4054–4059.
24 G. S. Grest and M. Murat, *Macromolecules*, 1993, **26**, 3108–3117.

25 D. Bedrov and G. D. Smith, *Langmuir*, 2006, **22**, 6189–6194.
26 P. Träskelin, T. L. Kuhl and R. Faller, *Phys. Chem. Chem. Phys.*, 2009, **11**, 11324–11332.
27 I. G. Elliot, T. L. Kuhl and R. Faller, *Macromolecules*, 2010, **43**, 9131–9138.
28 J.-M. Y. Carrillo and A. V. Dobrynin, *Langmuir*, 2010, **26**, 18374–18381.
29 Z. Benková, B. Szefczyk and M. N. D. S. Cordeiro, *Macromolecules*, 2011, **44**, 3639–3648.
30 O. A. Guskova and C. Seidel, *Macromolecules*, 2011, **44**, 671–682.
31 N. A. Kumar and C. Seidel, *Macromolecules*, 2005, **38**, 9341–9350.
32 G.-L. He, H. Merlitz, J.-U. Sommer and C.-X. Wu, *Macromolecules*, 2007, **40**, 6721–6730.
33 M. A. C. Stuart, W. T. S. Huck, J. Genzer, M. Müller, C. Ober, M. Stamm, G. B. Sukhorukov, I. Szleifer, V. V. Tsukruk, M. Urban, F. Winnik, S. Zauscher, I. Luzinov and S. Minko, *Nat. Mater.*, 2010, **9**, 101–113.
34 M. Motornov, R. Sheparovych, R. Lupitskyy, E. MacWilliams, O. Hoy, I. Luzinov and S. Minko, *Adv. Funct. Mater.*, 2007, **17**, 2307–2314.
35 F. Hofmeister, *Arch. Exp. Pathol. Pharmakol.*, 1888, **24**, 247–260.
36 Y. Zhang and P. S. Cremer, *Curr. Opin. Chem. Biol.*, 2006, **10**, 658–663.
37 W. L. Jorgensen and J. Tirado-Rives, *J. Am. Chem. Soc.*, 1988, **110**, 1657–1666.
38 H. J. C. Berendsen, J. R. Grigera and T. P. Straatsma, *J. Phys. Chem.*, 1987, **91**, 6269–6271.
39 D. van der Spoel, E. J. W. Wensink and A. C. Hoffmann, *Langmuir*, 2006, **22**, 5666–5672.
40 M. B. Gee, N. R. Cox, Y. Jiao, N. Bentenitis, S. Weerasinghe and P. E. Smith, *J. Chem. Theory Comput.*, 2011, **7**, 1369–1380.
41 J. G. Kirkwood and F. P. Buff, *J. Chem. Phys.*, 1951, **19**, 774–777.
42 D. van der Spoel, E. Lindahl, B. Hess, G. Groenhof, A. E. Mark and H. J. C. Berendsen, *J. Comput. Chem.*, 2005, **26**, 1701–1718.
43 B. Hess, C. Kutzner, D. van der Spoel and E. Lindahl, *J. Chem. Theory Comput.*, 2008, **4**, 435–447.
44 H. J. C. Berendsen, J. P. M. Postma, W. F. Van Gunsteren, A. Dinola and J. R. Haak, *J. Chem. Phys.*, 1984, **81**, 3684–3690.
45 M. Parrinello and A. Rahman, *J. Appl. Phys.*, 1981, **52**, 7182–7190.
46 S. Nosé and M. Klein, *Mol. Phys.*, 1983, **50**, 1055–1076.
47 S. Nosé, *Mol. Phys.*, 1984, **52**, 255–268.
48 W. G. Hoover, *Phys. Rev. A: At., Mol., Opt. Phys.*, 1985, **31**, 1695–1697.
49 T. Darden, D. York and L. Pedersen, *J. Chem. Phys.*, 1993, **98**, 10089–10092.
50 C. Seidel, *Macromolecules*, 2003, **36**, 2536–2543.
51 C. Ibergay, P. Malfreyt and D. J. Tildesley, *J. Phys. Chem. B*, 2010, **114**, 7274–7285.
52 A. Luzar, *J. Chem. Phys.*, 2000, **113**, 10663–10675.
53 H. A. Karimi-Varzaneh, P. Carbone and F. Müller-Plathe, *Macromolecules*, 2008, **41**, 7211–7218.
54 P. Jungwirth and D. J. Tobias, *J. Phys. Chem. B*, 2001, **105**, 10468–10472.
55 L. Tůma, D. Jeníček and P. Jungwirth, *Chem. Phys. Lett.*, 2005, **411**, 70–74.
56 M. Mucha, T. Frigato, L. M. Levering, H. C. Allen, D. J. Tobias, L. X. Dang and P. Jungwirth, *J. Phys. Chem. B*, 2005, **109**, 7617–7623.
57 P. Jungwirth and D. J. Tobias, *J. Phys. Chem. B*, 2002, **106**, 6361–6373.
58 F. G. Donnan, *Z. Elektrochem.*, 1911, 572–581.
59 Y. Shinagawa, *J. Theor. Biol.*, 1977, **64**, 551–556.
60 E. A. Algaer and N. F. A. van der Vegt, *J. Phys. Chem. B*, 2011, **115**, 13781–13787.

General discussion

Dr Goodman introduced the session: What information does a picture contain? A postcard of Cambridge University might be described as "Cambridge", but it is only a tiny part of the city. It might be labelled "King's College", if attention is focused on the highest peaks in the figure, ignoring the bulk of the data. It might have many other labels. Is the timescale hundreds of years (for the buildings) or seconds (for the tourists)? We need to know what phenomena we are investigating in order to interpret the data. Photographs are simple compared with the data in molecular dynamics simulations or DFT calculations. Why is an organic chemist like me interested in Hofmeister? The effects of ions are important in many areas of organic chemistry. For example, solubility; an important and familiar property that must be controlled to make useful pharmaceuticals and to get chemical processes to work effectively. However, we cannot calculate it accurately, and it is hard to measure it precisely (A. Llinas, R. C. Glen and J. M. Goodman, Solubility Challenge: Can You predict solubilities of thirty-two molecules using a database of one hundred reliable measurements?, *J. Chem. Inf. Model.*, 2008, **48**, 1289–1303.). Ions in the solution can have an important influence on solubility properties, and it is becoming possible to calculate the interactions between ions and biomolecular structures sufficiently accurately for the calculations to be related to readily interpretable experimental properties such as the extent to which oligopeptides form helices (for example, M. V. Fedorov, J. M. Goodman and S. Schumm, The effect of sodium chloride on poly-L-glutamate conformation, *Chem. Commun.*, 2009, 896–898). So we might hope that we can look at some molecular properties, such as size, charge, polarisibility, *etc.*, and go from these to the properties of solutions. But can all molecular properties be reduced to dials? Do we need a Hofmeister dial? Do we need to adjust size of the Hofmeister effect for different situations or even exchange it with a coulombic effect or other property? Are molecular structures much too complicated for this type of simplified analysis, and cannot be understood except by treating each molecule as an individual? In this session we will hear a series of presentations which describe the effects of ions on a wide range of other molecular systems

Professor Levin added: I think it is difficult to decouple Hofmeister from Coulomb. The two are intimately tied together.

The underlying equations that describe the electrostatics in these systems are highly non-linear; ionic size, polarizability, and electrostatic potential are all mixed up together. For some problems, such as surface and interfacial tensions, we can have a very accurate theory that is highly predictive.

For hydrophobic colloidal suspensions we can calculate critical coagulation concentrations, see: A. P. dos Santos and Y. Levin, Ion specificity and the theory of stability of colloidal suspensions, *Phys. Rev. Lett.*, 2011, **106**, 167801. I am not sure if it will be possible to do something with proteins, which are much more complex than simple colloidal particles. Nevertheless, I think the theory can lead us to better water and ion models, which can then be used to simulate protein–electrolyte solutions.

Professor Record responded: Hofmeister effects on processes that expose molecular surface to water result from specific short-range preferential interactions of ions with the surface that is exposed in the process. We find that, to a good approximation, Hofmeister effects on protein unfolding and DNA helix melting can be separated from coulombic effects at high salt concentration, where coulombic effects are minimized (L. M. Pegram *et al.*, *Proc. Natl. Acad. Sci. U. S. A.*, 2010, **107**(17), 7716–7721). For such situations the significance of the coulombic effect can be predicted computationally, (though the adequacy of the Poisson–Boltzmann equation at molar salt concentrations may be questioned), or can be determined experimentally if a salt without a significant Hofmeister preferential interaction is available. Hofmeister salt effects which apparently are relatively free of coulombic effects are observable for preferential interactions of salt ions with uncharged surfaces, though of course the accumulation or exclusion of the salt ion in the vicinity of that surface has some coulombic consequences. Contributions of cation and anion to the Hofmeister effect of a salt appear independent and additive, at least up to 1 molal (L. M. Pegram and M. T. Record, *Chem. Phys. Lett.*, 2008, **467**, 1–8; *J Phys Chem B.*, 2008, **112**, 9428–36). For processes involving polyelectrolytes at low to moderate salt concentration, Hofmeister and coulombic effects can be tightly coupled because the coulombic accumulation of counterions to high local concentrations in the vicinity of the polyion drives specific interactions of interactive counterions with the polyion surface, which can result in "inverse" Hofmeister effects where interactive counterions are stabilizing at low salt concentration but destabilizing at high salt concentration (see for example Y. Zhang and P. S. Cremer, *Proc. Natl. Acad. Sci. U. S. A.*, 2009, **106**(36), 15249–15253.

Mr Marslaek opened the discussion of the paper by D. T. Bowron: You are showing the pair correlation function of SPC/E water before any EPSR. Your result shows no structure beyond the first peak, which is inconsistent with the result one gets from the same model sampled using molecular dynamics. I am curious about the source of this discrepancy.

Dr Bowron answered: The damping of the structure beyond the first neighbour peak in Fig. 5 from the paper relates to the requirement for flexible water molecules if experimental data is to be correctly modeled. Standard Monte Carlo and Molecular Dynamics engines are generally focused on inter-molecular correlations, and consequently treat the water molecules as identical rigid structures. This prevents their models from generating accurately the intra-molecular terms that are unavoidably captured in experimental scattering data. EPSR restores the structure in the intermolecular correlations through the perturbation potential that it generates.

Professor Mason commented: I think what you have done in Fig. 5 is simply take TIP3P water which has the experimental geometry of water (104 degrees), and applied to it the SPC/E charges? It is documented that this will just look like TIP3P water, whereas if you take the SPC/E geometry (109 degrees) and apply the TIP3P charges, this looks like SPC/E water (P. E. Mason and J. W. Brady, *J. Phys. Chem. B*, 2007, **111**(20), 5669–5679, Fig. 2). Basically geometry plays a larger role than charge in determining the presence of the second peak in gOO(r). Further the presence of the second peak at 4.5 Å in gOO(r) is actually a relatively poor indicator of 'tetrahedral' structure, in that it can be shown that TIP3P water, despite having virtually no second peak in gOO(r) has a very similar three dimensional structure to SPC/E water. In essence all that Fig. 3 in the paper has shown is that EPSR can change a bond angle from 104 degrees to nearer 109 degrees.

Dr Bowron responded: The fundamental requirement of EPSR is to reproduce experimental total scattering data, which would not be possible if the simulation were performed using rigid molecules in the normal fashion of classical methods. All the molecules in an EPSR model are flexible and their flexibility is optimized to match the intrinsic level of quantum zero-point disorder present in the nuclear positions. As a result, it is not useful to obsess about fixed molecule geometries that are essential components of conventional classical simulations. EPSR is effectively a hybrid method, incorporating elements of classical, quantum and effective many-body interactions into the structures it generates. The method certainly does not change the H–O–H angle of the water molecules in its ensembles to 109 degrees, the mean water molecule geometry is specified based on intramolecular O–H and H–H distances of 0.98 Å and 1.55 Å respectively.

Mr Marslaek asked: Is the main reason for the introduction of flexibility the approximation of zero point disorder?

Dr Bowron responded: Yes. As the structure refinement method aims to reproduce experimental data, it must take into account the finite widths of the intra-molecular correlation features in the total radial distribution functions that are derived from the structure factors measured in the scattering experiments. Fixed geometry molecular configurations would result in unphysical delta functions in the radial distribution function at the intra-molecular correlation distances.

Mr Marslaek remarked: Is then perhaps the primary message that changing the geometry of the SPC/E model and introducing flexibility makes for a poor water model?

Dr Bowron answered: The structural disorder introduced into the molecular geometries to reflect the quantum zero point effects, leads to a distribution of dipole moments for local interactions. This is because we do not vary the partial charges assigned to the nuclear sites on each molecule and this results in the near neighbour inter-molecular correlations being disordered. In EPSR this problem for the inter-molecular interactions is then compensated for, by the empirical perturbation potential derived from the experimental data.

Professor Bain opened the discussion of the paper by Dor Ben-Amotz: Could you explain what assumptions you make *a priori* in your MCR approach about the nature of the component spectra? Do you assume that the spectra are a linear combination of two or three concentration-invariant spectra? Would you, for example, distinguish between the cases where the TBA molecules were in two states with and without a bound I^- ion, as opposed to a continuous shift in the C–H stretching frequency with increasing I^- concentration?

Professor Ben-Amotz answered: That is an excellent question. I believe that in these particular experiments the answer is that we could not tell the difference between the two cases you described, for the following reasons. Our experimental results suggest that the TBA molecules that do contain one I^- anion in their first hydration-shell produce a red-shift of about 5 cm^{-1} in the CH frequency of TBA (and the latter shift is consistent with our theoretically predicted shift of about 4 cm^{-1}). However, since this shift is significantly smaller than the width of the CH band (as well as the sub-peaks in the CH band), I would not expect to be able to distinguish whether our measured difference between the CH frequency of TBA in a particular NaI solution and that of TBA in pure water is due to a mixture of bound and unbound I^- ions or due to a continuous distribution of TBA–ion distances. More specifically, the results shown in Fig. 5 of our paper indicate that our measured TBA CH stretch frequency in 1 M NaI is red-shifted with respect to that of TBA in pure water by about 1 cm^{-1}. The latter red-shift may be due entirely to a two-state linear combination of TBA molecules which do or do not contain I^- in their first hydration-shell, or it could equally well be due to the average shift arising from a continuous distribution of I^- ions about TBA. Having said that, I also want to stress that the two-component SMCR analysis procedure that we have used in this work is appropriate, independent of the above two-state *vs.* continuous distribution issue. In other words, our two-component SMCR analysis only assumes that when TBA is added to a given salt solution, then the spectrum of the resulting solution is expressible as a linear combination of the spectra of the pure salt solution (with no TBA) and the TBA solute-correlated spectrum which contains features that reflect

TBA-induced spectral changes. That assumption should be valid as long as there are no significant TBA–TBA interactions at the experimental TBA concentrations. The validity of the latter assumption is confirmed by the fact that our TBA solute-correlated spectra are found to scale linearly with TBA concentration. The same goes for the results in Fig. 7 in our paper, whose red curve is the solute-correlated spectrum obtained when NaI is added to an aqueous solution of TBA. Thus, the latter red curve reveals how adding NaI has changed the Raman spectrum of aqueous TBA. We have again confirmed that the latter change is a linear function of the NaI concentration, which implies that there are no significant ion–ion interaction contributions to the solute-correlated spectrum of NaI in aqueous TBA.

Dr Ottosson commented: In your paper you show that iodide directly interacts with the hydrophobes, whereas fluoride does not, but that the local iodide concentration at the hydrophobe interface is only 60% of that of the bulk concentration. Since your observable in the solvent-correlated spectra is the shift of the C–H stretch peaks I wonder how you can distinguish between a situation where a low local concentration of iodide is binding strongly (thus giving a large shift per ion) and that where the a larger local concentration (which could be larger than the bulk concentration) binds weakly and causes a small shift per ion? In general, isn't it so that solvent correlated spectra (without additional input from simulations) only can give an estimate of the lower bounds of the number of affected molecules and ions due to the difficulty to distinguish between these kinds of scenarios?

Professor Ben-Amotz replied: Our experimental estimate that only about 60% of the TBA molecules are significantly perturbed by their interaction with iodide is in fact a lower bound, as you have noted, and so it could be larger. However, our Raman-MCR results, shown in Fig. 7, imply that some of the TBA molecules in the system are not significantly perturbed even in a 3 M NaI solution. That would seem to require that the local concentration of iodide in the first hydration-shell of TBA is indeed lower than the bulk iodide concentration. On the other hand, I would agree that there is more to be done in order to address these important issues. So, at this point, I would say that our evidence suggests, but does not yet definitively prove, that the local iodide concentration around TBA is lower than the surrounding bulk iodide concentration. Note that I have provided some further comments related to your question in my reply to Colin Bain's question above.

Dr Ottosson asked: You compare your solvent correlated spectra with simulations and get a near quantitative match. However, the shift in the experiments still results from an ensemble average of the changes induced by the solute so I wonder how straight-forward it is to directly compare with simulations. For example, did you perform an analysis of how the calculated spectral shift depends on the distance from the hydrophobic group? Even if you have a completely accurate description of this dependency, I feel there must still be some uncertainty in how the density profile of the iodide ions with respect to the distance from the hydrophobes must look, even if your calculations match the boundary conditions posed by your experiments – would you agree with that? Reasonably, a large number of qualitatively different types of density profiles should be able to cause the same solvent correlated spectrum.

Professor Ben-Amotz responded: Our calculated CH frequency shifts of TBA (given in Table 1 in the paper) are average values obtained from all of the iodide ions within the first hydration shell of TBA. Further work would be required in order to obtain sufficient statistics to accurately map the dependence of the CH frequency shift on the angular and radial location of the iodide ions. At this point, all we can say is that the magnitude of the CH frequency shift that we have measured, when compared with our calculated shifts, is consistent with our conclusion that iodide ions do penetrate significantly into the first hydration shell of TBA.

Mr Sharma said: How much does it effect the experimental spectra due to presence of any possible triiodide (triple anion) complexation with increasing concentration of NaI in aqueous solutions? Have you ever obtained any such evidence in any other monovalent ionic solutions?

Professor Ben-Amotz answered: We have not seen any evidence of triiodide, but I imagine that we would have been able to see it if it were present, as our solute-correlated spectra contain features arising both from the solute itself and from any molecules in the solution whose vibrational spectrum is perturbed by the solute. Thus, for example, if triiodide were formed then we should see its vibrational spectrum in our NaI solute-correlated spectrum (either in pure water or in water plus some other solute such as TBA).

Dr Gibb commented: From my perspective, as a flat surface is curved into a convex one, the number of non-covalent interactions that it can form with another entity decreases. Likewise, as a flat surface is made concave so the number of possible interactions that it can make increases (assuming that the other species can fit within the concavity). Isn't it the case that the interaction surface of *t*-butanol is simply too flat/small for any interaction with the iodide ion to be anything but transient? Replacing the *t*-butanol with a hydrophobe possessing a larger interface would lead to a stronger interaction that might be quantifiable using the C–H shift data.

Professor Ben-Amotz responded: Our experimental results suggest that iodide does penetrate into the first hydration shell of *t*-butanol, while fluoride does not. Moreover, our simulation results suggest that the iodide population is localized primarily around the higher curvature periphery of the three methyl groups, rather than near the flatter hydrophobic end of the molecule. It would be interesting to further explore the influence of surface curvature on the affinity of various ions for hydrophobic interfaces.

Dr Gibb asked: With the example of concave molecules we synthesize, we can measure – with excellent accuracy and precision – the association of chaotropes to the hydrophobic pocket. The values we measure are between 0.65 and 2.67 kcal mol^{-1} (iodide, 1.4 kcal mol^{-1}).

Professor Ben-Amotz answered: I would be interested in learning more about this and considering whether our experimental technique could be applied to such ion binding processes in hydrophobic pockets.

Professor Goodall said: Beyond simple ions to Raman active compounds, could you look at the Raman of more complex ions at the same time as your Raman-MCR measurements on water? Examples I would find particularly interesting are thiocyanate and nitrate.

Professor Ben-Amotz replied: Yes, we certainly can look at molecular ions, and we are currently doing just that. I expect that we will publish a follow-up paper describing the interactions between TBA and various ions.

Professor Record remarked: We've analyzed thermodynamic data for the effects of Hofmeister salts on hydrocarbon solubility to obtain a lower bound on the amount of "local" water at a molecular hydrocarbon surface (0.18 H_2O $Å^{-2}$) and local-bulk partition coefficients quantifying the individual distributions of the cation and the anion between this local water and bulk. How do your spectroscopically derived results for ion distributions in the vicinity of the *t*-butyl group compare with these thermodynamically derived results?

Professor Ben-Amotz answered: Thank you for this question. Your results are in fact more directly relevant to our measurements than I had realized. More specifically, your thermodynamic analysis[1] of hydrocarbon solubility in the presence of Hofmeister salts, as well as the partitioning of various ions between bulk water and either air–water or oil–water interfaces, are all generally consistent with our findings. For example, our spectroscopic measurements confirm that both sodium and fluoride ions are excluded from the hydrophobic hydration shell of TBA. Moreover, our additional unpublished measurements confirm that the same is true for sulfate ions, and that the affinity of halide ions for the first hydration shell of TBA increases with increasing size. The only apparent quantitative discrepancy between our conclusions and yours is that our results suggest that the concentration of iodide ions is comparable to, but slightly lower than, the concentration of the surrounding bulk aqueous salt solution. Although the latter conclusion is consistent with both our experimental and EFP simulation results, I would say that further work is required in order to verify this particular conclusion. I am more confident that our results do clearly show that I^- ions are present in the first hydration-shell of TBA, in contact with its hydrophobic groups, while F^- is not.

1. L. M. Pegram and M. T. Record, *J. Phys. Chem. B*, 2008, **112**, 9428–9436.

Dr Cremer asked: In direct relation to Tom Record's comments on the partitioning of iodide to *t*-butyl alcohol, would iodide partition better to a hydrophobic moiety that was slightly positively charged compared with a purely hydrophobic moiety such a *t*-butanol? Such hybrid hydrophobic/positively charged sites are present in the backbones of polypeptides.

Professor Ben-Amotz replied: Yes, one would clearly expect that a positively charged species would have a higher affinity for I^- than a neutral species. We have performed some measurements similar to those I have talked about, but using TMAO rather than TBA as the solute. As you know, TMAO looks quite similar to TBA but is zwitterionic, with a partially negative O atom and a partially positive N atom attached to three methyl groups. Thus, one might expect the methyl groups of TMAO to have a slightly positive net charge. Our preliminary measurements of TMAO in aqueous NaI suggest that the I^- ions do in fact interact more strongly with the methyl groups of TMAO than they do with the methyl groups of TBA. We plan to publish the results of these and other such measurements at some point.

Professor Bakker asked: For both I^- and F^- you observe a similar increase of the response of the OH dangling peak of the water molecules in the hydration shell of TBA, indicative of a slight distortion of the hydrophobic hydration shell by the ions. However, you also find that F^- is strongly excluded from the first hydration shell of TBA, and that the local concentration of I^- around TBA is nearly as large as in the surrounding bulk. Can you explain why nevertheless F^- and I^- lead to a similar increase of the response of the dangling OH?

Professor Ben-Amotz answered: The fact that both I^- and F^- produce a similar increase in intensity (and slight red-shift) of the dangling OH peak arising from the hydrophobic hydration-shell of TBA is surprising. Alhough we do not yet understand why the two ions produce such a similar effect on the dangling OH, we have speculated that the increase is related to the disruption of the tetrahedral water structure in the second hydration-shell of the anions.

Dr Tyrode asked: TBA tends to aggregate in aqueous solutions at concentration usually exceeding 2 M. These aggregates remain soluble in solution and may have long associated correlation lengths (see for example the paper presented in this *Faraday Discussion* by D. T. Bowron; or S. Paul and G. N. Patey, *J. Phys. Chem.*

B, 2006, **110**(21), 10514; or R. Gupta and G. N. Patey, *J. Chem. Phys.*, 2012, **137**(3), 034509). In the work presented here you carried out Raman-MCR measurements of varying concentrations of NaI and NaF at a fixed TBA concentration of 0.5 M. In contrast to NaF, at high NaI concentrations (3 M) you observed a 3 cm^{-1} red-shift in the CH stretching modes of TBA, which you interpreted as iodide ions present in the first hydration shell. Although you have confirmed in previous work (*e.g.* ref. 7 in your manuscript) that no TBA aggregates are formed at 0.5 M concentrations in PURE water, this may not necessarily be the case in the presence of NaI concentrations exceeding 1 M. I am concerned that red-shift detected as well as some of the other spectral changes, are a consequence of TBA aggregation triggered by an increase in the iodide concentration in solution (particularly above 1 M), rather than the anion adsorption to the first hydration shell, as described in the manuscript. In principle, this possibility could be discarded, collecting a series of spectra at different TBA concentrations. However, judging for what is described in the manuscript, it appears that this possibility was not explored.

Professor Ben-Amotz answered: This is a good question. Although we have not investigated this possibility in detail, we are quite convinced that TBA aggregation is not responsible for the salt-induced spectral changes that we have observed. The simplest evidence for that is that we do not observe any such spectral changes in NaF, which should have a greater "salting out" effect than NaI. We are currently pursuing additional studies aimed at extending our Raman-MCR methodology to the detection of hydrophobic aggregate formation. Those measurements indicate that there is essentially no direct contact TBA–TBA aggregate formation below a TBA concentration of 1 M.

Professor Harries added the general comment: I think that, as a community, our understanding of important biological processes that relate to proteins in solutions containing ions and many other cosolutes is becoming better over time. For example, over the past decade significant progress has been made in deciphering the way that the thermodynamic stability of small solute molecules, such as amino acids and peptides, is affected by salts and other cellular cosolutes such as osmolytes. The guiding idea is that understanding the effect on these smaller model compounds will afford a basis for learning about and predicting the effect of the same cosolutes on large molecules undergoing processes such as protein folding or self-assembly and aggregation (for example into amyloid fibers). This progressive, incremental strategy is helpful, as it allows a smoother transition between scales, is more easily accessible experimentally, and is easier to analyze. In addition, this approach has been used to successfully predict the changes in stability of several large proteins due to cosolutes.[1–3] Nevertheless, it is already clear that, in the process of going from small molecules to large, we will inevitably encounter considerations that apply to one scale but not to another. A full understanding of these factors is an essential part of our learning, and serves as another reason to improve our knowledge on all relevant length scales. In this respect, we are already making considerable progress in formulating predictive theories that will eventually allow quantification of the effect of cellular solutes on biologically relevant processes.

1. M. Auton and D. W. Bolen, *Proc. Natl. Acad. Sci. U. S. A.*, 2005, **102**, 15065.
2. L. M. Pegram, T. Wendorff, R. Erdmann, I. Shkel, D. Bellissimo, D. J. Felitsky and M. T. Record Jr, *Proc. Natl. Acad. Sci. U. S. A.*, 2010, **107**, 7716.
3. M. Auton, J. Rösgen, M. Sinev, L. M. F. Holthauzen and D. W. Bolen, *Biophys. Chem.*, 2011, **159**, 90.

Professor Mason continued the discussion of the paper by D. T. Bowron: In the second paragraph of your paper in the 'structure refinement' section you state:

"However it is important to recognize that any pair distribution functions not heavily constrained by the scattering data will primarily reflect the prior assumptions made for the underlying interatomic potential." And that "in spite of this, provided that these primary constraints are reasonable, the model can provide a useful guide for improving our understanding of how the structure of system gives rise to its observed physical and chemical properties". These two statements together suggest that if a structure factor of the solution is not heavily constrained by the neutron data, that the molecular mechanics will provide the insight. This is indistinguishable from the statement that molecular mechanics can provide insight into solution properties, and would render, in such a case the inclusion of neutron data in such a study irrelevant.

In the current case, if we take a look at Fig. 7 in the paper we find that all of these correlations contribute a very small amount to the total scattering. For instance, in the 1 molal TBA, 0.5 m AB electrolytes, each of the correlations shown in Fig. 7 contributes less than about one part in ten thousand to the total scattering data. To put this into perspective this is probably a couple of orders of magnitude smaller than the thickness of the line chosen to represent the residuals shown in Fig. 2. In this case it is probably reasonable to say that the empirical potential structural refinement technique is delivering the same as the molecular mechanics simulation. To me it looks like this point seems to have been all but lost in the conclusions of this paper. Do you think this is a reasonable criticism of the technique applied to this aspect of this system?

Dr Bowron answered: In an ideal world, where we have a perfect knowledge of the potentials of interaction between, atoms, ions and molecules, we should be able to rely on a molecular mechanics simulation to provide an accurate model of reality. Unfortunately we do not live in such a world, and we are dependent on experimental observations to establish the veracity of our knowledge. Empirical Potential Structure Refinement aims to generate configurations of atomic and molecular coordinates that are consistent with experimental scattering data and known physicochemical constraints such as density and molecular structure. The method starts from a baseline of our best guess at a set of interaction potentials, and then develops a set of perturbation potentials with a magnitude that scales in proportion to the structural contributions to the data that is provided for refinement. If the reference potentials were perfect and able to account for the provided data, the method would not change anything, but to date there are no perfect potentials and a perturbation function is always required to bring the model into good agreement with experiment. For example, existing classical potentials tend to overestimate the magnitude of polar group interactions between molecules.[1–2]

With regards to whether we learn anything about the dilute species interactions in an EPSR model, the answer is, yes we do. This is because in any condensed state system, atoms, ions and molecules do not live and act in isolation. The chemistry that takes place in a solvent does not occur on a structureless uniform background. What an EPSR model does, is optimize the nuclear positions in the structural ensembles it generates to be consistent with the best scattering information that we have about the *total* system, and this means that dilute species interactions that are to be accommodated in the model, have to do so against a background of the best solvent model that we can generate. In essence the method enforces rigorous, experimentally established, boundary conditions for the atomic and molecular interactions of interest. Although not presented in this work, chemically specific structural information from dilute species can be incorporated into the EPSR model through a requirement for the structural ensembles to reproduce X-ray absorption spectroscopy data, which can be sensitive to structural correlations at the parts-per-million level.

1. D. T. Bowron, J. L. Finney and A. K. Soper, *Mol. Phys.*, 1998, **93**, 531–543.
2. D. T. Bowron, J. L. Finney and A. K. Soper, *Mol. Phys.*, 1998, **94**, 249–251.

Dr Ball asked: I have a small question which is starting to seem like part of a bigger question. In your previous studies you reported finding salt bridges – halide ions – between the hydroxy groups of *t*-butyl alcohol. Is this something that survives your EPSR treatment? More generally, your results make me wonder how secure we can feel about using simple systems as analogues of more complex ones – tertiary alcohols as models of macromolecules with hydrophobic and hydrophilic regions, say. It seems that we might start to see qualitatively new phenomena appear (or vanish) as we alter the degree of complexity. We know that even for alcohols – even for methanol, in fact – there are effects such as clustering that might determine the hydration characteristics, which might not apply in the same way, or at all, to long-tail lipids or to protein side-chains. What would you say about that? How much continuity can we expect as we change the complexity of our solutes?

Dr Bowron replied: First, salt bridges do still survive in the EPSR simulations, but these studies performed with more powerful computers that allowed for better sampling of the experimentally consistent structural ensembles, suggest that they are more prevalent at higher alcohol concentrations. The earlier results in the most dilute system,[1–2] appear to have been given undue weight due to insufficient system equilibration.

The challenge of extending fundamental knowledge to the science of complex systems is extreme, but there is however the counter argument that without the fundamental understanding of the rich phenomena displayed by the toy models, we would not know where to begin to tackle the bigger problems. What is already known is that complexity can arise from very simple laws, so at the very least we need to study simple systems to establish these. Speaking as an experimental scientist, I actually think we should be very grateful for the fact that complexity can drive new phenomena as this means there is likely to be an almost endless supply of new and exciting things to discover, and good measurements will always have a role to play in their observation and characterization.

1. D. T. Bowron and J. L. Finney, *Phys. Rev. Lett.*, 2002, **89**, 215508.
2. D. T. Bowron and J. L. Finney, *J. Chem. Phys.*, 2003, **118**, 8357–8372.

Professor van der Vegt opened the discussion of the paper by Daniel Harries: Enthalpy and entropy variations contain contributions of changes in solvent–solvent interactions that do not affect the free energy. This is referred to as exact enthalpy–entropy compensation. Shouldn't we for this reason refrain from statements on enthalpy (or entropy) driven stabilisation?

Professor Harries replied: Regardless of the name we give these findings, the overall entropic and enthalpic parts of the free energy are important considerations or constraints on proposed molecular mechanisms for the action of cosolutes. It is true that the effect of cosolutes on the free energy of processes such as protein folding and aggregation can be much smaller than the underlying entropic and enthalpic terms, which often seem to compensate each other to a large extent. We simply call processes "enthalpically driven" if the net enthalpic contribution of cosolutes as measured in an experiment is the largest contribution to the relevant free energy. In addition, we find that at least for the peptide folding we have studied, the entropic and enthalpic contributions are in fact often of the same order of magnitude as the total change in folding free energy due to solute addition (for example, in the presence of PEG 2000 at 0.1 M, $\Delta\Delta G = -1.8$ kJ mol^{-1} while $\Delta\Delta H = -1.1$ kJ mol^{-1} and $T\Delta\Delta S = 0.7$ kJ mol^{-1}). Moreover, we know that these terms in the free energy can usually be further dissected into several contributions on the molecular scale. Take, for example, the process of protein folding in the presence of cosolutes. Sometimes the net effect is simply found to be dominated by entropy or enthalpy, while the other is smaller and acts to stabilize the protein (we and others have found indirect or direct evidence for this[1–2]). Even more dramatically, other cosolutes, such as polyols and sugars, can show stabilizing enthalpic dominance, while the net entropic contribution is *destabilizing*.[3] It is difficult to explain this last experimental observation by mechanisms that invoke entropy, such as excluded volume effects, as they would result in an entropic contribution of the wrong sign. While excluded volume effects undoubtedly contribute to the overall entropy, in this case they do not dominate the net action of these cosolutes. Therefore, the molecular mechanism that is responsible for this stabilizing effect necessarily includes contributions that are overwhelmingly enthalpic. Calling such mechanisms "enthalpically driven" serves as a mnemonic reminding us of the necessity to find new mechanisms that can account for the dominating enthalpic contributions, as well as the possible unfavorable entropy.

1. D. B. Knowles, A. S. Lacroix, N. F. Deines, I. Shkel and M. T. Record Jr, *Proc. Natl. Acad. Sci. U. S. A.*, 2011, **108**, 12699–704.
2. Y. Wang, M. Sarkar, A. E. Smith, A. S. Krois and G. J. Pielak, *J. Am. Chem. Soc.*, 2012, **134**(40), 16614–16618.
3. R. Politi and D. Harries, *Chem. Commun.*, 2010, **46**, 6449–51.

Dr Canchi remarked: It is very interesting to see that certain cosolutes can stabilize proteins by an enthalpic mechanism, even though stabilizing cosolutes are excluded from the vicinity of proteins. What do you think is the molecular picture for this effect?

Professor Harries replied: In models that describe the action of "depletion forces" or "molecular crowding", the solvent itself is often ignored.[1–2] Scaled particle theory, often used to describe the action of excluded cosolutes, typically parametrizes only the shape and size of the cosolute and the protein, and ignores the interactions of the solvent with these species.[3] However, water has strong, direct interactions with both protein and cosolutes, particularly in biological systems.[4] What we know about the enthalpic mechanism you ask about is based mostly on molecular dynamics simulations.[5] It seems that even in the absence of peptides or proteins, polyol osmolytes modify the hydrogen bonding network around them, so that the hydrogen bonds between water molecules becomes shorter and more linear, and hence stronger. Since the cosolute molecules are excluded from the vicinity of the peptide, a layer of hydration is formed between cosolutes and the peptide. This first hydration layer, too, is interconnected with stronger hydrogen bonds than in pure water, and as a consequence waters in this layer are less available for hydrogen bond formation with the peptide, forcing the peptide to find alternative internal hydrogen bonds. This translates into a stabilization of the peptide in the folded state. Overall, water mediates the interaction of cosolute with the peptide, and the concomitant changes in water hydrogen bonding are responsible for the enthalpic contributions. Further evidence of this mechanism is necessary. Already, recent experiments[6] have shown that the water–water hydrogen bonding in glycerol aqueous solutions is indeed altered in the way predicted from molecular dynamics simulations.[7]

1. S. Asakura and F. Oosawa, *J. Chem. Phys.*, 1954, **22**, 1255–1256.
2. H.-X. Zhou, G. Rivas and A. P. Minton, *Annu. Rev. Biophys.*, 2008, **37**, 375–397.
3. D. M. Hatters, A. P. Minton and G. J. Howlett, *J. Biol. Chem.*, 2002, **277**, 7824–7830.
4. G. N. Somero, C. B. Osmond and C. L. Bolis, *Water and life: comparative analysis of water relationships at the organismic, cellular, and molecular levels*, Springer-Verlag, 1992.
5. R. Gilman-Politi and D. Harries, *J. Chem. Theory Comput.*, 2011, 3816–3828.
6. J. J. Towey and L. Dougan, *J. Phys. Chem. B*, 2012, **116**, 1633–1641.
7. R. Politi, L. Sapir and D. Harries, *J. Phys. Chem. A*, 2009, **113**, 7548–7555.

Dr Henchman commented: You state that the salts tend to have an overall destabilizing effect on the model peptide and yet you include supposedly stabilizing salts such as Na_2SO_4. Could you explain why this is?

Professor Harries replied: The effect of salt on protein folding depends not only on the salt used, but also on the protein itself. For the peptide we have been studying, electrostatic interactions seem to play an important part,[1] so that at low concentration the screening of these interactions dominate, and the folded state is destabilized by salts. As salt concentrations increase, we find a stabilizing effect (compared to the initial destabilization) that is ion-specific. For example, in sodium sulfate the effect is much stronger than in sodium chloride at the same concentration. For this peptide, increasing the concentration of salt even further resulted in a tendency to aggregate. However, the theoretical predictions shown in Fig. 4 indicate that had the experiment been possible, salts would have stabilized the peptide in the folded state, compared to its state in pure water.

1. M. S. Searle, S. R. Griffiths-Jones and H. Skinner-Smith, *J. Am. Chem. Soc.*, 1999, **121**, 11615–11620.

Dr Gibb commented: Going way back to Philip Ball's comment, it seems that a natural next step for the field is to fill in the gap in molecular complicatedness between simple molecules such as *t*-butanol and complicated ones such as peptides or proteins. In that regard it is worth noting that over the last decade or so the field of supramolecular chemistry has developed considerably into the aqueous realm. In other words, there are many classes of highly symmetrical, water-soluble hosts that have a lot to offer the Hofmeister field in regards to identifying how small-scale effects come together synergistically to engender larger-scale Hofmeister phenomena.

Professor Tiddy continued the discussion of the paper by D. T. Bowron: Can you relate your measurements of the species present in solution to thermodynamic data such as water and TBA activities?

Dr Bowron replied: The Empirical Potential Structure Refinement process is performed in the constant NVT (canonical) ensemble. In consequence the calculation of thermodynamic parameters such as solution component activities requires the evaluation of the system's chemical potential through a method such as test particle insertion. At the current time, the method is not optimized to do this reliably as the simulation is parameterized to reproduce structural data over any grand-canonical thermodynamic observable.

Professor Ben-Amotz said: Am I correct in understanding that the small magnitudes of the ion–TBA association equilibrium constants that you have obtained imply that the local concentration of the ions around TBA is smaller than the total concentration of the corresponding ion and that your results imply that the affinity F^- for TBA is greater than that of Br^-? In a related comment, you indicated that your results indicate that there is significant TBA–TBA hydrophobic association at a mole fraction of 0.04, which is equivalent to a TBA concentration of over 2 M. At such a high concentration one might expect there to be significant TBA–TBA contacts even in a random mixture. So, it is not clear to me how you can distinguish such random contacts from contacts induced by a true hydrophobic interaction.

Dr Bowron answered: At the lowest investigated alcohol concentrations there is sufficient water in the solutions to independently hydrate the alcohol molecules, the cations and the anions. The coordination histograms tell us that the affinity of the ions for direct interactions with the alcohol molecules occurs at the 5% to 10% level in these systems whilst the equilibrium constants allow us to compare the relative propensities. Our results do show that there is a preference for alcohol interactions with fluoride ions compared to bromide, but with the caveat that the fluoride ions are always found to be hydrated and consequently the alcohol-ion interaction is solvent mediated. Solution crowding appears to become a significant factor as the concentration of the alcohol is increased. In the sodium chloride system, the apparent preference for alcohol–cation interactions at the lowest concentration is reduced and the anions start to play a more significant role. This correlates with the growth of a 9 Å alcohol–alcohol interaction feature (Fig. 9 in the paper), which had previously been assigned to salt bridged interactions.[1]

The concentration of the alcohol in the solutions investigated was deliberately chosen to bridge the 0.04 mole fraction composition as this was known to coincide with the maximum degree of hydrophobic effects displayed. The salt free systems have already been extensively characterized,[2–3] so this study was designed to allow us to establish how the salts drive the system away from the hydrophobic-interaction dominated state.

1. D. T. Bowron and J. L. Finney, *Phys. Rev. Lett.*, 2002, **89**, 215508.
2. D. T. Bowron and J. L. Finney, *J. Chem. Phys.*, 2003, **118**, 8357–8372.
3. D. T. Bowron, J. L. Finney and A. K. Soper, *J. Phys. Chem. B*, 1998, **102**, 3551–3563.

Professor Halling continued the discussion of the paper by Daniel Harries: This follows the previous comment about enthalpy–entropy compensation. There are those who argue that individual values of entropy and enthalpy change are of limited value in condensed phases, and especially aqueous solution. The argument distinguishes a direct process of interest, with a significant Gibbs energy change – and a secondary re-arrangement in the rest of the system, which makes almost no contribution to the Gibbs energy change, but has substantial and almost exactly balancing entropy and enthalpy changes. These will of course contribute to the total measured entropy and enthalpy changes. I am not convinced by this view, but would like to hear your argument. For example, on your enthalpy–entropy plot, classes of species often seem to lie quite close to a line for perfect enthalpy–entropy compensation, with almost the same Gibbs energy change. Thus you would define some class members as enthalpy driven, and others as entropy driven. Are you certain this really means a change in mechanism as we go through the series? An example is the line for polyethylene glycols.

Professor Harries answered: From a purely practical perspective, the entropic and enthalpic contributions that are found experimentally can be regarded as constraints to any mechanism that is offered to explain the overall changes in free energy. The value in making this dissection becomes clear once we notice that chemically distinct classes of cosolutes naturally fall into specific thermodynamic mechanisms (or "fingerprints") by which they change the free energy of a process. For example, we have shown that polymers (PEG and dextran) at low concentrations add to the favorable entropic component of peptide folding, see Fig. 1. At higher polymer concentrations this entropic dominance gives way to enthalpic dominance. Others have also observed a change in behavior at high PEG concentrations. Record and coworkers showed that while low PEG concentrations stabilize a DNA duplex, high concentrations destabilize it.[1] In contrast, the contributions of polyols and sugars to folding are much different, and consist of a favorable enthalpy that is counteracted by unfavorable entropy. Overall, it would be very hard to find a single mechanism that could explain the action of these two classes of cosolutes (polymers *versus* polyols) without directly rationalizing how the energetic contributions could be so different for the two cosolute classes. This has led us to propose that polymers and polyols likely act by different molecular mechanisms, where "mechanism" refers here to the collection of contributions to entropy and enthalpy that make up the total change in free energy. To conclude, similar changes in free energy could result from very different energetic or entropic contributions, and these could become hallmarks of specific underlying forces that determine the overall free energy change.

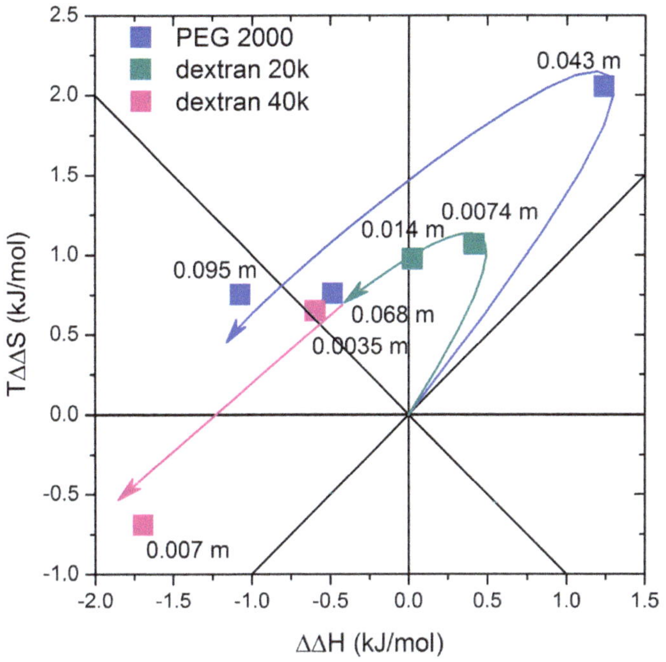

Fig. 1 Contributions of PEG and dextran on the entropic and enthalpic components of peptide folding. Cosolute concentrations are marked next to each of the data points, and arrows follow trends with increasing concentration.

1. D. B. Knowles, A. S. Lacroix, N. F. Deines, I. Shkel and M. T. Record Jr, *Proc. Natl. Acad. Sci. U. S. A.*, 2011, **108**, 12699–12704.

Mr Marslaek opened the discussion of the paper by Mikael Lund: If you have multiple binding constants, for different ions, how do you deal with sampling in the Monte Carlo simulation? Do you consider exchange moves at the same binding site, for example?

Professor Lund responded: In the presented scheme, ions can compete for binding sites and the site-to-bulk swap move is sufficient to explore all possible configurations. Site–site exchange moves are currently not implemented but may indeed improve sampling of tightly bound, competing ions.

Dr Vila Verde opened the discussion of the paper by Bob Eisenberg: How do we reconcile the fact that in biology, specific ion effects occur at ion concentrations as high as 20 M, when outside of a biological context they only occur for lower concentrations?

Dr Eisenberg replied: I do not know if the specific ion effects are in fact similar in detail.

Of course, many different mechanisms can produce the same result. An amplifier is an amplifier whether it is made of vacuum tubes (valves in UK English), bipolar transistors, or FETS.

Dr Goodman added: Bob Eisenberg has explained how his view of ion interactions is inconsistent with some other people's. However, as Philip Ball has pointed out, different phenomena emerge at different length scales and at different levels of complexity. If a molecular system is made ten times bigger, it may well be more than ten times as complicated, as co-operative effects which are impossible for simple systems become possible. For example, the discussion between Bruce Gibb and Dor Ben-Amotz about whether curvature affects binding led to the observation that small systems like tertiary butyl alcohol, cannot create concave binding surfaces and so generalisations based on the analysis of concave molecules may not be applicable to very simple systems. Observations on isolated carboxylates, or carboxylates in flat surfaces may well lead to different conclusions to multiple carboxylates carefully positioned to make ion channels, for which large synergistic effects lead to different behaviour. As a result, the apparent difference of opinion between Bob Eisenberg and Werner Kunz may really be consistent opinions about different systems. Therefore, you are both correct.

Dr Eisenberg replied: I agree.

Professor Levin continued the discussion of the paper by Mikael Lund: The simulation proposed uses the Yukawa potential which is based on linear Poisson–Boltzmann equation. How realistic is this? Near a charged protein residue the field can be quite large so that the non-linear effects can be very important.

Professor Lund replied: I agree that the Debye–Hückel approximation is simplistic and will break down at strong coupling conditions. However, the approach has previously been critically compared with explicit ion MC simulations (F. Luíf, S. B. Da Silva, B. Jönsson and R. Penfold, *Protein Sci.*, 2001, **10**, 1415–1425; and A. Kurut, B. A. Persson, T. Åkesson, J. Forsman and Mikael Lund, *J. Phys. Chem. Lett.*, 2012, **3**, 731–734) and for 1 : 1 salt and typical charge densities found in proteins, linearised PB seems fully applicable. In multivalent salt, however, the method can/will be qualitatively wrong.

Let me add that the idea of treating Hofmeister ion binding as a two-state process is not limited to implicit salt models, but can be readily incorporated in explicit salt MC simulations using a grand canonical scheme that does not suffer from the approximations made in PB theory, linearised or not (C. Labbez and B. Jönsson, *Applied Parallel Computing: State of the Art in Scientific Computing*, Springer, Berlin–Heidelberg, 2007,vol. 4699, pp. 66–72).

Professor Jungwirth asked: Your Monte Carlo model very nicely puts interactions of ions with proteins and amino acid protonation/deprotonation on the same conceptual footing. Of course, protonation is a chemical reaction and proton concentrations are typically orders of magnitudes lower than those of Hofmeister salts. Consequently, the (de)protonation effect effects are more dramatic than the Hofmeister effects, which may have led some early researchers (*e.g.*, Loeb and Kunitz, 1923) to disregard the latter.

Professor Lund replied: I have two comments on this:

(1) For reasonably short ranged, attractive potentials of mean force between the ion and a motif, a binding constant can be defined, regardless of the nature and depth of the minimum.

(2) On one hand pH – *i.e.* the proton concentration – can be used to tune the charge state of acids and bases at low ionic strengths. On the other hand, more weakly bound ions require higher bulk concentrations to promote a fully bound site. This consequently leads to screening of long range electrostatic interactions, whereby the charge change may not be as significant for intermolecular interactions as for charge changes induced by changes in the bulk proton concentration.

Mr Patko asked: I have a question in two parts mostly related to the treatment of anion binding as an analog to pH changes (*i.e.* site specific proton binding).

1. In your treatment of anions to be the complete analog to protons for your model of specific site binding for biological macromolecules, you assume the near instantaneous on/off kinetic behavior that is generally reserved for proton exchange and electron transfer reactions. Please explain why this assumption should also valid for anions, particularly those that are slow to lose their solvation shells? In short can anion binding truly be considered as a true fast on/off type analog as is the case for pH? 2. The use of the Monte-Carlo-type simulation mostly precludes the impact of the time-dependent solvent interactions upon the conformational changes and attributes them primarily to the impact of the site specific anion binding (ignoring any solvent shell and kinetics associated with each site binding itself as described in part one of my question). Can you please explain why these underlying assumptions needed to handle the actually more complex anion site-specific binding with this simplified on/off type binding mode are a good trade-off? The simplification is presumably made in the interests of computational efficiency and to develop a systematic reduced-model for macromolecular ion binding that can use some of the convenient and familiar language of pH dependent site binding.

Professor Lund replied: Statistical mechanics show that equilibrium or static properties can be rigorously separated from dynamics and, consequently, the kinetics of solvation or any time-dependent property have no influence on the studied thermodynamic properties. In the model, solvation is accounted for *implicitly* by the experimental ion-motif dissociation constant (pK_d), although we assume that solvation of a motif is unchanged from the model compound and when incorporated in a protein. This is often a good approximation (M. Lund, B. Jönsson and C. E. Woodward, *J. Chem. Phys.*, 2007, **126**, 225103), although for deeply buried sites, solvation changes may need to be considered as commonly done in more advanced Poisson–Boltzmann protein models – see for example the APBS project.

Professor Bakker opened the discussion of the paper by Nico van der Vegt: In your calculations of the salt bridges between the carboxylate groups and the different cations (overview in Table 2 from the paper) you find that for Li^+ the fraction of free ions is less and the fraction of mono-coordinated salt bridges is higher than for Na^+. From this systematic, one would expect that the fraction of di- and tri-coordinated salt bridges would also be larger for Li^+ than for Na^+. However, in your calculations you find that the fractions of di- and tri-coordinated salt bridges are smaller for Li^+ than for Na^+. Can you explain this? Could this observation be related to some steric effect, the mono-coordinated salt bridge with Li^+ somehow giving less opportunity for a second contact with a carboxylate than the mono-coordinated salt bridge with Na^+?

Dr Rodriguez-Ropero replied: As shown in Table 2, the population of mono-coordinated salt bridges is higher for Li^+ than for Na^+, but for di- and tri-coordinated salt bridges this situation is inverted and the population of such salt bridges is higher for Na^+ than for Li^+. This is mainly because of two reasons: (i) steric effects; the Na^+ cation is bigger than the Li^+ cation and it fits better between 2 or 3 carboxylate groups belonging to 2 or 3 neighbouring chains leading to the formation of, respectively, di- and tri-coordinated salt bridges. (ii) The hydration shell around the Li^+ cation is stronger than for Na^+, thus the energy penalty of losing one water molecule in Li^+ is higher. This higher energy penalty makes the formation of di- and tri-coordinated salt bridges slightly more difficult for Li^+ than for Na^+.

Professor Goodall continued the discussion of the paper by Bob Eisenberg: In your paper you presented the challenge to the community to identify a specific set of experimental measurements that operationally define "the Hofmeister Effect", to make a physical model of the setup of those experiments, and to compare with experiments in a range of solutions and concentrations. I think this is a really interesting proposal, and suggest that there would be enthusiasm from the biopharmaceutical community if you looked at the role of Hofmeister ions in stabilising antibody formulations. Every drug company has an interest in biotherapeutic proteins: these species are formulated at high concentration in solution, where they must be stabilised against aggregation and precipitation, so any insights from the Hofmeister community would be welcomed.

Dr Ball asked: You mentioned that finite-size effects are typically neglected in theoretical treatments. We know that effects such as crowding can be important in the cell, for example altering binding constants and kinetics. Is this the sort of thing you're referring to?

Dr Eisenberg replied: I have discussed these matters at length in my paper and in other publications[1–4] and it is best not to try to deal with all that here in brief. Crowding is easy to see and document for bio-ions like sodium, potassium, calcium, and chloride in and near nucleic acids, binding proteins, ion channels, enzymes and even near electrodes in electrochemical systems of our technology. Here extreme crowding producing number densities greater than 10 molar is easy to document. For reference, solid NaCl is 37 molar. Other forms of crowding produced by the general density of structures in the cytoplasm are obviously also important because they are likely to produce excess free energies that change all electrochemical properties of these ions. Since some of these ions are signalling ions, whose activity controls biological function the way a gas pedal controls the speed of a car; the consequences could be profound. This kind of crowding is harder to quantify and document but indeed may be very important, as Dr Ball hints.

1. B. Eisenberg, Crowded charges in ion channels, *Adv. Chem. Phys.*, 2007, 77–223; also available at http://arxiv.org as B. Eisenberg, Life's solutions are not ideal, http://arxiv.org/abs/1105.0184.
2. B. Eisenberg, Mass action in ionic solutions, *Chem. Phys. Lett.*, 2011, **511**, 1–6.
3. B. Eisenberg, Life's solutions. A mathematical challenge, 2012, http://arxiv.org/abs/1207.4737.
4. B. Eisenberg, Living devices: The physiological point of view, 2012, http://arxiv.org/abs/1206.6490.

Professor Halling added: Can I follow your suggestion that we should think what we can learn from observing the results of evolutionary selection, and suggest a possibly interesting example. These are the halophile proteins, which are adapted to work in about 4 M salts, usually KCl. One feature they share is an unusually large number of surface carboxyl groups. However, I also have to offer a warning. The halophile community state that enzymatic activity in 4 M salt is a special property of these enzymes. In fact however quite a few "normal" enzymes have been found to be active, even activated, in concentrated salts. So such activity is not a unique property of halophile proteins.

Dr Eisenberg replied: Thank you for the information.

Dr Konvalinka continued the discussion of the paper by Mikael Lund: It is well known that in biological systems (intracellular organels, budding virus particles *etc.*), the local concentration of ions could be very high. At the same time, in some of these compartments the local concentrations of polar groups of biomolecules could be extremely high as well and the number of available water molecules actually limited. Could a theory describe the behaviour of ions in such highly crowded environments, far away from ideal aqueous solutions?

Professor Lund replied: If an ion-binding site is exposed to a strong electric field from the surrounding or is significantly desolvated compared to the model compound representing the used pK_d value, then the current scheme will be less applicable as also discussed in the answer to Yan Levin's questions earlier.

We have however used this level of theory to study phase transitions in concentrated solutions of charged proteins (30 *v*/*v*%) and achieved semi-quantitative agreement with experiment (A. Kurut, B. A. Persson, T. Åkesson, J. Forsman and Mikael Lund, *J. Phys. Chem. Lett.*, 2012, **3**, 731–734).

Dr Rodriguez-Ropero added: Proteins are definitively a valid source of inspiration even in the design of synthetic systems like the diblock copolymer brush we are presenting today. In this sense we can, for instance, place the binding sites in specific positions within a polyelectrolyte brush so that their relative positions and local concentrations resemble those found in selective channel proteins or enzymes. In this way we could enhance the selectivity of the system towards a specific ionic species.

Professor Jungwirth commented: It is important to distinguish between (typically weak) Hofmeister salt effects and (typically strong) ion binding situations, *e.g.*, in enzymes or ion channels. The example brought from the audience concerning solubility and stability of strongly charged enzymes of halophilic bacteria falls into the former category and can be rationalized in terms of increasing protein charge leading to higher solubility in high salt and in terms of the higher affinity of Na^+ compared to K^+ to carboxylic groups at the protein surface. There may be situations lying at the fringes between the above two ion binding regimes and these may actually be very interesting to study.

Dr Goodman noted: A comment on Bob Eisenberg's abstract: "Chemistry is about chemical reactions". This is true, and chemical reactions are so complex that for many cases qualitative analyses may be all that is available. These are useful in themselves, even without a mathematical framework.

Dr Eisenberg replied: I agree.

Dr Goodman added: Sometimes I do not care how fast something goes, as long as A goes to B. This may represent a very complex and surprising transformation. Qualitative analyses can be very useful in the absence of an underlying mathematical theory.

Measuring the interaction between ions, biopolymers and interfaces – one polymer at a time

Sandra Kienle,[a] Tobias Pirzer,[ab] Stefanie Krysiak,[a] Michael Geisler[ac] and Thorsten Hugel*[ad]

Received 10th April 2012, Accepted 6th June 2012
DOI: 10.1039/c2fd20069d

Atomic force microscopy (AFM) based single polymer force spectroscopy allows to detect the interaction (energy) between single polymers and interfaces in aqueous environment. We use this method to delineate the effect of ions, pH, co-solutes and temperature on the adhesion of biopolymers onto solid substrates.

Introduction

Single molecule force spectroscopy is a technique based on the principle of the atomic force microscope (AFM) developed by Binnig, Quate and Gerber,[1] the invention of which has opened new perspectives for the investigation of surfaces at high lateral and vertical resolution. As the AFM can be operated in aqueous environments it has become possible to image biological surfaces such as cells and protein layers at nm-resolution under physiological conditions.[2] The AFM has not only a very high spacial resolution, but can also detect forces in the piconewton range in ambient environment. Therefore, it has evolved as a very sensitive tool to investigate properties of single (bio)polymers during the last 15 years, which is also termed dynamic force spectroscopy and utilized in the following.

The AFM consists of a cantilever with a sharp tip of about 20 nm radius at its free end, which is moved along three axes and positioned with nanometer accuracy relative to a sample surface with a piezoelectric actuator. The deflection of the cantilever is detected through an optical lever and a position sensitive photodiode. When the spring constant is determined[3] molecular forces down to the piconewton range can be measured. The recorded force–distance traces make intra-molecular stabilizing forces, structural energy barriers and mechanical properties of (bio)polymers readily accessible.[4,5] Single polymers covalently attached *via* a long flexible linker to a passivated AFM tip allow to determine ion specific (Hofmeister) effects,[6] co-solute effects[7] and temperature dependence.[8]

In the following we shortly summarize the fundamentals for this kind of single molecule measurement and present new results on how ions, pH, co-solutes and temperature affect the adhesion of biopolymers onto solid substrates.

[a]*IMETUM, Physics Department, CeNS, Technische Universität München, 85748 Garching, Germany. E-mail: thugel@mytum.de*
[b]*ZNN/WSI, Technische Universität München, 85748 Garching, Germany*
[c]*Center of Excellence Automated Infusion Systems, B.Braun Melsungen AG, 34212 Melsungen, Germany*
[d]*Center for Integrated Protein Science Munich, CIPSM, Germany*

AFM-based dynamic force spectroscopy

In force spectroscopy experiments the polymer under investigation is often tethered between cantilever tip and sample surface by unspecific attachment to the tip and substrate. This kind of bonding is sometimes strong enough to provide stability for several seconds and has been utilized to study for example the reversible unfolding of titin domains,[9] conformational transitions in polysaccharides,[10] and the elastic properties of engineered proteins.[11] As such physisorption is in many cases insufficient, the molecules under investigation are increasingly specifically bound to the substrate surface *via* chemical bonds in order to improve the long-term stability.[12] The attachment to the AFM tip that "picks up" the molecule is still unspecific in this case. This method suffers from collective effects arising from confinements and interactions between substrate-grafted chains.[12,13] Typically, less than 1% of the experimental recordings reflect true single molecule events due to abundant surface and multiple-molecule interactions.[14]

In other experiments, the AFM cantilever tip is covalently coated with polymers to measure the interaction strength of ligand-receptor bonds[15–17] or the adhesion forces from solid substrates.[18] In contrast to earlier measurements on pre-adsorbed polymer films, the substrate is no longer shielded by additional surface adsorbed polymer chains, and the interaction with the bare substrate can now be controlled on the few polymer level.[19] Such experiments yield interaction forces, but not free energies related to the thermodynamic adsorption equilibrium. One of the main challenges is that the number of interacting polymers is generally not known in such experiments. An application of such a tip modification with a uniform layer of chemical functional groups is chemical force microscopy.[20]

Despite more evolved preparations, AFM based force spectroscopy with a single covalently bound molecule at the tip apex *via* flexible linkers has proven advantages over former approaches[21] (Fig. 1A). For example, the detection of single molecule recognition events allowed receptor–ligand interactions to be characterized and mapped[22] and enabled the assembly of small molecules on the nanometer scale.[23] In addition, cysteine engineered polyprotein constructs for instance offer the possibility for site-directed force application and hence to study directional protein mechanics.[24] With stable covalent attachment of the polymers to the AFM-tips, it also became possible to investigate the adsorption process of the same set of molecules on a variety of surfaces and therefore deviations due to different sample preparations could be ruled out.[25] This AFM-based single molecule technique allows to measure forces and free energies in thermodynamic equilibrium, characterized by well defined plateaus in the force–distance traces[5] (Fig. 1B).

One of the indications for the thermodynamic equilibrium on the timescale of the experiments is a velocity independent plateau of constant force and length (Fig. 2). This method has meanwhile been extended to a wide range of interfaces ranging from solid–liquid, liquid–air, liquid–liquid and biointerfaces such as lipid bilayers.[26]

Interfacial adhesion

Charge effects

Single polymers covalently attached to an AFM-tip were used to investigate the equilibrium adhesion onto solid substrates and to determine their surface binding energies. In particular for polyelectrolytes on charged surfaces with charge density σ, the electrostatic part of the interaction is sensitive to the salt concentration, and the line charge density of the polymer τ_{ch} can be determined:[5]

$$F_{des}^{el} = F_0 + 4\pi l_B k_B T \sigma \kappa^{-1} \tau_{ch} \quad (1)$$

Here, l_B is the Bjerrum length and κ^{-1} the Debye length. F_0 is the part of the desorption force which is independent of charges. F_0 can be dominated by dispersion

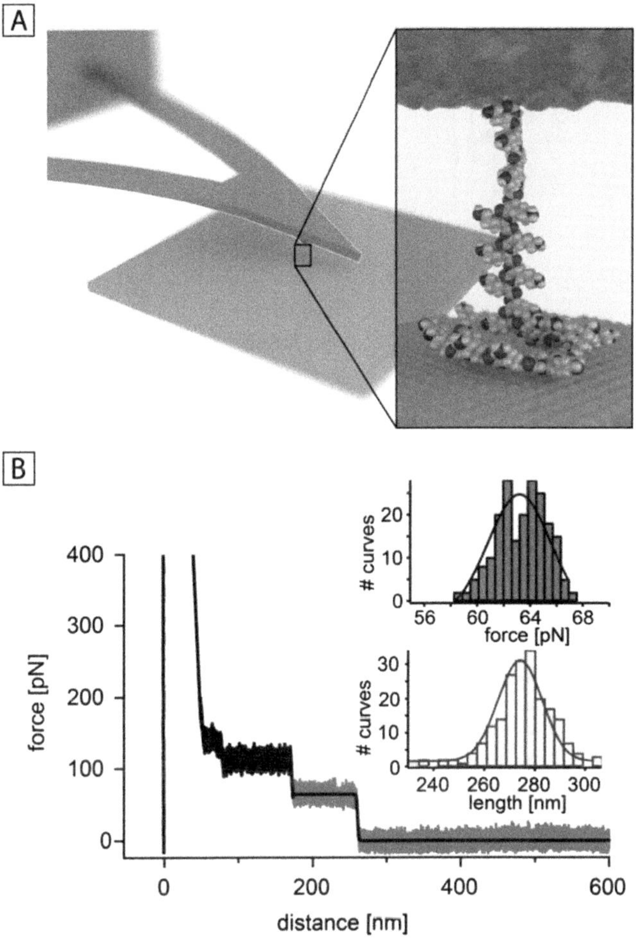

Fig. 1 (A) Scheme of an AFM cantilever with a polymer, which is covalently attached to the tip *via* a linker (here: PEG, first 30nm from the tip). (B) Force–distance curve measured with poly-L-lysine on a hydrophobic self-assembled monolayer (SAM) in water at room temperature. The plateau indicates the desorption of a single polymer from the surface. The fit (solid line) corresponds to a sigmoidal curve. The upper inset shows the histogram of the desorption force (*i.e.* the height of the plateau) and the bottom the detachment length (*i.e.* the length of the plateau) for one of the data points in Fig. 2. Both histograms were fitted with a Gaussian.

forces (*e.g.* van der Waals) or water structural (hydrophobic) forces. In fact we have shown that in single polymer adsorption on a variety of hydrophobic and hydrophilic surfaces the dispersion and water structural forces largely compensate each other.[27] In the following we describe how the adhesion force can be controlled by salt specific effects.

Salt effects

Fig. 3A shows the desorption force of poly-D-tyrosine from a hydrophobic SAM at various concentrations of the salt NaCl. The relative error of the measured force within the salt concentrations is around 2%, as they were all measured with one and the same cantilever. Consistent with other systems[6] (see Fig. 3B) we observed a linear dependency on the salt concentration. The increase in salt concentration of 1 M leads to an approximate increase in desorption force of about 5 pN. The

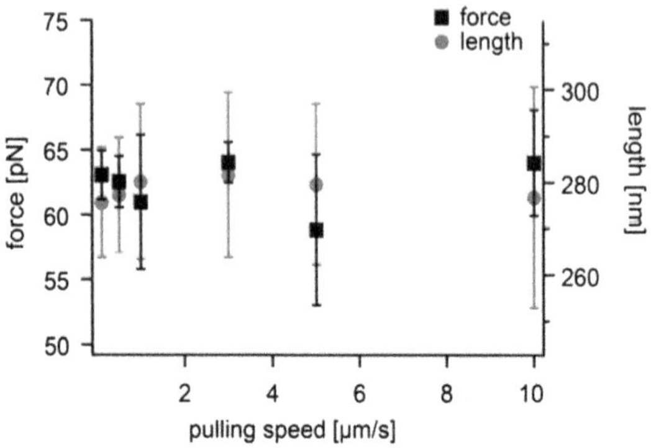

Fig. 2 The desorption force (squares) and the detachment length (circles) were measured for several pulling speeds. The measurements were performed with poly-L-lysine attached to the AFM tip and a hydrophobic SAM as a substrate.

offset F_0 (*i.e.* the force in pure water), which was determined from the fit to the data, is around 88 pN. The error for F_0 is about 13%, which is the calibration error for the spring constant of a cantilever, which is necessary to determine the absolute force.

Hofmeister effects

In Fig. 3B we delineate the effect of the Hofmeister sodium salts phosphate (NaH_2PO_4), chloride (NaCl) and iodide (NaI). NaH_2PO_4 is the most kosmotropic salt (mostly increases the hydrophobic effect) followed by NaCl, while NaI is already chaotropic (reduces the hydrophobic effect).

In analogy to salting out or denaturation experiments, we observe a linear dependence of the desorption force on salt concentration for the spider silk motive $(QAQ)_8NH3$.[6] The slopes of the linear fits to the data are given in Fig. 3B. They increase from NaI over NaCl to NaH_2PO_4 as expected from the Hofmeister series. Unexpected is the generally very shallow slope, which results in less than 10% change within the physiological relevant salt concentrations below 1 M compared to more than 500% in salting out experiments.[28] This might be due to the close to zero concentration of protein in our experiment or due to some kind of compensation of dispersion, water structure and co-solute forces at surfaces.

pH effects

Fig. 3C shows the dependence of the desorption force on pH for $(QAQ)_8NR3$ on a hydrophobic SAM; such a study is hampered by the fact that a change in pH either affects the valency of the ions or results in a non-buffered solution, *i.e.* a badly defined pH. Nevertheless, we show here the influence of the pH of a NaCl solution; the pH was adjusted by HCl and NaOH and found to be constant before and after the measurement. We used different cantilevers in the experiments and therefore have an uncertainty of 13%.

pH-values represent the concentration of H_3O^+ and OH^- ions. In the investigated pH regime, the concentration of these ions is less than 0.5 M. Therefore, pH effects could be interpreted as salt effects, which are also very small (<5 pN) for concentrations below 0.5 M.

Surface induced conformational transitions

It is text book knowledge that salt, pH or temperature affect the structure of proteins or polypeptides. In addition, surfaces or interfaces influence the aggregation of

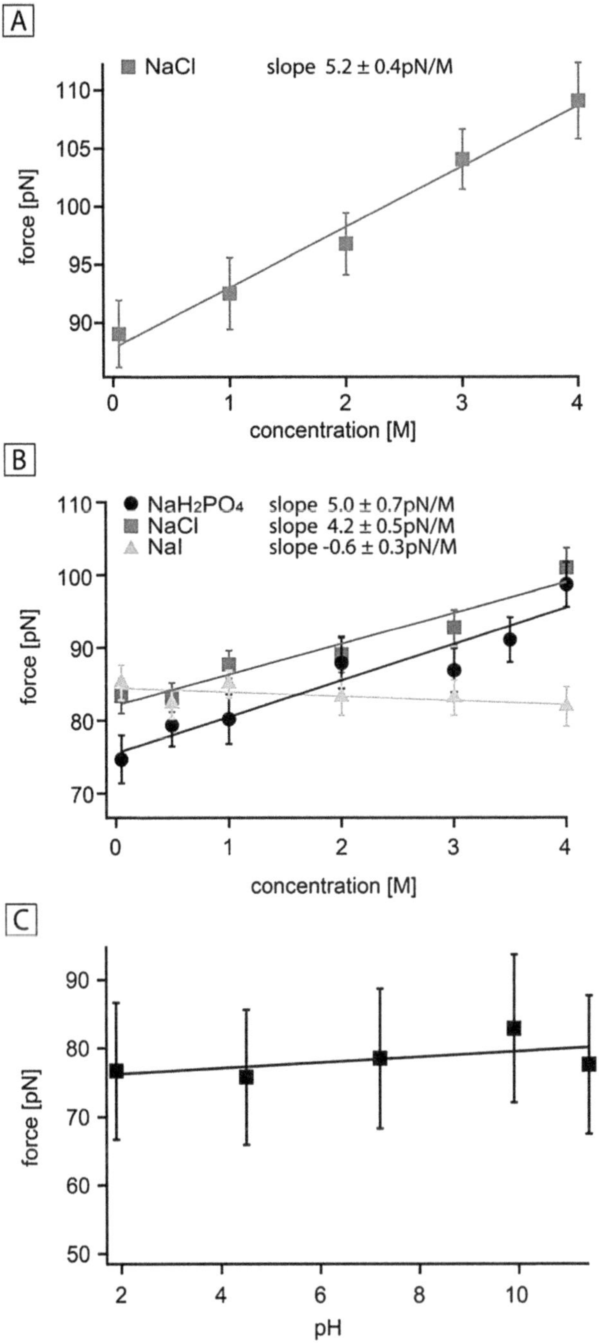

Fig. 3 (A) Desorption force for poly-D-tyrosine from a hydrophobic SAM for different concentrations of NaCl and a linear fit. The error shown corresponds to the standard deviation. (B) Desorption force for $(QAQ)_8NH3$ from a hydrophobic SAM for the three different salts NaI (triangles), NaCl (squares) and NaH_2PO_4 (circles). The results were fitted with a straight line and the error is the standard deviation.[6] (C) At five different pH values the desorption force was determined for $(QAQ)_8NH3$ on a hydrophobic SAM. To guide the eye, a linear fit is applied. Here, the error was set to 13% due to the use of two different cantilevers.

polypeptides into larger structures, which is likely the cause for various neurodegenerative diseases like Parkinson's disease or Alzheimer's disease.[29] Recently AFM based single molecule force spectroscopy enabled to show that interfaces can even cause a single polymer to adopt different folds. Geisler *et al.*[30] were able to follow the structure formation within a single engineered protein in real time by monitoring the transition from equilibrium force plateaus (Fig. 4A) to non-equilibrium unfolding patterns (Fig. 4B) by changing the temperature. The certainty that the populations of the different conformational states are not affected by possible oligomeric precursors from solution (because there was only a single polymer at the AFM tip and none in solution) ensured that this was a surface induced effect. It was shown that the transition from the unstructured native state into the final fold was induced in the vicinity of surfaces with high surface energy and can be prohibited in the presence of hydrophobic surfaces for the spider silk motive C16. This polymer consists of 16 identical repeats and exhibits random coil conformations in water at room temperature.

The saw-tooth like pattern on top of the plateau indicates the forced unfolding of secondary structures that had been formed in the vicinity of the surface at elevated temperature. Additional work has to be done in order to break intramolecular bonds and to stretch the protein to its extended conformation. The transition from the equilibrium to the non-equilibrium desorption is easy to observe and can be used to follow the conformational transition of the protein in real time.[30]

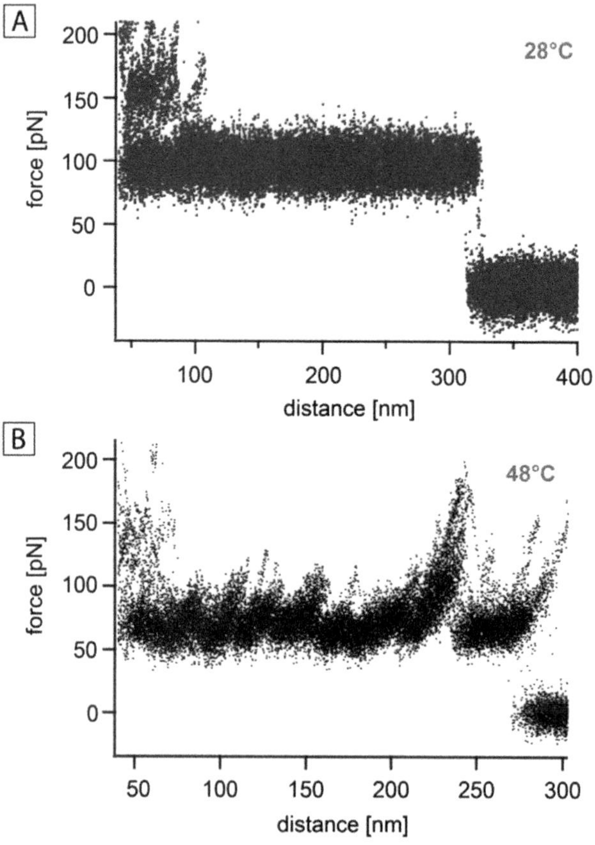

Fig. 4 Superposition of 20 force–distance traces for the desorption of a single C16 spider silk molecule in water from stainless steel at a temperature (A) of 28 °C and (B) 48 °C.

Finally, not only the formation of a structure can be observed, but even the details of the structure and its stability can be determined from the data, similar to typical non-equilibrium mechanical unfolding experiments.[14] C16 undergoes surface induced folding into a β-sheet rich secondary structure with parallel and anti-parallel strands.

Temperature dependence

On a hydrophobic diamond (lower surface energy compared to steel) no such structure formation was observed.[30] Therefore we did additional measurements with poly-D-tyrosine to investigate how the desorption force (the height of the force plateau) changes in such an equilibrium measurement with temperature. The only force that is involved in this desorption process is the hydrophobic attraction, because of the lack of ability of the surface to form hydrogen bonds or interact *via* electrostatic forces with the polymer.[8] The hydrophobic attraction was believed to be the main force that drives protein folding. The entropy convergence seemed to prove this fact, but general entropy convergence was recently refuted by Sedlmeier *et al.*[31] Nevertheless the hydrophobic attraction is one of the main forces in this folding process. While changing the temperature one expects to unfold proteins at high and at low temperatures. Therefore a maximum occurs in the free energy. The temperature dependent measurements with poly-D-tyrosine in water showed only a weak temperature dependence (Fig. 5A). The same was recently observed for poly-L-lysine and also in MD-simulations with polyglycine. This is due to a compensation mechanism between the various interactions during the desorption process (polymer–polymer, polymer–surface, polymer–water, surface–water, water–water).[8] To test the temperature dependence in combination with solvent specific effects, salt and ethanol were added to the solution (Fig. 5A, B). The measurements in 7% ethanol were done with the same cantilever which was used for the measurements in water (Fig. 5A). As previously shown the desorption force decreases after adding ethanol, but this does not change the temperature dependence. There might be a slight minimum in the free energy, but certainly again only a very weak temperature dependence. The same holds true for the measurement in 50 mM NaCl (Fig. 5B). Adding salt or ethanol in this concentration does not change the weak dependence of the desorption force on temperature.

This is another hint, that the hydrophobic attraction (which is the main contribution for the temperature dependence in desorption experiments) cannot be the only force dominating protein folding, which is strongly influenced by solvent properties.

Binding thermodynamics of small molecules

In addition to ions, co-solutes are present in the crowded environment of all biological systems which modify, or even determine, the interactions between macromolecules such as proteins, lipids and DNA. Single molecule sensors and equilibrium thermodynamics have been used to quantify the binding of co-solutes onto a macromolecule in solution as well as in its surface adsorbed state.[7] Therefore, desorption measurements were performed with poly(allyamin) at different glucose concentrations on two different surfaces. The change in force with increasing co-solute concentration was significantly different between the two surfaces. This desorption force F_{des} could then be translated into a line tension difference $\Delta\tau$:

$$F_{des}(a) = F_0 + \Delta\tau(a) \tag{2}$$

By fitting the following formula for $\Delta\tau$ to the concentration (activity, a) dependent desorption force, the values for the number of adsorbed co-solutes per length d and their association constants K were obtained[7]:

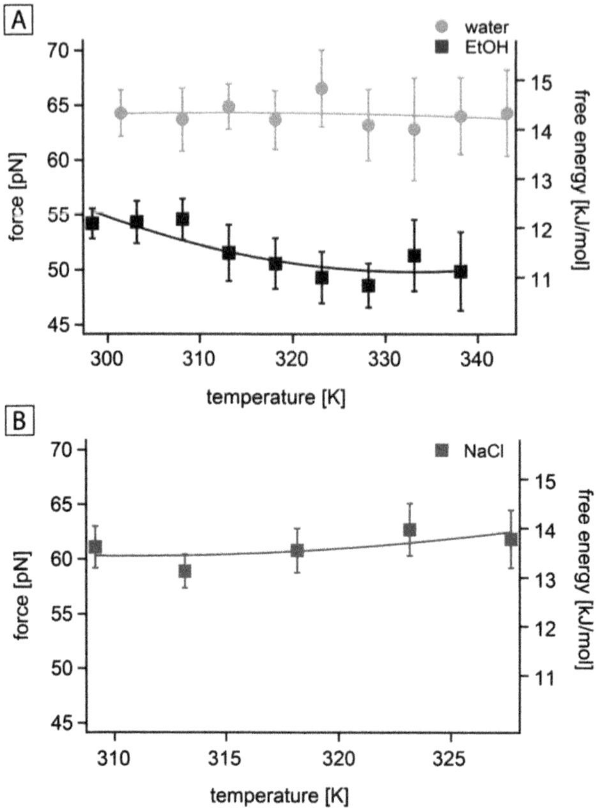

Fig. 5 The temperature dependence of the desorption force and the desorption free energy ΔG was determined for poly-D-tyrosine on a hydrophobic, hydrogenated diamond in (A) water (circles) and 7% ethanol (squares) and (B) with a different cantilever in 50 mM NaCl. The error shown here corresponds to the standard deviation.

$$\frac{\Delta\tau}{k_B T} = d_S^{-1}\ln(1 + aK_S) - d_b^{-1}\ln(1 + aK_b) \tag{3}$$

Here, the surface-adsorbed polymer state is characterized by the ligand association constant K_S and the adsorption site length d_S, and the bulk polymer state (which is the same for both surfaces) by the ligand association constant K_b and the adsorption site length d_b. $k_B T$ is the thermal energy.

The standard methods to date for the measurement of co-solute binding in bulk solution are densiometry, sedimentation and isothermal titration calorimetry (ITC). All these methods require a minimum amount of polymer to provide reliable free energy changes, and association constants, and the precision of the data increases with molecular concentration.[32] Thus, the availability of scarce and/or expensive polymer samples constitutes a certain limitation to this method. Moreover, self-assembly and aggregation restrict the application of ITC to less hydrophobic molecules.[32]

Such restrictions are not present in the AFM based method. But there is another restriction, namely that the simultaneous fit to the concentration dependent desorption force for different surfaces has to converge. This is only the case for systems in which the association constant of the investigated co-solute with the polymer changes considerably for the different surfaces (as in the above described example), which results in a different dependence of the desorption force on the concentration.

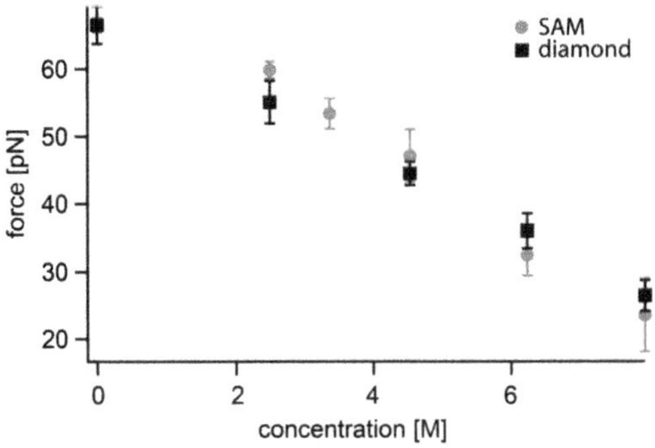

Fig. 6 Dependence of the desorption force on ethanol concentration for the desorption of poly-L-lysine from a hydrophobic SAM and a hydrogen terminated diamond.

Fig. 6 shows another example where this is obviously not the case. Here poly-L-lysine is desorbed from a hydrophobic SAM and an H-terminated diamond at various ethanol concentrations. Both series look very much the same and the (six) different fit parameters are therefore not sufficiently defined – even for a simultaneous fit to both traces. In order to determine the thermodynamic parameter for the interaction of ethanol with poly-L-lysine one would have to find a surface at which the concentration dependent desorption force deviates significantly from the data shown in Fig. 6. This would result in two additional parameters, but also in a strong restriction, because the parameters in bulk have to be the same for all three different surfaces.

Conclusions

During the last years the atomic force microscope (AFM) has been frequently used to measure the desorption force of single polymers from solid substrates in order to determine their adhesion free energy. Here we have shown that such desorption experiments do not only provide information on the interaction between a polymer and a substrate, but also allow to deduce further properties of biomolecular systems. Namely, the structure of biopolymers in contact with a solid substrate (surface induced folding), the effect of temperature on the hydrophobic attraction and thermodynamic parameters for the interaction of polymers with co-solutes (association constants and density of binding sites). Polymers at interfaces are affected by ions in a concentration and ion specific way, which results in different desorption forces and structures depending on the presence of ions. The presented measurements have been done for single polymers (small systems), therefore, their outcome is directly comparable to simulations. Together these experiments and simulations yield valuable information to understand molecular details of ion specific Hofmeister effects for polymers at interfaces.

Materials and methods

Surface preparation and contact angle measurements

As surfaces, a hydrophobic, hydrogen terminated polycrystalline diamond (ElementSix, Advancing Diamond Ltd., UK) of size 5 mm × 5 mm, a hydrophobic self-assembled monolayer (SAM) and stainless steel (316L)[30] were used.

The hydrogenation of the diamond was done similarly to a protocol which was published before.[33] First, it was heated in a vacuum chamber up to 700 °C to clean the sample, and then a hydrogen flow started (100 sccm, constant pressure of 10 mbar). It was activated by a microwave plasma reactor (Astex). After that the diamond was treated for 15 min at a pressure of 50 mbar until a soft shut down.

To finish the hydrogenation, the sample was cooled under hydrogen atmosphere (10 mbar, 100 sccm). Before using the diamond for the measurements it was cleaned by sonicating in acetone and 2-propanol.

For the preparation of the SAM a glass slide had to be coated with a thin gold layer. Therefore the glass slide was first cleaned by sonicating for 15 min in Hellmanex (Hellma GmbH und Co. KG, Mühlheim, Germany) and twice in ultra pure water (Biochrom, Germany). Additionally the slides were cleaned with RCA step 1 (v : v : v, 5 : 1 : 1, ultrapure water, ammonia (32%), hydrogen peroxide (35%)) at 70 °C for about 15 min and rinsed in ultrapure water. Afterwards they were dried and coated with a 10 nm layer of chrome nickel and a 100nm layer of gold using a vacuum coater (Edwards GmbH, Kirchheim, Germany). A day before the measurements the gold-coated slides were cleaned in RCA step 1 at 70 °C for about 15 min. After rinsing in ultrapure water and ethanol they were incubated overnight in a solution of 2 mM dodecanethiol (solved in ethanol). Before the measurement the slides were again rinsed in ultrapure water.

Before using any of the surfaces, its hydrophobicity was tested. Therefore the contact angle was determined using a home-built goniometer equipped with a CCD-camera. For the evaluation of the contact angles a Java based ImageJ plug-in (drop snake analysis) was used.[34]

AFM based force spectroscopy measurements

For the measurements a MFP-3D SA (Asylum Research, Santa Barbara, USA) was used with either a home-build fluid cell or a BioHeater™ as a closed fluid cell (for a well-defined temperature control). The cantilevers were silicon nitride cantilevers (MLCT, Bruker SPM probes, Camarillo, USA).

For the functionalization of the tip, the cantilevers were first activated in a plasma chamber with oxygen. After rinsing the cantilevers in acetone (anhydrous, ≥99.8%, VWR, Germany), they were incubated for 15 min in a Vectabond™ solution. Therefore 50 μl Vectabond were dissolved in 2.5 ml acetone. Then the cantilevers were again rinsed in acetone and chloroform (anhydrous, ≥99.8%, VWR, Germany) and incubated for 45 min in a chloroform-based solution of a 1 : 1500 mix of CH_3O-PEG-NHS (5 kDa) and PEG-α-ω-Di-NHS (6 kDa) (Rapp Polymere GmbH, Tübingen, Germany). The cantilevers were now rinsed in chloroform, ethanol (absolute, >99.9%, Merck, Germany) and borate buffer (pH 8–8.5) and incubated for at least 1 h in a 2 mg ml^{-1} polymer solution (dissolved in borate buffer) before rinsing in TRIS (molecular biology grade, AppliChem, Germany) and ultrapure water. As polymers either poly-L-lysine (Sigma Aldrich, 70 000–150 000 Da), poly-D-tyrosine (Sigma Aldrich, 40 000–100 000 Da), the spider silk motive C16 or $(QAQ)_8$NR3 (Thomas Scheibel, University Bayreuth) were used. $(QAQ)_8$NR3 includes the A and Q motifs from ADF-3 and C16 the C motif of ADF-4, both a genetically engineered variant of Araneus diadematus fibroin from the garden cross spider (see Geisler *et al.*[35] for details).

To calibrate the spring constant of the cantilevers the thermal noise method was used.[3] The same cantilever was used within one data set (except for the measurements with different pH values). For each velocity, pH, concentration or temperature the spring constant was determined separately and averaged afterwards. Therefore the inverse optical lever sensitivity (InvOLS) was determined five times before and after each data point and averaged for this data point (but not for the whole data set).

For each data point at least 100 force–distance curves were recorded. Therefore the cantilever was lowered with a constant velocity of 500 nm s^{-1} or 1 μm s^{-1} until it reached the surface. After a dwell time of 1 s, to let the polymer adsorb to the surface, the cantilever was retraced. During this retraction cycle the polymer was desorbed, which resulted in a constant force plateau in the force–distance curves (Fig. 1B). This plateau was then fitted with a sigmoidal curve to get information about the desorption force and the detachment length. All the obtained values for the force and the length were plotted in a histogram and fitted with a Gaussian. The maximum of the Gaussian was used as an average value for each data point shown here and the error corresponds to the standard deviation (SD), except for the measurements at different pH values. These measurements were performed with two different cantilevers and therefore the error was set to 13% due to the error in the calibration of the spring constant. For the free energy per amino acid (ΔG) the area under the force plateau was calculated and normalized by the length of one amino acid (assuming a length of 0.36 per amino acid).

Acknowledgements

We acknowledge helpful discussions with Bizan N. Balzer and funding from the German science foundation (SFB 863, and Hu 997/8).

Notes and references

1 G. Binning, H. Rohrer, C. Gerber and E. Weibel, *Phys. Rev. Lett.*, 1982, **49**, 57–61.
2 D. J. Muller and A. Engel, *Nat. Protoc.*, 2007, **2**, 2191–2197.
3 T. Pirzer and T. Hugel, *Rev. Sci. Instrum.*, 2009, **80**, 035110.
4 A. Janshoff, M. Neitzert, Y. Oberdorfer and H. Fuchs, *Angewandte Chemie-International Edition*, 2000, **39**, 3213–3237.
5 T. Hugel and M. Seitz, *Macromol. Rapid Commun.*, 2001, **22**, 989–1016.
6 T. Pirzer, M. Geisler, T. Scheibel and T. Hugel, *Phys. Biol.*, 2009, **6**, 025004.
7 M. Geisler, R. R. Netz and T. Hugel, *Angew. Chem., Int. Ed.*, 2010, **49**, 4730–4733.
8 S. Kienle, S. Liese, N. Schwierz, R. R. Netz and T. Hugel, *ChemPhysChem*, 2012, **13**, 982–989.
9 M. Rief, M. Gautel, F. Oesterhelt, J. M. Fernandez and H. E. Gaub, *Science*, 1997, **276**, 1109–1112.
10 A. F. Oberhauser, P. E. Marszalek, H. P. Erickson and J. M. Fernandez, *Nature*, 1998, **393**, 181–185.
11 Y. Cao and H. B. Li, *Nat. Nanotechnol.*, 2008, **3**, 512–516.
12 S. X. Cui, C. J. Liu and X. Zhang, *Nano Lett.*, 2003, **3**, 245–248.
13 L. Sonnenberg, Y. F. Luo, H. Schlaad, M. Seitz, H. Colfen and H. E. Gaub, *J. Am. Chem. Soc.*, 2007, **129**, 15364–15371.
14 H. Dietz and M. Rief, *Japanese Journal of Applied Physics Part 1 Regular Papers Brief Communications & Review Papers*, 2007, **46**, 5540–5542.
15 E. L. Florin, V. T. Moy and H. E. Gaub, *Science*, 1994, **264**, 415–417.
16 R. Merkel, P. Nassoy, A. Leung, K. Ritchie and E. Evans, *Nature*, 1999, **397**, 50–53.
17 C. Stroh, H. Wang, R. Bash, B. Ashcroft, J. Nelson, H. Gruber, D. Lohr, S. M. Lindsay and P. Hinterdorfer, *Proc. Natl. Acad. Sci. U. S. A.*, 2004, **101**, 12503–12507.
18 T. Hayashi, K. Sano, K. Shiba, Y. Kumashiro, K. Iwahori, I. Yamashita and M. Hara, *Nano Lett.*, 2006, **6**, 515–519.
19 M. Erdmann, R. David, A. Fornof and H. E. Gaub, *Nat. Nanotechnol.*, 2010, **5**, 154–159.
20 S. Akari, D. Horn, H. Keller and W. Schrepp, *Adv. Mater.*, 1995, **7**, 549–551.
21 C. Friedsam, A. D. Becares, U. Jonas, H. F. Gaub and M. Seitz, *ChemPhysChem*, 2004, **5**, 388–393.
22 P. Hinterdorfer and Y. F. Dufrene, *Nat. Methods*, 2006, **3**, 347–355.
23 S. K. Kufer, M. Strackharn, S. W. Stahl, H. Gumpp, E. M. Puchner and H. E. Gaub, *Nat. Nanotechnol.*, 2009, **4**, 45–49.
24 H. Dietz, M. Bertz, M. Schlierf, F. Berkemeier, T. Bornschlogl, J. P. Junker and M. Rief, *Nat. Protoc.*, 2006, **1**, 80–84.
25 M. Geisler, B. N. Balzer and T. Hugel, *Small*, 2009, **5**, 2864–2869.
26 T. Pirzer and T. Hugel, *ChemPhysChem*, 2009, **10**, 2795–2799.
27 D. Horinek, A. Serr, M. Geisler, T. Pirzer, U. Slotta, S. Q. Lud, J. A. Garrido, T. Scheibel, T. Hugel and R. R. Netz, *Proc. Natl. Acad. Sci. U. S. A.*, 2008, **105**, 2842–2847.
28 R. Vogel, *Curr. Opin. Colloid Interface Sci.*, 2004, **9**, 133–138.
29 F. Chiti and C. M. Dobson, *Annu. Rev. Biochem.*, 2006, **75**, 333–366.
30 M. Geisler, S. B. Xiao, E. M. Puchner, F. Grater and T. Hugel, *J. Am. Chem. Soc.*, 2010, **132**, 17277–17281.
31 F. Sedlmeier, D. Horinek and R. R. Netz, *J. Chem. Phys.*, 2011, **134**, 055105.
32 A. Velazquez-Campoy and E. Freire, *Nat. Protoc.*, 2006, **1**, 186–191.

33 M. Dankerl, S. Eick, B. Hofmann, M. Hauf, S. Ingebrandt, A. Offenhausser, M. Stutzmann and J. A. Garrido, *Adv. Funct. Mater.*, 2009, **19**, 2915–2923.
34 A. F. Stalder, G. Kulik, D. Sage, L. Barbieri and P. Hoffmann, *Colloids Surf., A*, 2006, **286**, 92–103.
35 M. Geisler, T. Pirzer, C. Ackerschott, S. Lud, J. Garrido, T. Scheibel and T. Hugel, *Langmuir*, 2008, **24**, 1350–1355.

Interactions of monovalent salts with cationic lipid bilayers

Sarka Pokorna,[a] Piotr Jurkiewicz,*[a] Lukasz Cwiklik,[ab] Mario Vazdar[bc] and Martin Hof[a]

Received 3rd May 2012, Accepted 21st May 2012
DOI: 10.1039/c2fd20098h

The influence of monovalent salts (NaF, NaCl, NaBr, $NaClO_4$, KCl) on the properties of lipid bilayers composed of binary mixtures of zwitterionic DOPC (dioleoylphosphatidylcholine) and cationic DOTAP (dioleoyltrimethylammoniumpropane) is experimentally measured and numerically simulated. Both approaches report a specific adsorption of the studied anions at the cationic bilayer. The adsorption is enhanced for higher content of DOTAP in DOPC/DOTAP mixtures and for larger anions (Br^- and ClO_4^-). The nonmonotonic dependence of the lipid headgroup mobility, determined using time-dependent fluorescence shifts of Laurdan located at the bilayer carbonyl level, on the content of cationic lipid is preserved in all examined salt solutions. Its maximum, however, is shifted towards higher DOTAP concentrations in the row: NaF < NaCl < NaBr. The same ordering of salts is found for the simulated area per lipid and the measured rigidification of pure DOTAP bilayers. Simulations reveal that Br^- strongly binds to the cationic headgroups of DOTAP neutralising the bilayer, which induces lateral inhomogeneities in the form of hydrophilic and hydrophobic patches at the membrane–water interface for pure DOTAP. In the equimolar DOPC/DOTAP mixture the neutralising effect of Br^- results in bending of the PC headgroups to a bilayer-parallel orientation. F^-, while attracted to the DOTAP bilayer, has an opposite effect to that of Br^-, *i.e.* it increases local mobility at the lipid carbonyl level. We attribute this effect to the disruption of the hydrogen-bonded structure of the molecules of lipids and water caused by the presence of the adsorbed F^-.

1 Introduction

Lipid bilayers are supramolecular aggregates with a complex, yet well-defined structure.[1] Being the fundamental structural units of all biological membranes they are often used as model systems in science. The self-assembly and the maintenance of the organization of the amphiphilic molecules of lipids in the bilayers is governed by the hydrophobic effect.[2] Yet, number of weak interactions with macromolecules as well as with salt ions affect the physical and biological properties of lipid membranes.[3] Specific ionic effects investigated in lipid membranes are often related to either the direct or reversed Hofmeister series.[4,5] The studies were usually limited

[a]*J. Heyrovský Institute of Physical Chemistry, Academy of Sciences of the Czech Republic, v. v. i., Dolejškova 3, 18223 Prague 8, Czech Republic*
[b]*Institute of Organic Chemistry and Biochemistry, Academy of Sciences of the Czech Republic, Center for Biomolecules and Complex Molecular Systems, Flemingovo nám. 2, 16610 Prague 6, Czech Republic*
[c]*Rudjer Bošković Institute, Division of Organic Chemistry and Biochemistry, POB 180, HR-10002 Zagreb, Croatia*

to the physiologically most abundant ions (Na^+, K^+, Cl^-, Ca^{++}, or Mg^{++}) and the most common lipids or lipid mixtures (neutral/zwitterionic and negatively charged phospholipids). As in many other biological systems, the specificity of the anions is more pronounced than that of the cations.[6] It is, however, possible to amplify the ionic effects by attracting the ions to the water/lipid interface by tuning the charge of the lipid bilayer.[6–8] This can be easily and precisely achieved by simply mixing an appropriate amount of charged and neutral lipids. We have recently shown that the addition of 20 mol% of negatively charged phosphatidylserine to the zwitterionic phosphatidylcholine (PC) membrane is enough to reveal a strong specificity of alkali metal chlorides (Na^+, K^+, and Cs^+) in their interactions with lipids.[9]

Cationic lipids, as opposed to those that have a negative charge, are very rare in nature. Synthetic cationic lipids are, however, successfully used in gene delivery, where they are usually associated with a nucleic acid polyanion to protect it and facilitate its transport into the cell.[10,11] Complexes of DNA and cationic lipids are used in experimental biology to modify cellular genetic make-up *in vitro* (*e.g.* to produce a protein of interest, like green fluorescent protein), but also in gene therapy.[10] There is no doubt that in the complexation of DNA with cationic lipids, the presence of the counterions, as well as the salt solutions in which the process occurs, cannot be ignored.

Lipid-based gene delivery systems often include a zwitterionic, so-called, helper lipid, which can accelerate the uptake of the lipid–DNA complex by the cells.[12] It has been shown that cationic and zwitterionic lipids mix anomalously and that the properties of lipid bilayers obtained from their mixtures depend non-monotonically on the molar content of the cationic lipid.[13–19] This was attributed to the complex electrostatic interactions between the cationic and zwitterionic headgroups of those lipids, which leads to (1) repulsion between cationic headgroups, and (2) reorientation of the P–N dipole of the zwitterionic lipid toward the aqueous phase.[13,18,20] Thus, when studying interactions of salts with cationic/zwitterionic lipid bilayers, one could expect not only the presence of the specific anionic effects, but also the influence of the ions on the above mentioned electrostatic interactions between the lipids.

The studies regarding interactions of ions with cationic lipid bilayers are scarce. Macdonald and Seelig, in their ^2H NMR, demonstrated, that the binding of chaotropic SCN^- to a neutral lipid bilayer can be further increased by the addition of dihexadecyldimethylammonium bromide.[7] Gurtovenko *et al.* studied the influence of NaCl on the structure of a dimyristoylphosphatidylcholine/dimyristoyltrimethylammoniumpropane (DMPC/DMTAP) mixture using MD simulations.[8] They found that at 6 mol% of the cationic DMTAP the bilayer behaves as a neutral one adsorbing Na^+ to the *sn*-2 carbonyl, which compresses the bilayer laterally, while Cl^- remains in the aqueous phase. At 50 and 75 mol% of DMTAP no Na^+ adsorption was observed, and while Cl^- was attracted closer to the bilayer surface, it still remained in the bulk. Authors concluded that cationic lipid membranes are almost insensitive to monovalent salt. Further contribution from the same group showed that also the lipid diffusion in DMPC/DMTAP bilayer is only slightly affected by NaCl solution.[21] Garcia-Celma *et al.* in their comprehensive investigation of ionic effects in hybrid alkanethiol/lipid membranes report that DOTAP binds Br^- and ClO_4^- stronger than PC does. DOTAP also amplified the anionic specific effects.[6] Woiterski *et al.* determined the amount of water bound to dioctadecyldimethylammonium (DODA) halide membranes using attenuated total reflection Fourier-transform spectroscopy.[22] They found highly structured water close to the DODAX bilayer, with 11 H_2O molecules bound per DODAF, 1–2 H_2O per DODACl and DODABr. No water was bound to DODAI.

In our previous studies, the time-dependent fluorescence shift (TDFS) method has been successfully applied to investigate fully hydrated free standing cationic/zwitterionic lipid bilayers. The relaxation time, which characterizes local mobility at the lipid glycerol level, was found to depend non-monotonously on the content of the cationic TAP lipids, with a clear maxima found at 30 and 45 mol% of TAP in DOTAP/DOPC[19] and DMTAP/DMPC[23] mixtures, respectively. In the present study, mixtures of cationic DOTAP and zwitterionic DOPC were used to study interactions of cationic lipid bilayers with monovalent ions. The two lipids possess the same aliphatic chains and are fully miscible, thus, the resulting lipid bilayers are well-defined and their charge can be controlled precisely. TDFS measurements were performed using the Laurdan fluorescence probe for the series of DOPC/DOTAP vesicles with varying lipid composition (0, 6, 15, 30, 50, 75, and 100 mol% of DOTAP) suspended in different monovalent salt solutions (NaCl, KCl, NaF, NaBr, $NaClO_4$). In addition molecular dynamics simulations were carried out for pure DOTAP and equimolar DOPC/DOTAP mixtures in the absence and in the presence of NaCl, NaF, and NaBr.

2 Methods

2.1 Materials and experimental methods

2.1.1 Materials. 1,2-Dioleoyl-*sn*-glycero-3-phosphocholine (DOPC) and 1,2-dioleoyl-3-methylammoniumpropane (DOTAP) were purchased from Avanti Polar Lipids, Inc. (Alabaster, USA). 6-Dodecanoyl-2-dimethylaminonaphthalene (Laurdan) was obtained from Molecular Probes (Eugene, USA). Salts were obtained from Sigma Aldrich (St. Louis, USA): NaCl, NaBr with purity > 99%, $NaClO_4$ with purity 98%, NaF and CsF > 99.99%; from Fluka (Buchs, Switzerland): KCl with purity ≥ 98%. Organic solvents of spectroscopic grade were supplied by Merck (Darmstadt, Germany). All chemicals were used without further purification.

2.1.2 Sample preparation. 150 mM or 1 M salt solutions were prepared in Mili Q (Millipore, USA) water. Appropriate volumes of stock solutions of lipids (in $CHCl_3$) and Laurdan (in CH_3OH) were mixed in glass tubes to yield a final concentration of 1.0 mM and 0.01 mM respectively. Solvents were evaporated under the stream of nitrogen, while heated, and left in the vacuum for at least 2 h in order to remove the remaining organic solvent. The dry lipid film was suspended in 1.5 ml of a salt solution or water and vortexed for 4 min. Large unilamellar vesicles (LUVs) were formed by extrusion through the 100 nm pore diameter membrane filters (Avestin, Ottawa, Canada).

2.1.3 Fluorescence measurements. All measurements were performed in a quartz cuvette at 10 °C to increase the extent of the relaxation process captured with the temporal resolution of the instrumentation. Each sample was equilibrated at least for 10 min before the measurement. Steady-state spectra were measured using Fluorolog-3 spectrofluorometer (model FL3–11; Jobin Yvon Inc., Edison, NJ, USA) equipped with a xenon-arc lamp. Time resolved fluorescence decays were recorded using a 5000U Single Photon Counting setup using a cooled Hamamatsu R3809U-50 microchannel plate photomultiplier. The data were collected for a series of wavelengths ranging from 400 nm to 540 nm with a 10 nm step. Each decay was fitted with a multi-exponential function using the iterative reconvolution procedure (IBH DAS6 software). Laurdan probe was excited at 373 nm with the IBH NanoLed 11 laser diode. Additionally, the emission >399 cutoff filter was used to eliminate the scattered light.

2.1.4 Time dependent fluorescence shift method. Time dependent fluorescence shift (TDFS) technique (called also fluorescence solvent relaxation) was described in detail previously[24–27] and has proved to be a reliable tool to determine local lipid mobility and hydration in model lipid membranes including those composed of cationic/zwitterionic lipid mixture.[19,23] The method is based on the observation of solvent reorientational motions (solvent relaxation) in the vicinity of the fluorescent probe. This rearrangement is a consequence of the rapid change in the charge distribution in the fluorophore upon the electronic excitation. Data are collected as the time resolved fluorescent decays for a series of wavelength spanning the emission spectra of the probe. These decays together with the steady-state spectrum are then used to reconstruct the time resolved emission spectra, TRES(t). TRES(t) are fitted with a log-normal function in order to determine the position of their maxima, $\nu(t)$, expressed in wavenumbers. From the analysis of $\nu(t)$ two parameters: $\Delta\nu$ and τ are obtained.

The first one, the total spectral shift: $\Delta\nu = \nu(0) - \nu(\infty)$ was shown to be directly proportional to the polarity of the medium.[24] Due to finite time resolution of every instrumentation, the so-called time-zero spectrum, TRES(0), is impossible to measure, thus the value of $\nu(0)$ cannot be obtained directly. TRES(0) can be, however, estimated using absorption and steady-state emission spectra measured in nonpolar solvent as described elsewhere.[28] The value obtained for Laurdan was equal to 23 800 cm^{-1} and was not affected by the changing lipid composition nor by the presence of ions. The origin of the polarity in lipid bilayers are mostly the dipoles of water molecules hydrating the bilayer. Thus, $\Delta\nu$ is often interpreted as the extent of membrane hydration. Although, if some other polar or charged moieties participate in the relaxation process, care should be taken not to misinterpret the results. Such is the case for some of the ions, which manifest their presence in the TDFS data in the form of ionic relaxation. For this reason in the present paper we discuss $\Delta\nu$ values rather in terms of polarity than hydration.

The second parameter provides information about the kinetics of the relaxation process. A convenient measure of the rate of the solvent relaxation is the mean integrated relaxation time, defined as:

$$\tau = \int_0^\infty \frac{\nu(t) - \nu(\infty)}{\Delta\nu} dt$$

Laurdan fluorophore is embedded at the carbonyl level of the lipid bilayer (*i.e.* close to the *sn*-1 carbonyl) (Fig. 1). All the water molecules hydrating this region are fully H-bonded to the lipids. Therefore, the nanosecond relaxation times

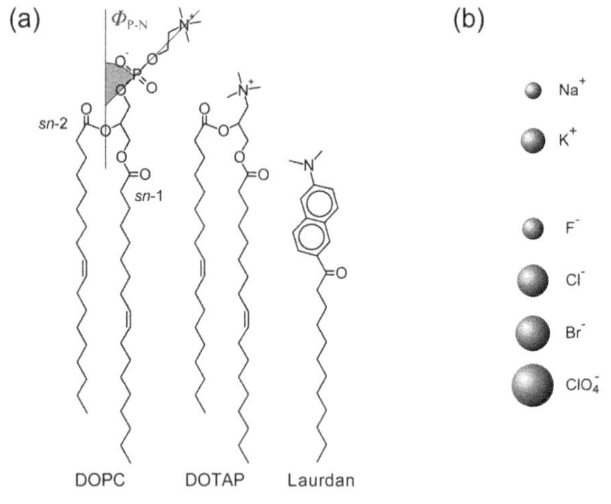

Fig. 1 (a) Chemical structures of cationic DOTAP, zwitterionic DOPC, and the fluorescent probe Laurdan. $\Phi_{P\text{–}N}$ angle between P–N dipole of DOPC and the membrane normal, *sn*-1 and *sn*-2 chains are marked. Laurdan fluorophore is located in accord with the results of parallax quenching,[19] and MD.[30] (b) Studied ions; the spheres are scaled according to the values of hydrodynamic radii taken from the literature.[31]

measured in lipid membranes reflect the local dynamics of the hydrated lipids rather than the motion of free water molecules.[29]

2.2 Simulation methods

Classical molecular dynamics (MD) simulations were performed for membranes build of 128 lipid molecules (64 in each leaflet). Two different bilayers were studied: 100 mol% DOTAP, and mixed 50 mol% of DOPC with 50 mol% of DOTAP. The bilayers were built based on a pre-equilibrated bilayer of 128 DOPC lipids with either all or 64 (32 in each leaflet) headgroups of DOPC changed to TAP^+ in order to form DOTAP molecules. Each system was hydrated by approximately 4600 molecules of water. In order to study the influence of salts, 13 monovalent anions (F^-, Cl^-, or Br^-) as well as 13 Na^+ cations were added to the water phase. Thus the concentration of salt measured in the water phase was equal to 150 mM. In each system, an additional number of anions (64 in the mixed DOTAP/DOPC and 128 in the pure DOTAP bilayer) was added to the water phase to neutralize the positive charge of the DOTAP molecules. Additionally, salt-free systems were simulated exclusively with the neutralizing Cl^- anions present. Test simulations were performed for a doubled ionic concentration (300 mM) for the pure DOTAP bilayer but no qualitatively new behavior of ions was observed and these systems will not be discussed here.

Interactions of lipid molecules were described using the non-polarizable united-atom Berger's force-field.[32] The parameters for DOTAP were derived based on the DOPC parameters set by removing the phosphate group and using the default Berger's parameters for the newly-formed headgroup. The SPC water model, for which the Berger's force-field was derived, was used throughout the simulations.[33] Force-field parameters for anions and sodium cation were taken from the GROMACS force field[34] with the exception of the fluoride anion where the parameterization introduced by Horinek *et al.* was employed.[35]

MD trajectories of 100 ns length were calculated applying periodic boundary conditions, with the initial box size of approximately 6.5 × 6.5 × 7.0 nm. Each system was shortly minimized, equilibrated for 50 ns, and the subsequent 50 ns of each trajectory were used for the analysis. The length of the equilibration period was chosen based on the stability of both the area per lipid and the number of ions adsorbed at the membrane–water interface. The cutoff of 1 nm was employed for both Lennard-Jones and short-range electrostatic interactions, while the long-range electrostatic forces were accounted for employing the particle-mesh Ewald method[36] with a Fourier spacing of 0.12 nm and the fourth-order approximation. Simulations were performed with the semi-isotropic pressure coupling, independently in the direction parallel and perpendicular with respect to the bilayer normal, employing the Parrinello–Rahman algorithm.[37] The pressure was set to 1 bar and the coupling constant of 2 ps was employed. The temperature of 310 K was controlled with the Nosé–Hoover thermostat,[38] independently for lipid and water phase, with the coupling constant of 1 ps. Bond lengths within the phospholipids were constrained using the LINCS algorithm.[39] Water molecules were kept rigid employing the SETTLE method.[40] Equations of motion were integrated using the leap-frog algorithm with the timestep of 2 fs. Molecular dynamics calculations were performed with program package GROMACS, version 4.0.7.[34]

3 Results and discussion

The interactions of monovalent ions with cationic lipid membranes were studied using TDFS experiments and MD simulations. This combination of methods has been recently successfully applied to a number of problems.[41] Below, we discuss the salt-induced changes in the properties of cationic lipid bilayers consisted of either pure DOTAP or DOPC/DOTAP and compared them with those of pure DOPC bilayer. First, the TDFS experimental results obtained for different salts and different DOPC/DOTAP mixtures are discussed. Next, the specific effects of anions in pure DOTAP bilayer and in the equimolar DOPC/DOTAP mixture that were observed in simulations are described.

3.1 Experimental measurements of DOPC/DOTAP mixtures

Large unilamellar vesicles (LUVs) formed from binary mixtures of lipids (cationic DOTAP and neutral DOPC) and labelled with Laurdan were measured using the TDFS method. The content of DOTAP was: 0, 6, 15, 30, 50, 75, and 100 mol%. The LUVs were suspended in different monovalent salt solutions. In all of the results discussed below TDFS parameters reflect local polarity and mobility in the vicinity of the lipid carbonyls where Laurdan is located. Chemical structures, hydrodynamic radii of ions, and the location of Laurdan are shown in Fig. 1.

3.1.1 Interactions between DOPC and DOTAP. In Fig. 2 the TDFS parameters obtained for DOPC/DOTAP vesicles suspended in water, NaCl, and KCl solutions are shown. In all of the media the relaxation kinetics described by the mean integrated relaxation time, τ, is a non-monotonic function of the mole fraction of DOTAP in the bilayer, Fig. 2a. The pronounced maximum of τ observed at 20–30 mol% DOTAP means that the moderate content of cationic lipid rigidifies the bilayer, which restricts the mobility of the hydrated carbonyls probed by Laurdan. Higher content of DOTAP lowers the values of τ, and the relaxation becomes faster again – local mobility is increased. This effect was observed in our earlier studies.[19,23] The reason of that unexpected bilayer rigidification was first identified in the NMR studies[13] of non-ideal mixing of cationic and zwitterionic lipids. Later it was also

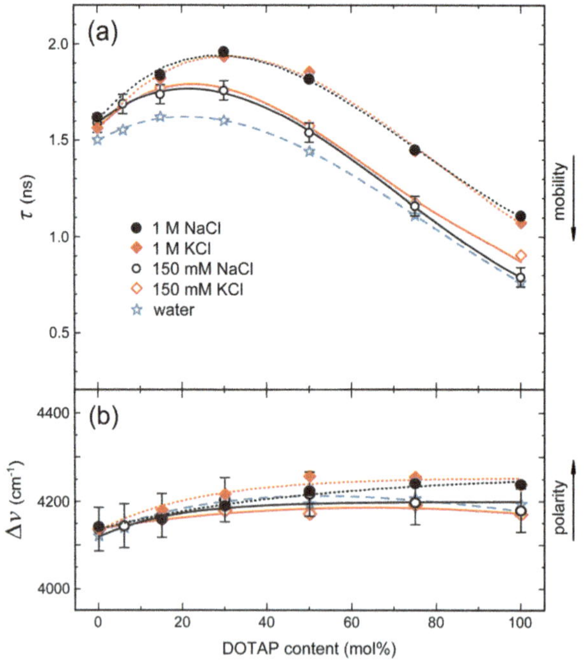

Fig. 2 Mean integrated solvent relaxation time, τ (a), and total spectral shift, $\Delta\nu$ (b) measured using TDFS of Laurdan in mixed DOPC/DOTAP vesicles as a function of DOTAP content. Data were measured at 10 °C for LUVs suspended in water, NaCl, and KCl. The lines represent polynomial fits and are shown as a guide to the eye only.

simulated in detail for DMPC/DMTAP mixture.[18] The molecular interpretation is as follows. TAP (trimethylammoniumpropane) cationic headgroup interacts electrostatically with the zwitterionic head of the phosphatidylcholine (PC), *i.e.* it attracts the phosphate anion and repels the NH_3^+. This leads to the straightening of the P–N dipole of PC molecule (Φ_{P-N} angle defined in Fig. 1 decreases) and the resulting lateral compression of the bilayer (simulated area per lipid decreases[18]). This effect was also observed in the present MD simulations for the equimolar DOTAP/DOPC system. For higher content of TAP^+, TAP-TAP repulsion starts to dominate and the bilayer expands again. This creates more space between the lipid molecules and makes the relaxation probed by Laurdan faster. The structural changes in DOPC/DOTAP bilayer are schematically depicted in Fig. 2a.

3.1.2 Water, KCl and NaCl – effect of salt concentration. The previously published TDFS experiments were performed for 10 mM Hepes pH = 7.4 with 100 mM NaCl.[19,23] In the present work DOPC/DOTAP vesicles suspended in water and in unbuffered solutions of 150 mM and 1 M NaCl and KCl were investigated. As it is visible in Fig. 2, while no major differences between KCl and NaCl were observed, the TDFS experiment was sensitive to the presence of salts. Increasing salt concentrations resulted in an elevated relaxation time (Fig. 2a) reporting a slowed down mobility at the carbonyl level. This effect was more pronounced for the cationic bilayers than for the pure neutral DOPC. Moreover, for lower DOTAP contents (up to 30 mol%) the τ values increased considerably upon addition of 150 mM salts, but did not change much when the salt concentration was further increased, *i.e.* Langmuir-like saturation was observed with the adsorption constant of Na^+ decreasing ~4.5 times when the DOTAP content increased from 0 to 30 mol %. On the contrary, the mobility of more cationic bilayers (*i.e.* ≥ 75 mol% of DOTAP) was very little affected by 150 mM salt solutions, but increased for the 1 M solutions. We believe that those two different types of salt concentration effects originate from the different dependence of the adsorption of metal and chloride ions to the DOPC/DOTAP bilayer on the DOTAP content.

It is known that Na^+ is able to adsorb not only to negative,[42] and neutral,[43–45] but also to slightly cationic[8,21] phospholipid bilayers, which usually restricts their mobility probed by TDFS. In the case of K^+, however, there is a disagreement between the results of our measurements and the MD simulations published. Except for the pure DOPC sample for which τ values for 1 M KCl and NaCl were 1.56 and 1.62 ns, respectively, Laurdan does not report any differences between KCl and NaCl, as for both salts the curves in Fig. 2a look almost identical. In previous MD simulations it was shown that, while Na^+ easily adsorbs to either phosphate[44] or carbonyl groups[43] of PC, K^+ is often reported to be only weakly associated with the bilayer.[44,46,47] This discrepancy can have a twofold reason. First, the type of K^+ adsorption and the location of Na^+ in the bilayer were shown to be force field-dependent,[47,48] thus, the effect of K^+ might be underestimated in MD. Second, it is also possible that TDFS is not sensitive enough to distinguish the effects of Na^+ and K^+ in neutral membranes. The substitution of 20 mol% of POPC with negatively charged POPS was found to be enough to reveal a strong cation specificity in the TDFS experiment.[42] Nevertheless, the present results show that both Na^+ and K^+ hinder mobility at the carbonyl level of DOPC/DOTAP bilayers with low DOTAP content.

For more positively charged bilayers cation adsorption is very unlikely, but chloride can associate with the bilayer more strongly. Chloride anions are usually only weakly associated with neutral lipid bilayers. They can be present at the phosphate level creating an electric double layer with deeper located sodium, but do not present any pronounced density peak.[44] For DOPC/DOTAP bilayers with high content of DOTAP, chloride is going to be attracted by the TAP^+ groups, which would hinder the headgroup mobility. This adsorption is apparently not strong enough to screen the electrostatic interactions between TAP^+ and PC headgroups, since the maximum of τ as a function of DOTAP content is preserved despite the salt presence.

The discussed above different salt concentration dependences for different DOTAP contents result in the change of the shapes of the τ curves presented in Fig. 2a and consequently the shift of the position of τ maximum. The τ maximum present for water at ~21 mol% of DOTAP shifts to ~23, and 31 mol% for 0.15 M and 1 M salt solutions, respectively.

The polarity at the carbonyl level of DOPC/DOTAP bilayers characterized by $\Delta\nu$ parameter was not much altered by the presence of NaCl nor KCl; see Fig. 2b. An increase above 15 mol% of DOTAP is subtle and within the experimental error (50 cm^{-1}). $\Delta\nu$ is usually the less sensitive TDFS parameter, which was also true in the case of the previous studies.[19]

3.1.3 Specific anion effects: NaF, NaCl, NaBr, $NaClO_4$. 150 mM sodium salt solutions (NaF, NaCl, NaBr, and $NaClO_4$) were used to study the interaction of monovalent anions with DOPC/DOTAP mixtures. The obtained TDFS parameters are collected in Fig. 3. On the contrary to the cations, for which the lack of specific effects was not surprising, the effects of the anions differ markedly for bilayers with higher content of DOTAP. Below the results of the integrated relaxation time (Fig. 3a) and later the ones of the total spectral shift (Fig. 3b) are discussed.

For pure DOPC, τ values are almost the same for all of the sodium halides, while the one for perchlorate is lower and close to that obtained for water. This agrees well with our previous studies,[45,49] where it was shown both by TDFS and MD simulations that large polarizable anions, such as I^-, ClO_4^-, and SCN^- penetrate easily into the DOPC bilayer headgroup region, where they increase polarity and mobility (1 M salt solutions were used). An addition of DOTAP causes a steep increase of τ values for $NaClO_4$ (Fig. 3a). Already at 6 mol% of DOTAP the effect observed in pure DOPC is cancelled. ClO_4^- is likely to attract TAP headgroups, which hinders

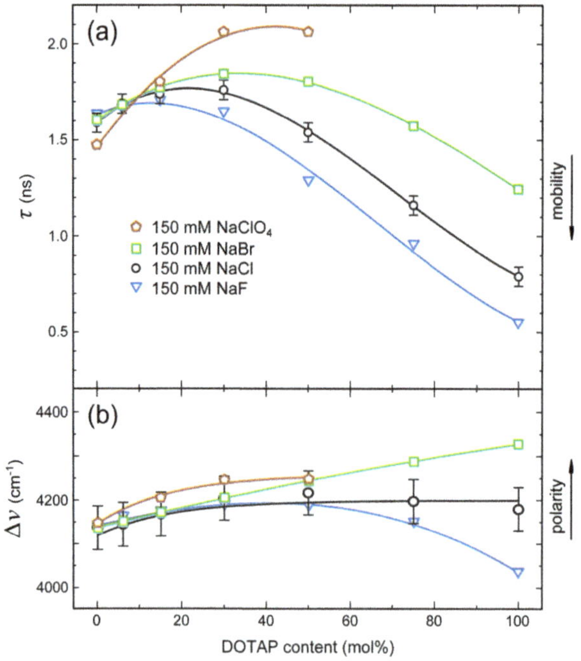

Fig. 3 Mean integrated solvent relaxation time, τ (a), and total spectral shift, $\Delta\nu$ (b) measured using TDFS of Laurdan in mixed DOPC/DOTAP vesicles as a function of DOTAP content. Data were measured at 10 °C for LUVs suspended in NaF, NaCl, NaBr, and $NaClO_4$. The lines represent polynomial fits and are shown as a guide to the eye only.

bilayer mobility. At 30 and 50 mol% of DOTAP the τ values for ClO_4^- are the highest. The interactions between perchlorate and DOTAP are so strong, that the formation of the vesicles with more than 50 mol% DOTAP was no longer possible, *i.e.* lipid film containing 75 and 100 mol% of DOTAP could not be suspended. Similar effects are expected for all soft anions (SCN^-, ClO_4^-, I^-, Br^-). They adsorb to the neutral bilayers due to their polarizability making the bilayer more mobile by repealing the negatively charged phosphates. On the contrary, they should bridge DOTAP molecules by attracting the TAP^+ groups, which would hinder the mobility of the DOPC/DOTAP bilayer. The effect of the stiffening of the cationic membrane is indeed observed here and follows the reversed Hofmeister series: $ClO_4^- > Br^- > Cl^- > F^-$. Thus, the introduction of the cationic lipid reverses the Hofmeister series. The more polarizable and larger the anion, the stronger are its interactions with the cationic bilayer. This effect is the most pronounced for pure DOTAP, where the substitution of fluoride by bromide slows down mobility at the carbonyl level more than 2 times. The above mentioned effect is also responsible for the shifting of the maximum of τ, which can be observed at ~13, ~23, and 34 mol% of DOTAP for NaF, NaCl, and NaBr, respectively.

Fluoride has an opposite effect to that of bromide and makes the DOTAP bilayer more mobile than it is in NaCl or in water. It is important to mention here, that the DOTAP used for the formation of liposomes had chloride as a counterion, so even in the "pure water" sample a 1 mM concentration of chloride was present. It is, however, unlikely that such a small concentration would produce the difference in τ for the DOTAP bilayer, which was 0.76 and 0.54 ns for water and 0.15 M NaF, respectively; especially that the chloride counterions were also present in the NaF solution. While it is well agreed that fluoride does not adsorb to the neutral lipid bilayers[50] the strong electrostatic interaction with pure cationic bilayer may easily change this rule; in fact, MD simulations point to the adsorption of F^- at the bilayer (see Section 3.2). Thus, it is also possible that the propensity of fluoride to form strong hydrogen bonds disturbs the structure of water hydrating the headgroups of DOTAP. To rule out a possible effect of pairing of F^- with Na^+ DOPC/DOTAP vesicles were suspended in 0.15 M solution of CsF. The obtained results (data not shown) were indistinguishable from those obtained for NaF, which demonstrates that the effects observed for fluoride are not counterion-specific.

In TDFS experiments in model lipid membranes $\Delta\nu$ usually corresponds to the hydration of the bilayer, since the main source of dipolar relaxation in the probe vicinity are the molecules of water. Thus, both TDFS parameters, $\Delta\nu$ and τ, are often related: the bilayer that is more hydrated is often more mobile, while dehydration restricts mobility. In the case of interactions of sodium halides with DOTAP this is clearly not the case (Fig. 3b). Bromide, which restricts mobility at the carbonyl level, at the same time increases $\Delta\nu$; fluoride has the opposite effect.

In the cases where ions can be present in the vicinity of the fluorophore, $\Delta\nu$ might not be interpreted in terms of bilayer hydration. It is known that the presence of ions can increase $\Delta\nu$ probed in a neat solvent, due to the so-called ionic relaxation, during which the ions undergo translational diffusion in the solvation shell of the fluorophore.[51] To test the hypothesis of the possible influence of the ionic relaxation on the increase of $\Delta\nu$ measured for bromide, a Laurdan anolog – Prodan – was used. Prodan, which is considerably more soluble in water than Laurdan, was dissolved in 1 M NaBr solution and in pure water. Indeed both the steady-state emission spectra and the TRES of Prodan were ~100 cm^{-1} red-shifted in NaBr solution. This demonstrated that the elevated $\Delta\nu$ values obtained for bromide can result from the ionic relaxation. Note, that the MD simulations show a strong clustering of TAP^+ groups induced by Br^- (see Section 3.2), which may be an additional source of the non-typical behaviour of $\Delta\nu$.

Possibility of Laurdan fluorescence quenching by Br^- was also tested. DOTAP vesicles containing 1 mol% of Laurdan were titrated with 2 M solutions of NaBr. A small reduction (about 6%) of the fluorescence lifetime was measured for 1 M NaBr concentration, but no changes were detected for 150 mM solution. Also the

Stern–Volmer quenching constant K_D measured for Prodan dissolved in water was $(0.25 \pm 0.03)\ M^{-1}$ for NaBr, which is relatively low in comparison to NaI, for which $K_D = (2.47 \pm 0.04)\ M^{-1}$.

Lowering of $\Delta\nu$ values obtained for DOTAP upon addition of NaF is difficult to interpret based on the experimental data alone. It is discussed in the simulation section (Section 3.2).

3.2 Molecular dynamics simulations

MD simulations were performed for pure DOPC and equimolar DOPC/DOTAP bilayers in the presence of 150 mM NaF, NaCl, and NaBr; in order to elucidate the molecular background of the strong specific anionic effects observed in TDFS experiments for monovalent atomic ions. The area per lipid (APL) and the membrane thickness calculated based on MD simulations are presented in Table 1. APL of the pure DOTAP bilayer in the absence of salts, *i.e.*, in the presence of Cl^- counterions only, is increased with respect to the purely zwitterionic DOPC bilayer, whereas under the same conditions in the mixed DOPC/DOTAP bilayer reduction of APL occurs. This is in accord with previous experimental and computational studies which have demonstrated that the lateral expansion of the purely-cationic membrane occurs due to electrostatic repulsion between DOTAP molecules whereas compression is observed in the mixed system resulting from the rising of DOPC headgroups interacting with the headgroups of DOTAP.[19,52]

In the case of the DOTAP bilayer, the presence of the considered 150 mM salt solutions causes the reduction of APL with respect to the system containing solely chloride counterions. APL in both NaCl and NaF solutions accounts for 86% of the value for the salt-free system while a significantly stronger reduction of APL is evident in the NaBr solution where APL equals 65% of that in the salt-free case. The bilayer thickness increases upon addition of salts. Similarly to the APL, the change of the thickness is relatively moderate in both NaCl and NaF solutions and strong in the presence of NaBr. The same effects, although less pronounced, occur in the case of the mixed DOTAP/DOPC bilayer. The nonmonotonic changes of the lateral lipid compression with the increasing DOTAP content resembles those of the relaxation time observed in the experiments (Fig. 2a) – in both salt-free and NaCl cases, APL decreases in the mixed DOTAP/DOPC bilayer, and significantly increases in the 100% DOTAP bilayer with respect to the DOPC one. The relation between the calculated APL and the experimental parameter τ is not direct, as demonstrated in our previous studies;[41] however, a qualitative agreement is expected.

The structural properties of the investigated bilayers along the direction of the membrane normal, as well as propensity of ions toward the lipid–water interface can be evaluated based on the density profiles presented in Fig. 4. Due to

Table 1 Area per lipid and bilayer thickness calculated in MD simulations. The data for the salt-free DOPC systems are taken from ref. 53. Membrane thickness is calculated as a distance between the maxima of the positions of phosphate groups

Lipid composition	Salt	APL/nm^{-2} ($\pm$0.01)	Thickness/nm ($\pm$0.05)
DOPC	—	0.68	3.87
DOTAP	—	0.74	3.40
	NaF	0.64	3.89
	NaCl	0.64	3.90
	NaBr	0.48	4.75
DOPC/DOTAP (1 : 1 mol/mol)	—	0.56	4.07
	NaF	0.56	4.05
	NaCl	0.56	4.18
	NaBr	0.52	4.42

electrostatic interactions between cationic DOTAP and anions, all considered negative ions exhibit affinity towards the DOTAP bilayer, as evident from the presence of the anion density profile peaks at the water–membrane interface (Fig. 4a–c). Br^- anions adsorb strongly at the lipid bilayer specifically in the region occupied by the TAP^+ groups: the two peaks overlap. Adsorption of both Cl^- and F^- is weaker: their peaks are broader and shifted ~0.2 nm outward with regard to the TAP^+ groups. The behaviour of sodium cations differs in the case of each salt. In the NaCl solution, Na^+ ions remain predominantly in the water phase, with only a small fraction penetrating toward the carbonyl groups of lipids. In the case of NaF, an enhancement of the Na^+ density at the membrane–water interface is observed, outward from the region occupied by F^-. In the NaBr solution a specific adsorption of Na^+ ions occurs; namely, some fraction adsorbs close to the carbonyl groups while another resides in the region occupied by both TAP^+ and Br^-. Note, that since the number of sodium cations accounts for merely 9% of the anions, the influence of the cations on the bilayer is negligible when compared with that of the anions.

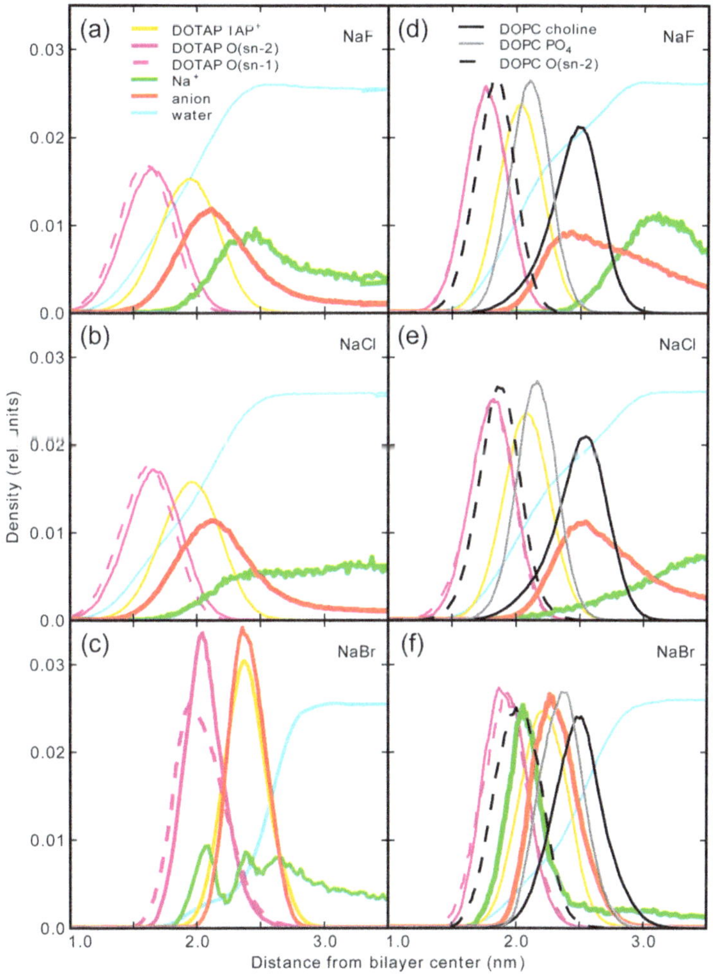

Fig. 4 Density profiles of ions, water, and selected components of lipid bilayers in different salt solutions for pure DOTAP (a–c) and mixed DOTAP/DOPC (d–f) bilayers. Densities from both bilayer leaflets are averaged. For clarity of presentation, the densities were renormalized. In particular, the ratio of the cation to anion density is enhanced.

The average positions of both TAP^+ and carbonyl oxygen atoms in the NaBr solution are shifted outward from the bilayer center with regard to the systems with the two other salts. This is a manifestation of the increase of the bilayer thickness in the NaBr solution, as discussed beforehand. Penetration of water into the membrane differs significantly between the NaBr and the two other salt solutions; in the former one, bilayer hydration is reduced, as evident from the decrement of water density in the region occupied by both TAP^+ and carbonyl groups of lipids. The diminished water penetration likely results from the reduced APL, since the decreasing average lipid–lipid distance causes steric hindrance for water molecules. In the NaBr solution, a significant difference between the shape of the density profiles of *sn*-1 and *sn*-2 carbonyl oxygens is present, suggesting alteration of the lipid configurations; this issue is addressed later on.

In the mixed DOTAP/DOPC system, no qualitative differences in the penetration of anions toward the bilayer are observed in comparison with the DOTAP bilayer, although their distributions are slightly broader and shifted outward from the bilayer center. In the case of Cl^- and F^- their maxima colocalize with those of DOPC choline. The behavior of the cations shows two dissimilarities. First, in both NaCl and NaF solutions cations are strongly repelled from the water–lipid interface. This is most likely due to electrostatic interactions with positively charged choline groups of DOPC; see the density profile peak of choline effectively separating the peak of cations from the membrane. Second, in the NaBr solution a deeper penetration of Na^+ toward carbonyl groups occurs. Such an attraction of Na^+ toward the bilayer results from the neutralization of the positive charge of the bilayer by massively adsorbed Br^- anions. Then, the adsorption of sodium in the carbonyl region occurs, in accord with the tendency of Na^+ to bind with carbonyl groups of phosphatidylcholines, which was demonstrated in previous computational studies.[9,47]

In Fig. 5 the distribution of P–N dipoles orientation calculated for DOPC in the mixed DOTAP/DOPC system is depicted. In all the systems, except the one containing Br^- ions, the P–N dipoles are pointing predominantly outward from the membrane, and the distribution is shifted toward lower angle values with respect to the distribution calculated in the salt-free DOPC bilayer. This agrees with similar observation for DOPC in the presence of salts.[54] In the case of Br^-, P–N dipoles predominantly orient in parallel with respect to the lipid–water interface. Such orientation is also evident based on the density profiles (see Fig. 4) where the positions of the peaks corresponding to the choline and phosphate groups of DOPC are much closer to each other in the presence of Br^- than in the two other systems. Such an orientation of DOPC headgroups in NaBr solution is remarkable and unexpected in light of the reduced APL in this system because, in general, an increased lateral packing of lipids can be expected to cause the rising of headgroups due to the steric effect. However, in the mixed DOPC/DOTAP bilayer molecules of DOTAP are located, on average, deeper in the membrane than molecules of DOPC (see the positions of the peaks of TAP^+ and DOPC phosphate in Fig. 4f). An inspection of MD trajectories revealed that the buried DOTAP molecules act as a spacer between DOPC, therefore reorientation of the DOPC headgroups is no longer hindered by the packing of lipids; the same is also valid in the case of the two other salts. It is, thus, electrostatics that governs the orientation of the P–N dipole. The presence of Br^- anions, which fully neutralize TAP^+ cations, allows the DOPC headgroups to return to their usual parallel conformation, as in a pure PC bilayer. In fact, the distribution of the Φ_{P-N} angle in DOPC/DOTAP with 150 mM NaBr discussed here is narrower than the one in pure PC bilayers.[9,18] Such a narrow distribution is likely the effect of attraction of the choline groups to the Br^- anions that reside below the choline groups of DOPC, which stabilizes P–N dipoles of DOPC in the bended conformation.

More specific analysis of the structural properties of the investigated bilayers can be gained from the analysis of the radial distribution functions ($g(r)$s) presented in

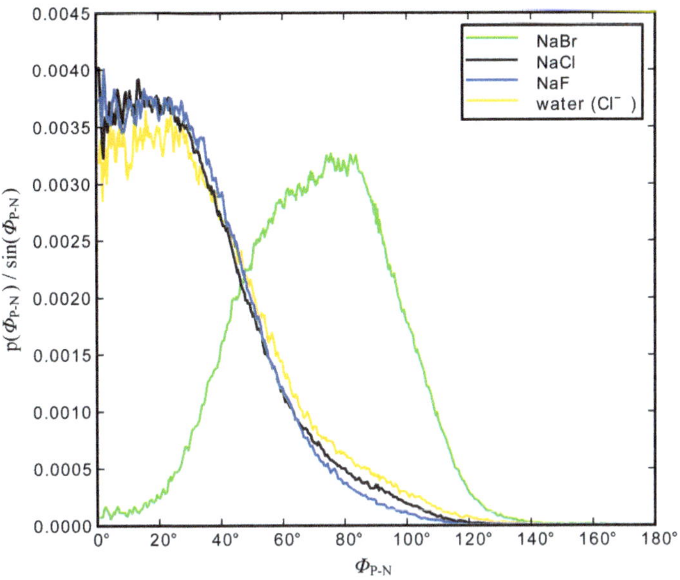

Fig. 5 Normalized distributions of Φ_{P-N} angles of DOPC in the mixed DOTAP/DOPC (50/50) bilayers; 0° represents P–N dipoles pointing outward of the membrane and 180° toward the membrane interior.

Fig. 6. TAP^+–water $g(r)$s depicted in Fig. 6a demonstrate that DOTAP headgroups are relatively well hydrated in the presence of both NaCl and NaF while they are dehydrated in the NaBr solution. This corroborates with the reduced hydration discussed beforehand based on the density profiles for the bilayer in the NaBr solution. The same hydration trend for the changing salt is visible in the radial distribution functions calculated for carbonyl groups and water (results not shown). The same tendency is also observed in the case of the mixed DOPC/DOTAP. In general, hydration of the studied bilayers for the simulated salt solutions decreases in the row: NaF ≈ NaCl > NaBr.

As evident from TAP^+-anion $g(r)$ functions in Fig. 6b, qualitative differences exist between the interactions of investigated anions with DOTAP headgroups. Namely, in the case of Br^- a very pronounced peak at ~0.35 nm is present, demonstrating a significant affinity of Br^- toward TAP^+ groups; in accord with the conclusion based on the density profiles presented in Fig. 4. Specific interactions of both Cl^- and F^- with TAP^+ also occur, as can be concluded from the presence of the $g(r)$ peaks at the distance of approximately 0.5 nm; however, these interaction are weaker than in the case of Br^-. Chloride ions are somewhat tighter bound to the DOTAP headgroups than F^-, as the $g(r)$ peak for Cl^- is more localized. The positions of the first peaks in Fig. 6b demonstrate that Br^-, while being the largest among the considered anions, resides from the nearest to the center of the TAP^+ groups, which indicates formation of contact ion pairs between Br^- and TAP^+. Both Cl^- and F^- remain at a relatively larger distance, predominantly above 0.4 nm, from the DOTAP headgroups, suggesting the presence of water-mediated anion-TAP^+ configurations.

Further structural features of the DOTAP bilayer in the investigated salt solutions can be inferred based on the lipid–lipid radial distribution functions. As shown in Fig. 6c, there is a well pronounced peak at 0.6 nm in the TAP^+-TAP^+ $g(r)$ for NaBr solution which demonstrates the presence of Br^-–induced clustering of DOTAP molecules *via* their headgroups. The first peak of the $g(r)$ corresponds to five TAP^+ groups neighboring an average TAP^+ group in the radius of 0.9 nm (based on the cumulative sum of the radial distribution function). The DOTAP headgroups

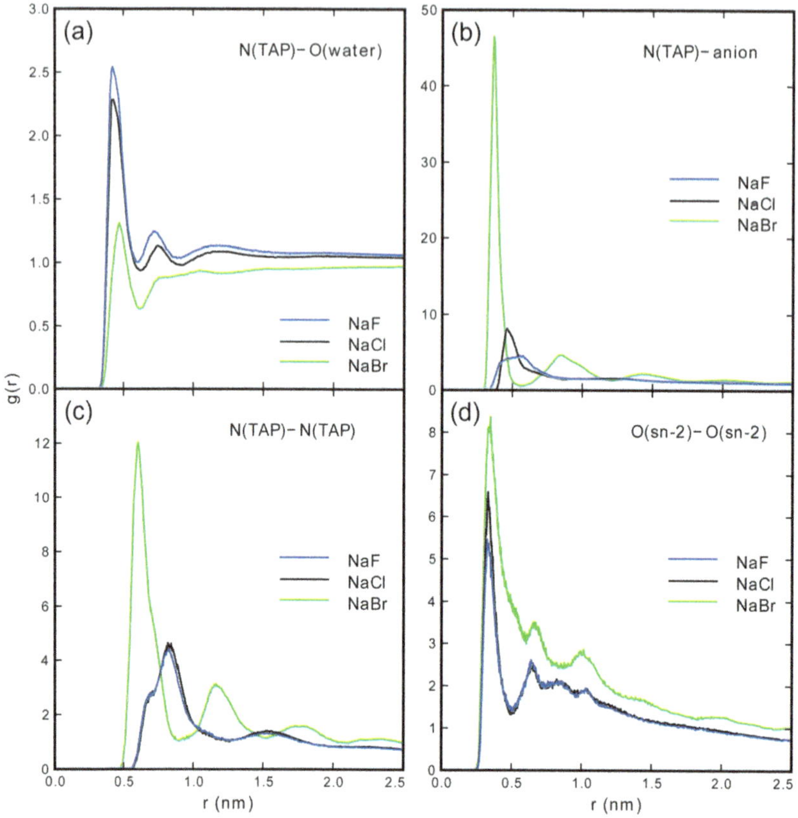

Fig. 6 Partial radial distribution functions calculated for the pure DOTAP bilayer between nitrogen atoms of TAP groups and oxygen atoms of water (a), nitrogen atoms of TAP groups and anions (b), nitrogen atoms of TAP groups between the lipids (c), and *sn*-2 carbonyl oxygen atoms between the lipids (d). For presentation purposes all curves are truncated at the distance of 2.5 nm.

clustering in both NaCl and NaF are weaker, the peaks are located at 0.8 nm and the number of five TAP^+-TAP^+ neighbors is reached only at the distance of ~1.1 nm. Note, that there is virtually no difference between the influence of Cl^- and F^- on the clustering of headgroup. Carbonyl–carbonyl radial distribution functions calculated for the *sn*-2 chains, presented in Fig. 6d, with well-pronounced peaks located at approximately 0.4 nm, show that clustering of lipid molecules occurs at the level occupied by carbonyl groups which is deeply buried in the bilayer. In this region, the differences between the salts are quantitative, with the tendency for lipid clustering increasing in the row: NaF < NaCl < NaBr. Note, that in contrast to the clustering of TAP^+ groups, some minor differences between the peaks calculated for Cl^- and F^- are present here. The ability to induce lipid clustering by the investigated anions is in accord with the changes in the experimentally measured local mobility at the carbonyl level (Fig. 3a). The strong headgroup clustering in the NaBr solution causes rigidification of the bilayer and hindrance of lipid motions. This effect is confirmed by the lateral diffusion coefficients calculated based on MD trajectories, equal to 1.5 ± 0.2, 1.4 ± 0.2, and $0.1 \pm 0.2 \times 10^{-7}$ cm^2 s^{-1} for the DOTAP bilayer in NaF, NaCl, and NaBr solutions, respectively. In the case of mixed DOPC/DOTAP bilayers, qualitatively the same trends in both the binding of ions and lipid clustering are observed based on the $g(r)$ functions (data not shown). Minor quantitative differences occur, mainly a stronger clustering of DOTAP lipids. This is because in the mixed bilayers, where the salt concentration was unchanged with regard to the simulations of purely-cationic membranes, the same number of anions interacts with a reduced number of positively charged DOTAP molecules.

A visual inspection of MD trajectories revealed that in the NaBr solution the clustering of DOTAP headgroups leads to formation of patches of TAP^+ groups bound together by Br^- anions, accompanied by the hydrophobic patches of hydrocarbon chains exposed to the water phase, as depicted in Fig. 7. Such lateral heterogeneities were not observed in the case of either NaCl or NaF. In the case of the mixed DOPC/DOTAP membrane, Br^- anions interact predominantly with DOTAP lipids forming tightly bound clusters of DOTAP molecules in a matrix of DOPC; hydrocarbon chains of lipids are not exposed to water in this system. In the case of the mixed bilayers, Cl^- and F^- interact predominantly with DOTAP molecules; however, interactions of anions with DOPC are also present.

As revealed in TDFS measurements, there are significant differences between the influence of Cl^- and F^- on the mobility and polarity of the cationic bilayer. These differences were measured at the level of *sn*-2 carbonyl and are the strongest for the pure DOTAP samples. They can be explained based on the molecular-level picture emerging from MD simulations. As evident from the density profiles, both Cl^- and F^- can penetrate into the bilayer, toward the carbonyl region of lipids. The negatively charged oxygen atoms of carbonyl groups do not interact with anions directly, as the $g(r)$ functions calculated for the *sn*-2 carbonyl oxygen and anions (not presented here) have first peaks at the distance of ~0.5 nm. Nevertheless, the presence of anions influences the hydrogen bonding of water in that region and thus, indirectly, affects the spatial correlations between carbonyl groups. F^- anions are relatively small and form a tight first solvation shell, thus, in the carbonyl region they strongly bind water molecules and cause a weakening of water-mediated carbonyl–carbonyl interactions. The solvation shell of Cl^- is less rigid and thus Cl^- binds water molecules in the carbonyl region less effectively. This leads to the increased local lipid mobility probed in TDFS experiment in NaF. On the other hand, as a consequence of the tight water binding by F^-, the water molecules which solvate fluoride anions should be less active in the solvent relaxation process since their rotational mobility is hindered. This would rationalize the experimentally

Fig. 7 Typical snapshots of the lipid/water interface (the view from the water phase) observed in the equilibrated MD trajectories for the pure DOTAP bilayer in NaCl (left) and NaBr (right). Color coding: yellow – TAP groups, gray – atoms of DOTAP not belonging to TAP, orange – Cl^-, green – Br^-. Water molecules and cations are not shown for clarity of presentation. Homogeneous localization of both TAP and anions is visible in NaCl, whereas formation of TAP clusters and hydrophobic patches (the latter seen as the grey areas) occurs in NaBr.

observed reduction of polarity in the vicinity of DOTAP carbonyl groups in the NaF solution.

4 Summary and conclusions

The aim of this paper was to explore specific interaction of anions (*i.e.* ClO_4^-, Br^-, Cl^-, and F^-) with lipid bilayers. As demonstrated in our previous study of the interactions of lipid bilayers with alkali cations,[9] the level of binding and the specificity of ions can be enhanced by introducing into the bilayer lipid species of opposite charge. In the present work we employ a binary mixture of neutral DOPC with cationic DOTAP, the two synthetic lipids that differ only in their polar heads. Significant differences between the effects induced by Br^-, Cl^-, and F^- on lipid bilayers have been observed in both TDFS measurements and MD simulations. For a detailed interpretation of these results it is essential to understand the interactions between the two lipids, first, in the absence of salts. We have previously demonstrated that the TDFS relaxation time, τ, which is a measure of local lipid mobility, displays a maximum as a function of DOTAP content at ~30 mol%; namely, with the increasing concentration of DOTAP in DOPC/DOTAP bilayer the local movements of lipid headgroups slow down in the range of 0–30 mol% of DOTAP, while for DOTAP contents > 30 mol% these movements become faster.[19] As concluded based on the experiments[13,14] and MD simulations,[18,20] this nonmonotonic relationship can be attributed to the reorientation of the P–N dipole of zwitterionic PC as a result of the electrostatic interactions between the headgroups of DOPC and DOTAP. The data presented herein demonstrate that most of the studied salts were not able to screen effectively these electrostatic interactions between TAP^+ and PC. Thus, the maximum of the relaxation time as a function of DOTAP content is qualitatively preserved for all of the salts, although its position shifts toward a larger DOTAP content: (a) with the increasing concentration of sodium or potassium chlorides, and (b) in the row of 150 mM salts: NaF < NaCl < NaBr, spanning the range from ~13 mol% to ~34 mol% of DOTAP for NaF and NaBr, respectively. Note in this context that at 50 mol% of DOTAP the vertical orientation of DOPC headgroups is stabilized by the presence of Cl^- anions adsorbed at the bilayer as demonstrated in simulations (Fig. 4e). Such orientation of headgroups should further rigidify the bilayer, and thus, hinder the mobility measured by Laurdan. In the view of the previous MD simulations, in which mixed DMPC/DMTAP bilayers were shown to be almost insensitive to the presence of up to 1 M NaCl solution, both in terms of their structure[8] and dynamics,[21] the kinetics of TDFS of Laurdan proves to be particularly sensitive to ions interacting with PC/TAP bilayer.

The results of TDFS of Laurdan in DOPC/DOTAP vesicles suspended in 150 mM sodium salt solutions indicate strong interactions between positively charged bilayers and the anions following reversed Hofmeister series of the form: ClO_4^- > Br^- > Cl^- > F^-. In the pure DOTAP bilayer Br^- penetrates to the vicinity of the Laurdan fluorescent probe, which is located at the carbonyl level, where it strongly hinders local lipid mobility. These experimental results are confirmed by our MD simulations in which Br^- is found to strongly and specifically adsorb in the region of DOTAP headgroups, inducing lipid clustering *via* formation of TAP^+–Br^- ion complexes. This is accompanied by the overall dehydration of the bilayer, the increased bilayer thickness and the reduced APL, as well as by reduction of the lateral mobility of lipids and formation of both hydrophillic and hydrophobic patches at the membrane–water interface (Fig. 7).

Notably, the effect of large soft anions on the relaxation time probed in a lipid bilayer is reversed upon the addition of cationic lipid. While in pure DOPC ClO_4^- increases bilayer mobility (similarly to SCN^- and I^-),[45] already at 30 mol% of DOTAP the effect is inverted. Large anions, like ClO_4^-, due to their significant polarizability, are incorporated deeply into the lipid bilayer, where they repel either DOPC carbonyls or phosphates, and attract either TAP^+ or DOPC cholines.

Somewhat more subtle effects occur in pure DOTAP bilayers in both NaCl and NaF solutions. As shown by MD simulations, Cl^- and F^- do not bind specifically to the headgroups of DOTAP, and thus, no lipid clustering at the headgroup level is observed. Nevertheless, even weakly associated Cl^- and F^- affect the mobility probed in the TDFS experiments in a distinct way for each anion. These differences can be rationalized based on the molecular-level picture emerging from MD simulations. Due to the strong affinity to water, the F^- anions, when attracted to the cationic bilayer, disturb its structure by breaking the network of hydrogen bonds created by molecules of lipids and water. This leads to a more fluid bilayer; in particular, the mobility probed by Laurdan at the carbonyl level of lipids is significantly increased. In the case of chloride anions, which penetrate the bilayer but form relatively flexible solvation shells, no strong influence on the mobility of lipids is observed.

Acknowledgements

We thank Prof. Pavel Jungwirth for introducing us to the topic of ionic effects, and for his valuable comments and stimulating discussions. Support from the Czech Science Foundation *via* grant P208/12/G016 (S. P., P. J., and M. H.) is acknowledged. Moreover, M. H. acknowledges the Praemium Academie Award (Academy of Sciences of the Czech Republic). L. C. acknowledges the support from the Granting Agency of the Czech Republic (ESF EUROMEMBRANE project MEM/09/E006).

References

1 M. Bloom, E. Evans and O. G. Mouritsen, *Q. Rev. Biophys.*, 1991, **24**, 293–397.
2 D. Chandler, *Nature*, 2005, **437**, 640–647.
3 D. Leckband and J. Israelachvili, *Q. Rev. Biophys.*, 2001, **34**, 105–267.
4 M. L. Berkowitz and R. Vacha, *Acc. Chem. Res.*, 2012, **45**, 74–82.
5 D. F. Parsons, M. Bostrom, P. L. Nostro and B. W. Ninham, *Phys. Chem. Chem. Phys.*, 2011, **13**, 12352–12367.
6 J. J. Garcia-Celma, L. Hatahet, W. Kunz and K. Fendler, *Langmuir*, 2007, **23**, 10074–10080.
7 P. M. Macdonald and J. Seelig, *Biochemistry*, 1988, **27**, 6769–6775
8 A. A. Gurtovenko, M. Miettinen, M. Karttunen and I. Vattulainen, *J. Phys. Chem. B*, 2005, **109**, 21126–21134.
9 P. Jurkiewicz, L. Cwiklik, A. Vojtískova, P. Jungwirth and M. Hof, *Biochim. Biophys. Acta*, 2012, **1818**, 609–616.
10 P. L. Felgner, T. R. Gadek, M. Holm, R. Roman, H. W. Chan, M. Wenz, J. P. Northrop, G. M. Ringold and M. Danielsen, *Proc. Natl. Acad. Sci. U. S. A.*, 1987, **84**, 7413–7417.
11 A. E. Regelin, S. Fankhaenel, L. Gürtesch, C. Prinz, G. von Kiedrowski and U. Massing, *Biochim. Biophys. Acta, Biomembr.*, 2000, **1464**, 151–164.
12 S. W. Hui, M. Langner, Y. L. Zhao, P. Ross, E. Hurley and K. Chan, *Biophys. J.*, 1996, **71**, 590–599.
13 P. G. Scherer and J. Seelig, *Biochemistry*, 1989, **28**, 7720–7728.
14 J. R. Silvius, *Biochim. Biophys. Acta*, 1991, **1070**, 51–59.
15 R. Zantl, L. Baicu, F. Artzner, I. Sprenger, G. Rapp and J. O. Radler, *J. Phys. Chem. B*, 1999, **103**, 10300–10310.
16 V. Matti, J. Säily, S. J. Ryhänen, J. M. Holopainen, S. Borocci, G. Mancini and P. K. J. Kinnunen, *Biophys. J.*, 2001, **81**, 2135–2143.
17 V. M. J. Säily, J.-M. Alakoskela, S. J. Ryhänen, M. Karttunen and P. K. J. Kinnunen, *Langmuir*, 2003, **19**, 8956–8963.
18 A. A. Gurtovenko, M. Patra, M. Karttunen and I. Vattulainen, *Biophys. J.*, 2004, **86**, 3461–3472.
19 P. Jurkiewicz, A. Olzynska, M. Langner and M. Hof, *Langmuir*, 2006, **22**, 8741–8749.
20 S. Bandyopadhyay, M. Tarek and M. L. Klein, *J. Phys. Chem. B*, 1999, **103**, 10075–10080.
21 M. S. Miettinen, A. A. Gurtovenko, I. Vattulainen and M. Karttunen, *J. Phys. Chem. B*, 2009, **113**, 9226–9234.
22 L. Woiterski, D. W. Britt, J. A. Käs and C. Selle, *Langmuir*, 2012, **28**, 4712–4722.
23 A. Olzynska, P. Jurkiewicz and M. Hof, *J. Fluoresc.*, 2008, **18**, 925–928.

24 M. L. Horng, J. A. Gardecki, A. Papazyan and M. Maroncelli, *J. Phys. Chem.*, 1995, **99**, 17311–17337.
25 R. Hutterer, A. B. J. Parusel and M. Hof, *J. Fluoresc.*, 1998, **8**, 389–393.
26 J. Sykora and M. Hof, *Cell. Mol. Biol. Lett.*, 2002, **7**, 259–261.
27 P. Jurkiewicz, J. Sykora, A. Olzynska, J. Humplickova and M. Hof, *J. Fluoresc.*, 2005, **15**, 883–894.
28 R. S. Fee and M. Maroncelli, *Chem. Phys.*, 1994, **183**, 235–247.
29 J. Sykora, P. Kapusta, V. Fidler and M. Hof, *Langmuir*, 2002, **18**, 571–574.
30 J. Barucha-Kraszewska, S. Kraszewski, P. Jurkiewicz, C. Ramseyer and M. Hof, *Biochim. Biophys. Acta, Biomembr.*, 2010, **1798**, 1724–1734.
31 Y. Marcus, *Chem. Rev.*, 1988, **88**, 1475–1498.
32 O. Berger, O. Edholm and F. Jähnig, *Biophys. J.*, 1997, **72**, 2002–2013.
33 H. J. C. Berendsen, J. P. M. Postma, W. F. van Gunsteren and J. Hermans, in *Intermolecular Forces*, 1981, pp. 331–342.
34 B. Hess, C. Kutzner, D. van der Spoel and E. Lindahl, *J. Chem. Theory Comput.*, 2008, **4**, 435–447.
35 D. Horinek, S. I. Mamatkulov and R. R. Netz, *J. Chem. Phys.*, 2009, 130.
36 U. Essmann, L. Perera, M. L. Berkowitz, T. Darden, H. Lee and L. G. Pedersen, *J. Chem. Phys.*, 1995, **103**, 8577–8593.
37 M. Parrinello and A. Rahman, *J. Appl. Phys.*, 1981, **52**, 7182–7190.
38 S. Nose, *Mol. Phys.*, 1984, **52**, 255–268.
39 B. Hess, H. Bekker, H. J. C. Berendsen and J. G. E. M. Fraaije, *J. Comput. Chem.*, 1997, **18**, 1463–1472.
40 R. W. Hockney, S. P. Goel and J. Eastwood, *J. Comput. Phys.*, 1974, **14**, 148–158.
41 P. Jurkiewicz, L. Cwiklik, P. Jungwirth and M. Hof, *Biochimie*, 2012, **94**, 26–32.
42 P. Jurkiewicz and a. al, *Biochim. Biophys. Acta, Biomembr.*, 2012.
43 R. A. Bockmann, A. Hac, T. Heimburg and H. Grubmuller, *Biophys. J.*, 2003, **85**, 1647–1655.
44 R. Vacha, S. W. I. Siu, M. Petrov, R. A. Bockmann, J. Barucha-Kraszewska, P. Jurkiewicz, M. Hof, M. L. Berkowitz and P. Jungwirth, *J. Phys. Chem. A*, 2009, **113**, 7235–7243.
45 R. Vacha, P. Jurkiewicz, M. Petrov, M. L. Berkowitz, R. A. Bockmann, J. Barucha-Kraszewska, M. Hof and P. Jungwirth, *J. Phys. Chem. B*, 2010, **114**, 9504–9509.
46 A. Cordomi, O. Edholm and J. J. Perez, *J. Phys. Chem. B*, 2008, **112**, 1397–1408.
47 A. A. Gurtovenko and I. Vattulainen, *J. Phys. Chem. B*, 2008, **112**, 1953–1962.
48 A. Cordomi, O. Edholm and J. J. Perez, *J. Chem. Theory Comput.*, 2009, **5**, 2125–2134.
49 M. Petrov, L. Cwiklik and P. Jungwirth, *Collect. Czech. Chem. Commun.*, 2011, **76**, 695–711.
50 E. Leontidis and A. Aroti, *J. Phys. Chem. B*, 2009, **113**, 1460–1467.
51 C. F. Chapman and M. Maroncelli, *J. Phys. Chem.*, 1991, **95**, 9095–9114.
52 W. Zhao, A. Gurtovenko, I. Vattulainen and M. Karttunen, *J. Phys. Chem. B*, 2012, **116**, 269–276.
53 L. Cwiklik and P. Jungwirth, *Chem. Phys. Lett.*, 2010, **486**, 99–103.
54 P. T. Vernier, M. J. Ziegler and R. Dimova, *Langmuir*, 2009, **25**, 1020–1027.

Ion specific effects of alkali cations on the catalytic activity of HIV-1 protease†

Jana Pokorná,‡[a] Jan Heyda‡[b] and Jan Konvalinka*[ac]

Received 1st May 2012, Accepted 8th May 2012
DOI: 10.1039/c2fd20094e

Human immunodeficiency virus 1 protease (HIV-1 PR), an important therapeutic target for the treatment of AIDS, is one of the most well-studied enzymes. However, there is still much to learn about the regulation of the activity and inhibition of this key viral enzyme. Specifically, the mechanism of activation of HIV-1 PR from the viral polyprotein upon HIV maturation is still not understood. It has been suggested that external factors like pH or salt concentration might contribute to regulation of this crucial step in the viral life cycle. Recently, we analyzed the activity of HIV-1 PR in aqueous solutions of sodium and potassium chloride by experimental determination of enzyme kinetics and molecular dynamics simulations. We showed that the effect of salt concentration is cation-specific [Heyda *et al.*, *Phys. Chem. Chem. Phys.*, 2009 (**11**), 7599]. In this study, we extended this analysis for other alkali cations and found that the dependence of the initial velocity of peptide substrate hydrolysis on the nature of the cation follows the Hofmeister series, with the exception of caesium. Significantly higher catalytic efficiencies both in terms of substrate binding (K_M) and turnover number (k_{cat}) are observed in the presence of K^+ compared to Na^+ or Li^+ at corresponding salt concentrations. Molecular dynamics simulations suggest that both lithium and sodium are attracted more strongly than potassium and caesium to the protein surface, mostly due to stronger interactions with carboxylate side chain groups of aspartates and glutamates. Furthermore, we observed a surprising decrease in the K_M value for a specific substrate at very low salt concentration. The molecular mechanism of this phenomenon will be further analyzed.

Introduction

The protease of human immunodeficiency virus 1 (HIV-1 PR) is indispensable for replication of the virus. Inactivation of HIV-1 PR or its chemical modification, inhibition, or mutation can block viral infectivity. Therefore, HIV-1 PR has become a prime target for antiretroviral drug discovery and a subject of research by both academia and industry. Today, nine HIV-1 PR inhibitors have been approved as anti-HIV drugs by the Federal Drug Administration, and this unprecedented research and development effort is a prime example of successful rational,

[a]*Gilead and IOCB Research Center Prague, Flemingovo nám. 2, 16610 Prague 6, Czech Republic. E-mail: konval@uochb.cas.cz*
[b]*F-I2 Soft Matter and Functional Materials, Helmholtz-Zentrum Berlin für Materialien und Energie, 14109 Berlin, Germany*
[c]*Department of Biochemistry, Faculty of Science of the Czech Republic, Charles University in Prague, Hlavova 8, 120 00 Prague 2, Czech Republic*

† Electronic supplementary information (ESI) available. See DOI: 10.1039/c2fd20094e
‡ These authors contributed equally to this work.

structure-based drug discovery (for reviews see ref. 1–3). However, due to the development of antiviral resistance and unwanted drug side-effects, the need for the development of novel, more potent HIV-1 PR inhibitors capable of blocking highly resistant viral species remains urgent.

HIV-1 PR is a dimeric aspartic protease that consists of two identical monomers, containing 99 amino acids each. In fact, HIV-1 PR is the smallest protease ever described. The enzyme dimerizes mostly *via* its N- and C- termini (residues 1–4 and 96–99), which form an intertwining four-stranded antiparallel β-sheet. Upon substrate or inhibitor binding, two flexible β-hairpins, called flaps, close over the active site cavity and form hydrogen bonds with the bound ligand, thus contributing to substrate/inhibitor recognition and binding. HIV-1 PR cleaves viral Gag and Gag-Pol polyproteins, primary products of the translation of viral mRNA, into structural proteins and replication enzymes that form progeny infectious viral particles.[1]

Like many other viral proteins, HIV-1 PR has a high pI value (8.66), which is caused by a higher number of positively charged basic amino acid residues (10 Lys and Arg residues per monomer) than acidic ones (8 Glu and Asp residues per monomer). Each HIV-1 PR monomer contributes one of the catalytic aspartates, which form a catalytic dyad in the active site of the PR dimer, activating a water molecule localized between them to attack the peptide bond of the substrate bound to the active cleft of the PR. The aspartate residues share a single proton that plays a role in the catalytic mechanism of the enzyme.

The substrate specificity of HIV-1 PR and its kinetics of peptide hydrolysis have been thoroughly studied during the last 25 years (for a review, see ref. 4). Several chromogenic and fluorogenic peptide substrates have been designed for continuous analysis of PR activity, and the subsite preferences, optimal pH, and buffer requirements for the enzymatic activity have been studied in depth.

Dependence on ionic strength or salt concentration is an important characteristic of all enzymatic activities. It has been shown that the presence of higher salt concentrations primarily affects the substrate binding (K_M) of HIV, although an effect on the catalytic rate (k_{cat}) has also been observed. The specific salt effect follows the Hofmeister series, *i.e* classification of cations and anions according to their ability to precipitate proteins (for review see ref. 5). Potassium salts activate the PR more effectively than equal concentrations of sodium salts.[6,7] The mechanism of this phenomenon is not clearly understood; it is usually attributed to the "salting out" of the substrate into the binding cleft of the enzyme. Szeltner and Polgar[8] suggest that the substrate affinity increase is the result of an entropic effect due to the increase in the conformational stability of HIV-1 PR.

Numerous effects of salts on enzyme kinetics have been described in the literature. Some enzymes are influenced very little by ionic strength: human T-cell leukemia virus 1 protease (HTLV-1 PR), an enzyme homologous to HIV-1 PR, does not change its activity with NaCl concentration ranging from 0 to 1.5 M,[9] while other enzymes require high ionic strength (thermolysin[10]) or even the presence of a specific cation for full enzymatic activity (thrombin and other serine proteases involved in the blood coagulation cascade; reviewed in ref. 11 & 12).

The effect of salt on the behaviour of proteins in solution is described by the lyotropic (Hofmeister) series. The mechanistic explanation underlying the Hofmeister series is not trivial. Traditionally, this effect is believed to be due to changes in the bulk water structure (for a review, see ref. 13). However, a series of recent studies seems to suggest that the "bulk water shell" is not affected by the monovalent ions in the solution and that the lyotropic series might be rather interpreted in terms of ion-protein interactions in the first hydration shell of the protein in question (for a review, see ref. 14).

We recently performed molecular dynamics (MD) simulations and experimental analysis of the higher affinity of Na^+ over K^+ to protein surfaces.[6,15] The MD calculations show that sodium binds twice as strongly to protein surfaces as potassium, mostly due to the higher affinity of Asp and Glu residues to the smaller cation. We suggest that the different affinities of individual ions to specific residues on protein surfaces can explain the Hofmeister series. In our most recent study, we analyzed the influence of sodium and potassium ions on the activity of HIV-1 PR. We found that: (i) the enzymatic activity of the protease increases with increasing salt concentration, and this effect is stronger for K^+ than for Na^+, as shown experimentally by higher catalytic efficiency (k_{cat}/K_M); and (ii) MD simulations show that Na^+ is more strongly attracted to the protein surface than K^+, primarily due to interactions with the COO^- side chain groups of Asp and Glu.[6]

In the present study, we set out to extend and complete our previous work by analyzing the influence of other alkali metal ions on the proteolytic activity of HIV-1 PR.

Methods

Molecular dynamics simulations

MD simulations were started from equilibrated structures of HIV-1 PR (PDB access number 3HVP)[16] in mixed Na-KCl solutions at 1 M concentration (1 M Na^+, K^+, and 2 M Cl^-). As in our previous work,[6] the protonation of individual amino acids corresponds to pH 6. For the pair of catalytic aspartates, the protonation of a single carboxylate was considered, based on recently published studies.[17,18] The system had dimensions of approximately $72 \times 76 \times 74$ Å^3 and contained 11458 SPC/E water molecules,[19] 439 molecules of salt, and five chloride anions to compensate for the total charge of the protein (+5e).

Li^+ cations were introduced by substituting the positions of Na^+ and K^+ cations, resulting in 2 M LiCl solution (439 ion pairs). A system with 2 M CsCl salt was prepared *per analogiam*. To control for the quantitative effect of salt concentration, we performed additional simulations with CsCl at concentrations of 1 M (200 ion pairs) and 0.5 M (100 ion pairs). NaCl, KCl, and CsCl were parameterized as in our previous studies.[6,20] LiCl was parameterized according to van der Vegt.[21]

MD simulations consisted of minimization, heating to 300 K, 1 ns of equilibration, and 10 ns production runs at constant temperature and pressure (at ambient conditions) for the simulations of HIV-1 PR for 2 M salt solutions, and 20 ns production runs for 1 M and 0.5 M salt solutions to improve statistics. For consistency with our earlier work, the computations were carried out using the AMBER11 simulation package[22] employing the parm99 forcefield[23] with nonbonded interactions cut off at 12 Å. Smooth particle mesh Ewald (PME) method accounted for the long range electrostatic interactions.[24] Bonds involving hydrogen atoms were constrained using the SHAKE algorithm.[25] Such a simulation protocol yields converged results with respect to binding of ions to protein surfaces on the timescale employed.[6,15]

Analysis of simulation data

The resulting trajectories were analysed by means of the spatial and proximal distribution of ions in the vicinity of HIV-1 PR. We also monitored the number of coordinated cations (1st solvent shell) and the number of cations within the 2nd solvent shell of HIV-1 PR.

The spatial distribution function was calculated as the ratio of local (spatially resolved) and bulk density of ions and reaches unity far from the protein, *i.e.*, in the bulk region. It is directly related to the spatial free energy [$F(x,y,z) = -RT \ln g(x,y,z)$], and can be integrated (over two angular coordinates) into the more common radial distribution function. The important feature is that locally, close to the surface of HIV-1 PR, it exhibits sharp peaks (even sharper than those in the radial distribution function). Employing an isocontour, *i.e.*, cut through these peaks, we can visualize spatial regions in the vicinity of HIV-1 PR where cations occur with significantly enhanced probability.

Like van der Vegt,[21] we focused both on direct binding of cations to the protein surface (contacts, or the 1st layer in our previous study[6]) and on solvent shared binding (2nd layer). Due to the spread of cation sizes, the exact positions are based on the 1st and 2nd minima of the distribution functions. The major reason for considering the 2nd layer is the investigation of Li^+, which exhibits significantly different behaviour from other alkali cations.

Enzyme activity determination

Buffers and reagents. Buffer solutions were prepared by dissolving the appropriate buffer powder in HPLC-quality water and adding different amounts of NaCl, KCl, or CsCl (Sigma-Aldrich, anhydrous, beads, 99.999%) or LiCl (Sigma-Aldrich, anhydrous, beads, 99.998%). The appropriate hydroxide (NaOH or KOH, Sigma-Aldrich, trace metals basis, 99.99%; $LiOH \cdot H_2O$ or CsOH, Sigma-Aldrich, trace metals basis, 99.95%) was used to adjust pH.

Enzymes. The expression, refolding, and purification of HIV-1 PR have been described elsewhere.[26] Briefly, a pET24a plasmid vector encoding HIV-1 PR and host strain *E. coli* BL21 (DE3) (Novagen) were used for heterologous expression. HIV-1 PR was purified from inclusion bodies by solubilization in acetic acid, followed by chromatography on Mono S-Sepharose (Amersham Pharmacia Biotech). The active-site concentration of HIV-1 PR in the final preparation was determined by titration with the specific tight-binding inhibitor brecanavir,[27] and frozen aliquots of the enzyme were stored at −20 °C.

Kinetic analyses. All kinetic assays were performed with the chromogenic peptide substrate KARVNle*NphEANle-NH_2 in 50 mM MES, HEPES, or phosphate buffer, pH 6.0. Kinetic assays were conducted as previously described[28] with some modifications. Typically, 16 nmol of substrate was added to 1 ml of buffer containing 6–7 nmol of PR and various concentrations of salt. Substrate hydrolysis was followed by the decrease in absorbance at 305 nm using a Unicam UV 500 spectrophotometer. All enzymatic reactions were performed in duplicate. The K_M and k_{cat} values for the peptide substrate were calculated using the Michaelis–Menten equation (GraFit 5 software).

Results

Molecular dynamics of alkali ion interactions with the HIV-1 PR surface

The interaction of HIV-1 PR with lithium, sodium, potassium, and caesium cations was investigated by MD simulations of the protein in aqueous salt solutions. At the timescale of our calculations, no changes in the secondary structure or root-mean-square-deviation were observed. Thus, the simulations sample the ion distribution in the vicinity of freely moving HIV-1 PR in its native state. Frequent exchanges of ions and water are observed that ensure ergodicity of the system and good convergence of distribution functions.

The results of our simulations are presented as spatial distribution functions of alkali cations in Fig. 1. The data show the distribution of lithium, sodium, potassium, and caesium cations and chloride anion at the surface of HIV-1 PR.

Although HIV-1 PR is positively charged (+5 charge), there are several interaction sites (hot spots) that are occupied by cations. The potentially most relevant for the enzyme activity are located near the aspartic dyad (*i.e.*, below the flaps) and close to the entrance to the HIV-1 PR binding cleft,[6] where negatively charged amino acids are most abundant.

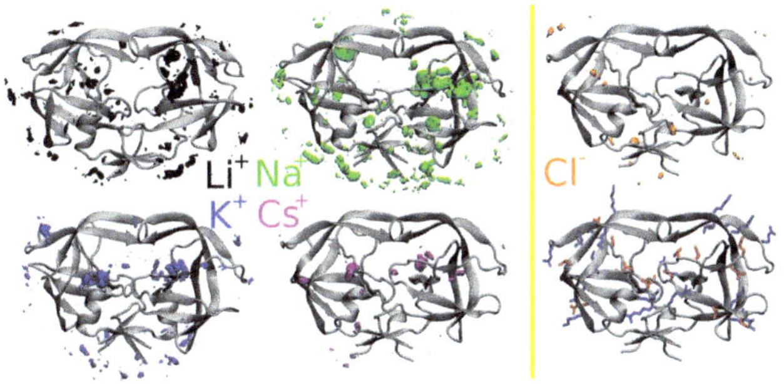

Fig. 1 Spatial distribution of lithium (black), sodium (green), potassium (blue), and caesium (magenta) cations and chloride anion (orange) at the surface of HIV-1 PR averaged over simulation runs. The PR surface is shown in ribbon structure representation, and the charged residues (positive in blue, negative in red) are visualized as thick lines in the bottom right of the Figure. The isocontours represent regions in which the ion concentration is equivalent to 10× that of the bulk value.

Li^+, Na^+, and to a lesser extent K^+ bind to the dangling carbonyls and terminal carboxyls located at the four-stranded antiparallel β-sheets formed by the intertwined N- and C- termini of HIV-1 PR.

The results summarized in Fig. 1 further suggest that the affinity of Li^+ and Na^+ is significantly stronger (larger clouds) and arguably also more complex (more clouds) than that of K^+ and Cs^+, which display smaller and less numerous isocontour clouds.

Cl^- does not exhibit strong binding to the surface of HIV-1 PR, even though the surface is overall positively charged. The very weak binding sites of Cl^- are spread over the HIV-1 PR surface and to a large extent correlate with positions of the positively charged amino acids.

In order to analyse the contribution of individual structural features of HIV-1 PR to the ion binding, we decomposed the surface, based on proximity criteria, into distinguishable groups (carboxyl groups, peptide bonds, and the remainder).[29] In Fig. 2, the running coordination numbers, the numbers of ions within distance r, of each cation are compared. The derivatives of coordination numbers in Fig. 2 reveal a clear difference between Li^+ and other alkali cations. Li^+ is the smallest cation studied, but there is only a negligible contribution of direct binding to the protease surface. Almost the entire contribution comes from the solvent-shared pairing. This feature is consistent with the strong hydration shell of Li^+ cation and is in agreement with recent studies.[21,30]

Sodium and larger cations are ordered in the Hofmeister series, and the contribution of the first solvent shell is more pronounced than that of the second. The number of sodium cations is roughly twice that of potassium, and caesium cations bind only slightly more weakly than potassium. Carboxyl and carbonyl groups are responsible for this ion specificity, while other functional groups do not distinguish significantly between the cations.

Two positions in the running coordination number plot are important, as they correspond to the number of ions in the 1st and 2nd solvent shells. The exact position differs for different cations (growing size), but can be identified as the minimum on the curve derivative. These positions, together with the number of cations within the 1st and 2nd shells, are summarized in Table 1. We again observe a marked difference between Li^+, which possesses almost no direct interaction with HIV-1 PR, and other cations studied, which are bound strongly (Na^+, 12 cations in the 1st solvent shell) or weakly (K^+, Cs^+ with 6 cations within 1st shell).

The coordination numbers reveal that Li^+ is the only cation that is found frequently at parts of the protein surface other than carbonyl and carboxyl groups.

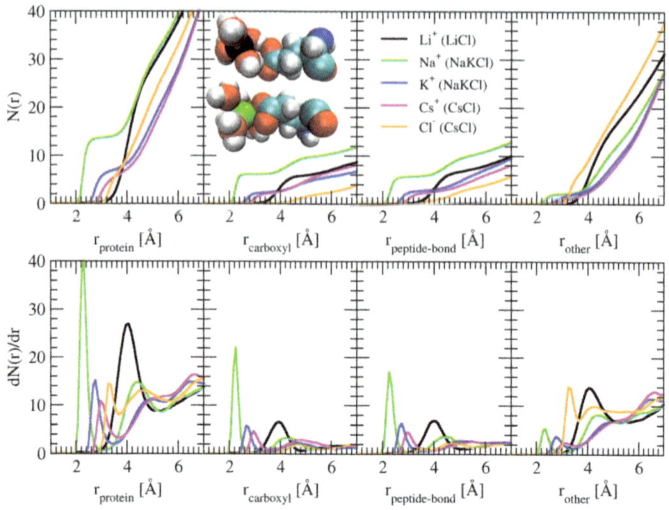

Fig. 2 The upper panel displays the running coordination numbers of alkali cations and chloride anions within distance *r* from the whole protein surface (far left). The dissection of the protein surface and corresponding coordination numbers of ions into three distinct groups – carboxyl (middle left), peptide-bond (middle right) and other groups (far right; positively charged, polar, and hydrophobic) – is also shown. The lower panel shows the simple derivative of the running coordination number and thus represents a not-normalized distribution function. Positions of the 1st and the 2nd solvation layer are defined by the 1st and 2nd minima in the distribution function. The inset shows solvent-shared pairing of lithium (black) and direct binding of sodium (green) to a glutamate side chain carboxyl group.

Table 1 Coordination number of cations involved in the direct and solvent-separated interaction with the protein surface and its components. Numbers are presented for cations within the 1st solvent shell (1SS), and within the 2nd solvent shell (2SS). The distances 2.8 Å and 5.2 Å were used for Li^+, 3.1 Å and 5.3 Å for Na^+, 3.4 Å and 5.5 Å for K^+, and 3.7 Å and 5.8 Å for Cs^+

	Li^+	Na^+	K^+	Cs^+
1SS – whole	0.2	13.6	6.7	6.3
2SS – whole	30.5	32.5	22.5	24.8
1SS – carboxyl	0.07	5.0	2.2	2.1
2SS – carboxyl	6.1	9.6	5.0	6.2
1SS – carbonyl	0.08	5.5	2.6	2.2
2SS – carbonyl	7.0	9.8	5.9	5.4
1SS – remainder	0	2.1	1.9	1.8
2SS – remainder	17.5	13.1	11.7	13

Activation of HIV-1 PR by alkali ions

The initial velocities of cleavage of a model chromogenic peptide substrate of HIV-1 PR were determined at various concentrations of sodium, potassium, lithium, and caesium ions. The peptide substrate used in this study is a consensus HIV-1 PR substrate derived from the processing site on the Gag polyprotein spanning the boundary between the capsid and nucleocapsid proteins. Due to the *para*-nitrophenyl residue introduced at the P1′ position of the substrate, the substrate's absorbance at 305 nm decreases upon hydrolysis by the protease. Comparison of the initial velocities of enzymatic hydrolysis of the substrate in 4-(2-hydroxyethyl)-1-piperazineethanesulfonic acid (HEPES) and phosphate buffers is given in Fig. 3. The plots show a monotonous increase in the initial velocities with increasing ionic strength for all the salts tested. KCl activates the enzymatic activity of HIV-1 PR most effectively, followed by NaCl and CsCl, which influence the enzymatic activity comparably. Somewhat surprisingly, lithium showed the least dependence of PR initial velocity on ion concentration. In order to analyze a possible substrate effect on the ion strength dependence, we performed similar analyses with another peptide substrate, 2-aminobenzoyl-TINleNphQR-NH_2. This fluorogenic substrate is based on the p2-nucleocapsid cleavage site of the HIV Gag polyprotein.[31] The initial velocities of this peptide hydrolysis follow a similar pattern as those shown in Fig. 3; namely, initial velocities observed in potassium-containing buffers are higher than those observed in the corresponding sodium-containing buffers.[6]

Very similar results were also obtained in 2-[bis(2-hydroxyethyl)amino]-2-(hydroxymethyl)-1,3-propanediol (BIS-TRIS) buffer, pH 6.0, as well as in acetate buffer, pH 4.7 (data not shown).

We observed different and somewhat surprising results in 2-(*N*-morpholino)ethanesulfonic acid (MES) buffer, pH 6.0 (Fig. 4). While the salt concentration

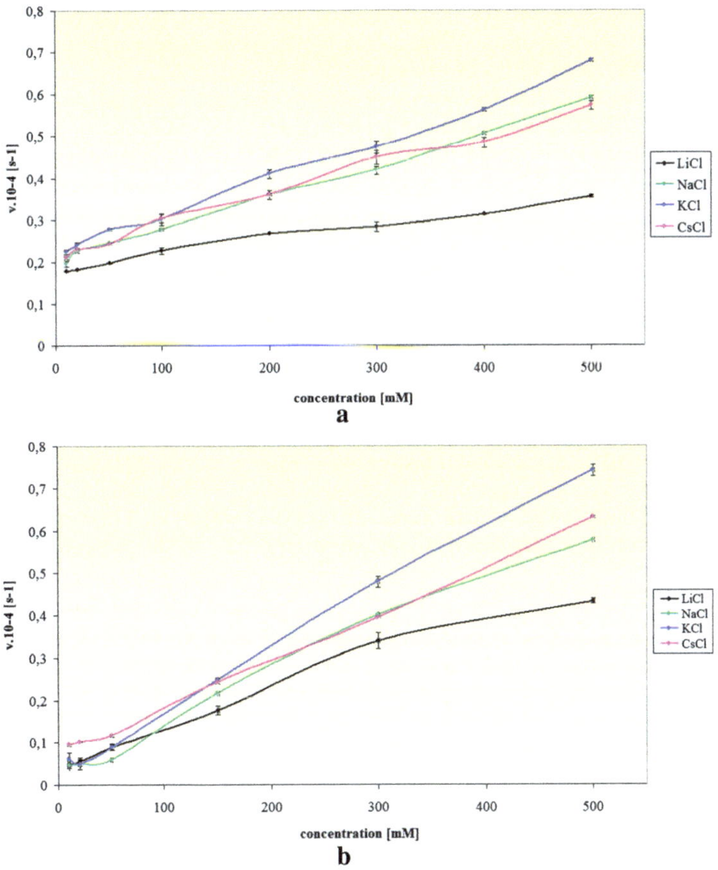

Fig. 3 Dependence of the initial velocities of the hydrolysis of the peptide substrate KARVNle*NphEANle-NH_2 by HIV-1 PR on increasing concentrations of alkali ions. The kinetic measurements were performed in 50 mM phosphate (a), or HEPES (b), pH 6.0. For details, see the Methods section.

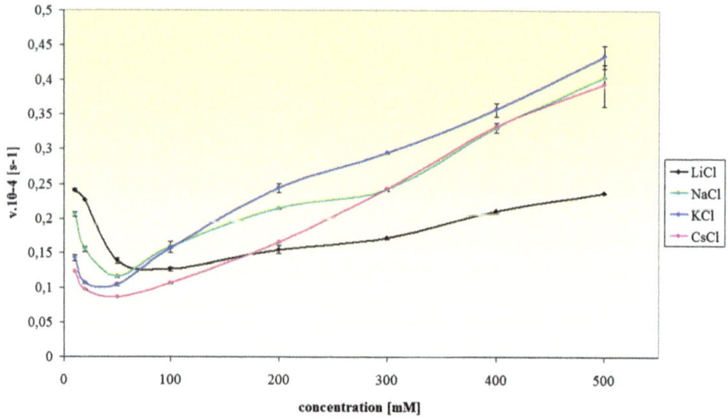

Fig. 4 Dependence of the initial velocities of hydrolysis of the peptide substrate KARVNle*NphEANle-NH_2 by HIV-1 PR on increasing concentrations of alkali ions. The kinetic measurements were performed in 50 mM MES, pH 6.0. For details, see the Methods section.

dependence in the higher concentration range follows a very similar pattern as in the other buffers, at very low salt concentrations (10–20 mM) we observed significant activation of the enzymatic activity of HIV-1 PR. No such activation was observed in any other buffer or with any other peptide substrate tested.

In order to quantify the dependence of peptide cleavage on the salt concentration, we determined the kinetic constants K_M and k_{cat} at various salt concentrations at pH 6.0 in phosphate (Table 2) and MES buffers (Table 3). The higher activity of HIV-1 PR in the presence of K^+ ions in comparison with other alkali ions was documented by higher catalytic efficiencies (k_{cat}/K_M) for both salt concentrations and both peptide substrates measured in phosphate buffer, pH 6.0. Increasing salt concentration leads to a drop in the K_M value for most ions analyzed. The higher catalytic efficiency in the presence of potassium compared with all other ions tested is due to both the substrate binding, as expressed by the lower K_M value, and a higher turnover number (k_{cat}). Interestingly, increasing the lithium concentration actually leads to an increase in K_M value, in contrast to all other cations analyzed.

The data summarized in Table 3 shed light on the mechanism of unexpected activation of HIV-1 PR in MES buffer at very low salt concentrations. Comparing the K_M values for 10 mM and 300 mM salt concentrations reveals that activation at low ionic strengths is entirely due to the dramatic increase in substrate binding, while the k_{cat} values actually increase with increasing salt concentration. The most profound activation effect at a very low salt concentration occurs in the presence of lithium

Table 2 Kinetic parameters of the cleavage of peptide substrate KARVNle*NphEANle-NH_2 by HIV-1 PR in phosphate buffer, pH 6.0, at various concentrations of alkali ions

Ion	K_M/μM	k_{cat}/s^{-1}	k_{cat}/K_M/s^{-1} mM^{-1}
LiCl 50 mM	170 ± 20	13 ± 1	79 ± 13
NaCl 50 mM	120 ± 20	10 ± 1	89 ± 14
KCl 50 mM	110 ± 20	16 ± 1	150 ± 30
CsCl 50 mM	160 ± 20	12 ± 1	73 ± 10
LiCl 300 mM	190 ± 20	17 ± 1	88 ± 14
NaCl 300 mM	86 ± 10	11 ± 1	130 ± 20
KCl 300 mM	41 ± 2	17 ± 1	400 ± 30
CsCl 300 mM	60 ± 3	12 ± 0.4	200 ± 10

Table 3 Kinetic parameters of the cleavage of peptide substrate KARVNle*NphEANle-NH_2 by HIV-1 PR in MES, pH 6.0, at various concentrations of alkali ions

Ion	K_M/μM	k_{cat}/s^{-1}	k_{cat}/K_M/s^{-1} mM^{-1}
LiCl 10 mM	2.5 ± 0.2	3.9 ± 0.1	1600 ± 200
NaCl 10 mM	10 ± 1	2.4 ± 0.1	240 ± 20
KCl 10 mM	19 ± 3	4.9 ± 0.7	260 ± 50
CsCl 10 mM	47 ± 6	6.1 ± 0.3	130 ± 20
LiCl 300 mM	180 ± 20	20 ± 2	110 ± 20
NaCl 300 mM	150 ± 10	16 ± 2	100 ± 20
KCl 300 mM	57 ± 5	14 ± 1	240 ± 30
CsCl 300 mM	67 ± 6	16 ± 1	240 ± 30

(note the approximately 70-fold difference in K_M value between 300 mM and 10 mM LiCl). On the other hand, the same ion is only a moderate activator of HIV-1 PR in the higher concentration range.

Discussion

Even though HIV-1 PR is one of the most well-studied enzymes, the regulation of its activity within the budding virus remains poorly understood. Specifically, the precise mechanism of activation of HIV-1 PR from the viral polyprotein upon HIV maturation is unclear. External factors like pH or salt concentration might contribute to regulation of this crucial step in the viral life cycle. Indeed, increased salt concentration improves the binding of peptide substrate to the HIV-1 PR binding cleft.[7,28] The common interpretation of this observation is that the substrate is "salted-out" to the rather hydrophobic binding cavity of the enzyme.[7,32,33]

In this study, we confirmed and extended our previous analysis of the influence of alkali cations on the activity of HIV-1 PR.[6] Our MD calculations confirm that the affinity of alkali cations to the protein surface follows the Hofmeister series. The experimental data (the dependence of the initial rates of HIV-1 PR substrate cleavage on alkali ion concentration) generally follow the reverse order of the Hofmeister series. This activity dependence is to a large extent due to improved substrate binding at higher concentrations of K^+ and Na^+ salts in comparison with Li^+. Interestingly, Cs^+ is an outlier: its influence on substrate cleavage is difficult to distinguish from that of Na^+. The explanation for this observation is not straightforward.

Fig. 1 and 2 point to the weaker affinity of chloride for the HIV-1 PR surface compared with sodium, even though the enzyme surface is positively (+5) charged. This phenomenon could be explained by two observations: (1) Cl^- exhibits weaker interactions with arginine or lysine residues compared with binding of alkali cations to aspartate or glutamate;[15,34] and (2) negatively charged amino acids are non-uniformly distributed on the protease surface and are often localized in pairs, which further strengthens the binding of cations. This does not violate the electroneutrality condition. However, the whole HIV-1 PR surface has to be taken into account. The running coordination numbers in Fig. 2 clearly show that in case of HIV-1 PR in CsCl salt solution, the electroneutrality condition is fulfilled already beyond 4 Å from the protein surface. Similar observations could be made for other salts (data not shown).

Na^+ and Li^+ occupy similar surface regions and show similar binding strengths (Fig. 1 and 2). However, the ion-protein affinities have different origins (Fig. 2 and Table 1). For sodium, direct binding dominates, while lithium gains its major contribution from solvent-shared pairs. The representative coordination for these two cases is shown in the inset of Fig. 2. This observation is in agreement with recent simulation results, which showed that solvent-shared pairs are the dominant contributors to the action of Li^+.[21] Note, however, that these conclusions depend to some extent on force field details, and it is more difficult to construct a satisfactory parameterization for Li^+ than Na^+ due to lithium's higher charge density and significant charge transfer to solvent. Lithium differs from other alkali cations in that it avoids less the remaining patches of HIV-1 PR (other than carboxyl/carbonyl groups). This may be due to the strongly bound water molecules that to some extent mask the high charge density of the small Li^+ cation. We can only speculate whether this behaviour of Li^+ is responsible for the experimental observation that increasing Li^+ concentration leads to weaker substrate binding, in contrast to the behavior of all other ions analyzed. Wondrak *et al.* reported a similar observation in their early work.[7]

The effect of caesium on the observed enzymatic activity is unexpected. The simulation results (Fig. 1) show that caesium exhibits the weakest binding to carboxylates in and near the binding cleft. The overall depletion (Fig. 2 and Table 1) of Cs^+ from the HIV-1 PR vicinity is, however, not weaker than that of potassium. The observed effects of caesium compared with potassium may be explained by the weak "salting-out" effect of Cs^+ together with its low affinity to carboxylates. These two counteracting effects explain why HIV-1 PR is most active in the presence of K^+. Na^+ and Cs^+ show comparable influences on the enzyme activity for different reasons, such as salting-out and binding to the protein surface. Interestingly, Wondrak *et al.*[7] also reported that the ability of caesium to activate HIV-1 PR lies in between those of potassium and lithium.

In our previous communication,[6] we proposed that the larger affinity of Na^+ over K^+ ions to the Asp dyad in the PR active site and at the entrance to the PR substrate binding cleft might lead to the modulation of the electrostatic potential of the PR surface at the active site entrance and thus influence the substrate/inhibitor recognition. Recent simulations by Friedman,[35] however, showed no appreciable binding of sodium cations to the singly protonated Asp dyad at physiological concentration (0.1 M) and only weak binding of sodium to the HIV-1 PR surface. The sodium binding reappeared only at elevated concentration (1 M). According to Friedman, this points to the importance of the concentration effect.

In this study, we performed calculations at non-physiological concentrations of 2 M in order to obtain converged data for the construction of the spatial distribution function. Because Friedman's work[35] opened the question of concentration effect on ion–protein binding, we set up additional calculations at 0.5 M and 1 M CsCl in order to analyze the role of salt concentration. The running coordination numbers and distribution functions are shown in Fig. S1† and suggest that the effect of the salt concentration is weak or negligible in the concentration range studied.

The direct comparison of our results with data obtained by Friedman[35] is not straightforward due to the choice of a different ion force-field and a different model of HIV-1 PR (2PC0 *vs.* our 3HVP). Friedman's model carries a charge of +7, while our model has a +5 charge. The Coulomb repulsion of cations from the surface increases with the growing net positive charge of the protein, which may to a large extent explain the different result obtained by Friedman. Moreover, the proper protonation state of the aspartic dyad as it is approached by a cation is disputable. Lund[36] recently suggested that the presence of a cation can induce deprotonation of the second aspartate (the pK_a is close to the experimental pH 6). In light of the present experimental and computational results, we propose that the stronger binding of cations to the negatively charged amino acid residues at the entrance of the PR substrate binding cleft and to the active site aspartate dyad might lead to the less favourable binding of the peptide substrate.

While analyzing the dependence of HIV-1 PR activity on the salt concentration, we observed an unexpected increase in the hydrolysis of the prototype peptide substrate KARVNle*NphEANle-NH_2 at very low salt concentration (below 30 mM) in MES buffer (Fig. 4). The activation is almost entirely caused by a dramatic, nearly 70-fold decrease in K_M, *i.e.*, tighter substrate binding. Very interestingly, the phenomenon is substrate- and buffer-specific: we did not observe similar activation for the hydrolysis of any other substrate and in any other buffer system investigated (compare Fig. 3). Even more surprisingly, the phenomenon could be observed only in the highest purity grade salts and buffers: the common p.a. buffers and salt preparations gave variable results at low concentrations. Very interestingly, the p.a. quality sodium chloride that shows monotonous salt concentration dependence for HIV-1 PR activity (*i.e.*, does not show an activity increase in the low salt concentration range) starts to show significant activation of HIV-1 PR after heat annealing (500 °C for 5 h). We speculate that MES forms a semistable complex with the KARVNle*NphEANle-NH_2 substrate in the HIV-1 PR binding cleft that significantly stabilizes the Michaelis complex at very low salt concentrations and that could be disrupted by higher salt concentration or by unidentified impurities in the commonly used grade of chemical reagents. The detailed mechanistic and structural explanation of this surprising observation will be the subject of further studies.

Acknowledgements

We would like to thank Pavel Jungwirth for inspiration and very fruitful discussions and the Grant Agency of the Czech Republic, projects P208/12/G016 and P207-11-1798 for support. The Institute of Organic Chemistry and Biochemistry is supported by Project RVO 61388963.

References

1 J. Pokorná, L. Machala, P. Řezáčová and J. Konvalinka, Current and novel inhibitors of HIV protease, *Viruses*, 2009, **1**(3), 1209–1239.
2 A. Wlodawer and J. W. Erickson, Structure-based inhibitors of HIV-1 protease, *Annu. Rev. Biochem.*, 1993, **62**, 543–585.
3 A. Wlodawer and J. Vondrasek, Inhibitors of HIV-1 protease: A major success of structure-assisted drug design, *Annu. Rev. Biophys. Biomol. Struct.*, 1998, **27**, 249–284.
4 J. Tözsér, Comparative studies on retroviral proteases: Substrate specificity, *Viruses*, 2010, **2**(1), 147–165.
5 W. Kunz, P. Lo Nostro and B. W. Ninham, The present state of affairs with Hofmeister effects, *Curr. Opin. Colloid Interface Sci.*, 2004, **9**(1–2), 1–18.
6 J. Heyda, J. Pokorná, L. Vrbka, R. Vácha, B. Jagoda-Cwiklik, J. Konvalinka, P. Jungwirth and J. Vondrášek, Ion specific effects of sodium and potassium on the catalytic activity of HIV-1 protease, *Phys. Chem. Chem. Phys.*, 2009, **11**(35), 7599–7604.
7 E. M. Wondrak, J. M. Louis and S. Oroszlan, The effect of salt on the Michaelis–Menten constant of the HIV-1 protease correlates with the Hofmeister series, *FEBS Lett.*, 1991, **280**(2), 344–346.
8 Z. Szeltner and L. Polgar, Conformational stability and catalytic activity of HIV-1 protease are both enhanced at high salt concentration, *J. Biol. Chem.*, 1996, **271**(10), 5458–5463.
9 J. J. Ha, D. A. Gaul, V. L. Mariani, Y. S. Ding, R. A. Ikeda and S. B. Shuker, HTLV-1 protease cleavage of p19/24 substrates is not dependent on NaCl concentration, *Bioorg. Chem.*, 2002, **30**(2), 138–144.
10 K. Inouye, K. Kuzuya and B. Tonomura, Effect of salts on the solubility of thermolysin: A remarkable increase in the solubility as well as the activity by the addition of salts without aggregation or dispersion of thermolysin, *J. Biochem.*, 1998, **123**(5), 847–852.
11 E. Di Cera, Thrombin: a paradigm for enzymes allosterically activated by monovalent cations, *C. R. Biol.*, 2004, **327**(12), 1065–1076.
12 M. J. Page and E. Di Cera, Role of Na^+ and K^+ in enzyme function, *Physiol. Rev.*, 2006, **86**(4), 1049–1092.
13 M. G. Cacace, E. M. Landau and J. J. Ramsden, The Hofmeister series: Salt and solvent effects on interfacial phenomena, *Q. Rev. Biophys.*, 1997, **30**(3), 241–277.
14 Y. J. Zhang and P. S. Cremer, Interactions between macromolecules and ions: The Hofmeister series, *Curr. Opin. Chem. Biol.*, 2006, **10**(6), 658–663.
15 L. Vrbka, J. Vondrasek, B. Jagoda-Cwiklik, R. Vacha and P. Jungwirth, Quantification and rationalization of the higher affinity of sodium over potassium to protein surfaces, *Proc. Natl. Acad. Sci. U. S. A.*, 2006, **103**(42), 15440–15444.

16 A. Wlodawer, M. Miller, M. Jaskolski, B. K. Sathyanarayana, E. Baldwin, I. T. Weber, L. M. Selk, L. Clawson, J. Schneider and S. B. H. Kent, Conserved Folding in Retroviral Proteases - Crystal-Structure of a Synthetic Hiv-1 Protease, *Science*, 1989, **245**(4918), 616–621.
17 K. Y. Nam, B. H. Chang, C. K. Han, S. G. Ahn and K. T. No, Investigation of the protonated state of HIV-1 protease active site, *Bull. Korean Chem. Soc.*, 2003, **24**(6), 817–823.
18 K. Wittayanarakul, S. Hannongbua and M. Feig, Accurate prediction of protonation state as a prerequisite for reliable MM-PB(GB)SA binding free energy calculations of HIV-1 protease inhibitors, *J. Comput. Chem.*, 2008, **29**(5), 673–685.
19 H. J. C. Berendsen, J. R. Grigera and T. P. Straatsma, The missing term in effective pair potentials, *J. Phys. Chem.*, 1987, **91**(24), 6269–6271.
20 R. Vácha, S. W. I. Siu, M. Petrov, R. A. Böckmann, J. Barucha-Kraszewska, P. Jurkiewicz, M. Hof, M. L. Berkowitz and P. Jungwirth, Effects of alkali cations and halide anions on the DOPC lipid membrane, *J. Phys. Chem. A*, 2009, **113**(26), 7235–7243.
21 B. Hess and N. F. A. van der Vegt, Cation specific binding with protein surface charges, *Proc. Natl. Acad. Sci. U. S. A.*, 2009, **106**(32), 13296–13300.
22 D. A. Case, T. A. D., T. E. Cheatham, III, C. L. Simmerling, J. Wang, R. E. Duke, R. Luo, R. C. Walker, W. Zhang, K. M. Merz, B. Roberts, B. Wang, S. Hayik, A. Roitberg, G. Seabra, I. Kolossvai, K. F. Wong, F. Paesani, J. Vanicek, J. Liu, X. Wu, S. R. Brozell, T. Steinbrecher, H. Gohlke, Q. Cai, X. Ye, J. Wang, M.-J. Hsieh, G. Cui, D. R. Roe, D. H. Mathews, M. G. Seetin, C. Sagui, V. Babin, T. Luchko, S. Gusarov, A. Kovalenko, and P. A. Kollman. *Amber 11*, University of California, San Francisco, San Francisco, 2010.
23 J. M. Wang, R. M. Wolf, J. W. Caldwell, P. A. Kollman and D. A. Case, Development and testing of a general amber force field, *J. Comput. Chem.*, 2004, **25**(9), 1157–1174.
24 U. Essmann, L. Perera, M. L. Berkowitz, T. Darden, H. Lee and L. G. Pedersen, A smooth particle mesh Ewald method, *J. Chem. Phys.*, 1995, **103**(19), 8577–8593.
25 J. P. Ryckaert, G. Ciccotti and H. J. C. Berendsen, Numerical-integration of Cartesian equations of motion of a system with constraints – molecular-dynamics of *N*-alkanes, *J. Comput. Phys.*, 1977, **23**(3), 327–341.
26 J. Weber, J. R. Mesters, M. Lepsik, J. Prejdova, M. Svec, J. Sponarova, P. Mlcochova, K. Skalicka, K. Strisovsky, T. Uhlikova, M. Soucek, L. Machala, M. Stankova, J. Vondrasek, T. Klimkait, H. G. Kraeusslich, R. Hilgenfeld and J. Konvalinka, Unusual binding mode of an HIV-1 protease inhibitor explains its potency against multi-drug-resistant virus strains, *J. Mol. Biol.*, 2002, **324**(4), 739–754.
27 R. Hazen, R. Harvey, R. Ferris, C. Craig, P. Yates, P. Griffin, J. Miller, I. Kaldor, J. Ray, V. Samano, E. Furfine, A. Spaltenstein, M. Hale, R. Tung, M. St. Clair, M. Hanlon and L. Boone, *In vitro* Antiviral activity of the novel, tyrosyl-based human immunodeficiency virus (HIV) type 1 protease inhibitor brecanavir (GW640385) in combination with other antiretrovirals and against a panel of protease inhibitor-resistant HIV, *Antimicrob. Agents Chemother.*, 2007, **51**(9), 3147–3154.
28 A. D. Richards, L. H. Phylip, W. G. Farmerie, P. E. Scarborough, A. Alvarez, B. M. Dunn, P. H. Hirel, J. Konvalinka, P. Strop, L. Pavlickova, V. Kostka and J. Kay, Sensitive, soluble chromogenic substrates for HIV-1 proteinase, *J. Biol. Chem.*, 1990, **265**(14), 7733–7736.
29 P. K. Mehrotra and D. L. Beveridge, Structural analysis of molecular solutions based on quasi-component distribution functions: Applications to [H_2CO]AQ at 25 °C, *J. Am. Chem. Soc.*, 1980, **102**(13), 4287–4294.
30 M. B. Gee, N. R. Cox, Y. Jiao, N. Bentenitis, S. Weeerasinghe and P. E. Smith, A Kirkwood–Buff derived force field for aqueous alkali halides, *J. Chem. Theory Comput.*, 2011, **7**(5), 1369–1380.
31 M. V. Toth and G. R. Marshall, A simple, continuous fluorometric assay for HIV protease, *Int. J. Pept. Protein Res.*, 1990, **36**(6), 544–550.
32 L. J. Hyland, T. A. Tomaszek and T. D. Meek, Human immunodeficiency virus-1 protease. 2. Use of pH rate studies and solvent kinetic isotope effects to elucidate details of chemical mechanism, *Biochemistry*, 1991, **30**(34), 8454–8463.
33 D. J. T. Porter, M. H. Hanlon, L. H. Carter, D. P. Danger and E. S. Furfine, Effectors of HIV-1 protease peptidolytic activity, *Biochemistry*, 2001, **40**(37), 11131–11139.
34 J. Heyda, T. Hrobarik and P. Jungwirth, Ion-specific interactions between halides and basic amino acids in water, *J. Phys. Chem. A*, 2009, **113**(10), 1969–1975.
35 R. Friedman, Ions and the protein surface revisited: Extensive molecular dynamics simulations and analysis of protein structures in alkali-chloride solutions, *J. Phys. Chem. B*, 2011, **115**(29), 9213–9223.
36 M. Lund, Electrostatic chameleons in biological systems, *J. Am. Chem. Soc.*, 2010, **132**(49), 17337–17339.

Dramatically stabilizing multiprotein complex structure in the absence of bulk water using tuned Hofmeister salts

Linjie Han, Suk-Joon Hyung and Brandon T. Ruotolo*

Received 3rd May 2012, Accepted 6th June 2012
DOI: 10.1039/c2fd20099f

The role that water plays in the salt-based stabilization of proteins is central to our understanding of protein biophysics. Ion hydration and the ability of ions to alter water surface tension are typically invoked, along with direct ion-protein binding, to describe Hofmeister stabilization phenomena observed for proteins experimentally, but the relative influence of these forces has been extraordinarily difficult to measure directly. Recently, we have used gas-phase measurements of proteins and large multiprotein complexes, using a combination of innovative ion mobility (IM) and mass spectrometry (MS) techniques, to assess the ability of bound cations and anions to stabilize protein ions in the absence of the solvation forces described above. Our previous work has studied a broad set of 12 anions bound to a range of proteins and protein complexes, and while primarily motivated by the analytical challenges surrounding the gas-phase measurement of solution-phase relevant protein structures, our work has also lead to a detailed physical mechanism of anion-protein complex stabilization in the absence of bulk solvent. Our more-recent work has screened a similarly-broad set of cations for their ability to stabilize gas-phase protein structure, and we have discovered surprising differences between the operative mechanisms for cations and anions in gas-phase protein stabilization. In both cases, cations and anions affect protein stabilization in the absence of solvent in a manner that is generally reversed relative to their ability to stabilize the same proteins in solution. In addition, our evidence suggests that the relative solution-phase binding affinity of the anions and cations studied here is preserved in our gas-phase measurements, allowing us to study the influence of such interactions in detail. In this report, we collect and summarize such gas-phase measurements to distill a generalized picture of salt-based protein stabilization in the absence of bulk water. Further, we communicate our most recent efforts to study the combined effects of stabilizing cations and anions on gas-phase proteins, and identify those salts that bear anion/cation pairs having the strongest stabilizing influence on protein structures *in vacuo*.

1. Introduction

Proteins are central molecular machines in the critical biological processes necessary for life. In many cases, these essential functions are regulated by protein structure, dynamics, and stability.[1] As such, a deep understanding of these properties has been sought by biochemists for well over a century. In that time, we have learned that many of the important biophysical properties of proteins can be influenced

Department of Chemistry, University of Michigan, 930 N. University Ave., Ann Arbor, MI 48109. E-mail: bruotolo@umich.edu; Fax: +734-615-3718; Tel: +734-615-0198

dramatically by the presence or absence of salts *in vivo*.[2–5] Indeed, such disparate biochemical properties as cell growth and protein crystallization have been directly linked to the influences of critical anions and cations.[5,6] In pioneering work, Hofmeister discovered that such salts can either stabilize or destabilize proteins differentially, and that empirical observations can be used to construct a generally predictive rank order.[7] Understanding the basic physical mechanism(s) that underlie the series that now bears his name has become an active area of research, due to its central importance in our understanding of protein biophysics.

While our knowledge of Hofmeister-type protein–salt interactions is still evolving, several experimental results, some of them fairly recent, have provided tremendous insight into the important aspects of their stabilization mechanism. Originally, the structure of bulk water, and its alteration through specific long range forces generated by anions and cations in solution, was thought to be critical for understanding Hofmeister-type protein stabilization.[8,9] Anions and cations were classified as either water structure makers (kosmotropes) or breakers (chaotropes), but recent experiments have indicated strongly that such structural effects are minimal at relevant solute concentrations, and thus have no direct causal relationship to Hofmeister-type protein stabilization.[10–14] Revised theories center on direct anion/cation interactions with proteins in three main ways.[15] First, anions and cations may directly interact with the protein backbone and side chains through ion pairing interactions.[16,17] Anions are known to have high affinity for amino functional groups within proteins,[18–21] and cations are likely to interact with an array of sites,[22,23] in many cases involving carboxylate groups.[20,24,25] Arguably more important, in light of current data, are the more-indirect interactions between Hofmeister ions and the layer of water closely associated with proteins. Anions and cations can alter both the surface tension and hydrogen-bonding network surrounding proteins significantly, such that hydration entropy and protein stability are dramatically affected.[26–29] Although canonical Hofmeister series are operative in many cellular processes, reversed Hofmeister series have also been observed, and rationalized through alterations to local water structures and direct protein–ion interactions, as above.[15,30–32] Thus, while local protein–water interactions have been deemed important in Hofmeister stabilization, the magnitude of their importance is a somewhat malleable concept and subject to change based on the specific process or proteins being studied.

Based on the potential importance of solvent in Hofmeister-type protein stabilization, a number of groups have undertaken experiments carried out in environments of rarefied solvation (*e.g.*, the gas-phase) to study both the local water structure surrounding small ions, and their interactions with proteins. Many of these experiments have been carried out using mass spectrometry (MS), where shifts in ion molecular mass can be interpreted relative to direct protein–counterion binding in solution.[20,21,33–36] For example, precise measurements of molecular mass allowed Kebarle and co-workers to define the binding mode of many anions and cations to specific proteins.[20] More recently, MS conditions have been identified whereby anion binding observed in the gas-phase can correlate precisely to the number of solvent accessible basic sites on a protein, thus mirroring bound populations in solution.[21] Similarly, wavelength-resolved action spectroscopy, using MS detection, has been utilized to deduce the relative population of charge-solvated and zwitterionic structures in a range of amino acid–cation complexes.[37–43] Similar measurements of anions and cations clustered with varying amounts of water have also been used to deduce ion specific effects on both local and bulk water structure.[44–51] While these data have demonstrated significant local water structure effects for specific ions, in some cases extending out to long ranges, they have also shown that bulk water structure is largely unaffected by cations and anions in solution at relevant concentrations.

The desolvated structures of proteins can also be measured using ion mobility (IM) spectrometry,[52,53]where ions are separated according to their orientationally-averaged collision cross sections (CCSs) by measuring protein transit times through a gas-filled chamber under the influence of a relatively-weak electric field. IM data has been used to study the general dependence of protein structure as a function of bound solvent, revealing both the complexity of protein structural states that exist in the gas-phase and the level of solvation necessary to compact Coulombically unfolded proteins.[54–56] In addition, IM data have recently been used to deduce the influence of a range of anions and cations on protein stability and structure in the absence of bulk solvent. For example, anion binding can result in the recovery of compact protein structures when the unbound gas-phase protein ions occupy unfolded states in the gas-phase.[35] Similarly, our group has demonstrated the differential effects on desolvated protein ion stability for various cations and anions, showing that the identity of condensed salt can stabilize gas-phase protein structure through mechanisms vastly different from those found in solution.[34,36] In all of these cases, gas-phase data have served both to inform on the importance of solvent in protein structure and also highlight examples where such structures can be conserved in the absence of bulk solvent.

The correlation between solution and gas-phase protein structure has, in part, driven the development and application of IM-MS in structural biology. For example, MS measurements have been used to study bioactive peptide aggregation,[57] membrane protein structure,[58,59] and protein stability changes upon ligand binding.[60,61] The application of IM and MS to protein quaternary structure has rapidly developed in recent years, enabling the determination of multiprotein stoichiometry, dynamics, and 3D topology.[62–67] In IM-MS experiments aimed at the determination of protein quaternary structure, detailed MS measurements of protein connectivity are combined with equally detailed IM measurements of gas-phase protein size in order to generate topology models.[67] As such, the above-mentioned correlation between solution and gas-phase protein size and structure has become a critical component in protocols aimed at deducing multiprotein topology. Our group aims to use Hofmeister salts to stabilize protein complex structure in the absence of solvent, and our efforts to deduce the mechanisms driving cation and anion-mediated protein stability in the gas phase have primarily been motivated by our work to develop general IM-MS methods for determining protein assembly structure.

Here, we describe and summarize our previous results where we have used Hofmeister anions and cations as gas-phase stabilizers for protein structure. After providing a detailed account of the IM-MS methods we have developed to measure multiprotein complex stability, we then discuss the evidence to date illuminating the operative mechanisms of cation and anion stabilizing proteins in the gas phase. Anions primarily stabilize gas-phase proteins through a 'dissociative-cooling' mechanism that requires bound anions to dissociate from complexes to carry away excess energy from the system,[34] and stabilizing cations often bind tightly to tether regions of the protein together and limit charge mobility.[36] While our previous results focus primarily on measuring the influence of isolated anions and cations on protein structure, we also describe new experiments in which we seek to combine the influence of both bound anions and cations on gas-phase protein structure simultaneously. These results demonstrate that both anions and cations can be used in concert to stabilize protein structures in the absence of bulk solvent to an extent not previously accessible using either alone. We conclude by discussing the general relevance of gas-phase protein stability measurements for Hofmeister stabilization in solution, and emphasize the role that such measurements can play in elucidating the role of solvent in protein–salt interactions.

2. The IM-MS approach for measuring the stability of desolvated protein structure

2.1 A multidimensional measurement of protein stability

Our experimental approach for measuring the stability of gas-phase multiprotein complexes employs IM-MS to make rapid serial measurements of protein size and mass as a function of experimental variables. The measurements described in this report were carried out on a Synapt G2 HDMS (Waters, Milford, MA) platform.[68–70] The instrument is equipped with a quadrupole mass analyzer, a travelling-wave ion mobility separator, and a time-of-flight (ToF) mass analyzer arranged in tandem.[69] In order to generate useful information regarding the structure of proteins, the non-covalent contacts and course-grained structures of protein ions must to be maintained during their transfer to the gas phase. Nano-electrospray ionization (nESI, typically using ~5 μL of sample, operating in the positive ion mode) is used in our experiments for both its broad tolerance to aqueous salt concentrations and its comparatively direct route in generating gas-phase protein ions from such solutions. Emitters were prepared as previously described,[71] and ions were generated using capillary voltages ranging from 1.4–1.7 kV. The sampling cone voltage and source pressures were also optimized in order to transmit intact protein complexes and to preserve non-covalent interactions, as described previously.[71]

Ionized proteins and complexes are introduced into the vacuum of the IM-MS instrument *via* a series of differentially pumped regions, each having tightly controlled pressures and temperatures such that volatile solution additives are gradually removed to limit unwanted ion heating. Upon arrival at the quadrupole mass filter, ions of interest can be selected prior to IM-MS analysis based on their *m/z*. In this tandem-MS mode, ions in a narrow *m/z* range are selected and accumulated in an ion trap region situated in front of the IM separator (containing 3.3×10^{-2} mbar of argon gas). Each mass-selected ion can be activated by increasing the acceleration voltage used to introduce ions into the ion trap just prior to the mobility separator. Increases in this acceleration cause controlled ion heating, which in turn causes multiprotein complex ions to first unfold and then dissociate into both highly-charged protein monomers and complexes stripped of individual subunits. The monomers produced by this collision induced dissociation (CID) process are highly-charged primarily because of the relatively unfolded transition states occupied by the heated ions during the preceding collision induced unfolding (CIU) process.[72,73] For the CID and CIU data shown in this report, trap collision voltage (the accelerating potential described above) was varied in a stepwise fashion while keeping other settings constant in order to construct collision energy-dependent datasets.

The IM separator within the instrument utilizes a series of low-voltage 'waves' that 'travel' at relatively low speeds by altering the potentials on a series of individually addressable electrode rings within the device (typical values are 40 V wave amplitudes, traveling at 800–1000 m s^{-1}).[68] Under high-vacuum conditions, all ions are transported through the device at the velocity of these traveling waves. To undergo IM separation, the device is pressurized with ~3.5 mbar N_2 gas such that the drift times of large ions, which collide frequently with the neutral gas, are slowed relative to ions having a smaller size, or CCS. A single IM separation typically takes ~30 milliseconds during which 200 ToF datasets can be recorded over a broad *m/z* range. Drift time measurements using a traveling-wave type IM separator are typically calibrated externally using a series of proteins having known gas-phase sizes to generate CCS values for unknowns. It should be noted that the general approach described above is not exclusive to the Synapt instrument geometry. In fact, many previous reports of monomeric protein unfolding have been described in the literature using both drift-tube ion mobility spectrometry (DT-IMS) [74–76] and field asymmetric IM (FAIMS)[77] coupled to MS detectors.

The IM-MS approach that we have used to measure the stability of desolvated proteins and their complexes is illustrated in Fig. 1, which shows a basic schematic diagram of the inner-workings of the IM-MS instrumentation. Our illustration depicts two simplified ion classes: those ions that lack stabilizing salts bound to their exterior (blue), and those proteins bound to a large numbers of such stabilizers (red). After creation through nESI, IM analysis of stabilized protein complexes exhibit narrow, monomodal IM drift time distributions (Fig. 1B, red). In contrast, less stable protein complex ions unfold prior to IM analysis and exhibit broad, multimodal drift time spectra (Fig. 1B, blue). The clear differences in the IM arrival time distributions recorded for these two ion classes allows us to infer that protein ions better maintain their compact states when bound to stabilizing salts (Fig. 1A and 1B). MS spectra for these same two protein ion classes, recorded on a ToF mass analyzer (*m*/*z* range of 800–15000 operated at a pressure of 1.6 $\times$ 10^{-6} mbar), can be used to determine the composition of the complex and verify protein binding stoichiometry. When observed at relatively low activation energies, the measured masses of protein ions bound to salt adducts are significantly larger than the known sequence mass of the assembly, as illustrated by the higher *m*/*z* values observed for those ions in Fig. 1A and 1C. While this excess can be quantified using precise mass measurements and comparison with control samples, ion heating can also be used to drive the evaporation of salt adducts from protein ions, reducing their masses toward that predicted by sequence alone.[54]

2.2 Preparing multiprotein–salt complexes for IM-MS measurements

Our approach involves the protection of protein complex structure through the addition of salts in solution prior to nESI. We screened a series of Hofmeister-type cations and anions for their ability to stabilize multiprotein complexes in the absence of bulk solvent. The proteins studied herein include the protein tetramers avidin (egg white), transthyretin (TTR, human), concanavalin A (ConA, jack bean), and alcohol dehydrogenase (ADH, yeast) as well as the β-lactoglobulin A dimer (BLA, bovine) purchased from Sigma (St. Louis, MO, USA). All protein samples were buffer exchanged into 100 mM ammonium acetate at pH 7 using Micro Bio-Spin 6 columns (Bio-Rad, Hercules, CA) and prepared to a final concentration of 5 μM (avidin, TTR, ConA, ADH tetramers) or 10 μM (BLA dimer).

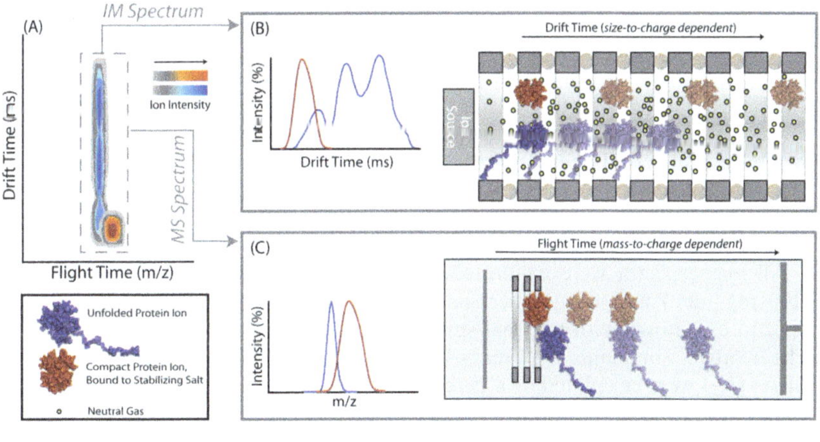

Fig. 1 A schematic diagram of the IM-MS experiment, in which collisional heating of two protein complexes has generated different CIU responses. A) IM-MS data is multi-dimensional, consisting of IM drift time (in ms), *m*/*z* (measured in a time-of-flight mass analyzer), and ion intensity (shown on a color axis, different for the two ion populations). Signal corresponding to stabilizing salt-bound protein is illustrated on a red color axis, and a blue color axis corresponds to signal from protein ions in the absence of stabilizing salt adducts. B) Equally-charged protein ions are separated in a travelling-wave type IM separator based on their conformational differences following CIU. Protein ions bound to stabilizing salts travel more quickly in the IM separator when compared to non-stabilized protein ions due to their more-compact, folded geometry. IM data for unfolded proteins typically reveal a multimodal drift time profile, indicative of multiple unfolded intermediate states, stable on the time scale of the experiment, created during the CIU process. C) Subsequent mass analysis of ions using a time-of-flight mass analyzer reveals that salt molecules remain bound to protein ions during the analysis, resulting in an increased molecular mass for stabilized complexes.

We began our experiments by assessing the stabilizing effect of isolated cations and anions on gas-phase protein structure by adding salts where cations (acetate anion with ammonium, tetramethylammonium (TMA), sodium, potassium, rubidium, lithium, Tris (2-Amino-2-hydroxymethyl-propane-1,3-diol), calcium, barium, and magnesium counterions) or anions (ammonium cation with acetate, fluoride, chloride, nitrate, citrate, thiocyanate, bicarbonate, tartrate, iodide, hydrogen phosphate, sulfate and perchlorate counter-ions) were altered specifically so that their stabilizing influence could be evaluated individually. In later experiments, salts of varying anion and cation content are added to solution to observe the combined effects of simultaneous anion and cation adduction on protein ion stability. All salts were purchased from Sigma (St. Louis, MO, USA) and prepared in aqueous solution prior to nESI. In order to study the influence of different salts on protein stability without significantly altering buffer capacity or solution pH, the salts were prepared as stock solutions in 100 mM ammonium acetate at a concentration of 20 mM, each of which was then added to protein solutions. Final solutions contained added salt concentrations of less than or equal to 5 mM (2 mM for avidin, TTR, ConA, ADH and 0.5 mM for BLA samples for cation-based studies, 5 mM for avidin, TTR, ConA, ADH and 0.2 mM for BLA samples for anion-based studies). While these concentrations are much lower than that used for deriving Hofmeister-ion effects on proteins in solution (>300 mM), and selected for our experiments primarily to avoid ion suppression effects, the mechanics of nESI desolvation allows all the samples tested to achieve very high relative salt concentrations prior to ionization.

2.3 IM-MS data analysis techniques for the measurement of desolvated protein stability

Our IM-MS measurements of protein complex stability are all performed in a tandem-MS (quadrupole selection) mode. Ions were selected for stability measurements in the quadrupole mass filter at an *m/z* corresponding to the 16^+ charge state of Avidin, the ConA 19^+ state, the TTR 14^+ state, the ADH 24^+ state for the tetramers and the 11^+ state for BLA dimers. Charge states were chosen based on their intensity across each solution state studied, and control IM data were screened for evidence of overlapping non-tetrameric or non-dimeric ions at the same *m/z* values respectively. Each of these mass-selected ions was activated by increasing the trap collision voltage. For all the protein–salt systems investigated here, trap collision voltages were incremented in 5 V steps. Upper voltage limits were considered to be where no further CID was observed.

All mass spectra were calibrated externally using a solution of cesium iodide (100 mg ml^{-1}) and were processed with Masslynx 4.1 software (Waters, UK). Spectra are shown with minimal smoothing and without background subtraction. The relative abundances of mass-selected tetrameric ions (or dimeric ions for BLA) (I_{tet}) were calculated as a percentage of the total intensity for all the signals observed in the MS data corresponding to either intact protein complexes or identified product ions (*i.e.*, monomer or stripped protein complexes). The relative abundances of the compact form observed for tetrameric (or dimeric for BLA) ions separated in IM experiments (I_f) were calculated as a percentage of the total intensity of the peaks observed in the arrival time distribution at a selected *m/z* value corresponding to intact tetramer (or dimer for BLA):

$$I_{tet}(\%) = \frac{I_{tet}}{I_{tet} + I_{mon}} \times 100 \quad (1)$$

$$I_f(\%) = \frac{I_{folded}}{\sum I_{conformers}} \times 100 \quad (2)$$

The average relative standard deviation for the determination of either I_{tet} (%) or I_f (%) is 2–4%. Some figures contain axes labeled in collision energy (units of eV*). This axis is a normalized version of ion kinetic energy, which takes into account both the ion charge and the reduced mass of the ion-neutral collision complex, for making comparisons across large mass ranges.[34]

Typical protein CIU and CID data are shown in Fig. 2, using the 64 kDa avidin tetramer as an example. The IM-MS dataset at low activation energies shows well-resolved peaks primarily corresponding to intact and folded tetrameric protein complex ions having 14^+ to 17^+ charges (Fig. 2A, red). When the most intense charge state, the 16^+ tetramer, is selected in the quadrupole mass filter, the ions are resolved in the IM separator as a single narrow peak requiring ~20 ms of drift time to traverse the IM separator (Fig. 2C, red). When the ions are activated by increasing the trap collision energy to 65 V, the IM-MS dataset exhibits substantially different features. For example, the IM data for the same tetramer ions are now resolved as multiple conformational species populating significantly longer drift times (Fig. 2C, blue). The appearance of new features is also prominent at both lower and higher *m*/*z*, corresponding to the 7^+ to 9^+ charge state avidin monomer ions and 8^+ to 10^+ charge state avidin trimer ions respectively (Fig. 2B, blue). Thus, CIU and CID information can be obtained in a single dataset, and carefully monitoring this information over a wide range of collision energies enables a quantitative measure of protein quaternary structure stability (through CID) and subunit tertiary structure stability (through CIU).

To ease the interpretation of desolvated protein stability, we typically reduce IM-MS measurements to histograms that plot the relative structural integrity of the complexes observed against additives used for protein stabilization. This procedure is illustrated with data acquired from measurements of the avidin tetramer incubated in three different buffers (control/100% 100 mM ammonium acetate,

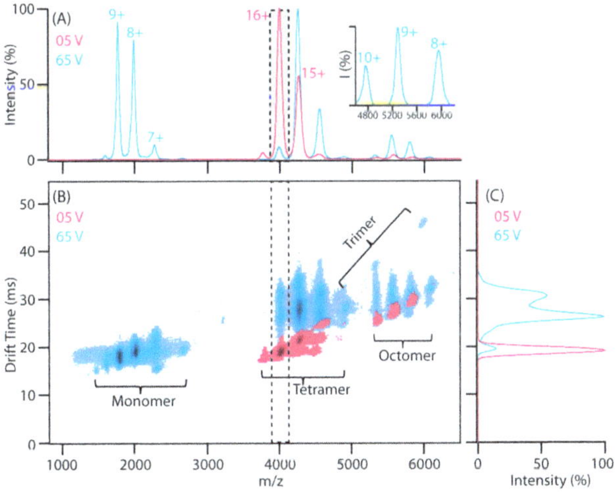

Fig. 2 Example CIU and CID data for the avidin tetramer. A) Mass spectra of avidin acquired using instrument conditions that either preserve (trap collision voltage: 5 V, red), or activate protein complex ions (65 V, blue). Peaks corresponding to the 14–17^+ charge states of the avidin tetramer and the 7–10^+ charge states of the monomer are shown. The inset shows peaks corresponding to the 8–10^+ charge states of the avidin trimer. B) Contour plots of IM drift time *versus m*/*z* acquired at a trap collision voltage of 5 V (red) and 65 V (blue). Major peaks from the charge state series corresponding to monomeric and tetrameric avidin are labeled. C) IM drift time data for only the selected 16^+ tetramer ions reveals compact ions at low levels of ion activation (red), and extended tetrameric ions generated under activating conditions (blue).

100 mM ammonium acetate with 5 mM added ammonium fluoride, and 100 mM ammonium acetate with 5 mM added ammonium nitrate). MS for avidin bound to nitrate anions reveal a broadened mass profile consistent with a statistical distribution of bound adducts (Fig. 3A). In contrast, the MS data for avidin incubated in non-stabilizing fluoride-based salts is indistinguishable from control data acquired from sample buffers comprised entirely from salts containing highly volatile acetate anions. This result is consistent with previous observations indicating that nitrate binds to protein in large numbers.[34] The tandem MS data for 16^+ avidin tetramer, acquired under elevated trap collision energy, shows that avidin incubated with stabilizing nitrate anions yields significantly less monomer signal upon activation, compared to both fluoride-added and control datasets (Fig. 3B). Likewise, the IM arrival time distribution of the same charge state reveals that protein incubated with nitrate consists primarily of a relatively narrow peak at 20 ms. By contrast, avidin incubated with fluoride anions populates longer drift times (~25 ms), indicating that these ions adopt significantly extended conformations by comparison (Fig. 3C). Thus, this data demonstrates how anion binding can dramatically influence the stability of protein complexes in gas phase.

For quantitative measurements of protein stability, the trap collision voltage at which ions undergo CIU and CID is monitored, and plots of collision voltage *versus* the intensity observed for compact (I_f) and intact (I_{tet}) tetramer ions are recorded

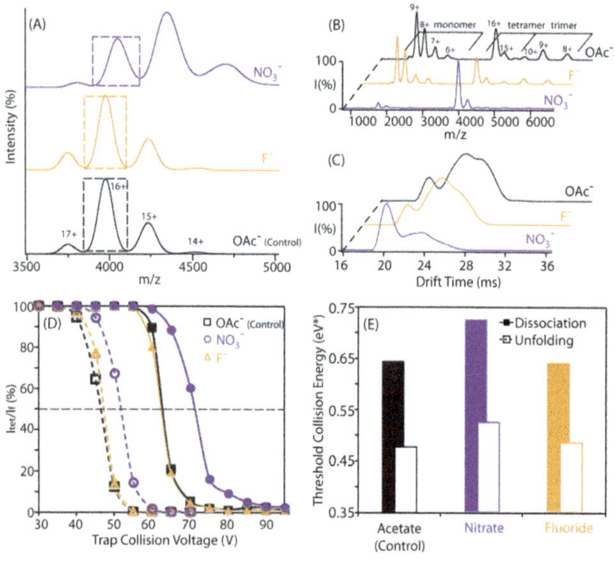

Fig. 3 Addition of anions in solution alters the dissociation and unfolding profiles of protein assemblies. A) Example MS data for the avidin tetramer incubated with anions (nitrate, fluoride, acetate) acquired at instrument conditions that preserve the tetrameric assembly. The peak corresponding to 16^+ avidin is marked with a dashed box, and corresponds to those ions selected from this primary MS dataset for CID (B) and CIU (C) stability analysis, which use energy resolved datasets from MS and IM respectively to deduce the influence of selected anions on protein complex stability in the gas-phase. D) Plots of the relative intensities for intact (I_{tet}, solid lines) and compact (I_f, dashed lines) avidin tetramer 16^+ ions are shown as a function of trap collision voltage. Avidin ions were generated using solutions containing either added nitrate (circle), fluoride (triangle), or acetate anions (control, square). The energy at which the relative intensity of I_{tet} or I_f reduces to 50% of its original value is marked with a horizontal dashed line. E) A histogram showing the energy at which 50% dissociation (filled) and unfolding (open) occurs for avidin tetramers generated from solutions having the indicated additives.

(Fig. 3D) while keeping the other parameters in the experiment constant. Both I_f and I_{tet} decrease as the collision voltage is increased, and I_f is observed to decrease at lower voltages than I_{tet} for ions generated from all three buffer compositions. We also observe that the addition of ammonium nitrate to sample solutions (circles) increases the voltage values at which I_f and I_{tet} decrease, and that this change is significantly greater than that observed for the addition of ammonium fluoride (triangles), relative to control samples (squares). This result is more-clearly illustrated through a comparison of the normalized collision energy at which I_f and I_{tet} decrease to 50% of their original intensity (Fig. 3E). This general workflow is used throughout our experiments in order to quantitatively compare the influence of Hofmeister anions and cations on the stability of desolvated protein ions in the gas phase.

3. Gas-phase proteins and their complexes are stabilized by cationic and anionic additives through different mechanisms and to different extents

Using the protocol illustrated in Fig. 3, we have developed a classification system that allows us to generally rank cations and anions in terms of the stability they afford to gas-phase protein ions (Fig. 4A). Our data suggests that cations (red) influence the unfolding and dissociation processes of protein complexes to similar degrees. Bound cations increase the threshold dissociation/unfolding energy in the order: $NH4^+ \approx TMA^+ < Rb^+ < K^+ < Na^+ < TrisH^+ < Ba^{2+} \approx Li^+ < Ca^{2+} < Mg^{2+}$, progressively stabilizing gas-phase proteins to greater degrees. In contrast to cations, bound anions (blue) tend to stabilize protein complex CID and CIU to different extents. Therefore rather than a clearly-defined rank order, our IM-MS data reveals that anions can be categorized into three distinct groups. Tartrate^{2-}, Cl^-, citrate^{2-} and NO_3^- are among the most efficient stabilizers of protein unfolding and dissociation, populating a 'highly-stabilizing' cluster. In contrast, HCO_3^-, I^- and ClO_4^- provide little or no additional stability to protein complex ions, and are thus regarded as 'weakly-stabilizing' salts in the gas-phase. The remaining anions including SO_4^{2-}, HPO_4^{2-}, SCN^-, and F^- form a final 'medium-stabilizing' cluster.

For a more quantitative comparison of the correlation between gas-phase unfolding and dissociation found in our IM-MS data, linear regression analysis was used to derive residual plots and correlation coefficients for the anion and cation data shown in Fig. 4A. Fig. 4B and 4C show these data, and support our earlier assertion regarding the superior linearity of cation-based IM-MS stability data when compared to anion data, resulting in average residual sum of squares (RSS) values of 0.004 and 0.057 respectively. Furthermore, this analysis is also consistent with the correlation coefficients calculated for the same data (Fig. 4B, caption), therefore suggesting a link between local and global protein stabilization in the case of cation additives that is absent for anions. In addition, cationic additives seem to stabilize protein complex ions against CID to a greater extent, on average, than equivalent anions while anionic adducts are, in general, better stabilizers of gas-phase protein unfolding.

The above differences between anionic and cationic stabilizers allow us to construct separate mechanistic descriptions of their action (Fig. 4D). For anion-based stabilization, we can classify adducts into three categories based on both their protein binding affinity and ability to dissociate from proteins complexes following activation in the gas-phase. It is important to point out that these three anion classes, while related, are not the same as those presented in Fig. 4A. The class that includes Cl^-, tartrate^{2-} and NO_3^- exhibits a strong stabilizing influence on protein structure (green track, Fig. 4D). They can bind in large numbers and readily dissociate from the protein surface after relatively minimal activation. The observed dissociation of

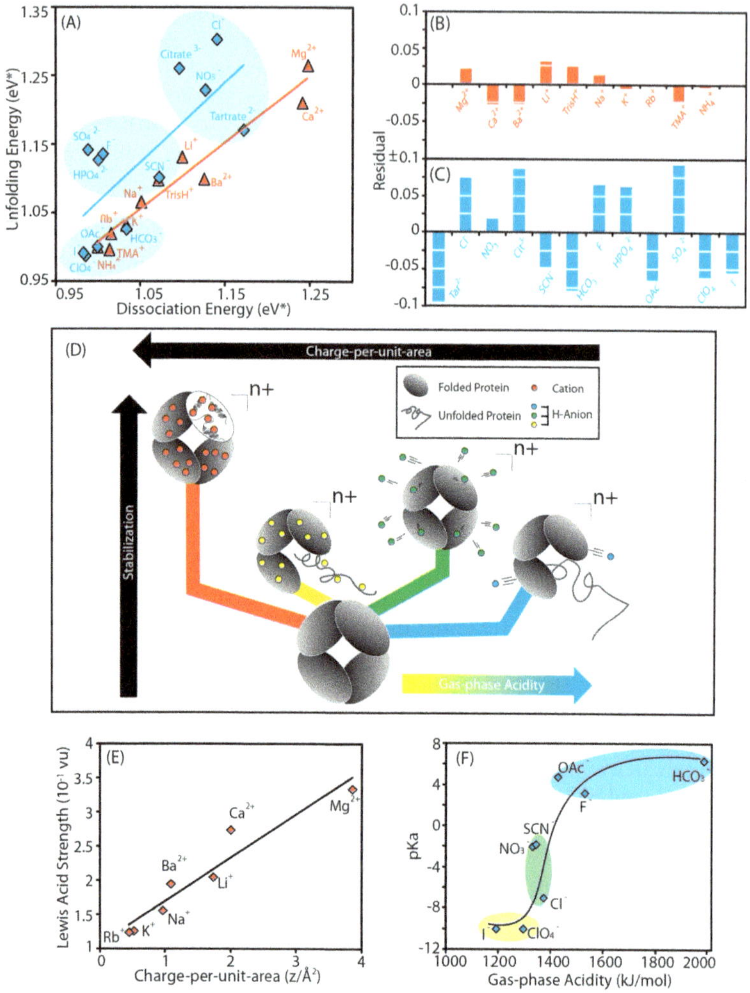

Fig. 4 Comparison of anion and cation-mediated stabilization afforded to gas-phase protein ions. A) A plot of the average CID *versus* average CIU energies (eV*) for the 5 protein complexes studied herein for cation (red) and anion (blue) additives. The plot reveals that the anions can be categorized into three distinct groups (blue background color), according to their ability to stabilize protein complexes relative to the control data set (OAc^-). Both B) and C) show residual plots for stability data from best-fit linear relationships between CIU and CID energies, for cation and anion datasets respectively. The correlation between unfolding and dissociation stabilization is much higher for our cation data ($R^2 = 0.97$) relative to equivalent data recorded for anion additives ($R^2 = 0.55$). D) A diagram depicting our current mechanistic understanding of gas-phase protein structure stabilization through bound cations and anions. Anions with mid-range acidities bind the protein with high affinity and are released upon dissociation leading to high protein structural stability in the gas phase through dissociative cooling (green track). Anion additives that do not bind, nor dissociate from gas-phase protein ions do not provide significant structural stabilization (blue and yellow track). In contrast, high charge-per-unit-area cations, having much greater surface charge densities than any of the anions tested to date, bind in large numbers to protein complexes and remain bound to become less-mobile as charge carriers, thus affording increased stability enhancement to gas-phase protein structure (red track). E) A plot of the charge-per-unit-area of cations against Lewis acid strength reveals a high level of correlation to the best fit line (shown). F) A plot of the gas-phase acidity of anions against pK_a reveals a positive correlation between the two parameters. The data are color-coded to indicate the membership of each anion in the three relevant mechanistic tracks depicted in Fig. 4D, and a trend line is added to guide the eye.

anion-based adducts corresponding to [H-"anion"] type neutrals acts to carry away excess rotational and vibrational energy from gas-phase protein ions, thus abating any dramatic internal energy increases for the protein and allowing it to retain a compact, native-like structure.[34] In contrast, anions that do not bind (HCO_3^- and F^-), nor dissociate from gas-phase protein ions (I^- and ClO_4^-) do not provide significant structural stabilization (blue and yellow tracks respectively, Fig. 4D). As shown in Fig. 4D and 4F, the gas-phase acidity of the conjugate acid form of the anion (equivalent to anion proton affinity) correlates well with these three classes of stabilizing anions, and in turn with their relative binding affinities. This correlation is apparent when singly-charged anions are considered, where anions with low gas-phase acidity fall into the yellow track, those with intermediate affinity in the green track, and those with the highest values are in the blue track. Multiply-charged anions are more-difficult to place *a priori* due to likely multi-dentate protein interactions.

Whereas anions studied to date perform optimally as stabilizers when relying entirely on the dissociative-cooling process described above, optimal cationic stabilizers are those that remain bound to the protein assembly, even following extensive activation in the gas phase (red track, Fig. 4D). These highly-stabilizing cations correlate with those that have larger charge-per-unit-area values, such as Mg^{2+} and Ca^{2+} shown in Fig. 4E. The higher surface charge density of these cations, much in excess of any anions tested, gives these adducts access to modes of stabilization that rely either upon multidentate interactions within the protein, enabling them to tether regions of its structure, or by replacing highly-mobile proton charge carriers with less-mobile cationic charge carriers that restrict charge mobility and frustrate the Coulombic unfolding of subunits within the complex.[36] Note that anions with low gas-phase acidity (*e.g.*, I^- and ClO_4^-), despite having higher protein binding strengths when compared to the other singly-charged anions studied here, do not reach the charge-per-unit-area values of stabilizing cations (Mg^{2+} and Ca^{2+}). Armed with this knowledge, we have endeavored to use tailored anion/cation pairs for protein stabilization in order to make use of both stabilization mechanisms simultaneously.

4. Drastic differences observed for gas-phase protein stabilities in the presence of cationic/anionic additives

On first inspection, drastic differences are apparent between the rank order determined by our data in the gas-phase (Fig. 4A) and the well-known Hofmeister series of cations and anions describing their influence upon protein stability in solution: $TMA^+ > NH_4^+ > Rb^+ > K^+ > Na^+ > Li^+ > Ca^{2+} > Mg^{2+} > Ba^{2+}$ and $SO_4^{2-} > HPO_4^- > F^- > OAc^- > Citrate^{3-} > HCO_3^- > Cl^- > NO_3^- > I^- > ClO_4^- > SCN^-$. For example, NO_3^-, SCN^- and Cl^- are all protein stabilizers in the gas-phase, but act as either neutral agents or destabilizers in bulk solvent. Similarly, cations stabilize gas-phase proteins following a nearly reversed order relative to most measurements of protein stability carried out in solution. Multiple reports have shown that Hofmeister-type protein stabilization in solution depends upon ion hydration and the ability of ions to alter water surface tension, along with direct ion–protein binding.[3,15,78] In our stability measurements of desolvated proteins, the influence of both bulk solvent and local hydration layers are absent, and our data are instead dominated by overall protein–counterion binding affinity, adduct dissociation, and Coulombic effects incumbent upon protein unfolding in the gas-phase (Fig. 4D). Thus, it is likely that the lack of protein solvation contributes substantially to the differences we observe between the stabilizing influence of Hofmeister salts in the gas-phase and in solution.

In spite of the clear differences between protein stabilization in the gas-phase and in solution noted above, it is potentially instructive to mine our current dataset for any information that may suggest critical links between our data and those collected in the condensed phase. To attempt this, we first note the strong correlation between gas-phase acidity and pK_a measurements acquired in solution for the anions studied here (Fig. 4F). As stated above, anion proton affinity seems likely to govern the relative amount of anions bound to gas-phase proteins in our data (Fig. 4D). We also note that the surface charge density of the cations studied here have a high correlation to their Lewis acid strengths in solution. Mirroring the above anion-based correlation, we found cation charge-per-unit-area to be an able predictor of protein–cation complex stability (Fig. 4E). Both correlations, while not representing demonstrable proof, provide some evidence that the relative binding affinities observed for protein–counterion complexes in our gas-phase measurements may mirror those in solution.

To further probe the potential correlations between our gas-phase data and protein–salt interactions in solution, we used MS to quantify the anions bound to proteins for a broad range of interacting pairs, as previous data had found a strong correlation between perchlorate binding observed by MS and the number of surface accessible basic sites on a given protein.[21] It is important to note that all the data and analysis shown in Fig. 4 is derived by normalizing the stabilization effects observed to the number of adducts bound to protein ions in the first instance. Fig. 5A shows MS data for 7^+ CYC monomers generated from solutions containing a range of anion additives. Unlike MS data for more-massive multiprotein complexes, where only average numbers of [H-"anion"] type adducts can be extracted from the average mass shifts observed relative to control (Fig. 5B), the mass resolving power achieved in our CYC experiments is sufficient to resolve individual bound populations for the anion–protein complexes observed. The total number of anions bound to each protein is indicated in the blue histograms in Fig. 5C–F, under which the black column represents the number of basic sites already occupied by charge carriers based on the ion charges state observed. Our data show that for anions known to have strong protein interactions (including nitrate, chloride, iodide and perchlorate anions), the quantity of binding observed by MS, when added to those sites occupied by unpaired charge carriers, correlates reasonably well the maximum expected number of binders based on the known surface assessable binding sites in solution. We attribute those cases where we observe excess binding (values exceeding the black dashed line) to anion condensation during the nESI process. Taken together with protein–ligand binding studies carried out where intensity values in MS measurements can be directly correlated to binding strengths between proteins and small molecules,[79–82] it is highly likely that our data represent a direct measure of protein–anion binding affinities highly correlated to those in solution. As such, it is likely that our observations, and the rank orders of stabilizing/destabilizing salts extracted from our data, serve primarily to highlight the critical importance of protein solvation in the Hofmiester stabilization of proteins in solution.

The rank order describing the stabilizing influence of protein salt interactions in the absence of solvent also exhibits reasonable agreement with a number of studies that have observed a so-called 'reversed' Hofmeister series, primarily for positively charged proteins at low salt concentrations.[30–32] Spectroscopic data indicates that these inversed datasets result from the differing strengths of anion associations with positively charged protein surfaces, are well-correlated with the sizes of hydrated anions, and are related directly to the hydration free energy of stabilizing anions.[83] While our gas-phase data correlate more-strongly to these 'reversed' Hofmeister series, a direct mechanistic correlation between the two rank orders is unlikely given what we currently understand of our gas-phase data. It is likely that continuing studies that highlight differences between gas-phase and 'reversed' Hofmeister series may further-pinpoint the role of anion hydration in solution-phase protein stabilization.

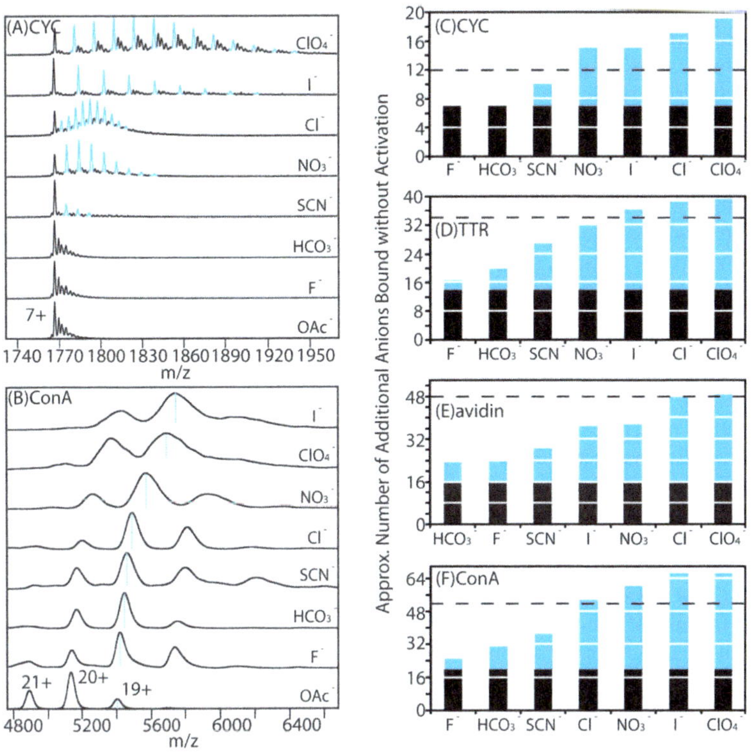

Fig. 5 Correlations between our gas-phase data and protein–salt interactions in solution. A) nESI-MS data of the 7^+ charge state of CYC (10 μM) generated from solutions containing 100 mM ammonium acetate (control) and a series of solutions containing both 100 mM ammonium acetate and 1 mM salts (ammonium-based salts with different anions) reveal a distribution of adducts (indicated as blue peaks) resolved by MS without collisional activation. Peaks corresponding to adducted ions arising from sodium, potassium, and sodium + potassium-binding are shown in black, and observed to dominate in our control (acetate), fluoride, and bicarbonate datasets. B) nESI-MS data for tetrameric ConA (10 μM) obtained from a solution containing 100 mM ammonium acetate (control) and a series of solutions containing both 100 mM ammonium acetate and 5 mM salts (ammonium-based salts with different anions). Each spectrum was obtained using identical instrumental conditions without collisional activation. The centroid *m/z* values corresponding to the 19^+ charge state of protein–anion complexes are indicated by blue dashed lines. C–F) Histogram plots showing the approximate number of residual anions bound to protein complex ions (shown as blue columns) stacked on the number of sites occupied by unpaired positive charges (shown as black columns), for CYC, TTR, avidin, and ConA, respectively. The black dashed line represents the number of surface solvent-accessible basic sites, as determined by the DEPTH program.[90]

5. Tailored anion/cation combinations enhance the stability of gas-phase proteins and multiprotein complexes

In our previous experiments, we found that the stabilization mechanism accessed by cations and anions are not mutually exclusive, and can therefore be accessed simultaneously to enhance the gas-phase stability of protein structure through tailored salts. Data demonstrate that Ca^{2+}, Mg^{2+}, tartrate^{2-}, Cl^-, citrate^{2-}, and NO_3^- are strongly stabilizing for protein structure in the gas-phase, and their combination leads to 8 potential salts that may be useful for IM-MS measurements of native-like protein structure. However, if we consider the binding of both free anions and cations in solution a prerequisite for the enhanced stabilization of proteins in the gas phase, then salts containing tartrate and citrate must be excluded from our list, as they can act as strong chelators for suppressing protein–metal interactions during nESI.[84] Consequently, our preliminary list of highly-stabilizing salt additives is reduced to four potential choices: $Ca(NO_3)_2$, $CaCl_2$, $Mg(NO_3)_2$ and $MgCl_2$.

In this report, we use $Ca(NO_3)_2$ to demonstrate such combined stabilizing effects for gas-phase protein structure. To ensure that both Ca^{2+} and NO_3^- can simultaneously bind to protein complexes and are subsequently carried into the gas-phase, we performed preliminary experiments on CYC monomers, where we are able to use MS to resolve individual protein-bound populations. Data for 7^+ CYC is shown in Fig. 6A, where monomer ions are generated from four different buffer compositions (control/100% 100 mM NH_4OAc, 100 mM NH_4OAc with added NH_4NO_3, 100 mM NH_4OAc with added $Ca(OAc)_2$, 100 mM NH_4OAc with added $Ca(NO_3)_2$). The resolved adduct populations corresponding to [H-'NO_3^-'] type adducts (blue) together with Ca^{2+} adducts (red) can be observed in our CYC monomer dataset following incubation with $Ca(NO_3)_2$. Following these proof-of-principle experiments, we extended our data to include a number of tetrameric protein complexes prepared under the same conditions. In these data, we observe large shifts in ion mass, indicative of simultaneous Ca^{2+} and NO_3^- binding when compared with negative (samples containing pure ammonium acetate buffer) and positive (samples containing either NH_4NO_3 or $Ca(OAc)_2$ additives) control data (Fig. 6B).

In order to investigate the stabilizing effects of the complexes created above, we performed CIU and CID stability measurements on monomer (CYC) and tetramer ions (TTR, avidin and ConA) created from solutions containing added $Ca(NO_3)_2$. Fig. 6C and 6D show histogram plots of the normalized collision energy (eV*) at which I_{tet} and I_f for these ions decrease to 50% of their original values as a function of buffer composition. Generally, we observe that cations are stronger stabilizers of gas-phase protein structure than anions, as they act by remaining bound to the assembly at relatively high internal temperatures. Most importantly, it is clear from our data that the simultaneous presence of both stabilizing cations and anions causes a significant increase in protein complex stability relative our control datasets, resulting in 8% and 10% average increases in gas-phase protein quaternary and tertiary structure stabilities respectively.

To further validate the joint stabilization provided by combined cations and anions, and provide insight into the stabilization mechanism at work, we constructed CIU unfolding 'fingerprints' for protein complex ions derived from solutions containing added $Ca(NO_3)_2$ as well as control samples containing its constituent anion and cation components. Changes in the protein ion tertiary/secondary structures are induced during the CIU process, leading to several structural ensembles that are stable on the millisecond timescale and can be resolved in both IM drift time and collision energy. For clarity, CIU fingerprint data is projected as a contour plot (Fig. 6E) where intensities for the features observed are denoted by a color-based axis. Careful analysis of CIU fingerprint data allows the nature of protein stabilization to be identified by noting the conformational features that are stabilized (elongated on the collision energy axis) relative to control data.[34,56,60,85]

Control fingerprints for the TTR tetramer reveal three major conformational families (I, II, III) observed under the conditions used for our experiments here. These features are easily resolved in drift time and have distinct patterns as a function of collision voltage (Fig. 6E). For example, fingerprint data acquired for TTR with added NO_3^- shows that the most compact conformer (I) is observed at substantially higher collision voltages when compared to those ions generated from pure ammonium acetate solutions, indicating that the stabilization observed in our experiments is due primarily to the enhanced stability of this compact conformer. In contrast, TTR incubated with added Ca^{2+} displays a substantially different CIU fingerprint, where both the most compact and partially unfolded forms of the complex are stabilized. As discussed above, the differences in the fingerprints recorded for Ca^{2+}-bound and NO_3^--bound TTR are indicative of the separate stabilization mechanisms operative for these two adduct populations. Specifically, Ca^{2+}

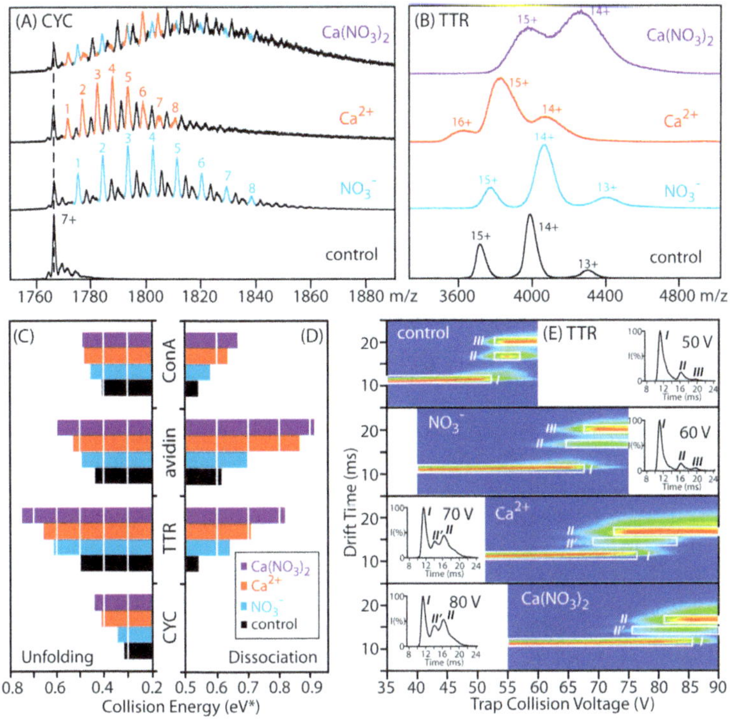

Fig. 6 The combined influence of tailored Hofmeister anions and cations on gas-phase protein stability. A) nESI-MS data of the 7^+ charge state for CYC obtained from ammonium acetate-based solutions containing 10 µM CYC and no added salt (control), 400 µM ammonium nitrate, 200 µM calcium acetate, or 200 µM calcium nitrate. The resolved populations of cation and anion additives are denoted by red and blue colors, respectively. The other peaks (black) except for the apo-CYC correspond to adducts arising from sodium or potassium-binding. B) nESI-MS data for TTR (5 µM) generated from ammonium acetate-based solutions with no added salt (control, black), 4 mM ammonium nitrate (blue), 2 mM calcium acetate (red), or 2 mM calcium nitrate (purple). C & D) Histogram plots of collision energy (eV*) required to dissociate or unfold 50% of the population of monomeric CYC, tetrameric avidin, ConA, and TTR are shown for the four buffer conditions mentioned above. E) A CIU fingerprint contour plots are shown for 14^+ TTR generated from ammonium acetate-based solutions with no added salt (control), 4 mM ammonium nitrate, 2 mM calcium acetate, or 2 mM calcium nitrate, where ion trap collision voltage is charted against IM drift time, and the ion intensities are denoted by a color-coded axis (blues correlated to low ion intensity, whereas reds indicate high ion intensity). The conformational forms for the tetramer (I, II, II′ or III) are labeled as described in the text, and the two most compact forms of the protein observed are highlighted (white box). The inset figures show the IM arrival time distributions of 14^+ TTR ions acquired at the indicated trap collision voltages, with the peak centroids corresponding to the intermediate structural families labeled.

stabilizes the complex through tight binding, such that intermediately unfolded conformers of the protein are partially stabilized. In contrast, NO_3^- stabilizes the complex through dissociative cooling, and therefore cannot, by definition, stabilize partially unfolded protein conformers. Additionally, a new intermediate unfolded species (II′, inset Fig. 6E) emerges in the Ca^{2+} fingerprint. This observation is also consistent with our mechanism, in which strongly-bound cations modulate the collisional unfolding process through tethering flexible regions within proteins or by limiting charge migration by replacing highly mobile proton charge carriers. Critically, fingerprint data collected from ions incubated with $Ca(NO_3)_2$ displays elements from both the fingerprints of its constituent components, exhibiting a highly stabilized compact (I) state, a stabilized II state, and the appearance of a II′ state. Thus, our CIU fingerprint data supports the joint stabilization of protein complexes through simultaneous binding of both Ca^{2+} and NO_3^- adducts, resulting in enhanced protein stability when compared to their individual effects.

6. The future of gas-phase protein structure measurements and their potential impact on our understanding of Hofmeister stabilization

Measurements of desolvated protein stability have a clear place in supporting the emergent field of gas-phase structural biology, which seeks to use solvent-free structural information to develop native-state protein models.[86] This challenging endeavor hinges on a strong correlation between solution and gas-phase protein structures, and such a correlation has been measured for a broad array of proteins and their complexes.[62] At the same time, a number of cases and specific structural elements within proteins have been observed to substantially rearrange in the absence of solvent.[65,87,88] Stabilized proteins and multiprotein complexes could potentially avoid many of these effects, facilitating the use of gas-phase measurements in the construction of native-state protein models. It is also likely that the mechanistic aspects of anion and cation stabilization discussed here have significant implications for the use of MS in the analysis of membrane-bound protein complexes. In these experiments, detergent micelles are used as stabilizing agents to transport hydrophobic proteins and protein complexes into the gas-phase from aqueous solutions by nESI, and collisional activation is used to remove bound detergent from the protein ions following desolvation.[58,89] Understanding the mechanistic principles that govern the action of the relatively simplified stabilizing additives studied here will likely provide an enhanced ability to utilize IM and MS data for membrane protein complexes by avoiding unwanted unfolded protein conformations, thus enabling the rapid determination of protein topologies critical in our understanding of multiple cellular processes and disease states.[59]

The direct implications of the gas-phase data shown in this report for Hofmeister stabilization in solution depends upon both the drastically different stabilization mechanisms observed, and the similarities detected in the numbers of bound anions in our experiments to those expected to associate with the protein in solution. IM-MS data clearly show that anions and cations can stabilize protein complexes through separate mechanisms, and that these modes of action are unique to the gas-phase. Furthermore, it is likely that the number of bound anions and cations observed by MS in our data bear a strong correlation to the bound populations present in solution. As such, the dramatically different rank orders observed in gas-phase experiments when compared with solution can be taken as evidence for the critical importance of solvation effects in the mechanism of Hofmeister stabilization in solution. Finally, in addition to providing enhanced stabilizing additives for the gas-phase measurement of native-like protein structures, it is clear that continuing measurements of protein-counterion complexes will provide a useful tool for quantifying bound cation and anion populations in support of solution-phase measurements of Hofmeister protein stabilization.

Acknowledgements

This work is supported by the National Institutes of Health (1-R01-GM-095832-01) and by University of Michigan startup funds.

References

1 C. V. Robinson, A. Sali and W. Baumeister, *Nature*, 2007, **450**, 973–982.
2 L. Vrbka, P. Jungwirth, P. Bauduin, D. Touraud and W. Kunz, *J. Phys. Chem. B*, 2006, **110**, 7036–7043.

3 Y. J. Zhang, S. Furyk, D. E. Bergbreiter and P. S. Cremer, *J. Am. Chem. Soc.*, 2005, **127**, 14505–14510.
4 R. L. Baldwin, *Biophys. J.*, 1996, **71**, 2056–2063.
5 W. Kunz, *Curr. Opin. Colloid Interface Sci.*, 2010, **15**, 34–39.
6 K. D. Collins, *Methods*, 2004, **34**, 300–311.
7 W. Kunz, J. Henle and B. W. Ninham, *Curr. Opin. Colloid Interface Sci.*, 2004, **9**, 19–37.
8 K. D. Collins and M. W. Washabaugh, *Q. Rev. Biophys.*, 1985, **18**, 323–422.
9 F. Vanzi, B. Madan and K. Sharp, *J. Am. Chem. Soc.*, 1998, **120**, 10748–10753.
10 A. W. Omta, M. F. Kropman, S. Woutersen and H. J. Bakker, *Science*, 2003, **301**, 347–349.
11 M. F. Kropman and H. J. Bakker, *J. Am. Chem. Soc.*, 2004, **126**, 9135–9141.
12 J. D. Batchelor, A. Olteanu, A. Tripathy and G. J. Pielak, *J. Am. Chem. Soc.*, 2004, **126**, 1958–1961.
13 M. Bostrom, D. R. M. Williams and B. W. Ninham, *Phys. Rev. Lett.*, 2001, **87**.
14 J. D. Smith, R. J. Saykally and P. L. Geissler, *J. Am. Chem. Soc.*, 2007, **129**, 13847–13856.
15 Y. J. Zhang and P. S. Cremer, in *Annual Review of Physical Chemistry, Vol 61*, ed. S. R. Leone, P. S. Cremer, J. T. Groves, M. A. Johnson and G. Richmond, 2010, vol. 61, pp. 63–83.
16 M. Lund, R. Vacha and P. Jungwirth, *Langmuir*, 2008, **24**, 3387–3391.
17 M. Lund, L. Vrbka and P. Jungwirth, *J. Am. Chem. Soc.*, 2008, **130**, 11582.
18 H. Prakash, B. T. Kansara and S. Mazumdar, *Int. J. Mass Spectrom.*, 2010, **289**, 84–91.
19 J. L. Stephenson and S. A. McLuckey, *Anal. Chem.*, 1997, **69**, 281–285.
20 U. H. Verkerk and P. Kebarle, *J. Am. Soc. Mass Spectrom.*, 2005, **16**, 1325–1341.
21 T. G. Flick, S. I. Merenbloom and E. R. Williams, *Anal. Chem.*, 2011, **83**, 2210–2214.
22 U. H. Verkerk, M. Peschke and P. Kebarle, *J. Mass Spectrom.*, 2003, **38**, 618–631.
23 N. Felitsyn, M. Peschke and P. Kebarle, *Int. J. Mass Spectrom.*, 2002, **219**, 39–62.
24 O. Timofeev, M. M. Zhu and M. L. Gross, *Int. J. Mass Spectrom.*, 2004, **231**, 113–117.
25 P. Kebarle and U. H. Verkerk, *Mass Spectrom. Rev.*, 2009, **28**, 898–917.
26 L. M. Pegram and M. T. Record, *J. Phys. Chem. B*, 2007, **111**, 5411–5417.
27 L. M. Pegram and M. T. Record, *Proc. Natl. Acad. Sci. U. S. A.*, 2006, **103**, 14278–14281.
28 L. M. Pegram and M. T. Record, *J. Phys. Chem. B*, 2008, **112**, 9428–9436.
29 L. M. Pegram and M. T. Record Jr, *Chem. Phys. Lett.*, 2008, **467**, 1–8.
30 Y. J. Zhang and P. S. Cremer, *Proc. Natl. Acad. Sci. U. S. A.*, 2009, **106**, 15249–15253.
31 M. Bostroem, D. F. Parsons, A. Salis, B. W. Ninham and M. Monduzzi, *Langmuir*, 2011, **27**, 9504–9511.
32 M. Bostrom, F. W. Tavares, S. Finet, F. Skouri-Panet, A. Tardieu and B. W. Ninham, *Biophys. Chem.*, 2005, **117**, 217–224.
33 T. G. Flick, S. I. Merenbloom and E. R. Williams, *J. Am. Soc. Mass Spectrom.*, 2011, **22**, 1968–1977.
34 L. J. Han, S. J. Hyung, J. J. S. Mayers and B. T. Ruotolo, *J. Am. Chem. Soc.*, 2011, **133**, 11358–11367.
35 S. I. Merenbloom, T. G. Flick, M. P. Daly and E. R. Williams, *J. Am. Soc. Mass Spectrom.*, 2011, **22**, 1978–1990.
36 L. Han, S. J. Hyung and B. T. Ruotolo, *Angew Chem Int Ed Engl*, 2012.
37 J. S. Prell, T. G. Flick, J. Oomens, G. Berden and E. R. Williams, *J. Phys. Chem. A*, 2010, **114**, 854–860.
38 M. F. Bush, J. Oomens, R. J. Saykally and E. R. Williams, *J. Am. Chem. Soc.*, 2008, **130**, 6463–6471.
39 R. C. Dunbar, A. C. Hopkinson, J. Oomens, C. K. Siu, K. W. M. Siu, J. D. Steill, U. H. Verkerk and J. F. Zhao, *J. Phys. Chem. B*, 2009, **113**, 10403–10408.
40 R. C. Dunbar, N. C. Polfer and J. Oomens, *J. Am. Chem. Soc.*, 2007, **129**, 14562.
41 J. T. O'Brien, J. S. Prell, J. D. Steill, J. Oomens and E. R. Williams, *J. Phys. Chem. A*, 2008, **112**, 10823–10830.
42 M. F. Bush, J. T. O'Brien, J. S. Prell, R. J. Saykally and E. R. Williams, *J. Am. Chem. Soc.*, 2007, **129**, 1612–1622.
43 P. B. Armentrout, M. T. Rodgers, J. Oomens and J. D. Steill, *J. Phys. Chem. A*, 2008, **112**, 2248–2257.
44 R. S. Walters, E. D. Pillai and M. A. Duncan, *J. Am. Chem. Soc.*, 2005, **127**, 16599–16610.
45 J. W. Shin, N. I. Hammer, E. G. Diken, M. A. Johnson, R. S. Walters, T. D. Jaeger, M. A. Duncan, R. A. Christie and K. D. Jordan, *Science*, 2004, **304**, 1137–1140.
46 D. J. Miller and J. M. Lisy, *Journal of Chemical Physics*, 2006, 124.
47 M. F. Bush, J. T. O'Brien, J. S. Prell, C. C. Wu, R. J. Saylkally and E. R. Williams, *J. Am. Chem. Soc.*, 2009, **131**, 13270–13277.
48 T. Iino, K. Ohashi, K. Inoue, K. Judai, N. Nishi and H. Sekiya, *Journal of Chemical Physics*, 2007, 126.

49 J. H. Choi, K. T. Kuwata, Y. B. Cao and M. Okumura, *J. Phys. Chem. A*, 1998, **102**, 503–507.
50 W. H. Robertson and M. A. Johnson, *Annu. Rev. Phys. Chem.*, 2003, **54**, 173–213.
51 M. F. Bush, R. J. Saykally and E. R. Williams, *J. Am. Chem. Soc.*, 2007, **129**, 2220.
52 T. Wyttenbach and M. T. Bowers, *Top. Curr. Chem.*, 2003, **225**, 207–232.
53 J. A. Loo, B. Berhane, C. S. Kaddis, K. M. Wooding, Y. M. Xie, S. L. Kaufman and I. V. Chernushevich, *J. Am. Soc. Mass Spectrom.*, 2005, **16**, 998–1008.
54 J. Freeke, C. V. Robinson and B. T. Ruotolo, *Int. J. Mass Spectrom.*, 2010, **298**, 91–98.
55 S. I. Merenbloom, T. G. Flick, M. P. Daly and E. R. Williams, *Journal of the American Society for Mass Spectrometry*, 2011, 1–13.
56 J. Freeke, M. F. Bush, C. V. Robinson and B. T. Ruotolo, *Chem. Phys. Lett.*, 2012, **524**, 1–9.
57 N. Carulla, M. Zhou, E. Giralt, C. V. Robinson and C. M. Dobson, *Acc. Chem. Res.*, 2010, **43**, 1072–1079.
58 N. P. Barrera, S. C. Isaacson, M. Zhou, V. N. Bavro, A. Welch, T. A. Schaedler, M. A. Seeger, R. N. Miguel, V. M. Korkhov, H. W. van Veen, H. Venter, A. R. Walmsley, C. G. Tate and C. V. Robinson, *Nat. Methods*, 2009, **6**, 585–U549.
59 M. Zhou, N. Morgner, N. P. Barrera, A. Politis, S. C. Isaacson, D. Matak-Vinković, T. Murata, R. A. Bernal, D. Stock and C. V. Robinson, *Science*, 2011, **334**, 380–385.
60 S. J. Hyung, C. V. Robinson and B. T. Ruotolo, *Chem. Biol.*, 2009, **16**, 382–390.
61 J. T. S. Hopper and N. J. Oldham, *J. Am. Soc. Mass Spectrom.*, 2009, **20**, 1851–1858.
62 J. L. P. Benesch and B. T. Ruotolo, *Curr. Opin. Struct. Biol.*, 2011, **21**, 641–649.
63 M. F. Bush, Z. Hall, K. Giles, J. Hoyes, C. V. Robinson and B. T. Ruotolo, *Anal. Chem.*, 2010, **82**, 9557–9565.
64 J. A. Leary, M. R. Schenauer, R. Stefanescu, A. Andaya, B. T. Ruotolo, C. V. Robinson, K. Thalassinos, J. H. Scrivens, M. Sokabe and J. W. B. Hershey, *J. Am. Soc. Mass Spectrom.*, 2009, **20**, 1699–1706.
65 A. Politis, A. Y. Park, S. J. Hyung, D. Barsky, B. T. Ruotolo and C. V. Robinson, *Plos One*, 2010, 5.
66 B. T. Ruotolo, J. L. P. Benesch, A. M. Sandercock, S. J. Hyung and C. V. Robinson, *Nat. Protoc.*, 2008, **3**, 1139–1152.
67 Y. Zhong, S.-J. Hyung and B. T. Ruotolo, *Expert Rev. Proteomics*, 2012, **9**, 47–58.
68 K. Giles, S. D. Pringle, K. R. Worthington, D. Little, J. L. Wildgoose and R. H. Bateman, *Rapid Commun. Mass Spectrom.*, 2004, **18**, 2401–2414.
69 K. Giles, J. P. Williams and I. Campuzano, *Rapid Commun. Mass Spectrom.*, 2011, **25**, 1559–1566.
70 Y. Y. Zhong, S. J. Hyung and B. T. Ruotolo, *Analyst*, 2011, **136**, 3534–3541.
71 H. Hernandez and C. V. Robinson, *Nat. Protoc.*, 2007, **2**, 715–726.
72 N. Felitsyn, E. N. Kitova and J. S. Klassen, *Anal. Chem.*, 2001, **73**, 4647–4661.
73 J. C. Jurchen and E. R. Williams, *J. Am. Chem. Soc.*, 2003, **125**, 2817–2826.
74 J. W. Li, J. A. Taraszka, A. E. Counterman and D. E. Clemmer, *Int. J. Mass Spectrom.*, 1999, **185**, 37–47.
75 K. B. Shelimov, D. E. Clemmer, R. R. Hudgins and M. F. Jarrold, *J. Am. Chem. Soc.*, 1997, **119**, 2240–2248.
76 S. J. Valentine, J. G. Anderson, A. D. Ellington and D. E. Clemmer, *J. Phys. Chem. B*, 1997, **101**, 3891–3900.
77 A. A. Shvartsburg, F. M. Li, K. Q. Tang and R. D. Smith, *Anal. Chem.*, 2007, **79**, 1523–1528.
78 Y. Zhang, S. Furyk, L. B. Sagle, Y. Cho, D. E. Bergbreiter and P. S. Cremer, *J. Phys. Chem. C*, 2007, **111**, 8916–8924.
79 J. A. Loo, *Int. J. Mass Spectrom.*, 2000, **200**, 175–186.
80 L. Liu, D. Bagal, E. N. Kitova, P. D. Schnier and J. S. Klassen, *Journal of the American Chemical Society*, 2009, **131**.
81 L. Deng, N. Sun, E. N. Kitova and J. S. Klassen, *Anal. Chem.*, 2010, **82**, 2170–2174.
82 E. N. Kitova, N. Soya and J. S. Klassen, *Anal. Chem.*, 2011, **83**, 5160–5167.
83 Y. Marcus, *Ion properties*, Wiley Online Library, 1997.
84 J. X. Pan, K. Xu, X. D. Yang, W. Y. Choy and L. Konermann, *Anal. Chem.*, 2009, **81**, 5008–5015.
85 K. Pagel, S. J. Hyung, B. T. Ruotolo and C. V. Robinson, *Anal. Chem.*, 2010, **82**, 5363–5372.
86 S. J. Hyung and B. T. Ruotolo, *Proteomics*, 2012, **12**, 1547–1564.
87 C. J. Hogan, B. T. Ruotolo, C. V. Robinson and J. F. de la Mora, *J. Phys. Chem. B*, 2011, **115**, 3614–3621.
88 E. Jurneczko and P. E. Barran, *Analyst*, 2011, **136**, 20–28.
89 N. P. Barrera, N. Di Bartolo, P. J. Booth and C. V. Robinson, *Science*, 2008, **321**, 243–246.
90 K. P. Tan, R. Varadarajan and M. S. Madhusudhan, *Nucleic Acids Res.*, 2011, **39**, W242–W248.

General discussion

Dr Hassenkam opened the discussion of the paper by Thorsten Hugel: Looking at Fig. 3A in your paper, what, in your view, is going on between molecules, ions, the surface and water as you change the NaCl concentration?

Professor Dr Hugel responded: Upon desorption of a polymer from a surface in ionic aqueous environment there will be changes in all nine interactions (polymer–water, polymer–polymer, polymer–ion, polymer–surface, water–water, water–ion, water–surface, ion–surface, ion–ion). We believe that the general picture is the same as in the absence of ions (like *e.g.* in Fig. 7A), where all the contributions largely compensate to give a small free energy change (which is comparable to the protein folding problem, where large entropic and enthalpic contributions compensate to yield a small free energy change between the folded and unfolded state of a polymer). Changing the salt concentration itself will considerably affect the four interactions involving ions. In total we measure again only a small free energy change in between the salt concentrations (up to several molar), therefore we think that there is again a compensation mechanism. Simulations to get insights into the molecular details for that system have not been done yet.

Professor Ben-Amotz asked: It is not clear to me how you can distinguish whether the force plateau that you have measured represents the force associated with transferring a monomer into the aqueous solution from the solid surface or from the interior of the collapsed polymer (protein) chain on the surface. I would also like to understand better the connection between your results and those obtained in similar experiments by Li and Walker.[1] Note that the latter experiments revealed a non-monotonic temperature dependence of the plateau force, and the resulting transfer free energy, whose monomer size dependence is apparently consistent with hydrophobic crossover predictions.[2]

1. I. T. S. Li and G. C. Walker, *Proc. Natl. Acad. Sci. U. S. A.*, 2011, **108**, 16527–16532.
2. S. Garde and A. J. Patel, *Proc. Natl. Acad. Sci. U. S. A.*, 2011, **108**, 16491–16492.

Professor Dr Hugel answered: Distinguishing between transfer from the surface or the interior of the polymer is indeed not trivial. We have therefore tested many different surfaces and seen significant differences in the height of the force plateau (see for example ref. 1 & 2). For our experiments we therefore have an effect of the surface, which excludes that the monomer is only transferred from the interior of the collapsed polymer. In addition, imaging of the polymers suggests no spherical objects (the height of the polymers is always significantly less than their width) in contradiction to the interpretation by Li and Walker, interestingly the AFM images by Li and Walker also show no spherical objects (see their *PNAS* supplement). Finally, in our opinion the pressure in the interior of a collapsed polymer will change with size and the plateau should therefore show some slope (which it does not). Our interpretation is that the polymer has significant interaction with the substrate and we measure a combination of the solvation free energy and the surface interaction free energy (which is not included by Li and Walker), which largely compensate each other. This is reminiscent of the protein folding problem, where large entropic and enthalpic contributions compensate to yield only small free energy changes between the folded and unfolded state. Considering the non-monotonic temperature dependence, we can tell that this might happen (*e.g.* for lysine on glass, Fig. 3A), but that this is not a universal feature. This is perfectly in line with the recent theoretical treatment from the Netz group (Sedlmeier *et al.*, *J. Chem. Phys.*, 2011, **134**, 055105). This publication even explains why previous simulations suggested the maximum in free energy as a universal feature. Li and Walker draw their conclusions from a system where the polymer is unspecific attached (spin coated) onto the substrate and then an untreated AFM-tip is brought in contact with the polymer sometimes making contact at undefined position(s). This system is therefore not as well defined as our systems, in their setup the polymer might for example slide off the highly curved AFM-tip.

1. T. Pirzer and T. Hugel, *ChemPhysChem*, 2009, **10**(16), 2795.
2. M. Geisler, B.N. Balzer and T. Hugel, *Small*, 2009, **5**(24), 2864.

Professor Vegt asked: In interpreting the temperature dependent measurements with poly-D-tyrosine you discuss an analogy with protein folding where a maximum occurs in the free energy. You explain the weak temperature dependence observed in your experiments with the compensation of various types of interactions. This point is not entirely clear as it seems to me that similar compensation effects, excluding those involving the surface, occur in protein folding.

Professor Dr Hugel replied: In the beginning we were also puzzled by our experimental results. Meanwhile they are well understood also from a theoretical point of view (Sedlmeier *et al.*, *J. Chem. Phys.*, 2011, **134**, 055105). As you state correctly, we have (entropy and enthalpy) compensation mechanisms occurring in protein folding leading to a maximum in the free energy. This could to a first approximation be compared to dissolving the protein upon desorption (which might result in a maximum), however for the desorption we have to consider additional contributions, namely changes at the surface. In sum, this yields a maximum, minimum or no temperature dependence of the desorption free energy depending on the system (Fig. 3A, Fig. 7A,B).

Dr Roeselova opened the discussion of the paper by Piotr Jurkiewicz by stating: I have two questions regarding the mobility information that you obtain from the fluorescence measurements: First, the characteristic time scale of the solvent relaxation presented in Fig. 2 and 3 is of the order of units of nanoseconds. What types of motion in the system (lipids, ions) does this correspond to? Second, what timescales are available with your technique, *i.e.*, what types of motion can you probe? Can you, for example, probe lateral motion of the lipid molecules (2D diffusion)?

Dr Jurkiewicz responded: Fluorescence solvent relaxation, when measured in lipid membranes at room temperature, is about 1000 times slower than in neat solvents like water, where it comprises mainly of librational and rotational motions of water dipoles. In lipid bilayers, at the level we probe them, water molecules are hydrogen bonded to lipids. Thus, we believe, these are collective motions of hydrated lipids or their parts that we observe locally in the vicinity of the fluorescent polarity probe. In such a complex environment it is difficult to attribute the relaxation time to a defined single type of movement. It is, however, quite clear that lipid rotations and lateral diffusion should participate in the relaxation response. When comparing the results of TDFS technique with, *e.g.*, *z*-scan FCS that gives lateral diffusion, we usually see a qualitative agreement, although we found TDFS to be much more sensitive. The timescale is limited by the temporal resolution of the instrumentation and by the fluorescence lifetime of the probe. In our case this gives a range of about 20 ps to 20 ns.

Professor Tiddy added: I have two questions. 1. If there were small changes in the location of the probe (say 2–3 Ångströms) in the presence of the different salts, would this influence the results?

2. How long do you think it takes for the lipids to reach their equilibrium conformations, and how does this compare to the simulation time?

Dr Jurkiewicz answer: 1. Yes – certainly. There is a very steep gradient of the relaxation times along the membrane normal. As a result, a well-defined and stable location of the fluorescent probe is a prerequisite for TDFS method. We know the mean location of the probe from parallax-quenching experiments and from simulations of the probe both in its ground and excited states; so we know exactly what region of the bilayer we investigate. Also we test whether the probe changes its location during the experiment using a water soluble quencher (*e.g.* acrylamide). No such changes were observed for the studied salt solutions.

Dr Cwiklik added: In the case of charged membranes, we typically calculate 100-ns-long trajectories, where the final 50 ns are taken for analysis. Such an equilibration time is sufficient for studying interactions of ions with charged lipid, where interactions with the headgroup region of a membrane are of a main interest. Note that in the case of neutral zwitterionic membranes, interactions with monovalent ions, such as sodium or potassium cations, are much weaker and thus equilibration for at least 150 ns would be required. Regarding the membranes studied in our paper, in the case of bromide anions, strong lateral inhomogeneities and lipid ordering are observed at the membrane surface. Therefore, it is possible that this particular system is not fully relaxed due to an insufficient size of the simulation box in the direction parallel to the membrane.

Mr Dudgeon asked: Have you simulated or seen any changes in the binding and interactions of the ions depending on the phase (lamella/gel) of the bilayer (on change in temperature)?

Dr Jurkiewicz replied: No, we have neither simulated nor measured it. It would be interesting. Unfortunately, relaxation in the gel phase is considerably slower than the fluorescence lifetime, so the TDFS method would show only the initial part of the relaxation – too little to analyze ionic effects. It should be possible, however, to measure those effects for liquid ordered phase. We have not done it yet.

Dr Cremer opened the discussion of the paper by Brandon T. Ruotolo: Please explain in more detail the relative differences between the interactions of anions and cations with proteins in the gas phase *vs.* in aqueous solution.

Dr Ruotolo replied: One of the central points that we to make in our report here, is that there are probably significant correlations between the binding sites and relative affinities for anions/cations observed from our gas-phase data and those same bound populations in solution. For decades, many researchers have used mass spectrometry, and tandem mass spectrometry, to identify the binding sites of various cations and anions within peptides and proteins (P. Kebarle and U. H. Verkerk, *Mass Spectrom. Rev.*, 2009, **28**, 898–917). More recently, electron-mediated dissociation techniques, in combination with tandem mass spectrometry, have been able to provide detailed binding site information for specific metals within intact multiprotein complexes (H. Zhang, W. D. Cui, J. Z. Wen, R. E. Blankenship and M. L. Gross, *Anal. Chem.*, 2011, **83**, 5598–5606). Fig. 5 in our report here shows a general correlation between the mass-shifts and observed in our gas-phase data, with the expected binding affinities for those same elements in solution. Differences between anion/cation populations bound to protein ions in the gas-phase and their analogs in solution include: (1) the increased strength of electrostatic interactions and salt-bridges in a solvent-free environment; and (2) the absence of any specific water-mediated or bridged interactions in the gas-phase data. We also note that, because of the polarity of the ions probed in our experiments, cation binding observed in our experiments may be skewed by the protein ionization process. In the future, we plan to perform similar experiments using negative ion mass spectrometry, and record any differences observed in anion/cation binding.

Professor Jungwirth asked: Can you in your electrospray experiments change the conditions such that you would be gradually adding water molecules to your complexes, which would allow to bridge the gas phase and liquid phase conditions?

Dr Ruotolo replied: While we do not demonstrate this in our work, many reports have previously shown that the step-wise addition of water molecules to gas-phase biomolecules can be performed in combination with structural probes like IM-MS, and that these additions can significantly alter gas-phase protein structure. For example, early IM-MS results on protein-water clusters (J. Woenckhaus, R. R. Hudgins and M. F. Jarrold, *J. Am. Chem. Soc.*, 1997, **119**, 9586–9587) indicate that for some proteins ions, the addition of only a few water molecules is necessary to alter the gas-phase protein structure to one that better-resembles that found in solution. While the ultimate aims of those experiments are somewhat different from our own, we are certainly interested in investigating gas-phase water addition experiments in the future as a bridge between our measurements and those performed in solution.

Professor Goodall enquired: Is there any consideration of what happens with proteins at the droplet interface during the ESI desolvation process? Rheology studies for solutions shows that proteins denature at the air–water interface. This denaturation can be prevented by using a small amount of surfactant (*e.g.* Tween) to compete for the interface. Would there be any benefit in having a non-ionic surfactant present with the protein in the MS experiment?

Dr Ruotolo noted: Based on what we know about the ionization mechanisms operative during our experiments, surfactant additives such as Tween are probably not necessary to avoid significant denaturation of native-like proteins at the air/water interface of the charged droplets produced by electrospray. We know from a number of different datasets that the amount of charging experienced by native-like proteins from aqueous solvent (J. F. de la Mora, *Anal. Chim. Acta*, 2000, **406**, 93–104), as well as the sizes those ions adopt in the gas-phase (J. L. P. Benesch and B. T. Ruotolo, *Curr. Opin. Struct. Biol.*, 2011, **21**, 641–649), that significant structural rearrangements consistent with denaturation do not take place during the electrospray ionization of proteins from aqueous solvents. This evidence, combined with the knowledge that the oligomeric states for proteins are well-preserved during desolvation and ionization, suggests that droplet surface effects are likely negligible influences on the structures of the protein ions produced. That having been said, the mechanism of electrospray ionization is an area of active research, and details of the physical processes at work during protein desolvation and ionization are being continuously revealed by cutting-edge experiments. For example, recent models suggest that denatured proteins, which take on large amounts of charge during electrospray ionization, are likely produced by an entirely different mechanism from their native analogs (L. Konermann, A. D. Rodriguez and J. Liu, *Anal. Chem.*, 2012, **84**, 6798–6804). Added surfactant stabilizers have, however, been observed recently to dramatically stabilize membrane-based proteins during electrospray, allowing their oligomeric state to be captured by mass spectrometry experiments (N. P. Barrera, N. Di Bartolo, P. J. Booth and C. V. Robinson, *Science*, 2008, **321**, 243–246). While these datasets continue to evolve, and the influence of added surfactant on the electrospray mechanism has yet to be completely elucidated, it is clear that surfactant molecules are absolutely essential for accurate stoichiometry measurements of membrane protein complexes by mass spectrometry.

Professor Goodall added: It would be really nice if you could use your gas-phase protein structure measurement technique to look at well-studied proteins such as BSA, for which there is a massive amount of solution-phase data on ion effects and ion binding, to look for correlation with the Hofmeister series.

Dr Ruotolo responded: We absolutely agree, and we have attempted to construct our current dataset to include measurements for well-studied multimeric (*i.e.*, avidin tetramer) and monomeric (*i.e.*, cytochrome c) proteins so that conclusions made regarding the anion and cation-based stability differences we observe can be directly correlated to those same protein systems in solution. Arguably, there is even more Hofmeister-based stability data for bovine serum albumin (BSA), as you rightly point out, and we certainly plan to include studies of this protein in our upcoming work. In addition, we also plan to extend our investigations to include a greater number of multiprotein systems, especially those that contain sufficient internal cavity regions, in order to investigate the influence of quaternary structure on the stability differences we observe amongst gas-phase protein ions.

Professor Halling opened the discussion of the paper by Jan Konvalinka: Could I suggest two factors that might help explain some of the strange effects you observe. First the different salt effects in different buffers. You use a fixed buffer concentration of 50 mM, but the buffer will contribute very different cation concentrations at pH 6, about 60 mM for phosphate, about 20 mM for MES and about 3 mM for HEPES. These are very significant differences compared with the added salt concentrations you are studying. Secondly, I wonder if some findings are due to salt effects on the activity coefficient of the substrate, which carries multiple charges. These will lead to a change in observed affinity for the enzyme, without any direct effects on the enzyme itself – and this would most likely appear as a change in K_m value, as you observe. Such effects also amount to a challenge to physical chemistry – to predict the effects of salt on the activity coefficients of organic species like enzyme substrates.

Dr Konvalinka replied: This point is very well taken and I would like to thank Professor Halling for raising it. Of course, the actual ionic strengths in individual buffers do differ according to the p*K* of the buffer used. While the contribution of the alkali counterion in case of HEPES is negligible, the phosphate buffer might indeed overcome the effect of the added salt in the very low concentration region. It could be speculated that the ion concentration is already too high for the activation to occur even without any added salt.

In order to address this possibility, we analyzed the influence of the sodium, lithium, potassium and cesium ion concentration on the enzyme activity in low ion strength phosphate buffer (yielding 15 mM final alkali ion concentration without additional salt). We have not seen any activation effect in the low salt range, in agreement with the results observed for the MES buffer (data not shown). Therefore, we believe that the conclusions of the paper are valid.

As for the analysis of the salt effects on the activity coefficient of the peptide substrate, this is an interesting idea but it would require additional experimentation that would be beyond the scope of this paper.

Dr Vila Verde asked: Can the ions influence the opening-and-closing motion of your protease? Changes in small-amplitude (*i.e.* short timescale, such as ns) protein motion often correlate with changes in large-amplitude motion. Would it be possible to characterize the short timescale motion of your protein using simulations? Would it be possible to probe the opening-and-closing protein motion experimentally, using NMR?

Dr Konvalinka replied: This might be a possible explanation. The dynamics of flap opening/closure is indeed implicated in the binding of ligands to the HIV protease binding site. However, in our MD simulations (performed for 10 or 20 ns) no change in the overall structure connected with alkali ions have been observed. NMR is, of course, a method of choice for the study of the protein dynamics. However, NMR analysis of HIV PR is especially demanding due to the low solubility and insufficient stability of HIV protease in the high concentrations necessary for the measurements.

Dr Roberts asked: In your paper you are investigating the effect of three different buffers on the initial velocities of the enzymatic hydrolysis of the substrate. You mention further two buffers investigated in your study. Could you please comment on the buffer(s)' effect at low salt concentrations and why there is such a difference in measurements with MES buffer?

Dr Konvalinka replied: As described in the paper, in order to analyze a possible pH and buffer effect on the enzymatic hydrolysis of the substrate in the presence of different alkali cations, we performed analyses with another three buffers: acetate buffer (pH 4.7); 2-[bis(2-hydroxyethyl)amino]-2-(hydroxymethyl)-1,3-propanediol (BIS-TRIS) buffer (pH 6.0); and ethanolamine (pH 6.0). The initial velocities of substrate cleavage in these buffers show a monotonous increase with increasing alkali ion concentration and follow a similar pattern to those observed in 4-(2-hydroxyethyl)-1-piperazineethanesulfonic acid (HEPES) or phosphate buffers that are shown as representative in the paper. This effect is more pronounced in KCl containing buffers, followed by NaCl and LiCl containing buffers (the influence of CsCl was not analyzed in these additional experiments).

As we discuss in the paper, it is difficult to explain the unexpected and significant activation of the enzymatic activity of HIV-1 PR in 2-(*N*-morpholino)ethanesulfonic acid (MES) buffer (pH 6.0) at very low salt concentrations of alkali ions (10–20 mM). However, since this activation is clearly due to a dramatic increase in substrate binding, as shown in Table 3 of the paper, we suggest that MES might form a semi-stable complex with the peptide substrate in the enzyme active site that stabilizes the Michaelis complex at very low salt concentrations. Direct experimental proof of this hypothesis requires further NMR and/or X-ray structure analyses and will be very demanding.

Professor Goodall asked: In your paper you mentioned that the effects could only be seen using the highest purity grades of salts and buffers, and that common buffers and salts gave variable results at low concentrations. Because the effects of polyvalent metal ions often confound measurements of univalent cation binding studies to anionic proteins and polynucleotides, I wonder whether you know the content of the cationic impurities in your salts? Also, whether you have tried using EDTA as an additive to compete for binding of any polyvalent metal ion impurities?

Professor Jungwirth added: At low salt conditions the salt concentration is comparable to that of the buffer. Shouldn't we than include the buffer salts into the Hofmeister series on a similar footing as we do with the added salt ions?

Dr Konvalinka replied: It is important to stress that the buffers used for the kinetic analysis have been prepared from the free acids and corresponding alkalic hydroxides. Consequently, the only cations present in the buffers were those analysed in the appropriate experiments. As for the anions, their influence on activity of various enzymes, including HIV protease, was thoroughly studied elsewhere (see the paper for references) and were not the focus of the current study.

Professor Dr Hugel addressed the meeting: I need to clarify a couple points regarding our interpretation of plateau forces compared to the interpretation in Li and Walker, *Proc. Natl. Acad. Sci. U. S. A.*, 2011, **108**, 16527:

Li and Walker use the three polymers: polystyrene, poly(4-*tert*-butylstyrene) and poly(4-vinylbiphenyl) adsorbed onto a Si substrate. They assume that in their systems the polymer remains a compact spherical globule in bad solvent, even in contact with the Si substrate. They then assign the plateau of constant force to pulling the polymer out of its globule with negligible impact of the substrate, directly reflecting the solvation free energy.

This interpretation is not valid for any of our systems showing plateau forces. We have shown significant impact of the solid substrate on the plateau forces. In addition, we have pulled polymers horizontally under a tension of several 10 pN for many micrometers, which we consider impossible in the case of only a small (point) contact of the polymer with the substrate. Finally, imaging of the polymers suggest no spherical objects (the height of the polymers is always significantly less than the width). Our interpretation is therefore that the polymer has significant interaction with the substrate and we measure a combination of the solvation free energy and the surface interaction free energy, which largely compensate each other. This is reminiscent of the protein folding problem, where large entropic and enthalpic contributions compensate to yield only small free energy changes between the folded and unfolded state.

Dr Vazdar then asked: How do you attach the investigated polymer to the AFM cantilever tip? What type of covalent bonds are present and how stable they are? Additionally, are there any differences in measured forces depending on the choice of the linker, *i.e.* how robust is the setup presented in the paper?

Professor Dr Hugel replied: The polypeptides are attached *via* their N-terminal amine *via* an NHS-ester to a previously amino-silanized AFM-tip (for details see the Experimental and computational methods section). We therefore get siloxane bonds and peptide bonds. Both have been shown to be stable up to more than 1.5 nN. The setup is extremely robust once you got a single polymer attached to the AFM-tip. This can be seen by the often several thousand force–extension traces which can be taken with one and the same molecule in different solutions and on different surfaces. It just takes some time to get all the conditions right to have exactly a single polymer attached to the apex of the AFM-tip. The desorption force is not affected by the choice of the linker. There is for example no difference in force comparing a coupling of the polymer by an NHS ester or a maleinimide to the AFM tip.

Dr Jurkiewicz spoke to the meeting: I would like to add something to the discussion on the adsorption of small cations on the surface of liquid-crystalline layer of polymer, like the one studied by Professor Kunz, or lipid bilayers in our case. We have lately investigated properties of negatively charged lipid bilayers formed from zwitterionic phosphatidylcholine and negatively charged phosphatidylserine, which is a model of the inner leaflet of the plasma membrane, suspended in alkali metal chloride solutions. Both experiments and simulations, some of them we recently published (Jurkiewicz *et al.*, *Biochim. Biophys. Acta*, 2012, **1818**, 609) show deep penetration of sodium and lithium to the lipid carbonyl groups instead of the predicted adsorption on the carboxylic groups of the phosphatidylserine, which are more exposed to the solution and thus more accessible for cations. This, in my eyes, reveals the complexity of lipid bilayers, and in general biological systems, the specificity of which is often governed by the interplay of electrostatics, hydrophobicity, and structural rearrangement.

Dr Hassenkam addressed Thorsten Hugel: You mention in your paper that you had to use a dwell time of 1 s for the polymer to reach the surface, did you do any experiments where you changed the dwell time?

Professor Dr Hugel replied: To keep our recent results comparable to previous work, we used a dwell time of 1 s. This is not necessary for the polymer to reach the surface. We tested dwell times in the range of 10 ms to 10 s and they do not affect the basic results.

Dr Cremer asked: Do you not see a difference between different chemical species as they pull away from the surface along the polymer chain? Please explain why the chemistry of the polymer doesn't play a greater role in your *z*-pulling experiments.

Professor Dr Hugel replied: The differences between the different chemical species within a polymer are well resolved in MD-simulations on polypeptides. Experimentally, we only see differences in between polymers of different species (*e.g.* polytyrosine in general has a higher desorption force compared to polylysine), but we do not see the 'sequence' of a polypeptide because the long (about 30 nm) flexible PEG linker 'averages' these sub-nanometer differences. The small differences with the chemistry of the polymer for the *z*-pulling experiments are puzzling on first sight. A couple of years ago we learned with the help of MD-simulations by the Netz group that this is caused by a compensation mechanism. Dispersion and water structural forces, which both have to be overcome upon desorption of the polymers (Horinek *et al.*, *Proc. Natl. Acad. Sci. U. S. A.*, 2008, **105**, 2842), largely compensate each other. When the force to dissolve the polymer becomes higher (*e.g.* for a hydrophobic polymer in water), the dispersion forces (between polymer and surface) become lower. This has been confirmed during the last years. Finally, recent MD-simulations by the Netz group show that the water structure at the surface and not the polymer determines single polymer adhesion, consistent with our results.

Dr Ruotolo enquired: Many groups have used AFM data to study the unfolding of multidomain proteins tethered to surfaces. Your work clearly shows that such measurements can also be used to determine the influence of anions and cations on the desorption of simple polymers from surfaces. If the two experiments were combined, one might be able to deduce Hofmeister-based effects on multidomain protein unfolding and surface adhesion. What, if any, 'cooperative' Hofmeister effects would you expect to see in such an experiment, and what would be the difficulties in the collection/interpretation of such data?

Professor Dr Hugel replied: We are currently investigating in this direction. There are two cases: (1) the domains of the protein stay folded in contact with the substrate and upon desorption (weak adhesion); or (2) the surface affects the protein structure (strong adhesion). In the first case the multidomain protein shows plateaus of constant force similar to the polypeptides and polymers we have investigated before. In the second case, data interpretation becomes tricky, as neither a plateau of constant force nor a unique unfolding pattern is observed. We are hoping to learn from the (in this case non-equilibrium) total work necessary for the desorption, which requires experiments with different force loading rates – which are on their way. I am not sure what to expect and looking forward to my students' data.

Professor Jungwirth addressed Brandon Ruotolo: You said that the "leftover species" in your systems is not water but most likely buffer. How about then including the effect of the buffer to the effect of the salt ions into your considerations?

Dr Ruotolo replied: The nomenclature we use in our report is probably slightly different from much of the Hofmeister community. We begin our experiments with a base *buffer* of 100 mM ammonium acetate. While this solution has little buffer capacity at neutral pH, it provides a base ionic strength for our experiments and ensures that the total solution pH does not tend toward great acidity during the electrospray process. Ammonium acetate is also a highly volatile salt, and for this reason it is a favorite buffer among mass spectrometrists who analyze proteins from aqueous solvent. To that solution, we add 20 mM of various *salts* which we intend to use as stabilizers. Note the concentration difference between the *salts* and *buffer* used in our experiments, neither within the concentration range for typical Hofmeister-type effects, and thus both components might be referred to as *buffer* by some experimentalists in the Hofmeister community. During electrospray ionization, our solvent evaporates, along with all of the volatile components of our buffer. Thus the effective concentration of all of the involatile components in solution is increased dramatically during what can best be described as a *desolvation jump* on the millisecond timescale. Therefore, what could have been called *buffer* prior to the electrospray process, becomes *added salt* during desolvation. Our overall goal is to do just what you suggest, tailor the composition of low-concentration buffer components such that the *leftover species* that remain after electrosrpay ionization are beneficial for gas-phase protein stability. In our report here, we examine the influence of both anion and cation species acting in concert toward this end, rather than in isolation, for the first time.

Dr Cremer asked: In water one can often observe a reversal of the Hofmeister series when the surface charge is changed from negative to positive. What is the nature of series reversal in the absence of water?

Dr Ruotolo replied: The generally-reversed rank order we observe for the ability of anions and cations to stabilize protein ions in the absence of solvent is a product of the global stabilities that we record for intact gas-phase protein complex ions. These stability measurements depend largely upon phenomenology specific to gas-phase protein ions, including: charge migration upon thermal unfolding of complexes, Coulombic unfolding, and eventual dissociation of the complex *via* a pathway that ejects unfolded constituent proteins from multi-unit complexes. As stated previously, the micro-scale stabilities and binding affinities we find in our gas-phase measurements, specifically those related to local anion and cation binding, are likely more-directly correlated with the canonical Hofmiester series.

Dr Canchi addressed the meeting with a short presentation entitled "Preferential exclusion of TMAO from proteins": I thank Prof. Jungwirth for giving me the opportunity to share new results on the protecting effect of TMAO on protein stability from my doctoral research in the lab of Professor Angel Garcia at Rensselaer Polytechnic Institute. As discussed previously by Professor Record and Professor Harries, proteins can be stabilized or destabilized by the addition of small organic molecules, known as cosolvents, to the solution. Urea is a commonly used denaturant, *i.e.*, it destabilizes proteins. Trimethylamine *N*-oxide (TMAO) is a naturally occurring osmolyte that stabilizes proteins, and can counteract the destabilizing effect of urea in a 2 : 1 concentration ratio of urea to TMAO. We are interested in understanding the molecular mechanism of these phenomena using all-atom Molecular Dynamics (MD) simulations. We have previously reported folding equilibrium of the Trp-cage miniprotein in presence of urea using Replica Exchange MD, and obtained *m*-values in good agreement with experiment. However, we could not capture the stabilizing effect of TMAO in our simulations (Fig. 1).

This prompted us to reparametrize the popularly used Kast model for TMAO based on solution thermodynamics, in our case osmotic pressure data. Using the recently proposed method by Luo and Roux to calculate osmotic pressure in MD simulations, we find that the Kast model shows osmotic pressure greater than the van't Hoff law, but underestimates the osmotic pressure of TMAO solutions when compared to experiment (Fig. 1) The interpretation here is that the water-mediated repulsive interaction among two TMAO molecules in the Kast model is repulsive, but not to the extent suggested by experiment. We have proposed a new model for TMAO, the Osmotic model, that captures experimental osmotic data for TMAO solutions by making two changes to the Kast model: (i) scale charges uniformly by a factor of 1.2; (ii) use a non-default mixing rule to reduce TMAO–TMAO L-J interaction parameter, ε, by 25%. We next examine the preferential interaction behavior of these two models of TMAO with protein surfaces. For a wide variety of proteins, we find that the Osmotic model shows greater exclusion from proteins than the Kast model (Fig. 2).

This behavior, qualitatively consistent with experiment, is a consequence of incorporating realistic TMAO–water interactions through osmotic pressure measurements. It is instructive to examine the interactions in the two-component TMAO–water

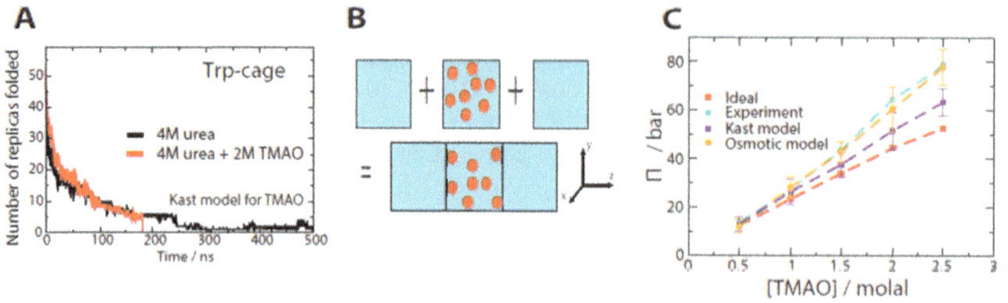

Fig. 1 Model development for TMAO. (A) REMD simulations of Trp-cage in 2 : 1 urea–TMAO mixture show similar decay to equilibrium as the simulation with urea alone, indicating that the Kast model is unable to capture the stabilization of proteins by TMAO. (B) Scheme to calculate osmotic pressure in MD simulation. TMAO molecules (red) are confined to the central region while water (blue) can diffuse freely. Osmotic pressure is obtained from the force required to confine TMAO to the central region. (C) The Osmotic model for TMAO is obtained by matching the osmotic pressure of TMAO from simulation to the experimental value, over a range of TMAO concentrations.

system that leads to change in preferential exclusion from proteins. We demonstrate that varying the polarity of TMAO by systematically scaling the charges has a significant effect; it's preferential interaction with proteins, shown here for Trp-cage. (Fig. 3).

By scaling down the charges or making the molecule hydrophobic, we increase the attraction between TMAO molecules and reduce the osmotic pressure of the TMAO–water system. The preferential interaction of TMAO with protein increases concomitantly with scaling down of the charges. Scaling up the charges on TMAO leads to an increase in repulsion between the TMAO molecules, increases the osmotic pressure and increases the preferential exclusion from the protein. Qualitatively similar results are obtained for the case of urea. (Fig. 3) This intimate connection between preferential exclusion and osmotic pressure provides us physical insight into the role of solute–solute interaction in determining preferential accumulation or exclusion of co-solvents from proteins. Denaturants such as urea have weak favorable interactions with protein surfaces and a significant concentration of cosolvent is required to denature the proteins. For this to happen, we speculate that denaturants, in addition to favorable interactions with proteins, must also be able to admit their own excess concentration in the vicinity (local domain) of the protein, *i.e.* show a negative deviation from ideality with osmotic coefficient < 1. Osmolytes on the other hand must have repulsive interactions with each other in addition to repulsive interactions with the protein, which provides a factor that contributes to making its accumulation in the vicinity of the protein unfavorable. One can think of denaturants and osmolytes as lying on opposite sides of the ideal osmotic pressure curve given by the van't Hoff law. (Fig. 4).

From experimental osmotic pressure data, this scenario is true for the case of urea and TMAO, as well as for other osmolytes such as betaine, trehalose, proline and glycerol in the concentration range reported.

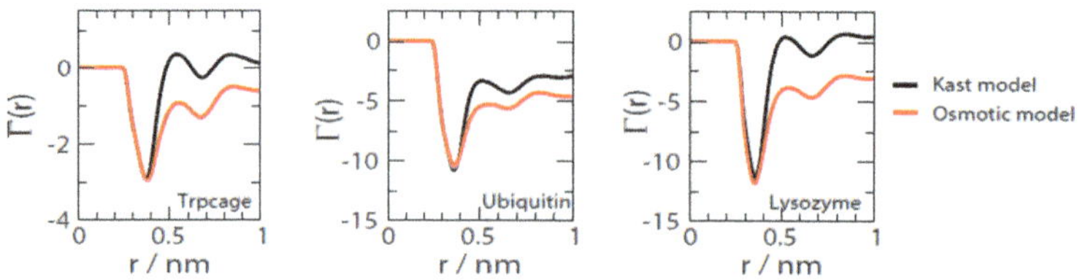

Fig. 2 The Osmotic model for TMAO shows greater exclusion from protein surfaces over the Kast model, as indicated by more negative values of $\Gamma(r)$. This is a consequence of the osmotic pressure parameterization.

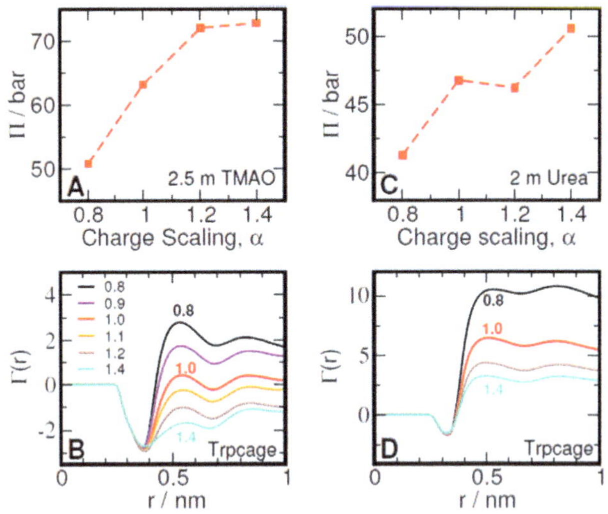

Fig. 3 Osmotic pressure and preferential interaction. (A) Osmotic pressure of 2.5 M TMAO solution as a function of scaling of TMAO charges. $\alpha = 1.0$ corresponds to the Kast model. (B) Preferential interaction of Trp-cage with TMAO as a function of scaling of TMAO charges. (C) Osmotic pressure of 2.5 M urea solution as a function of scaling of urea charges. $\alpha = 1.0$ corresponds to the KBFF model for urea. (B) Preferential interaction of Trp-cage with urea as a function of scaling of TMAO charges.

The exercise of model development for TMAO has given us insight into the interactions that govern preferential interaction of cosolutes with proteins. We expect that the use of the Osmotic model for TMAO will enable simulations of folding equilibrium of proteins that captures the stabilizing effect of TMAO, and provide a detailed molecular mechanism of the phenomenon.

Professor Record asked: The effectiveness of solutes as osmolytes *in vivo* is a result both of their unfavorable preferential interactions with cell biopolymers and their unfavorable self interactions relative to interactions with water. For example, glycine betaine (GB) is a very effective osmolyte *in vivo* (Cayley and Record, *Biochemistry*, 2003, **42**(43), 12596–12609) in large part because it lacks hydrogen bond donor groups and so cannot compete effectively with water to interact with nucleic acid and membrane phosphates and with protein carboxylates and amide oxygens, which together make up a significant fraction of the surface of native biopolymers in the cell (Capp *et al.*, *Biochemistry*, 2009, **48**, 10372–10379). In addition, GB is found to interact less well than water than with aliphatic hydrocarbon, which makes up about 50% of the surface of native proteins, though GB interacts more favorably than water with aromatic hydrocarbon. The lack of hydrogen bond donor capability means that GB cannot complete with water to interact with its own carboxylate oxygens, resulting in a large positive osmotic coefficient for a two component GB–water solution (Capp *et al.*, *op. cit.*). Similar reasoning presumably applies to TMAO, which chemically is quite analogous to GB. Other *E. coli* osmolytes like proline, potassium glutamate and trehalose are less effective than GB in allowing the osmotically stressed cell to retain water and maintain a high growth rate (Cayley *et al.*, *J. Bacteriol.*, 1992, **174**, 1586–1595). We are quantifying the interactions of these osmolytes with the different functional groups of cell biopolymers; results to date indicate that these interactions with cell biopolymers, and not their self-interactions, are the primary determinant of their effectiveness as *in vivo* osmolytes. A comment about TMAO: while an effective *in vivo* osmolyte, this solute is not nearly as good choice for *in vitro* experimental studies as the equally

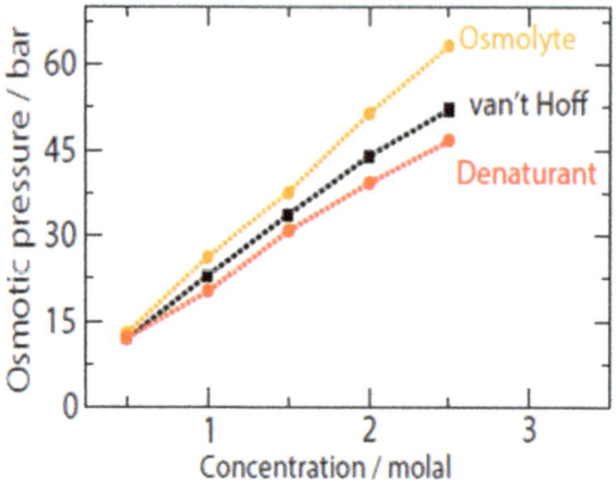

Fig. 4 Osmotic behavior of cosolvents. We hypothesize that osmolytes (orange) show a positive deviation from the van't Hoff law (black), while denaturants (red) show a negative deviation from the same.

effective *in vivo* osmolyte GB. Not only is TMAO chemically unstable, decomposing to the volatile trimethylamine which complicates vapor presssure osmometry, but also its solutions are quite alkaline and neutralization converts TMAO to $TMAOH^+$ which has a hydrogen bond donor capability and hence must compete better with water to interact with anionic and amide oxygen than TMAO does. Solutions of GB, on the other hand, are chemically stable and near-neutral pH.

Dr Canchi replied: Based on our results, we are trying to convey that osmolyte self-interaction plays an important role in determining the magnitude of exclusion from protein surfaces. From experimental data, we note that stabilizing/excluded osmolytes show positive osmotic coefficient for two-component solution. (Courtenay *et al. Biochemistry*, 2000, **39**(15), 4455; Fig 1). However, at the same two-component osmolality, it is shown that the ΔOsm upon addition of the protein varies depending on the identity of solute. (Courtenay *et al.*; Fig 2). This shows that chemical nature of the osmolyte matters too, and it is not just the self-interaction that determines it's preferential interaction with the protein. We are suggesting that cosolute self-interaction is a part of nature's strategy in selecting which molecules are used as osmoprotectants (or denaturants). This is a hypothesis that can be tested by synthesizing a library of molecules similar to a given osmolyte, and characterizing osmotic and exclusion behavior as well as cell growth rates for this library. Consider the case of urea and methylated urea derivatives, as discussed by Professor Bakker. Urea has a tendency to aggregate, and shows a negative deviation from the van't Hoff law. Methylated ureas are known to be stronger denaturants than urea – we propose that this is due to two effects: (i) the direct L-J interaction of methylated ureas with a protein surface is stronger than urea; (ii) methylated ureas are expected to show a larger self-interaction than urea, *i.e.* a smaller osmotic pressure. Both these factors are expected to act in concert to increase the preferential accumulation of methylated ureas as compared to urea.

The point about difficulty of using TMAO in experimental studies is well taken. It will indeed be useful to develop MD simulation parameters for GB, for which a wealth of experimental data is available.

Professor Mason enquired: I have a concern that salt bridges are over-emphasized in non-polarizible MD simulations following our recent study on potassium carbonate (*J. Phys. Chem. B*, 2012, **116**(28), 8145–8153). Do you have any thoughts on the idea that the simulation may have been giving anomalous results due to the salt bridge in Trp-cage, and the possibility that the reparameterization of TMAO that you have performed here is just compensating for this shortcoming?

Dr Canchi replied: While it may true that non-polarizable MD forcefields may overemphasize salt-bridges, I believe our results are not affected by that artifact. For the calculation of preferential interaction of TMAO with Trp-cage, at various charge-scaling factors for TMAO, the protein was simulated in its folded state and remains folded for the entire course of the simulations. The variation in preferential interaction comes from the change in strength of interaction of TMAO with water. This behavior is quite general, and we have observed it in similar charge-scaling simulations of TMAO (as well as urea) with other proteins too.

Dr Ottosson addressed Thomas Record: If I understood right, in your two-state model of the air–water interface you have to "calibrate" it by assuming a certain thickness and a concentration of ions for some extreme limiting cases – in this case sodium sulfate; first then can you from the surface tension data for other salts predict an ion partitioning for these? My question is then; how sensitive are your model predictions for other salts to the assumption that both sodium and sulfate are fully depleted in the interface layer? Similarly, how influential is the assignment of the interface layer thickness in the case of Na_2SO_4 on the overall results?

Professor Record replied: Thank you for the question. The molecular thermodynamic analysis of a surface tension increment for a solute $d\gamma/dm_2$ or of a chemical potential derivative $d\mu_2/dm_3$ for the interaction of a solute (3) with the surface of a biopolymer or model compound (2) in solution interprets these quantities in terms of the product $(K_p - 1)b_1$ where b_1 is the number of water molecules per Å^2 of surface and K_p is the microscopic partition coefficient (the ratio of the local concentration of the solute at the surface to its bulk concentration). We assume b_1 is a property of the surface, and obtain a lower bound value of b_1 using the solute with the most positive values of $d\gamma/dm_2$ or $d\mu_2/dm_3$, which is the most excluded solute and hence the solute with the smallest value of K_p. For the air–water interface and molecular hydrocarbon surface in aqueous solution, Na_2SO_4 has the most positive values of these derivatives. Assigning $K_p = 0$ for both Na^+ and SO_4^{2-} ions, we calculate $b_1 = 0.18$–0.19 H_2O Å^{-2} for both these nonpolar surfaces, which is two layers of water at bulk density. From analysis of $d\mu_2/dm_3$ for interactions of the excluded solute glycine betaine with amides and carboxylates, we find that glycine betaine is almost completely excluded from this same amount of local water at amide O and carboxylate O groups. This is a lower bound value of b_1 for these nonpolar and polar surfaces; if Na_2SO_4 and glycine betaine are less completely excluded ($K_p > 0$) then b_1 is larger than 0.18 H_2O Å^{-2}. We assume that $b_1 = 0.18$ H_2O Å^{-2} for other molecular surfaces. (The only possible exception to date may be anionic phosphate oxygen, where we determine a larger value of b_1 using glycine betaine.) We assume $b_1 = 0.18$ H_2O Å^{-2} is generally applicable, and calculate values of K_p for other salt ions and solutes based on this b_1. If b_1 is larger because of incomplete exclusion or if b_1 is not the same for other functional groups (*e.g.* amide or cationic N, hydroxyl O, *etc.*) as for the ones we have investigated, K_p values for solutes or salt ions and these functional groups will change accordingly, but it seems unlikely that the semiquantitative conclusions like the rank order of interactions of salt ions or solutes with a given surface or functional group obtained from the order of K_p values will be greatly affected.

Dr Ottosson addressed Professor Jungwirth: The two-state ion partitioning model presented by Tom Record made use of the fact that sodium sulfate seems to be an extreme case where both ions are strongly repelled from the interface and the calculation of the partitioning of other ions was made using the assumption that both sodium and sulfate are fully depleted in the interface layer (and if I remember right the thickness of the this layer was on the order of a nanometer or so?) Tom made references to the MD simulations of Pavel Jungwirth which also showed this behavior, nearly quantitatively. My question is therefore: how well does the density profiles from similar simulations of other salts which are less depleted (or even adsorbed) at the interface compare with the predictions from Tom's model? Can we learn something interesting from such comparisons?

Professor Jungwirth replied: The assumption of the partitioning model that sodium and sulfate are both fully depleted from the water surface is a reasonable one (although simulations indicate that sulfate is more depleted than sodium). Overall, the agreement between the partitioning model and results from MD simulations is quite good. It cannot be perfect also since simulations are often done at high salt concentrations and the partitioning model treats the ions separately, excluding ion–counterion interactions.

Mr Faraci asked: It might be of interest to the delegates the reading of the works published by Jacques Loeb between 1920 and 1924. The book *Proteins and the theory of colloidal behavior*[1] in particular, contains a collection of the author's previous findings in the field of colloid chemistry, together with a long explanation of the results obtained. The book can be downloaded freely from the internet. Note: A second edition of the book exists which appeared in 1924 and which contains two years more of experiments. This is however not available online, at least to my knowledge.

The phenomena Loeb dealt with include the specific effects of several acids, alkalis, and neutral salts, on swelling, osmotic pressure, as well as viscosity of protein solutions; the proteins used are gelatin, albumin, and casein.

By using the pH of the protein solution as a basis for the comparison between the effects of the acids, alkalis, and salts used, instead of their concentration, Loeb showed that for a fixed pH the influence of ions on the properties studied depended on their sign and valency alone.

It is worth noting that Loeb voluntarily avoided the study of precipitation phenomena, as protein aggregation would have prevented their stoichiometric combination with the compounds used. The results in ref. 1 are indeed expected to hold only to the extent that no constitutional change of the protein molecules occurs. The understanding of the precipitating action of salts was nevertheless suggested to be possible after that of their effects on the properties mentioned above. Although Loeb's findings do not provide *per se* a solution to the questions regarding ion specificities, and in particular do not explain how the pH is altered specifically by each different ion, the perspective adopted (namely that of studying the effects observed by plotting them against the pH instead of against concentrations) helps clarifying the relations between the role of H^+ and that of other ions in influencing the behaviour of colloids. At least with respect to the effects and conditions studied by Loeb, which is worth noting are of primary importance in physiological preparations, it appears that the specific influences of ions manifest themselves through an alteration of the pH.

1. J. Loeb, *Proteins and the theory of colloidal behavior*, 1922, McGraw-Hill Book Company.

Professor Jungwirth addressed the meeting: There may be two ways how to look at the specific ion effects – in the context of overall effects of ions on proteins or in the context of a particular biological function (enzymatic reaction, ion channeling, *etc.*). The former, which concern protein salting in/out and (de)stabilization are the traditional Hofmeister effects, where the interactions of salt ions are relatively weak. But many interesting questions for future "Hofmeister investigations" may concern situations involving stronger ion bindings; situations toward those characteristic for biological function.

Professor Lund added: Pavel Jungwirth asked if there is a fundamental difference between weakly and strongly bound ions – *i.e.* Hofmeister ions *vs.* protons and if this would matter when treating such binding with a two-state, thermodynamic model. Two comments to this:

(1) For reasonably short ranged, attractive potentials of mean force between the ion and a motif, a binding constant can be defined, regardless of the depth of the minimum.

(2) On one hand pH – *i.e.* the proton concentration – can be used to tune the charge state of acids and bases at *low* ionic strengths. On the other hand, more weakly bound ions require *higher* bulk concentrations to promote a fully bound site. This consequently leads to screening of long range electrostatic interactions, whereby the charge change may not be as significant for intermolecular interactions as for charge changes induced by changes in the bulk proton concentration.

Professor Levin added: I agree with Mikael's comment. Our philosophy is very similar. Since we do not have the full QM theory for proton water interaction, we write a very simple square well potential for hydronium at the interface. The range of the potential is one hydrogen bond, and its depth is adjusted to produce the correct surface tension of HCl. The potential depth is the only adjustable parameter in our theory. Once it is fixed we use the same potential to calculate surface and interfacial tensions of many different acids, obtaining a good agreement with experiments, see: A. P. dos Santos, and Y. Levin, Surface tensions and surface potentials of acid solutions, *J. Chem. Phys.*, 2010, **133**, 154107.

Professor Canongia Lopes continued: My main scientific research interests lie in the area of ionic liquids. One of the reasons I have attended this conference on the Hofmeister series is that more-and-more people from the ionic liquid community are interested in studying systems of aqueous solutions of ionic liquids interacting with different solutes, namely polymers or proteins. The formation and use of aqueous biphasic systems (ABSs) is a hot topic in this context; cellulose processing is another.

On the other hand, I think the "Hofmeister series community" can also have some feedback from the ionic liquids area: in order to find new salts that are liquid at room temperature many "exotic" anions and cations have been developed during the last decade. For instance, many anions like halides, thiocyanate or methyl sulfate are shared between the two communities, but anions such as bistriflamide, $N(CF_3SO_2)_2$, that exhibits not only charge dispersion between the central nitrogen atom and its four oxygens but also different conformations, could be useful ways to extend the Hofmeister series to novel, uncharted territories.

Professor Dr Fedorov concluded: I would like to attract some attention to ion effects in non-aqueous solvents and nanoparticle dispersions. Indeed, water is not the only one polar solvent where considerable amounts of ions can be dissolved. These days the interest to non-aqueous polar solvents (*e.g.* acetonitrile, pyrrolidone-based solvents, ionic liquids *etc.*) is rapidly increasing because of a number of applications of these liquids in electrochemistry (non-aqueous electrolytes), solubilization of nanoparticles, cellulose processing *etc.* As such, it has been recently found that contrary to water, pyrollidone-based solvents (*e.g.* *N*-methylpyrollidone, NMP) can disperse significant amounts of carbon nanoparticles without using any extra dispersing agents (*e.g.* surfactants of polymers). Because NMP is a rather inexpensive solvent it opens new avenues towards mass-scale manufacturing of carbon nanomaterials. However, the number of works on fundamental physical–chemical mechanisms of ion solvation and ion effects in these solvents is still very small compared to the amount of works on ions in aqueous solvents. I would like to note that ions could show quite interesting effects in polar non-aqueous solvents. For instance, we recently found that even millimolar concentrations of an inorganic salt (NaI) can have very significant effects on the stability of carbon nanotube dispersions in NMP and the mechanisms of carbon nanotube bundle formation in these dispersions (see Fedorov *et al.*, *Phys. Chem. Chem. Phys.*, 2011, 13(27), 12399–12402; Frolov *et al.*, *Chem. Sci.*, 2012, **3**(2), 541–548). Overall, I would like to suggest that as a community we would pay somewhat more attention in the future to ion effects in non-aqueous solvents and nanoparticle dispersions because these effects can be important for many application areas in energy, biopolymer processing and nanoscience.

Concluding remarks: *Cum grano salis*

Philip Ball*

Received 2nd October 2012, Accepted 4th October 2012
DOI: 10.1039/c2fd20126g

Hofmeister effects are part of a larger story – one in which the devil is perhaps in the details, but which promises to give us a much deeper understanding of how the solvent is a part of cell and molecular biology.

1 Introduction

A meeting of this sort seems to presuppose almost by definition that everyone knows why we're here. Many of the papers in this volume dispense with the usual justifying preliminaries, because, I presume, of an unspoken agreement among the participants that ion-specific effects in biological, colloid and polymer chemistry – the effects Franz Hofmeister discovered in the late nineteenth century – are interesting and important.

But I've observed that in the discussion sessions this silent consensus began to unravel. There the unresolved questions started to arise: what exactly *is* it that we're here to talk about? Where are the boundaries of this topic? What is the ultimate goal? Is this chemistry, or physics, or molecular biology, or physiology?

This diversity of views is, however, also not so far from the surface in this collection of papers. One of them might imply, for example, that understanding salt effects is important for the practical reason of precipitating and crystallizing biological macromolecules. Another might remind us that the topic is important because, after all, living cells are salty. Or it is important because it bears on broader questions of protein stability and folding. It could prove significant for devising drug formulations. It could be relevant to the ion-dependent catalytic function of all manner of proteins. Or perhaps we want to understand the air–electrolyte interface for reasons to do with atmospheric chemistry, or mineralogy, or electrochemistry. Or we want to understand industrial chemical formulations such as dispersants, foams and surface coatings. Some contributions, meanwhile, very reasonably take it for granted that Hofmeister effects are worth elucidating for their own sake, as an issue in pure (bio)physical chemistry.

A part of what interests me about this topic is that, as Thomas Record has mentioned, it is an aspect of biology that tends to be swept under the rug. Isn't that odd? For while it is easy to understand how Hofmeister's experiments on the addition of salts to egg white might be regarded as a historical curiosity of slight relevance to a remote corner of protein chemistry, it rapidly becomes clear – indeed, you barely have to think about it for a moment – that the behaviour of proteins and other biological macromolecules in salt solution is an ubiquitous phenomenon. In fact, one might even argue that "the behaviour of proteins and other biological macromolecules in salt solution" is perhaps as good a working definition of what life is as any other. If we don't understand this, how can we hope to understand anything in biochemistry and molecular biology?

But Hofmeister effects are of course not the only aspect of molecular science that tends to get swept under the rug in cell biology. In fact, the whole idea that molecular

18 Hillcourt Road, East Dulwich, London, UK SE22 0PE; Tel: +44 208 693 6336

biology needs to be based in sound physical chemistry is one that, with some notable exceptions such as muscle action and nerve signalling, seems to have been steadily eroded over the second half of the twentieth century. I believe one can locate the beginnings of that process to 1953, when the primary sequence of DNA became seen as the carrier of all biological information worth knowing, and proteins became apparently slaved to the nucleic-acid master. The message that an evolutionarily shaped genetic sequence implicitly acknowledges the environment in which it will be expressed gradually gave way to a notion that the environment barely warranted any consideration at all. I used to tell lay audiences that, without water, proteins don't generally fold and function, and DNA loses its beautiful double helix. I now realise I shall hereafter have to add that, without a bit of salt around, it wouldn't much matter whether they did or not.

In any case, I feel it is worth mentioning a few more of these neglected physio-chemical aspects of molecular and cell biology. It is not necessarily that these issues are never acknowledged in biology textbooks, but that the impression that emerges from the standard discourse of the field is that they are either well understood or peripheral, and can be relegated to the background. Here are a few:

- The structure and dynamics of pure liquid water[1,2]
- The hydration of biomolecules[3,4]
- The hydrophobic effect[5,6]
- Crowding and the effects of very high local concentrations[7,8]
- The mechanics of DNA[9,10]
- The packaging of chromatin and its relation to gene function[11,12]
- The phase behaviour of complex biological fluids[13,14]

To take just one example from this list: the hydrophobic effect becomes more elusive, or at least more polyvalent, the more closely it is interrogated. One common view is now that the hydration of small hydrophobes is qualitatively different from that of large ones.[15] In the former case, water molecules in the hydration shell might be able to adapt their orientation so as to preserve hydrogen-bonding around the cavity represented by the hydrophobic entity,[16] while for larger, extended surfaces, breaking of hydrogen bonds is unavoidable, and the picture is then dominated by the interfacial energies. When two such surfaces come close together, at some point the water remaining between these depletion layers might simply evaporate. If the gap between the surfaces becomes dry, the surface tension of the menisci at the edges of the space will pull them into contact. This is called a dewetting transition, and Lum *et al.* predict that it might be common between two hydrophobic surfaces when they are just 1 nm or so across.[15] However, it now looks as though dewetting is probably rather rare as an assembly mechanism for hydrophobic domains of proteins, since these surfaces are typically not hydrophobic enough, or geometrically conducive, to make dewetting happen.[17–20]

But this does not mean that it is of only marginal relevance to biochemistry. Chandler and coworkers have argued that dewetting does not demand the spontaneous appearance of a cavity but draws instead on the intrinsic fluctuations of water density at the water–hydrophobe interface.[21] In this view, proteins that do aggregate by dewetting[17] lie on one side of the fluctuation-driven transition, and others that do follow this mechanism[18] lie on the other side.

Similarly, while the discourse around the hydrophobic effect in ligand binding has long been dominated by the assumption that there is a single explanatory picture involving the expulsion of hydration water in the binding cleft, in reality there may be many different types of interaction and structural change involved whose details depend on the specifics of the situation.[22] At a broad level the 'hydrophobic effect' surely does involve changes in water structure close to solute surfaces, relative to that in the bulk phase, but a detailed balance of entropic and enthalpic contributions to substrate binding and water release – which must take into account also desolvation of the substrate – is likely to vary on a case-by-case basis that can be understood only by detailed analysis.

One could doubtless add more instances to the above list of neglected questions in 'the physical chemistry of molecular biology'. It is an interesting sociological question why certain communities in science decide that particular aspects of a problem are worth devoting a great deal of attention to while others become minority concerns, if not in fact regarded as somewhat suspect and disreputable. Why would we place so much emphasis, for example, on determining crystal structures of proteins and relatively little on a deep understanding of the forces (of which hydrophobic interactions are a major component) that hold that structure together and that enable it to change and flex so that the molecule can do its job?

Some reasons for this relative neglect of the detailed *principles* governing biomolecular structure, dynamics and function might be fairly obvious. It looks more easy to understand a system like that in Fig. 1a than in Fig. 1b. But there is now good reason to believe that all those solvent molecules in the halo of the latter are not background noise but may play a central part in the molecule's function.[4] This overlooking of the 'active environment' is starting to change, however. A recent structural determination of the entire photosystem II complex, for example, presented

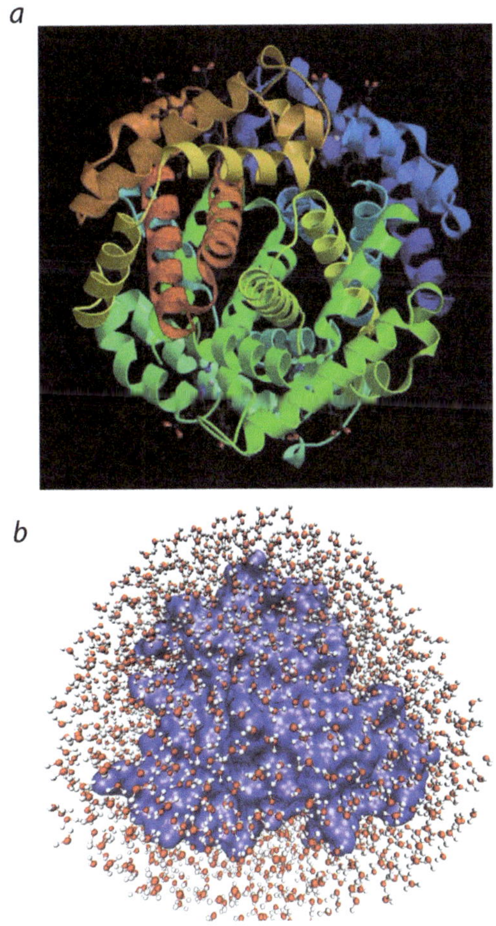

Fig. 1 Biomolecules such as proteins tend to be displayed as if *in vacuo* (a). A consideration of the detailed solvation environment complicates the picture (b), but may ultimately be the only way to understand *in vivo* structure and function [(b) from H. Frauenfelder *et al.*, *Proc. Natl Acad. Sci. U. S. A.*, 2009, **106**, 5129].

a detailed view of the hydration shell of around 1700 water molecules, some of which were revealed to be arranged in a highly specific and functional manner (Fig. 2).[23] (Even here, however, one must still ask where the salt goes!).

Clearly one can't expect to handle all this complexity at once. And the question that raised its head at this meeting is that of which parts of the phase space of these excessively complicated systems one might legitimately suppress in order to make progress. One option is to simplify the environment, perhaps by suppressing the co-solutes and coarse-graining or eliminating the solvent (Fig. 3a). Another is to keep the complexity of the solvent, perhaps along with its ions and small molecules, and to reduce the complexity of the solute (Fig. 3b). In the other direction, one can work towards the ultimate goal – a living organism – along various pathways: to acknowledge the heterogeneity and spatial constraints of the cell, for example, or to insert the macromolecules into a topological web of interactions – an interactome or genetic network (Fig. 3c, d).

There's no obviously best way to do it. Each choice has its virtues; each will miss something likely to be significant. In particular, we need to recognize that each of these pathways is in general unlikely to be continuous and incremental, but will have its own emergent phenomena that manifest at different stages of the idealization. The trick is always to know which of the phenomena at one level need to be carried over, perhaps in some parametrized form, to the next, and which can be left behind. All this is really by way of saying that all of the investigations of ion-specific effects that have been presented at this gathering have a valid claim to occupying some part of this picture.

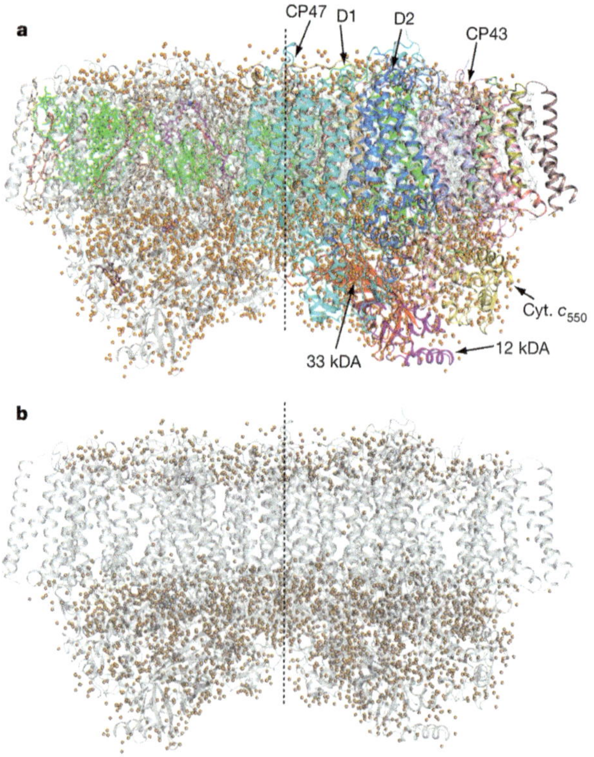

Fig. 2 The structure of photosystem II, with the hydration shell highlighted in (b). Reprinted with permission from ref. 23.

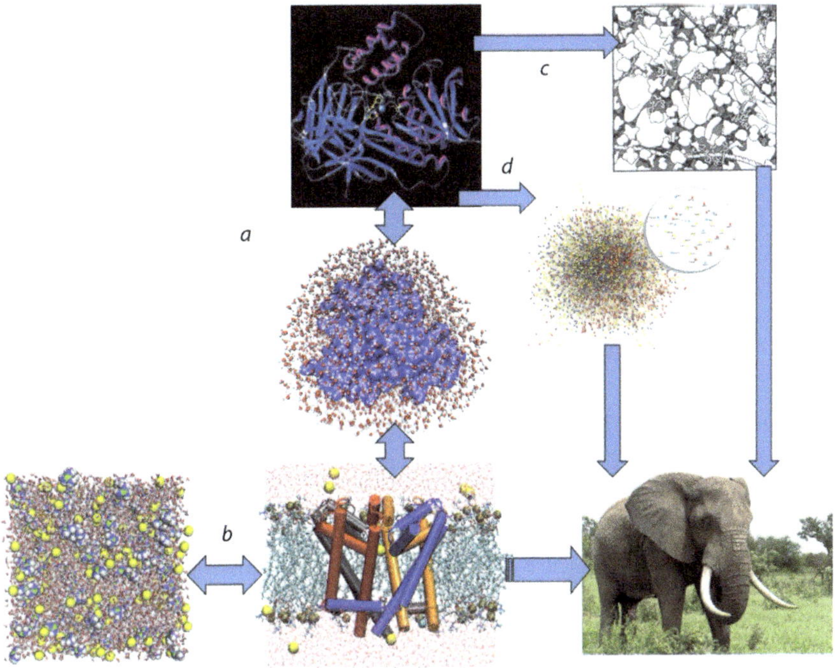

Fig. 3 To understand the complexity of the cell at the molecular level, we have to start by simplifying – somehow or another (a, b). Just as there is no unique prescription for that, so there is no unique way to progress from this level of detail to the organismal scale (c, d).

2 What's in a series?

I feel that I have been led to a better comprehension of what Hofmeister effects are. I don't mean that I now understand why Hofmeister's series looks the way it does. Rather, I am left suspecting that the series is in itself perhaps somewhat arbitrary. After all, it is merely a series in the first place: not a quantitative or even a qualitative correlation, but a list of how a diverse collection of reagents happen to rank in terms of their effect on protein solubility. Writing it as a series invites the inference that there is a unifying mechanism. My impression from the discussion at this meeting is that there is unlikely to be such a mechanism. Certainly, a steady progression in size and polarizability of halides might be reasonably expected to elicit a comparably gradual trend in Hofmeister activity. But once you add multivalent ions, polyatomic ions, hydrogen-bonders and so forth, there's no reason at all to expect any unity in the nature of the interactions involved,[24] or any preservation of the ranking when other variables (such as pH) are modified (Fig. 4). Are the observations of partial or complete reversals of the series telling us that there are several, sometimes conflicting influences behind it?[25] It remains possible, one must presume, that general principles might be identified if, for example, Coulombic effects dominate the interactions. But even in the case of alkali-metal cations it remains unclear whether one should necessarily expect monotonic trends when the nature of the hydration shell itself might not change in monotonic fashion.[26,27]

So just as it seems to be increasingly accepted that the hydrophobic effect is in fact a collection of several, perhaps many different phenomena with diverse origins, so too it appears that we may have to talk about Hofmeister effects in the plural, and to recognize that this bit of physical chemistry is too complicated and many-factored to yield to simple, all-encompassing explanations. In other words (and putting it a little too simplistically), this is a question of chemistry, not physics.

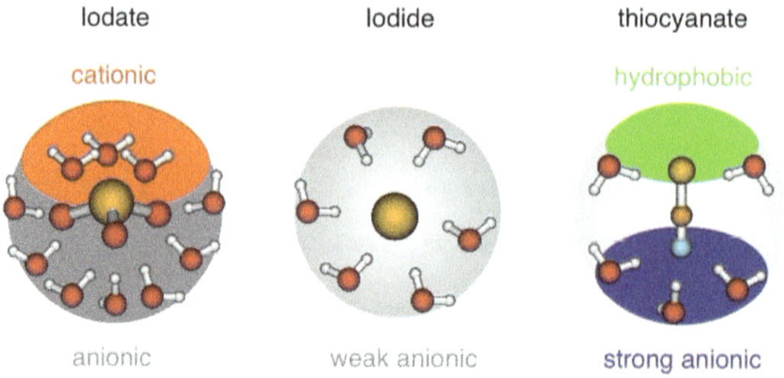

Fig. 4 The Hofmeister series: one effect or several? Courtesy of Marcel Baer.[24]

And this is perhaps after all how chemistry sometimes differs from physics, in that it often moves not towards unification but towards phase separation and clearer systematization of difference. Recall that five hundred years ago all salts themselves were considered to be expressions of some generic 'principle of salt', a useful myth that was relinquished when better ideas of similarity and difference were put in place.

Nonetheless, it makes sense to have at least some notion of what is and is not a Hofmeister phenomenon, as opposed to a biochemical or colloidal phenomenon in which particular ions play particular roles. I don't presume to suggest what that definition should be, but I would suggest both that it is worth formulating it and that it is important not to make it too prescriptive or dogmatic.

Among earlier attempts to provide some unified understanding of the Hofmeister series, the alleged effects of ions on 'water structure' have offered a seductive and now tenacious theoretical framework. This traditional explanation invokes the concept of 'structure-making' and 'structure-breaking' ions. The basic idea is that large, low-charge ions such as I^- and NH_4^+ disrupt 'water structure' – they are structure-breakers – while small or highly charged ions such as F^- and Mg^{2+} are structure-makers, imposing order on the hydrogen-bonded network. Then salting-out and salting-in of proteins are explained on the basis of entropic changes induced in their hydration shells by the addition of ions, or alternatively, of a reduction in the strength of hydrogen bonding of water molecules complexed to dissolved ions. The classical hypothesis is that salting-out arises from a competition for solvation between the salt and the protein, in which an ion's ability to sequester hydration water is somehow connected to its effect on bulk water structure. Thus the structure-making effect of small or highly charged ions depletes proteins of hydration water and causes precipitation.

In this view, ions are classified in terms of their ability to enhance or degrade the short-ranged order in the liquid: they are called either kosmotropes or chaotropes, respectively. By allegedly increasing the local order, kosmotropes render water more ice-like, with a consequent decrease in its ability to dissolve electrolytes. Chaotropes, meanwhile, promote a denser, more 'liquid-like' water structure. Thus, chaotropes and kosmotropes may set up gradients in chemical potential and solute concentration – a hypothesis that has been advanced to account for a range of cell functions extending well beyond the original Hofmeister effects.[28]

Yet there is little consensus – and sometimes plain contradiction – about what structure-making and -breaking actually entail. For example, does structure-making render water denser, or does it, in making water more ice-like, actually reduce the density? I feel that a survey of the literature of this topic rather encourages one to concur with Felix Franks,[29] who said 20 years ago that 'much has been written about structure-making and structure-breaking, and much of it is misleading.'

Perhaps more to the point, experimental studies of the structure of electrolyte solutions offer no real evidence that significant changes to the bulk hydrogen-bonded network of water really do occur in the presence of salts (at least, not at moderate concentrations).[30] The current view is that 'structure-making' and 'structure-breaking' in bulk water are of little relevance to the Hofmeister effect, and that one should instead focus on the nature of direct interactions between the ions and the macromolecule, and the ways in which ions are partitioned at interfaces.[31,32]

Whereas traditionally ions have been considered to be excluded from the air–water interface because electrolytes increase surface tension, recent studies show that the picture is not so simple.[33–35] While it may hold for hard (non-polarizable) ions such as sodium and fluoride, large soft ions such as iodide (and to a lesser extent, bromide and chloride) can accumulate preferentially at the surface.

An example of the changes in intermolecular interactions that might stem from the inhomogeneity of ion distributions near interfaces was provided by Zangi and Berne, who considered how ions interact with small hydrophobic particles 0.5 nm across[36] (Fig. 5). They found that preferential absorption or exclusion of the ions at the particle surfaces does not produce any simple, monotonic relation between ionic charge density, hydrophobic interactions, and aggregation. For example, low-charge-density ions can have either a salting-out or a salting-in effect, depending on their concentration. The mechanisms are somewhat subtle, but have no need to invoke vague notions of making and breaking 'water structure'.

I have heard the words 'chaotrope' and 'kosmotrope' mentioned only a very few times at this meeting – and then, as far as I could see, as mere labels to denote the placement of ions in the series rather than as descriptors of what they are thought to do to water. This is striking. If you compare a widely cited paper on Hofmeister effects from the 1980s[37] with a more recent one,[38] the first has plentiful allusions to structure-making and -breaking while the second makes no mention of these things. I appreciate that terminology can sometimes get frozen in long after the original meaning has become obsolete, but I offer the humble suggestion, *pace* Franks, that the concepts of structure-making and breaking have already created so much confusion and misdirection that it might be useful to let them now retire gracefully.

3 The bigger picture

The presentations at this meeting have shown that part of the value of studying Hofmeister effects is that it forces one to think about questions of much broader relevance. A few of these are:

- How is water structured around solutes?
- What are the roles of molecular polarizability?

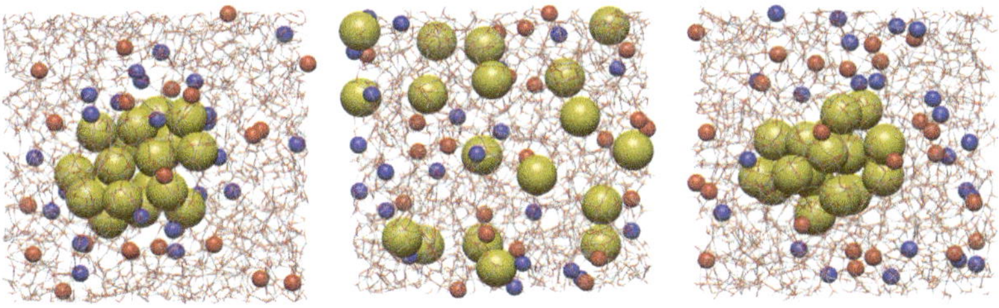

Fig. 5 The distribution of ions around hydrophobic (Lennard-Jones) particles in water. The hydrophobes are yellow, positive ions are red and negative ions are blue. Low-charge-density (q) ions (left) are adsorbed preferentially at the particle surfaces, leading to micelle-like clusters of hydrophobic particles surrounded by ions, which prevents further aggregation and precipitation. High-q ions (right) tended to be depleted at the particle surfaces, which again leads to the formation of clusters. In the intermediate-q case (centre) there is neither adsorption nor depletion, and the hydrophobes remain individually dispersed. Reprinted with permission from ref. 36.

- What are the effects of solvent dynamics over a wide range of length and time scales?
- How may theory and experiment handle complex and non-ideal mixtures?
- What is the relationship between gas-phase and solution-phase structures?
- How do interfaces affect solvation?

More generally, as Tom Record indicated at the outset,[39] Hofmeister effects seem to be now acknowledged as a subset of phenomena in which biomolecules are influenced by other species in solution, which could be regarded, depending on the concentration, as either cosolutes or cosolvents. Some of the same considerations apply to osmolytes, cryoprotectants, denaturants, protein stabilizing agents, polyols and polyelectrolytes.

There is now a growing view that all types of denaturation are intimately connected to changes in the way proteins are hydrated.[40,41] In other words, the opening-up of the native protein is not an intrinsic property of the polypeptide chain, but comes about in collaboration with the solvent. And to this outsider's eye, this story has shifted significantly in the past several years from the one that seemed to predominate for decades previously. As is the case with Hofmeister effects, no longer is there an idea that these cosolutes are somehow altering the structure of water in a global sense, nor that there is a simple equilibrium involving competition for hydration water. Rather, an understanding all these effects probably needs to focus to a large extent on the direct interactions between the ions or cosolutes, the biomolecule, and its first and perhaps second hydration shell.

For instance, it was long supposed that denaturants such as urea or guanidinium chloride (GdmCl) somehow perturb water's bulk structure in a way that destabilizes the folded protein, perhaps by altering the hydrophobic interactions that keep insoluble residues buried.[42] This idea now looks increasingly flawed.[43] Instead, we need to look at how denaturants affect the protein's hydration shell. For lysozyme, urea displaces water from the hydration shell and penetrates into the hydrophobic core, suppressing the native fold in favour of a swollen, rather disordered 'molten globule'.[44] But GdmCl appears to behave differently: the molecules stick to the hydrophobic surfaces and give them a less water-repelling veneer.[45] Here the denaturant is acting somewhat like a surfactant, stabilizing the interface between the surface and the solvent. Moreover, urea not only alters hydrophobic interactions but also disrupts the hydration of hydrophilic parts of a protein, sticking there *via* hydrogen bonds.[46] In effect, urea can usurp the hydrogen bonds that otherwise may help to bind the native state together. While these ideas are still being debated, the emerging picture is one in which denaturants exert their influence through direct interactions with the solute, not by restructuring the bulk solvent.

Some other small molecules, such as glycerol, have the opposite effect of stabilizing a native protein against denaturation, most probably again *via* direct interactions.[47] Glycerol, and also ethylene glycol and propylene glycol, protects proteins against mechanically induced unfolding, possibly *via* the formation of 'solvent bridges' that pin together hydrogen-bonding sites on the polypeptide backbone.[48]

Whether it is protein stabilization, precipitation, multimolecular assembly, or even the additive-induced conformational transitions of nucleic acids, these influences of small cosolutes are in an important sense a part of the same story. And that story is about the true chemical complexity and sophistication of biology, and indeed of biotechnology. It goes beyond the narrative of proteins and DNA to explore what truly makes those molecules tick, what sustains and constrains them, and how we might intervene in their function.

4 Conclusions

Pavel Jungwirth suggested at the start of this meeting that the study of Hofmeister effects might now be reaching the point at which we know at least which questions to ask. I have a sense that this might well be true, but also that this is not going to be a cumulative process in which the questions are answered sequentially starting from the most fundamental levels and proceeding steadily to the more complex. This is one of the delightful aspects of the meeting. You ask why ions may or may not segregate to an air–water interface, and immediately offer several answers, all appealing, none obviously mutually exclusive: it is because of capillary waves, or polarizability, or the cost of cavitation, or surfactant-like behaviour, or maybe some or all of the above and more.

That's not to suggest for a moment that the truth is you haven't got a clue. Perhaps you have too many clues, too many techniques suggesting different things. But this is liberating. Some might find it shocking that, 90 years after Debye–Hückel, it is possible to say, as Tom Record did, that "a general theory of ionic hydration does not exist". But I feel it is nothing short of exhilarating to dispense with the cosy but wrong textbook metaphors. And of course science never works in that bottom-up way. We don't need string theory before we can start working on nuclear physics and so on up the scale. We have now sequenced many hundreds of human genomes, but don't exactly agree on what a gene is. Roald Hoffmann has argued that the beauty of chemistry is that it makes do with imprecision, with imperfect definitions and understanding that can be bridged with metaphors.[49] He's right of course, but in fact physics and biology are like that too. And so I think that we do need metaphors. But we also need to remember not to start believing them, and to be ready to throw them out when they are no longer helpful. Bob Eisenberg [private correpondence] argues that we might use them for hypothesis-building while considering them dispensable, even obstructive, for hypothesis-testing.

This is also why I have mixed feelings about the idea floated at this meeting of identifying Grand Challenges for the study of Hofmeister effects. The ideas proposed, such as stabilizing antibody solutions or understanding what is special about halophiles, are potentially wonderful projects. But I believe researchers in this area can afford to trust that grand challenges already exist *at each levels of the hierarchy of questions*, from formulating a theory of ionic hydration to figuring out how salts might promote HIV inhibition.[50] Compared to genomics or dark energy, you might fear that you are working in a deeply unfashionable field. But on the contrary, this general area of *solvation science* should and will be a growth area. That is indicated, to take just one example, by the recent approval of the German Cluster of Excellence called RESOLV, centred at the Ruhr University of Bochum.[51] This recognizes that solvation processes, sharing some common principles, are important to areas ranging from electrocatalysis and green chemistry to molecular biology – and, as we have heard at this meeting, one might add for example supramolecular chemistry, polymer chemistry and liquid crystallinity. If you don't sense this already, I suspect that, at the next meeting on Hofmeister effects, when questions might really start to become answers, you will find yourselves to be a part of a much wider community.

References

1 F. Franks (ed.), *Water: A Comprehensive Treatise* Vol. 1, 1972, Plenum, New York.
2 H. E. Stanley, P. Kumar, L. Xu, Z. Yan, M. G. Mazza, S. V. Buldyrev, S.-H. Chen and F. Mallamace, *Phys. A*, 2007, **386**, 729.
3 V. Marakov, B. M. Pettitt and M. Feig, *Acc. Chem. Res.*, 2002, **35**, 376.
4 P. Ball, *Chem. Rev.*, 2008, **108**, 74.
5 C. Tanford, *The Hydrophobic Effect*, 2nd edn, 1980, Wiley, New York.
6 W. Blokzijl and J. B. F. N. Engberts, *Angew. Chem., Int. Ed. Engl.*, 1993, **32**, 1545.
7 R. J. Ellis and A. P. Minton, *Nature*, 2003, **425**, 27.
8 R. J. Ellis, *Trends Biochem. Sci.*, 2001, **26**, 597.
9 A. A. Travers and J. M. T. Thompson, *Philos. Trans. R. Soc. London, Ser. A*, 2004, **362**, 1265.
10 C. Bustamante, S. B. Smith, J. Liphardt and D. Smith, *Curr. Opin. Struct. Biol.*, 2000, **10**, 279.
11 B. Fierz and T. W. Muir, *Nat. Chem. Biol.*, 2012, **8**, 417.

12 A. Németh and G. Längst, *Briefings Funct. Genomics Proteomics*, 2004, **2**, 334.
13 D. Lingwood and K. Simons, *Science*, 2009, **327**, 46.
14 A. A. Hyman and K. Simons, *Science*, 2012, **337**, 1047.
15 K. Lum, D. Chandler and J. D. Weeks, *J. Phys. Chem. B*, 1999, **103**, 4570.
16 M. Montagna, F. Sterpone and L. Guidoni, *J. Phys. Chem. B*, 2012, **116**, 11695.
17 P. Liu, X. Huang, R. Zhou and B. J. Berne, *Nature*, 2005, **437**, 159.
18 R. Zhou, X. Huang, C. J. Margulis and B. J. Berne, *Science*, 2004, **305**, 1605.
19 L. Hua, X. Huang, P. Liu, R. Zhou and B. J. Berne, *J. Phys. Chem. B*, 2007, **111**, 9069.
20 N. Giovambattista, C. F. Lopez, P. J. Rossky and P. G. Debenedetti, *Proc. Natl. Acad. Sci. U. S. A.*, 2008, **105**, 2274.
21 A. J. Patel, P. Varilly, S. N. Jamadagni, M. F. Hagan, D. Chander and S. Garde, *J. Phys. Chem. B*, 2012, 2107523.
22 P. W. Snyder, J. Mecinovic, D. T. Moustakas, S. W. Thomas III, M. Harder, E. T. Mack, M. R. Lockett, A. Héroux, W. Sherman and G. M. Whitesides, *Proc. Natl. Acad. Sci. U. S. A.*, 2011, **108**, 17889.
23 Y. Umena, K. Kawakami, J.-R. Shen and N. Kamiya, *Nature*, 2011, **473**, 55.
24 M. D. Baer and C. J. Mundy, *Faraday Discuss.*, **160**, DOI: 10.1039/C2FD20113E (this volume; paper 5).
25 A. Salis, F. Cugia, D. F. Parsons, B. W. Ninham and M. Monduzzi, *Phys. Chem. Chem. Phys.*, 2012, **14**, 4343.
26 S. Imberti, A. Botti, F. Bruni, G. Cappa, M. A. Ricci and A. K. Soper, *J. Chem. Phys.*, 2005, **122**, 194509.
27 S. Varma and S. B. Rempe, *Biophys. Chem.*, 2006, **124**, 192.
28 P. M. Wiggins, *Microbiol. Rev.*, 1990, **54**, 432.
29 F. Franks, *Water*, 1983, Royal Society of Chemistry, London.
30 J. L. Finney and A. K. Soper, *Chem. Soc. Rev.*, 1994, **23**, 1.
31 Y. Zhang and P. S. Cremer, *Curr. Opin. Chem. Biol.*, 2006, **10**, 658.
32 Y. Zhang and P. S. Cremer, *Annu. Rev. Phys. Chem.*, 2010, **61**, 63.
33 S. J. Stuart and B. J. Berne, *J. Phys. Chem.*, 1996, **100**, 11934.
34 P. Jungwirth and D. J. Tobias, *Chem. Rev.*, 2006, **106**, 1259.
35 P. B. Petersen and R. J. Saykally, *Annu. Rev. Phys. Chem.*, 2006, **57**, 333.
36 R. Zangi and B. J. Berne, *J. Phys. Chem. B*, 2006, **110**, 22736.
37 K. D. Collins and M. W. Washabaugh, *Q. Rev. Biophys.*, 1985, **18**, 323.
38 W. Kunz, P. Lo Nostro and B. W. Ninham, *Curr. Opin. Colloid Interface Sci.*, 2004, **9**, 1.
39 T. Record and C. J. Mundy, *Faraday Discuss.*, **160** (this volume; paper 1).
40 V. P. Denisov, B.-H. Jonsson and B. Halle, *Nat. Struct. Biol.*, 1999, **6**, 253.
41 M. Davidovic, C. Mattea, J. Qvist and B. Halle, *J. Am. Chem. Soc.*, 2009, **131**, 1025.
42 G. G. Hammes and J. C. Swann, *Biochemistry*, 1967, **6**, 1591.
43 J. D. Batchelor, A. Olteanu, A. Tripathy and G. J. Pielak, *J. Am. Chem. Soc.*, 2004, **126**, 1958.
44 L. Hua, R. Zhou, D. Thirumalai and B. J. Berne, *Proc. Natl. Acad. Sci. U. S. A.*, 2008, **105**, 16928.
45 J. L. England, V. S. Pande and G. Haran, *J. Am. Chem. Soc.*, 2008, **130**, 11854.
46 B. J. Bennion and V. Daggett, *Proc. Natl. Acad. Sci. U. S. A.*, 2003, **100**, 5142.
47 E. J. Guinn, L. M. Pegram, M. W. Capp, M. N. Pollock and M. T. Record, *Proc. Natl. Acad. Sci. U. S. A.*, 2011, **108**, 16932.
48 L. Dougan, G. Z. Genchev, H. Lu and J. M. Fernández, *Proc. Natl. Acad. Sci. U. S. A.*, 2011, **108**, 9759.
49 J. Kovac and M. Weisberg (ed), *Roald Hoffmann on the Philosophy, Art, and Science of Chemistry*, 2012, Oxford University Press, New York.
50 J. Pokorná, J. Heyda and J. Konvalinka, *Faraday Discuss.*, **160**, DOI: 10.1039/C2FD20094E (this volume; paper 20).
51 See http://www.ruhr-uni-bochum.de/solvation/Resolv.html.

Poster titles

A comparison of the effects of ionic and non-ionic additives on the self-association and chromatic mesophase formation of Edicol sunset yellow, **J. W. Jones, H. Wheatcroft, A. P. Ormerod and G. J. T. Tiddy**, *University of Manchester, UK*

Hofmeister chaotropic anions and their affinity for hydrophobic concavity, **B. C. Gibb, R. Carnegie, V. Santhosh, H. Gan and C. L. D. Gibb**, *Tulane University, USA*

Intrinsic switching thermodynamics and kinetics of ultrathin thermoresponsive polymer brushes in sodium halide solutions, **C. A. Naini, M. Thomas, S. Franzka, S. Frost, M. Ulbricht and N. Hartmann**, *Universität Duisburg-Essen, Germany*

Shifts in the temperature of maximum density of water induced by ionic liquids, **M. Tariq, J. M. G. S. Esperanca, J. N. Canongia Lopes and L. P. N. Rebelo**, *Universidade Nova de Lisboa, Portugal*

Reversible and irreversible thermal denaturation of lysozyme at varying pH at low ionic strength, **A. Blumlein and J. J. McManus**, *National University of Ireland, Maynooth, Republic of Ireland*

Hydration and ion binding of phosphate ions $H_2PO_4^-$, HPO_4^{2-} and PO_4^{3-}, **A. Eiberweiser and R. Buchner**, *Universität Regensburg, Germany*

Hydration of the osmolyte ectoin in water and aqueous NaCl, **A. Eiberweiser, A. Nazet, T Rück, A. Stoppa and R. Buchner**, *Universität Regensburg, Germany*

Manipulating self-assembly of small organic molecules using specific ion effects, **S. M. Mooi and B. Hayne**, *University of Calgary, Canada*

Terahertz solvation study of transition metal halides in aqueous solutions, **V. Sharma, G. Schwaab, T. K. Roy, E. Edengeiser, S. Bureekaew, R. Schmid, H. Weingartner and M. Havenith**, *Ruhr University Bochum, Germany*

Terahertz hydration study of aqueous lanthanide halide solutions, **G. Schwaab, V. Scharma, M. Seitz, F. Böhm, E. Edengeiser and M. Havenith,** *Ruhr University Bochum, Germany*

Ions at the ice/vapour interface, **I. Gladich and M. Roeselova**, *Academy of Sciences of the Czech Republic, UOCHB, Czech Republic*

Like-charge ion pairing in water: An *ab initio* molecular dynamics study of aqueous guanidinium cations, **M. Vazdar**, *Institute of Organic Chemistry and Biochemistry, ASCR, Czech Republic*

Ion pairing of Li^+ in water by classical and *ab initio* molecular dynamics, and neutron scattering, **E. Pluharova P. E. Mason and P. Jungwirth**, *Institute of Organic Chemistry and Biochemistry, ASCR, Czech Republic*

Effect of ions on the mechanical and conformational properties of zwitterionic polymers, **Z. Zhang*, M. Moxey, M. Mears, A. J. Morse, S. P. Armes, A. L. Lewis, M. Geoghegan and G. J. Leggett**, *University of Sheffield, UK*

Crystallographic analysis of weaker counter-ion interactions with proteins, **M. Cianci, M. A. L. Martinez, J. Helliwell and P. Halling**, *University of Strathclyde, UK*

Specific ion effects on intermolecular interactions between lysozyme molecules, **M. M. Nuhu, Y. Lo, D. Roberts and R. Curtis**, *University of Manchester, UK*

Accurate hydrogen-bond definition sharply resolves the structure and dynamics of water and aqueous solutions, **R. H. Henchman and S. J. Irudayam**, *University of Manchester, UK*

Specific ion effects on intermolecular interactions between homopolypeptides, **R. Curtis**, *University of Manchester, UK*

Comparison between the preferential adsorption of ions to selected hydrophobic and polar surfaces: a vibrational sum frequency study, **E. Tyrode**, *University of Manchester, UK*

Influence of ion type, pH and ionic strength on interactions between protein and polyelectrolyte, **D. Roberts, A. Oyefoso and R. Curtis**, *Manchester Institute of Biotechnology, UK*

How friendly are ionic liquids to enzymes?, **N. Ayixiamuguli, N. R. Thomas and G. Stephens**, *University of Nottingham, UK*

Dynamics of the hydration shell: spatially decomposed time correlation approach, **O. Marsalek, P. Jungwirth and N. Matubayasi**, *Academy of Sciences of the Czech Republic, Czech Republic*

Molecular mechanisms of ion-specific effects on proteins, **J. Paterová, J. Heyda, K. Rembert, C. Hilty, P. Cremer and P. Jungwirth**, *IOCB AS CR, Czech Republic*

Ion effects on guest-host binding of aminobenzoic acids with cyclodextrins, **A. Romanova and M. V. Fedorov**, *University of Strathclyde, UK*

Adsorption kinetics of surfactants to solid–solution interfaces, **A. P. Dudgeon, C. D. Bain and P. Starck**, *Durham University, UK*

Water reorientation in acidic and salty electrolytes, **N. Ottosson and H. Bakker**, *AMOLF, The Netherlands*

New interaction parameters for charged amino acid side chains in the GROMOS force field, **M. Reif, P. Hünenberger and C. Oostenbrink**, *Institut für Molekulare Modellierung und Simulation, Wien, Austria*

The Skinner Prize for the best poster was awarded to Miss Eva Pluharova of the Institute of Organic Chemistry and Biochemistry, ASCR, Czech Republic, for her poster on the ion pairing of Li^+ in water by classical and *ab initio* molecular dynamics and neutron scattering.

List of participants

Professor Vicente Aguilella, *Universitat Jaume I, Spain*
Mr Crispin Amiri Naini, *University of Duisburg-Essen, Germany*
Professor Michael Ashfold, *University of Bristol, UK*
Dr Marcel Baer, *Pacific Northwest National Laboratory, USA*
Professor Colin Bain, *Durham University, UK*
Dr Huib J. Bakker, *AMOLF, The Netherlands*
Dr Philip Ball, *Freelance Science Writer, UK*
Professor Dor Ben-Amotz, *Purdue University, USA*
Ms Alice Blumlein, *NUIM, Republic of Ireland*
Dr Daniel Bowron, *ISIS Facility, UK*
Dr Deepak Canchi, *Intel, USA*
Mr Saso Cebasek, *University of Ljubljana, Slovenia*
Professor Ian Collins, *BP Exploration Operating Company Ltd, UK*
Dr Paul Cremer, *Texas A&M University, USA*
Dr Robin Curtis, *The University of Manchester, UK*
Dr Lukasz Cwiklik, *J Hetrovsky Institute of Physical Chemistry, Czech Republic*
Dr Liem Dang, *Pacific Northwest National Laboratory, USA*
Professor Anthony Davis, *University of Bristol, UK*
Mrs Susanne Kerstin Dengler, *University of Regensburg, Germany*
Miss Angeliki Ditsiou, *University of Dundee, UK*
Mr Alexander Dudgeon, *Durham University, UK*
Mr Andreas Eiberweiser, *University of Regensburg, Germany*
Mr Robert Eisenberg, *Rush University, USA*
Mr Federico Faraci, *Max Planck Institute for Dynamics and Self-Organization, Germany*
Prof Dr Maxim Fedorov, *University of Strathclyde, UK*
Professor Fabio Herbst Florenzano, *Universidade Federal de Alfenas - UNIFAL-MG, Brazil*
Professor Yi Qin Gao, *Peking University, China*
Professor Phillip Lewis Geissler, *University of California, Berkeley, USA*
Dr Bruce Gibb, *Tulane University, USA*
Dr Ivan Gladich, *Academy of Sciences of The Czech Republic, Czech Republic*
Dr David Goodall, *University of York, UK*
Dr Jonathan Goodman, *University of Cambridge, UK*
Kirsty Gould, *Royal Society of Chemistry, UK*
Dr Jason Hallett, *Imperial College London, UK*
Professor Peter Halling, *University of Strathclyde, UK*
Professor Daniel Harries, *Hebrew University Jerusalem, Israel*
Dr Tue Hassenkam, *Nano-Science center, University of Copenhagen, Denmark*
Dr Richard Henchman, *The University of Manchester, UK*
Prof Dr Thorsten Hugel, *TU Munchen, Germany*
Mr Klaus Juhl, *NanoGeoScience, Denmark*
Professor Pavel Jungwirth, *Academy of Sciences of The Czech Republic, Czech Republic*
Dr Piotr Jurkiewicz, *J Hetrovsky Institute of Physical Chemistry, Czech Republic*
Mr Andras Komaromy, *Monash University, Australia*
Dr Jan Konvalinka, *Institute of Organic Chemistry and Biochemistry, Czech Republic*
Professor Werner Kunz, *University of Regensburg, Germany*
Dr Michel Laguerre, *IECB - CBMN UMR CNRS 5248, France*
Professor Yan Levin, *Instituto De Fisica, UFRGS, Brazil*

Professor Jose Nuno Canongia Lopes, *Universidade Técnica de Lisboa, Portugal*
Professor Mikael Lund, *Lund University, Sweden*
Professor Ruth Lynden-Bell, *University of Cambridge, UK*
Professor Paul Madden, *University of Oxford, UK*
Mr Ondrej Marslaek, *IOCB AS CR, Czech Republic*
Professor Phil Mason, *Cornell University, USA*
Mr Jesper Matthiesen, *NanoGeoScience, Denmark*
Ms Sara Mooi, *University of Calgary, Canada*
Mrs Ayixiamuguli Nueraimaiti, *The University of Nottingham, UK*
Miss Mariam Nuhu, *The University of Manchester, UK*
Dr Niklas Ottosson, *FOM institute AMOLF, The Netherlands*
Professor Christopher Outhwaite, *University of Sheffield, United Kingdom*
Ms Jana Paterova, *IOCB AS CR, Czech Republic*
Mr Thomas Patko, *Advanced Sensor Technologies, USA*
Miss Eva Pluharova, *Institute of Organic Chemistry and Biochemistry, Czech Republic*
Mrs Jana Pokorna, *Institute of Organic Chemistry and Biochemistry, Czech Republic*
Ms Sarka Pokorna, *J Hetrovsky Institute of Physical Chemistry, Czech Republic*
Dr Karen Marie Polizzi, *Imperial College London, UK*
Dr Gleb Priimov, *Unilever R & D, UK*
Matthew Reaveley, *Royal Society of Chemistry, UK*
Professor Thomas Record, *University of Wisconsin - Madison, USA*
Dr Maria Reif, *University of Natural Resources and Life Sciences, Vienna, Austria*
Dr Dorota Roberts, *The University of Manchester, United Kingdom*
Dr Francisco Rodriguez-Ropero, *TU Darmstadt, Germany*
Dr Martina Roeselova, *Academy of Sciences of The Czech Republic, Czech Republic*
Dr Brandon Ruotolo, *University of Michigan, USA*
Dr Gerhard Schwaab, *Ruhr University Bochum, Germany*
Mr Vinay Sharma, *Physical Chemistry II, Ruhr University Bochum, Germany*
Professor Uri Sivan, *Technion-Israel Institute of Technology, Israel*
Laura Smart, *Royal Society of Chemistry, UK*
Dr Michael Spencelayh, *Royal Society of Chemistry, UK*
Miss Isabella Stoecker, *BP Exploration Operating Company Ltd, UK*
Dr Mohammad Tariq, *ITQB, Portugal*
Mrs Rachel Thompson, *Royal Society of Chemistry, UK*
Emer Prof Gordon Tiddy, *The University of Manchester, UK*
Professor Douglas Tobias, *University of California - Irvine, USA*
Professor Nico van der Vegt, *Center of Smart Interfaces, TU Darmstadt, Germany*
Dr Mario Vazdar, *Institute of Organic Chemistry and Biochemistry, Czech Republic*
Dr Ana Vila Verde, *Max Planck Institute of Colloids and Interfaces, Germany*
Dr Helen Wheatcroft, *The University of Manchester, UK*
Dr Bernd Winter, *Helmholtz Centre Berlin, Germany*
Dr Zhenyu Zhang, *University of Sheffield, United Kingdom*

Index of contributors*

*The page numbers in **bold** type indicate papers submitted for discussion.

